CON

MECHANISMS AND MECHANICAL DEVICES SOURCEBOOK

Fifth Edition

NEIL SCLATER

McGraw-Hill
New York • Chicago • San Francisco • Lisbon • London • Madrid
Mexico City • Milan • New Delhi • San Juan • Seoul
Singapore • Sydney • Toronto

The McGraw·Hill Companies

Copyright © 2011, 2007, 2001, 1996, 1991 by The McGraw-Hill Companies, Inc. All rights reserved. Except as permitted under the United States Copyright Act of 1976, no part of this publication may be reproduced or distributed in any form or by any means, or stored in a database or retrieval system, without the prior written permission of the publisher.

1 2 3 4 5 6 7 8 9 0   QDB/QDB   1 7 6 5 4 3 2 1

ISBN   978-1-265-63178-9
MHID   1-26-563178-6

Sponsoring Editor Larry S. Hager	**Copy Editor** Susan Fox Greenberg
Editing Supervisor Stephen M. Smith	**Proofreader** Constance Blazewicz
Production Supervisor Pamela A. Pelton	**Art Director, Cover** Jeff Weeks
Acquisitions Coordinator Michael Mulcahy	**Composition** Cenveo Publisher Services
Project Manager Tania Andrabi, Cenveo Publisher Services	

McGraw-Hill books are available at special quantity discounts to use as premiums and sales promotions, or for use in corporate training programs. To contact a representative, please e-mail us at bulksales@mcgraw-hill.com.

Information contained in this work has been obtained by The McGraw-Hill Companies, Inc. ("McGraw-Hill") from sources believed to be reliable. However, neither McGraw-Hill nor its authors guarantee the accuracy or completeness of any information published herein, and neither McGraw-Hill nor its authors shall be responsible for any errors, omissions, or damages arising out of use of this information. This work is published with the understanding that McGraw-Hill and its authors are supplying information but are not attempting to render engineering or other professional services. If such services are required, the assistance of an appropriate professional should be sought.

CHAPTER 5   LINKAGES: DRIVES AND MECHANISMS   89

CHAPTER 6   GEARS: DEVICES, DRIVES, AND MECHANISMS   131

This is the fifth edition of a one-of-a-kind engineering reference book covering the past, present, and future of mechanisms and mechanical devices. It includes clear illustrations and straightforward descriptions of specific subjects rather than the theory and mathematics found in most engineering textbooks. You will find that this book contains hundreds of detailed line drawings that will hold your interest regardless of your background in mechanical engineering. The text accompanying the illustrations is intended to help you to understand the basic concepts of subjects that may or may not be familiar to you.

You will find drawings and illustrations that are simply interesting and informative and perhaps others that could spur your creativity and prompt you to recycle them into your new designs or redesigns. They may offer solutions you had not previously considered because they were not visible inside contemporary products unless the product is disassembled. Solid state electronics and computer circuitry have displaced many earlier mechanical solutions, no doubt improving product reliability and efficiency while reducing their price.

Nevertheless, many of those displaced mechanical components have lives of their own and may very well turn up in other products in different form performing different functions after undergoing dimensional and material transformations.

Classical, proven mechanisms and mechanical devices may seem to disappear only to reappear in other forms and applications. Anyone who believes that all mechanisms will be replaced by electronics need only examine the sophistication of the latest self-winding mechanical watches, digital cameras, gyro-stabilized vehicles, and navigational systems.

This book illustrates the ongoing importance of classical mechanical devices as well as the latest mechatronic devices formed by the merger between mechanics and electronics. It is a must addition to your personal technical library, and it offers you a satisfying way to "get up to speed" on new subjects or those you may have studied in the past but have now faded from your memory. Moreover, it is hoped that this book will encourage you to refresh your knowledge of these and other topics that interest you by accessing the many related Web sites on the Internet.

What's New in This Book?

This fifth edition contains three new chapters: Chapter 3, Stationary and Mobile Robots, Chapter 4, Mechanisms for Renewable Power Generation, and Chapter 17, 3D Digital Prototypes and Simulation. Chapter 18, Rapid Prototyping, has been updated and completely revised, and new articles have been added to Chapters 5 through 16 that make up the archival core of the book. Five new articles have been added to Chapter 13, Motion-Specific Devices, Mechanisms, and Machines, which is part of the archival core. Also, five new articles have been added to Chapter 19, New Directions in Mechanical Engineering.

A Quick Overview of Some Chapters

Chapter 1 on basic mechanisms explains the physics of mechanisms including inclined planes, jacks, levers, linkage, gears, pulleys, genevas, cams, and clutches—all components in modern machines. A glossary of common mechanical terms is included.

Chapter 2 on motion control explains open- and closed-loop systems with diagrams and text. Described and illustrated are the key mechanical, electromechanical, and electronic components that comprise modern automated robotic and mechatronic systems, including actuators, encoders, servomotors, stepper motors, resolvers, solenoids, and tachometers. It includes a glossary of motion control terms.

Chapter 3, a new discussion of robots, includes an overview of stationary industrial robots and a wide range of mobile robots. Drawings and text explain the geometry of industrial robots and leading specifications are given for four of the newest robots. Seven mobile robots are described accompanied by their illustrations and leading specifications. They operate on Mars, on Earth, in the air, and under the sea. Other articles describe innovative NASA robots that climb, crawl, hop, and rappel down cliffs. In addition, a glossary defines common robot terms.

Chapter 4, a new addition, describes the leading contenders for generating carbon-free renewable power, which happen to be mechanical in nature. They are driven by the free energy of the wind, sun, and natural water motion. Examples described and illustrated include wind turbines and their farms, four different solar thermal farm concepts, and proposed methods for tapping ocean tidal and wave energy. Both the upsides and downsides of these plants are stated. Attention is given to location, efficiency, public acceptance, backup power sources, and connections to the power grid. Included is a glossary of wind turbine terms.

Chapter 17, also new, explains the features of the latest computer software making it possible to design new or revise old products in 3D right on the computer screen, taking advantage of features including the ability to manipulate, "slice and dice," and re-dimension the virtual model in a range of colors to finalize the design complete with manufacturing data. Compatible simulation software permits a model to be subjected to virtual mechanical and multiphysics stresses to verify its design and choice of materials without the need to build a physical model for testing. Included in the chapter is a glossary of CAD/CAE terms.

Chapter 18, an update of an earlier chapter on rapid prototyping, explains and illus-trates innovations and new additions to the many commercial additive and subtractive processes for building 3D solid prototypes. They are being made from soft or hard mate-rials for "hands-on" evaluation. Some prototypes are just for display while others are built to withstand laboratory stress testing. However, the newer applications include the fabrication of replacement parts for older machines, specialized metal tools, and molds for casting.

Chapter 19 is an update of a collection of articles discussing cutting-edge topics in mechanical engineering. These include the latest developments in microelectromechani-cal devices (MEMS) and progress in developing practical applications for the carbon allotropes, nanotubes, and graphene in products ranging from transparent sheets, strong fiber, cable, capacitors, batteries, springs, and transistors. Other topics include electron microscopes for R&D and a proposed long-range Moon rover.

The central core of the book, Chapters 5 through 16, contains an encyclopedic col-lection of archival drawings and descriptions of proven mechanisms and mechanical devices. This revised collection is a valuable resource for engineers, designers, teachers, and students as well as enthusiasts for all things mechanical. New entries describe a pre-cision linear actuator, polygon connections, slip clutches, shape memory alloy latches, and an energy exchanger for making desalination more efficient.

A complete Index makes it easy for readers to find all of the references to specific mechanisms, mechanical devices, components, and systems mentioned in the book.

Engineering Choices to Examine

Renewable Energy versus Fossil Fuel for Power Generation

The chapter on renewable power generation discusses three of the most promising mechanical methods for generating carbon-free, grid-compatible electric power. Wind turbine farms and concentrating solar thermal (CST) plants are the most likely candidates for government subsidies. These technologies are described and illustrated, and their upsides and downsides are explained. Electricity can also be generated by ocean waves and tides, but these technologies lag far behind wind and solar thermal plants.

The U.S. government is offering financial incentives for building electrical generating plants fueled by renewable energy, primarily for reducing atmospheric carbon dioxide (CO_2) emissions, considered by some to be the principal source of manmade global warming. The administration has set the goal of increasing the number of carbon-free, non-hydro power plants from about 3 percent today to 20 percent by 2020. Wind and solar thermal power plants have the best chance of meeting this goal, but many worry that the building of these plants and eliminating many fossil-fueled plants could endanger the utility industries' efforts to meet the nation's growing demand for low-cost, readily avail-able electric power.

Renewable energy power sources are handicapped by the inability of the overbur-dened power grid to transport electricity from remote parts of the country where most of these installations will be located to metropolitan areas where electricity demand is highest. When the wind dies or after sunset, these plants must be able to provide backup genera-tion or energy storage to meet their power commitments to the grid. This backup could include banks of batteries, heat stored in molten salt vats, and natural gas-powered steam generators, but the optimum choices have not been resolved because of variables such as plant power output and climate.

Digital 3D versus Rapid Prototyping

Recently introduced computer software makes it possible to design a product in a 3D for-mat from concept sketch to shop documentation on a computer. This process, 3D *digital prototyping* or *modeling,* can begin as an original design or be imported from another source. The software permits a 3D image to be disassembled and its dimensions, materi-als, and form changed before being reassembled as a new or modified product design on the same screen. The designer can work cooperatively with other specialists to merge valuable contributions for the achievement of the most cost-effective design. Changes can easily be made before the design is released for manufacture.

Virtual simulation software permits the 3D digital prototype to be given one or more virtual stress tests with the results appearing graphically in color on the computer screen. These simulations can include both mechanical and physical stress, and their results correlate so closely with actual laboratory tests results that, in many cases, these tests can be omitted. This saves time and the expense of ordering physical prototypes and can accelerate the whole design process and reduce time-to-market

There are, however, many reasons why physical models are desired. These include the advantages of having a solid model for "hands on" inspection, giving all persons with responsibilities for its design and marketing an opportunity to evaluate it. However, some products require mandatory laboratory testing of a physical model to determine its compliance with industry and consumer safety standards. Rapid prototyping has gained more acceptance as the cost of building prototypes has declined.

Solid prototypes can be made from wax, photopolymers, and even powdered metals, but those built for laboratory testing or as replacement parts can now be made from powdered metal fused by lasers. After furnace firing they gain the strength to match that of machined or cast parts. Rapid prototyping depends on dimensional data derived from a CAD drawing for the preparation of software that directs all additive and subtractive rapid prototyping machines.

The Origins of This Book

Many of the figures and illustrations in the archival Chapters 4 through 16 originally appeared in foreign and domestic engineering magazines, some 50 or more years ago. They were originally collected and republished in three McGraw-Hill reference books dating back to the 1950s and 1960s. The author/editor of those books, Douglas C. Greenwood, was then an editor for McGraw-Hill's *Product Engineering* magazine. The late Nicholas Chironis, the author/editor of the first edition of this book, selected illustrations and text from these books that he believed were worthy of preservation. He saw them as a collection of successful design concepts that could be recycled for use in new and modified products and would be a resource for engineers, designers, and students.

New illustrations and text were added in the subsequent four editions of this book. The older content has been reorganized, redrawn as necessary, and in some cases deleted. All original captions have been edited for improved readability and uniformity of style. All illustrations are dimensionless because they are scalable to suit the intended application. References to manufacturers and publications that no longer exist were deleted but, where available, the names of inventors were retained for readers wishing to research the status of the inventors' patents. All government and academic laboratories and manufacturers mentioned in this edition have Internet Web sites that can be explored for further information on specific subjects.

About the Illustrations

With the exception of illustrations obtained from earlier publications and those contributed by laboratories or manufacturers, the figures in this book were drawn by the author on a desktop computer. The sources for these figures include books, magazines, and Internet Web sites. The author believes that clear 3D line or wireframe drawings with callouts communicate engineering information more rapidly and efficiently than photographs, which often contain extraneous or unclear details.

Acknowledgments

I wish to thank the following companies and organizations for granting me permission to use selected copyrighted illustrations and providing other valuable technical information by various means, all useful in the preparation of this edition:

- ABB Robotics, Auburn Hills, Michigan
- Sandia National Laboratories, Sandia Corporation, Albuquerque, New Mexico
- SpaceClaim Corporation, Concord, Massachusetts

—Neil Sclater

ABOUT THE EDITOR

Neil Sclater began his career as a microwave engineer in the defense industry and as a project engineer at a Boston consulting engineering firm before changing his career path to writing and editing. He was an editor for *Electronic Design* magazine and later McGraw-Hill's *Product Engineering* magazine before starting his own technical communications firm.

He served clients by writing and editing marketing studies, technical articles, and new product releases. His clients included manufacturers of light-emitting diodes, motors, switching-regulated power supplies, and lithium batteries. During this 30-year period he contributed many bylined technical articles to various engineering publications on subjects ranging from semiconductor devices and servomechanisms to industrial instrumentation.

Mr. Sclater holds degrees from Brown and Northeastern Universities. He is the author or coauthor of 12 books including 11 engineering reference books published by McGraw-Hill's Professional Book Group. The subjects of these books include microwave semiconductor devices, electronics technology, an electronics dictionary, electrical power and lighting, and mechanical subjects. After the death of Nicholas P. Chironis, the first author/editor of *Mechanisms and Mechanical Devices Sourcebook*, Mr. Sclater became the author/editor of the four subsequent editions.

CHAPTER 1
BASICS OF MECHANISMS

INTRODUCTION

Complex machines from internal combustion engines to helicopters and machine tools contain many mechanisms. However, it might not be as obvious that mechanisms can be found in consumer goods from toys and cameras to computer drives and printers. In fact, many common hand tools such as scissors, screwdrivers, wrenches, jacks, and hammers are actually true mechanisms. Moreover, the hands and feet, arms, legs, and jaws of humans qualify as functioning mechanisms as do the paws and legs, flippers, wings, and tails of animals.

There is a difference between a *machine* and a *mechanism*: All machines transform energy to do work, but only some mechanisms are capable of performing work. The term *machinery* means an assembly that includes both machines and mechanisms. Figure 1a illustrates a cross section of a machine—an internal combustion engine. The assembly of the piston, connecting rod, and crankshaft is a mechanism, termed a *slider-crank mechanism*. The basic schematic drawing of that mechanism, Fig. 1b, called a *skeleton outline*, shows only its fundamental structure without the technical details explaining how it is constructed.

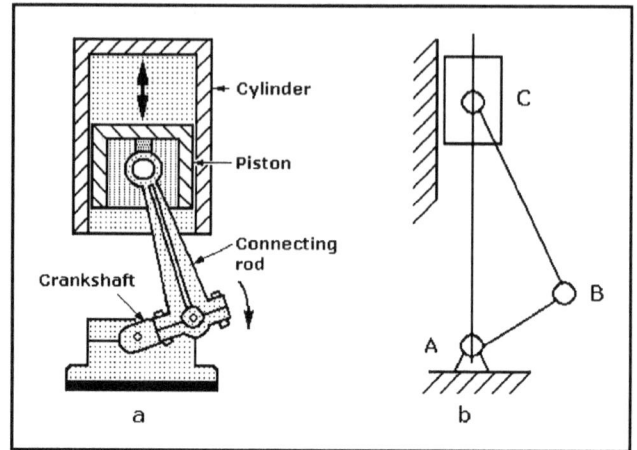

Fig. 1 Cross section of a cylinder of an internal combustion engine showing piston reciprocation (a), and the skeleton outline of the linkage mechanism that moves the piston (b).

PHYSICAL PRINCIPLES

Efficiency of Machines

Simple machines are evaluated on the basis of efficiency and mechanical advantage. While it is possible to obtain a larger force from a machine than the force exerted upon it, this refers only to force and not energy; according to the law of conservation of energy, *more work cannot be obtained from a machine than the energy supplied to it*. Because work = force × distance, for a machine to exert a larger force than its initiating force or operator, that larger force must be exerted through a correspondingly shorter distance. As a result of friction in all moving machinery, the energy produced by a machine is less than that applied to it. Consequently, by interpreting the law of conservation of energy, it follows that:

Input energy = output energy + wasted energy

This statement is true over any period of time, so it applies to any unit of time; because power is work or energy per unit of time, the following statement is also true:

Input power = output power + wasted power

The *efficiency of a machine is the ratio of its output to its input*, if both input and output are expressed in the same units of energy or power. This ratio is always less than unity, and it is usually expressed in percent by multiplying the ratio by 100.

$$\text{Percent efficiency} = \frac{\text{output energy}}{\text{input energy}} \times 100$$

or

$$\text{Percent efficiency} = \frac{\text{output power}}{\text{input power}} \times 100$$

A machine has high efficiency if most of the power supplied to it is passed on to its load and only a fraction of the power is wasted. The efficiency can be as high as 98 percent for a large electrical generator, but it is likely to be less than 50 percent for a screw jack. For example, if the input power supplied to a 20-hp motor with an efficiency of 70 percent is to be calculated, the foregoing equation is transposed.

$$\text{Input power} = \frac{\text{output power}}{\text{percent efficiency}} \times 100$$

$$= \frac{20 \text{ hp}}{70} \times 100 = 28.6 \text{ hp}$$

Mechanical Advantage

The *mechanical advantage* of a mechanism or system is the ratio of the load or weight W, typically in pounds or kilograms, divided by the effort or force F exerted by the initiating entity or operator, also in pounds or kilograms. If friction has been considered or is known from actual testing, the mechanical advantage, MA, of a machine is:

$$\text{MA} = \frac{\text{load}}{\text{effort}} = \frac{W}{F}$$

However, if it is assumed that the machine operates without friction, the ratio of W divided by F is called the *theoretical mechanical advantage*, TA.

$$TA = \frac{\text{load}}{\text{effort}} = \frac{W}{F}$$

Velocity Ratio

Machines and mechanisms are used to translate a small amount of movement or distance into a larger amount of movement or distance. This property is known as the *velocity ratio*. It is defined as the ratio of the distance moved by the effort per second divided by the distance moved by the load per second for a machine or mechanism. It is widely used in determining the mechanical advantage of gears or pulleys.

$$VR = \frac{\text{distance moved by effort/second}}{\text{distance moved by load/second}}$$

INCLINED PLANE

The *inclined* plane, shown in Fig. 2, has an incline length l (AB) = 8 ft and a height h (BC) = 3 ft. The inclined plane permits a smaller force to raise a given weight than if it were lifted directly from the ground. For example, if a weight W of 1000 lb is to be raised vertically through a height BC of 3 ft without using an inclined plane, a force F of 1000 lb must be exerted over that height. However, with an inclined plane, the weight is moved over the longer distance of 8 ft, but a force F of only $3/8$ of 1000 or 375 lb would be required because the weight is moved through a longer distance. To determine the mechanical advantage of the inclined plane, the following formula is used:

$$F = W \sin \theta \qquad \sin \theta = \frac{\text{height } h}{\text{length } l}$$

where height h = 3 ft, length l = 8 ft, $\sin \theta$ = 0.375, and weight W = 1000 lb.

$$F = 1000 \times 0.375$$
$$F = 375 \text{ lb}$$

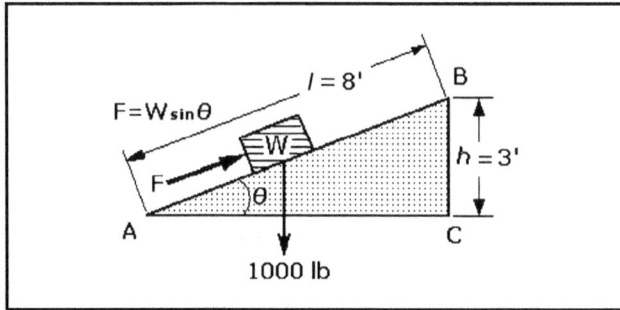

Fig. 2 Diagram for calculating mechanical advantage of an inclined plane.

$$\text{Mechanical advantage MA} = \frac{\text{load}}{\text{effort}} = \frac{W}{F} = \frac{1000}{375} = 2.7$$

PULLEY SYSTEMS

A single pulley simply changes the direction of a force so its mechanical advantage is unity. However, considerable mechanical advantage can be gained by using a combination of pulleys. In the typical pulley system, shown in Fig. 3a, each block contains two pulleys or sheaves within a frame or shell. The upper block is fixed and the lower block is attached to the load and moves with it. A cable fastened at the end of the upper block passes around four pulleys before being returned to the operator or other power source.

Figure 3b shows the pulleys separated for clarity. To raise the load through a height h, each of the sections of the cable A, B, C, and D must be moved to a distance equal to h. The operator or other power source must exert a force F through a distance $s = 4h$ so that the velocity ratio of s to h is 4. Therefore, the theoretical mechanical advantage of the system shown is 4, corresponding to the four cables supporting the load W. The theoretical mechanical advantage TA for any pulley system similar to that shown equals the number of parallel cables that support the load.

Fig. 3 Four cables supporting the load of this pulley combination give it a mechanical advantage of 4.

SCREW-TYPE JACK

Mechanisms are often required to move a large load with a small effort. For example, a car jack allows an ordinary human to lift a car which may weigh as much as 6000 lb, while the person only exerts a force equivalent to 20 or 30 lb.

The *screw jack*, shown in Fig. 4, is a practical application of the inclined plane because a screw is considered to be an inclined plane wrapped around cylinder. A force F must be exerted at the end of a length of horizontal bar l to turn the screw to raise the load (weight W) of 1000 lb. The 5-ft bar must be moved through a complete turn or a circle of length $s = 2\pi l$ to advance the load a distance h of 1.0 in. or 0.08 ft equal to the pitch p of the screw. The pitch of the screw is the distance advanced in a complete turn. Neglecting friction:

$$F = \frac{W \times h}{s}$$

where $s = 2\pi l = 2 \times 3.14 \times 5$, $h = p = 0.08$, and $W = 1000$ lb

$$F = \frac{1000 \times 0.08}{2 \times 3.14 \times 5} = \frac{80}{31.4} = 2.5 \text{ lb}$$

Mechanical advantage MA $= \dfrac{\text{load}}{\text{effort}} = \dfrac{2\pi l}{p} = \dfrac{31.4}{0.08} = 393$

Fig. 4 Diagram for calculating the mechanical advantage of a screw jack.

LEVERS AND MECHANISMS

Levers

Levers are the simplest of mechanisms; there is evidence that Stone Age humans used levers to extend their reach or power; they made them from logs or branches to move heavy loads such as rocks. It has also been reported that primates and certain birds use twigs or sticks to extend their reach and act as tools to assist them in obtaining food.

A lever is a rigid beam that can rotate about a fixed point along its length called the *fulcrum*. Physical effort applied to one end of the beam will move a load at the other end. The act of moving the fulcrum of a long beam nearer to the load permits a large load to be lifted with minimal effort. This is another way to obtain *mechanical advantage*.

The three *classes of lever* are illustrated in Fig. 5. Each is capable of providing a different level of mechanical advantage. These levers are called *Class 1*, *Class 2*, and *Class 3*. The differences in the classes are determined by:

- Position along the length of the lever where the effort is applied
- Position along the length of the lever where the load is applied
- Position along the length of the lever where the fulcrum or pivot point is located

Class 1 lever, the most common, has its fulcrum located at or about the middle with effort exerted at one end and load positioned at the opposite end, both on the same side of the lever. Examples of Class 1 levers are playground seesaw, crowbar, scissors, claw hammer, and balancing scales.

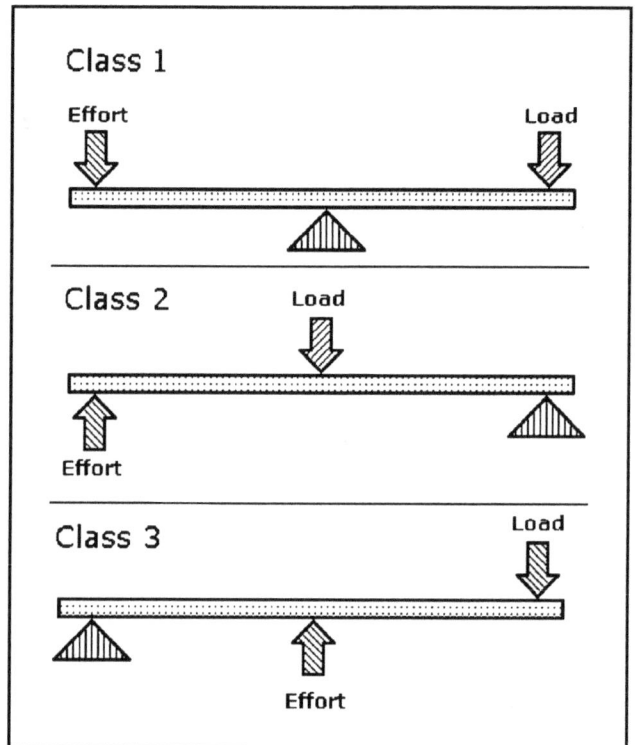

Fig. 5 Three levers classified by the locations of their fulcrums, loads, and efforts.

Class 2 lever has its fulcrum at one end; effort is exerted at the opposite end, and the opposing load is positioned at or near the middle. Examples of Class 2 levers are wheelbarrow, simple bottle openers, nutcracker, and foot pump for inflating air mattresses and inflatable boats.

Class 3 lever also has its fulcrum on one end; load is exerted at the opposite end, and the opposing effort is exerted on or about the middle. Examples of Class 3 levers are shovel and fishing rod where the hand is the fulcrum, tweezers, and human and animal arms and legs.

The application of a Class 1 lever is shown in Fig. 6. The lever is a bar of length AB with its fulcrum at *X*, dividing the length of the bar into parts: l_1 and l_2. To raise a load *W* through a height of *h*, a force *F* must be exerted downward through a distance *s*. The triangles AXC and BXD are similar and proportional; therefore, ignoring friction:

$$\frac{s}{h} = \frac{l_1}{l_2} \text{ and mechanical advantage MA} = \frac{l_1}{l_2}$$

Fig. 6 Diagram for calculating the mechanical advantage of a simple lever for raising a weight.

Winches, Windlasses, and Capstans

Winches, windlasses, and capstans are machines that convert rotary motion into linear motion, usually with some mechanical advantage. These machines are essentially Class 1 levers: effort is applied to a lever or crank, the fulcrum is the center of the drum, and the load is applied to the rope, chain, or cable.

Manually operated windlasses and capstans, mechanically the same, were originally used on sailing ships to raise and lower anchors. Operated by one or more levers by one or more sailors, both had barrels or drums on which rope or chain was wound. In the past, windlasses were distinguished from capstans; windlasses had horizontal drums and capstans had vertical drums. The modern term *winch* is now the generic name for any manual or power-operated drum for hauling a load with cable, chain, or rope. The manually operated winch, shown in Fig. 7, is widely used today on sailboats for raising and trimming sails, and sometimes for weighing anchors.

Ignoring friction, the mechanical advantage of all of these machines is approximately the *length of the crank* divided by the *diameter of the drum*. In the winch example shown, when the left end of the line is held under tension and the handle or crank is turned clockwise, a force is applied to the line entering on the right; it is attached to the load to perform such useful work as raising or tensioning sails.

Fig. 7 Diagram for calculating the mechanical advantage of a manually operated winch for raising anchors or sails.

LINKAGES

A *linkage* is a mechanism formed by connecting two or more levers together. Linkages can be designed to change the direction of a force or make two or more objects move at the same time. Many different fasteners are used to connect linkages together yet allow them to move freely such as pins, end-threaded bolts with nuts, and loosely fitted rivets. There are two general classes of linkages: *simple planar linkages* and more complex *specialized linkages*; both are capable of performing tasks such as describing straight lines or curves and executing motions at differing speeds. The names of the linkage mechanisms given here are widely but not universally accepted in all textbooks and references.

Linkages can be classified according to their primary functions:

- *Function generation*: the relative motion between the links connected to the frame
- *Path generation*: the path of a tracer point
- *Motion generation*: the motion of the coupler link

Simple Planar Linkages

Four different simple planar linkages shown in Fig. 8 are identified by function:

- *Reverse-motion linkage*, Fig. 8a, can make objects or force move in opposite directions; this can be done by using the input link as a lever. If the fixed pivot is equidistant from the moving pivots, output link movement will equal input link movement, but it will act in the opposite direction. However, if the fixed pivot is not centered, output link movement will not equal input link movement. By selecting the position of the fixed pivot, the linkage can be designed to produce specific mechanical advantages. This linkage can also be rotated through 360°.

- *Push-pull linkage*, Fig. 8b, can make the objects or force move in the same direction; the output link moves in the same direction as the input link. Technically classed as a four-bar linkage, it can be rotated through 360° without changing its function.

5

Fig. 8 Functions of four basic planar linkage mechanisms.

- *Parallel-motion linkage*, Fig. 8c, can make objects or forces move in the same direction, but at a set distance apart. The moving and fixed pivots on the opposing links in the parallelogram must be equidistant for this linkage to work correctly. Technically classed as a four-bar linkage, this linkage can also be rotated through 360° without changing its function. Pantographs that obtain power for electric trains from overhead cables are based on parallel-motion linkage. Drawing pantographs that permit original drawings to be manually copied without tracing or photocopying are also adaptations of this linkage; in its simplest form it can also keep tool trays in a horizontal position when the toolbox covers are opened.

- *Bell-crank linkage*, Fig. 8d, can change the direction of objects or force by 90°. This linkage rang doorbells before electric clappers were invented. More recently this mechanism has been adapted for bicycle brakes. This was done by pinning two bell cranks bent 90° in opposite directions together to form tongs. By squeezing the two handlebar levers linked to the input ends of each crank, the output ends will move together. Rubber blocks on the output ends of each crank press against the wheel rim, stopping the bicycle. If the pins which form a fixed pivot are at the midpoints of the cranks, link movement will be equal. However, if those distances vary, mechanical advantage can be gained.

Specialized Linkages

In addition to changing the motions of objects or forces, more complex linkages have been designed to perform many specialized functions: These include drawing or tracing straight lines; moving objects or tools faster in a retraction stroke than in an extension stroke; and converting rotating motion into linear motion and vice versa.

The simplest specialized linkages are four-bar linkages. These linkages have been versatile enough to be applied in many different applications. Four-bar linkages actually have only three moving links but they have one fixed link and four pin joints or pivots. A useful mechanism must have at least four links but closed-loop assemblies of three links are useful elements in structures. Because any linkage with at least one fixed link is a mechanism, both the parallel-motion and push-pull linkages mentioned earlier are technically machines.

Four-bar linkages share common properties: three rigid moving links with two of them hinged to fixed bases which form a *frame*. Link mechanisms are capable of producing rotating, oscillating, or reciprocating motion by the rotation of a crank. Linkages can be used to convert:

- Continuous rotation into another form of continuous rotation, with a constant or variable angular velocity ratio
- Continuous rotation into oscillation or continuous oscillation into rotation, with a constant or variable velocity ratio
- One form of oscillation into another form of oscillation, or one form of reciprocation into another form of reciprocation, with a constant or variable velocity ratio

There are four different ways in which four-bar linkages can perform inversions or complete revolutions about fixed pivot points. One pivoting link is considered to be the *input* or *driver* member and the other is considered to be the *output* or *driven member*. The remaining moving link is commonly called a *connecting link*. The fixed link, hinged by pins or pivots at each end, is called the *foundation link*.

Three inversions or linkage rotations of a four-bar chain are shown in Figs. 9, 10, and 11. They are made up of links AB, BC, CD, and AD. The forms of the three inversions are defined by the position of the shortest link with respect to the link selected as the foundation link. The ability of the driver or driven links to make complete rotations about their pivots determines their functions.

Drag-link mechanism, Fig. 9, demonstrates the first inversion. The shortest link AD between the two fixed pivots is the foundation link, and both driver link AB and driven link CD can make full revolutions.

Crank-rocker mechanism, Fig. 10, demonstrates the second inversion. The shortest link AB is adjacent to AD, the foundation link. Link AB can make a full 360° revolution while the opposite link CD can only oscillate and describe an arc.

Double-rocker mechanism, Fig. 11, demonstrates the third inversion. Link AD is the foundation link, and it is opposite the shortest link BC. Although link BC can make a full 360° revolution, both pivoting links AB and CD can only oscillate and describe arcs.

The fourth inversion is another *crank-rocker mechanism* that behaves in a manner similar to the mechanism shown in Fig. 10,

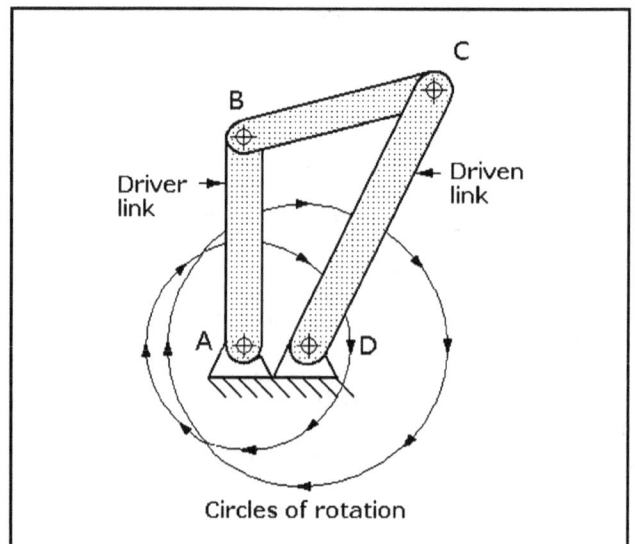

Fig. 9 Four-bar drag-link mechanism: Both the driver link AB and driven link CD can rotate through 360°. Link AD is the foundation link.

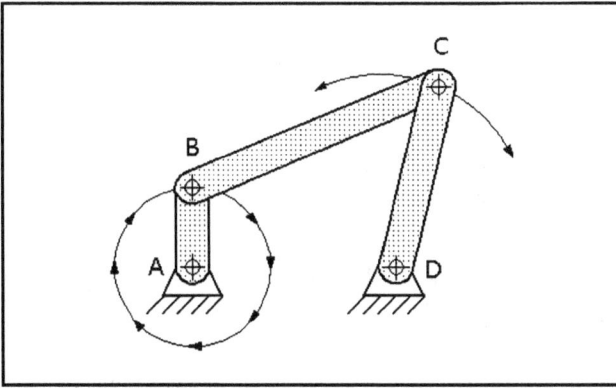

Fig. 10 Crank-rocker mechanism: Link AB can make a 360° revolution while link CD oscillates with C describing an arc. Link AD is the foundation link.

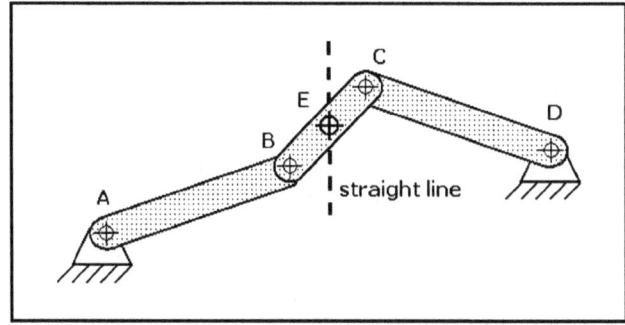

Fig. 11 Double-rocker mechanism: Short link BC can make a 360° revolution, but pivoting links AB and CD can only oscillate, describing arcs.

but the longest link, CD, is the foundation link. Because of this similarity between these two mechanisms, the fourth inversion is not illustrated here. A drag-link mechanism can produce either a nonuniform output from a uniform input rotation rate or a uniform output from a nonuniform input rotation rate.

Straight-Line Generators

Figures 12 to 15 illustrate examples of classical linkages capable of describing straight lines, a function useful in many different kinds of machines, particularly machine tools. The dimensions of the rigid links are important for the proper functioning of these mechanisms.

Watt's straight-line generator, illustrated in Fig. 12, can describe a short vertical straight line. Equal length links AB and CD are hinged at A and D, respectively. The midpoint E of connecting link BC traces a figure eight pattern over the full mechanism excursion, but a straight line is traced in part of the excursion because point E diverges to the left at the top of the stroke and to the right at the bottom of the stroke. This linkage was used by Scottish instrument maker, James Watt, in a steam-driven beam pump in about 1769, and it was a prominent mechanism in early steam-powered machines.

Scott Russell straight-line generator, shown in Fig. 13, can also describe a straight line. Link AB is hinged at point A and pinned to link CD at point B. Link CD is hinged to a roller at point C which restricts it to horizontal oscillating movement.

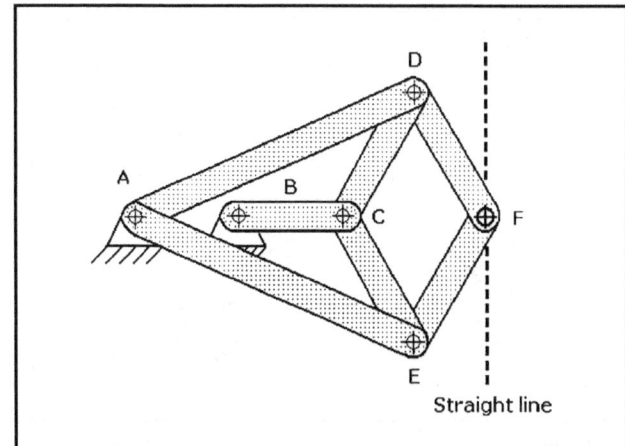

Fig. 12 Watt's straight-line generator: The center point E of link BC describes a straight line when driven by either links AB or CD.

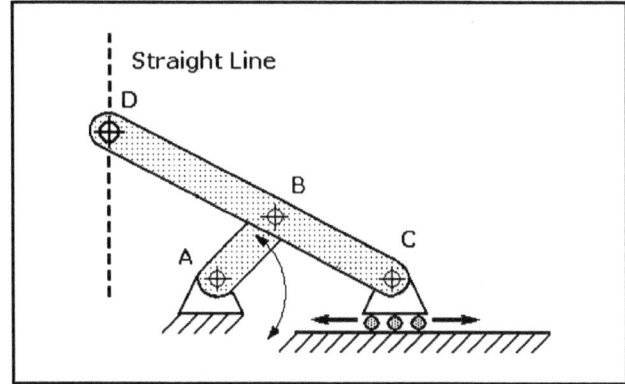

Fig. 13 Scott Russell straight-line generator: Point D of link DC describes a straight line as driver link AB oscillates, causing the slider at C to reciprocate left and right.

This configuration confines point D to a motion that traces a vertical straight line. Both points A and C lie in the same horizontal plane. This linkage works if the length of link AB is about 40 percent of the length of CD, and the distance between points D and B is about 60 percent of the length of CD.

Peaucellier's straight-line linkage, drawn as Fig. 14, can describe more precise straight lines over its range than either the Watt's or Scott Russell linkages. To make this linkage work correctly, the length of link BC must equal the distance between points A and B set by the spacing of the fixed pivots; in this figure, link BC is 15 units long while the lengths of links CD, DF, FE, and EC are equal at 20 units. As links AD and AE are moved,

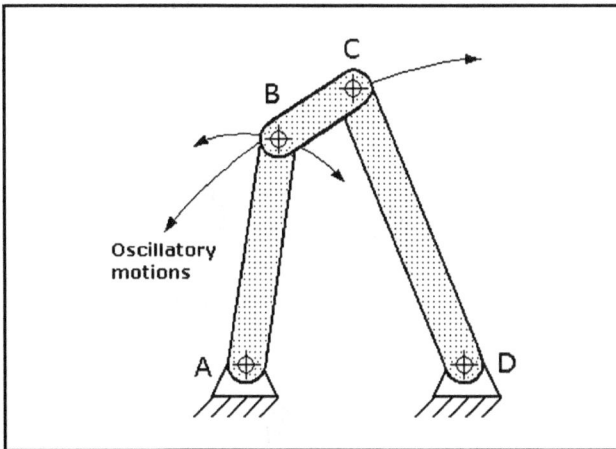

Fig. 14 Peaucellier's straight-line generator: Point F describes a straight line when either link AD or AE acts as the driver.

7

point F can describe arcs of any radius. However, the linkage can be restricted to tracing straight lines (infinite radiuses) by selecting link lengths for AD and AE. In this figure they are 45 units long. This linkage was invented in 1873 by the French engineer, Captain Charles-Nicolas Peaucellier.

Tchebicheff's straight-line generator, shown in Fig. 15, can also describe a horizontal line. Link CB with E as its midpoint traces a straight horizontal line for most of its transit as links AB and DC are moved to the left and right of center. To describe this straight line, the length of the foundation link AD must be twice the length of link CB. To make this mechanism work as a straight-line generator, CB is 10 units long, AD is 20 units long, and both AB and DC are 25 units long. With these dimensions, link CB will assume a vertical position when it is at the right and left extremes of its travel excursion. This linkage was invented by nineteenth-century Russian mathematician, Pafnuty Tchebicheff or Chebyshev.

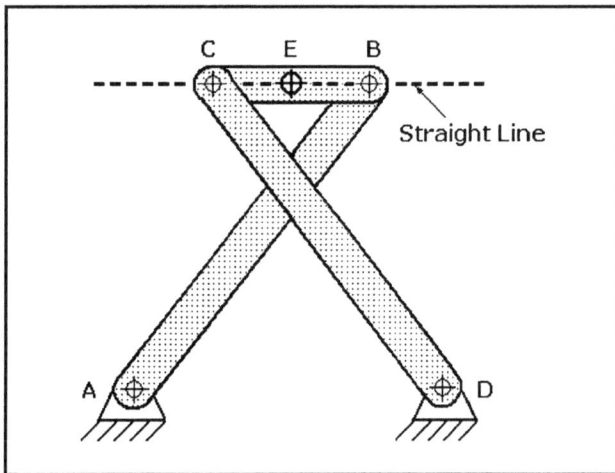

Fig. 15 Tchebicheff's straight-line generator: Point E of link CB describes a straight line when driven by either link AB or DC. Link CB moves into a vertical position at both extremes of its travel.

Rotary/Linear Linkages

Slider-crank mechanism (or a simple crank), shown as Fig. 16, converts rotary to linear motion and vice versa, depending on its application. Link AB is free to rotate 360° around the hinge while link BC oscillates back and forth because point C is hinged to a roller which restricts it to linear motion. Either the slider or the rotating link AB can be the driver.

This mechanism is more familiar as the piston, connecting rod, and crankshaft of an internal combustion engine, as illustrated in Fig. 1. The piston is the slider at C, the connecting rod is link BC, and the crankshaft is link AB. In a four-stroke engine, the piston is pulled down the cylinder by the crankshaft, admitting

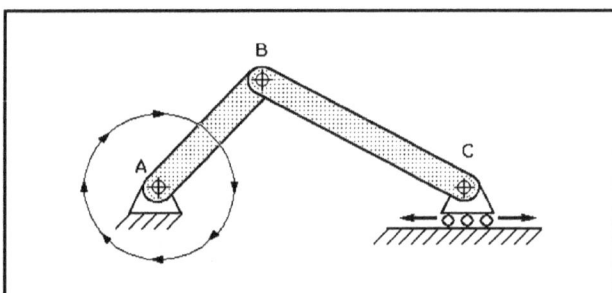

Fig. 16 Slider-crank mechanism: This simple crank converts the 360° rotation of driver link AB into linear motion of link BC, causing the slider at C to reciprocate.

the air-fuel mixture; in the compression stroke the piston is driven back up the cylinder by the crankshaft to compress the air-fuel mixture. However, the roles change in the combustion stroke when the piston drives the crankshaft. Finally, in the exhaust stroke the roles change again as the crankshaft drives the piston back to expel the exhaust fumes.

Scotch-yoke mechanism, pictured in Fig. 17, functions in a manner similar to that of the simple crank mechanism except that its linear output motion is sinusoidal. As wheel A, the driver, rotates, the pin or roller bearing at its periphery exerts torque within the closed yoke B; this causes the attached sliding bar to reciprocate, tracing a sinusoidal waveform. Part a shows the sliding bar when the roller is at 270°, and part b shows the sliding bar when the roller is at 0°.

Rotary-to-linear mechanism, drawn in Fig. 18, converts a uniform rotary motion into an intermittent reciprocating motion. The three teeth of the input rotor contact the steps in the frame or yoke, exerting torque 3 times per revolution, moving the yoke with attached bar. Full linear travel of the yoke is accomplished in 30° of rotor rotation followed by a 30° delay before returning the yoke. The reciprocating cycle is completed 3 times per revolution of the input. The output is that of a step function.

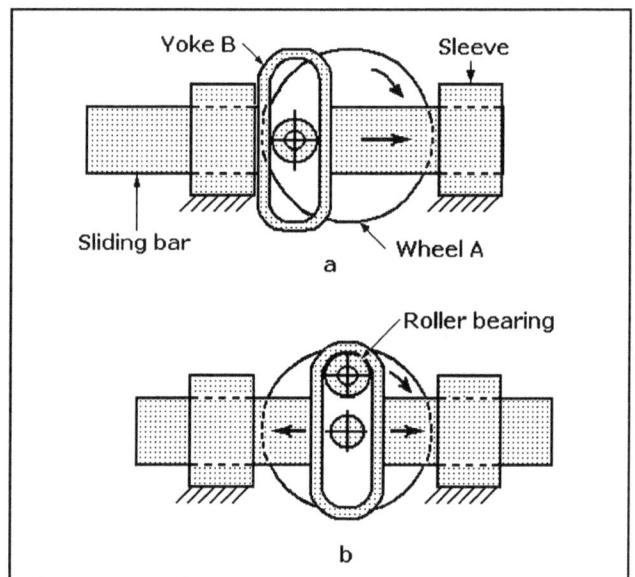

Fig. 17 Scotch-yoke mechanism translates the rotary motion of the wheel with a peripheral roller into reciprocating motion of the yoke with supporting bars as the roller exerts torque within the yoke. The yoke is shown in its left (270°) position in (a) and in its center (0°) position in (b).

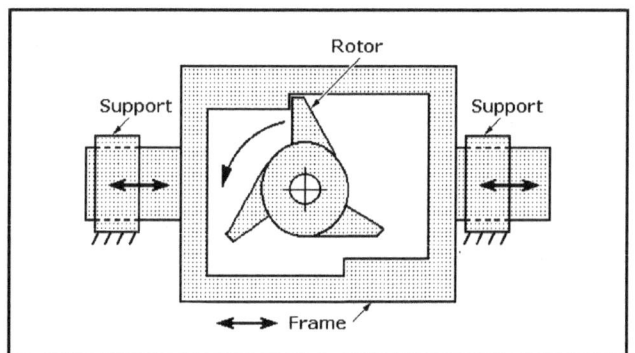

Fig. 18 Rotary-to-linear mechanism converts the uniform rotation of the 3-tooth rotor into a reciprocating motion of the frame and supporting bars. The reciprocating cycle is completed 3 times per rotor revolution.

SPECIALIZED MECHANISMS

Geneva wheel mechanism, illustrated in Fig. 19, is an example of intermittent gearing that converts continuous rotary motion into intermittent rotary motion. Geneva wheel C makes a quarter turn for every turn of lever AB attached to driving wheel A. When pin B on lever AB turns clockwise, it enters one of the four slots of geneva wheel C; the pin moves downward in the slot, applying enough torque to the geneva wheel to turn it counterclockwise $1/4$ revolution before it leaves the slot. As wheel A continues to rotate clockwise, it engages the next three slots in a sequence to complete one geneva wheel rotation. If one of the slots is obstructed, the pin can only move through part of the revolution, in either direction, before it strikes the closed slot, stopping the rotation of the geneva wheel. This mechanism has been used in mechanical windup watches, clocks, and music boxes to prevent overwinding.

Fig. 20 Swing-arm quick-return mechanism: As drive link AB rotates 360° around A, it causes the slider at B to reciprocate up and down along link CD, causing CD to oscillate though an arc. This motion drives link DE in a reciprocating motion that moves the rolling slider at E slowly to the right before returning it rapidly to the left.

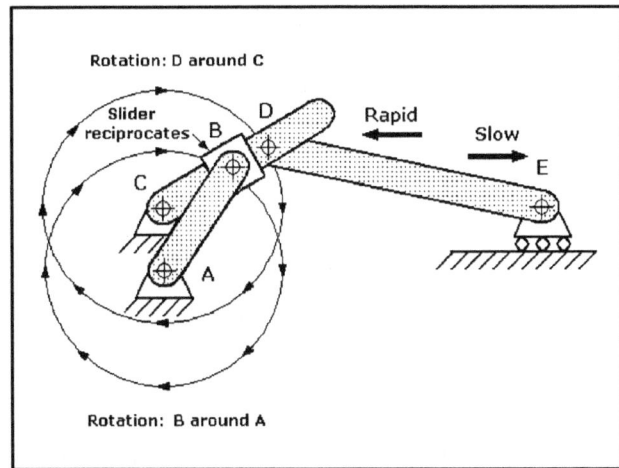

Fig. 19 Geneva wheel escapement mechanism: Pin B at the end of lever AB (attached to wheel A) engages a slot in geneva wheel C as wheel A rotates clockwise. Pin B moves down the slot, providing torque to drive the geneva wheel counterclockwise $1/4$ revolution before it exits the first slot; it then engages the next three slots to drive the geneva wheel through one complete counterclockwise revolution.

Fig. 21 Whitworth's quick-return mechanism: As drive link AB rotates 360° around A, it causes the slider at B to reciprocate back and forth along link CD, which, in turn causes CD to rotate 360° around C. This, motion causes link DE to reciprocate, first moving rolling slider at E slowly to the right before returning it rapidly to the left.

Swing-arm quick-return mechanism, drawn as Fig. 20, converts rotary motion into nonuniform reciprocating motion. As drive link AB rotates 360° around pin A, it causes the slider at B to reciprocate up and down along link CD. This, in turn, causes CD to oscillate left and right, describing an arc. Link DE, pinned to D with a rolling slider pinned at E, moves slowly to the right before being returned rapidly to the left.

Whitworth quick-return mechanism, shown as Fig. 21, converts rotary motion to nonuniform reciprocating motion. Drive link AB rotates 360° about pin A causing the slider at B to reciprocate back and forth along link CD; this, in turn, causes link CD to rotate 360° around point C. Link DE is pinned to link CD at D

and a rolling slider at E. The slider at E is moved slowly to the right before being returned rapidly to the left. This mechanism, invented in the nineteenth century by English engineer, Joseph Whitworth, has been adapted for shapers, machine tools with moving arms that cut metal from stationary workpieces. A hardened cutting tool attached at the end of the arm (equivalent to point E) advances slowly on the cutting stroke but retracts rapidly on the backstroke. This response saves time and improves productivity in shaping metal.

Simple ratchet mechanism, drawn as Fig. 22, can only be turned in a counterclockwise direction. The ratchet wheel has many wedge-shaped teeth that can be moved incrementally to turn an oscillating drive lever. As driving lever AB first moves clockwise to initiate counterclockwise movement of the wheel, it drags pawl C pinned at B over one or more teeth while pawl D prevents the wheel from turning clockwise. Then, as lever AB reverses to drive the ratchet wheel counterclockwise, pawl D is released, allowing the wheel to turn it in that direction. The amount of backward incremental motion of lever AB is directly proportional to pitch of the teeth: smaller teeth will reduce the degree of rotation while larger teeth will increase them. The contact surfaces of the teeth on the wheel are typically inclined, as shown, so they will not be disengaged if the mechanism is subjected to vibration or shock under load. Some ratchet mechanisms include a spring to hold pawl D against the teeth to assure no clockwise wheel rotation as lever AB is reset.

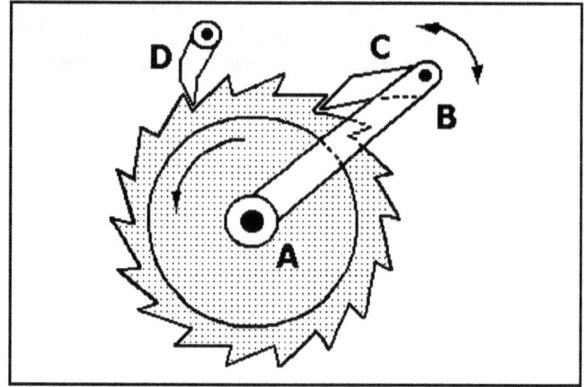

Fig. 22 This ratchet wheel can be turned only in a counterclockwise direction. As driving lever AB moves clockwise, it drags pawl C, pinned at B over one or more teeth while pawl D prevents the wheel from turning clockwise. Then as lever AB reverses to drive the ratchet wheel counterclockwise, pawl D is released allowing the wheel to turn it in that direction.

GEARS AND GEARING

A *gear* is a wheel with evenly sized and spaced teeth machined or formed around its perimeter. Gears are used in rotating machinery not only to transmit motion from one point to another, but also for the mechanical advantage they offer. Two or more gears transmitting motion from one shaft to another is called a *gear train,* and *gearing* is a system of wheels or cylinders with meshing teeth. Gearing is chiefly used to transmit rotating motion but can also be adapted to translate reciprocating motion into rotating motion and vice versa.

Gears are versatile mechanical components capable of performing many different kinds of power transmission or motion control. Examples of these are

- Changing rotational speed
- Changing rotational direction
- Changing the angular orientation of rotational motion
- Multiplication or division of torque or magnitude of rotation
- Converting rotational to linear motion, and its reverse
- Offsetting or changing the location of rotating motion

The teeth of a gear can be considered as levers when they mesh with the teeth of an adjoining gear. However, gears can be rotated continuously instead of rocking back and forth through short distances as is typical of levers. A gear is defined by the number of its teeth and its diameter. The gear that is connected to the source of power is called the *driver*, and the one that receives power from the driver is the *driven gear*. It always rotates in a direction opposing that of the driving gear; if both gears have the same number of teeth, they will rotate at the same speed. However, if the number of teeth differs, the gear with the smaller *r* number of teeth will rotate faster. The size and shape of all gear teeth that are to mesh properly for working contact must be equal.

Figure 23 shows two gears, one with 15 teeth connected at the end of shaft A, and the other with 30 teeth connected at the end of shaft B. The 15 teeth of smaller driving gear A will mesh with 15 teeth of the larger gear B, but while gear A makes one revolution gear B will make only $\frac{1}{2}$ revolution.

The number of teeth on a gear determines its diameter. When two gears with different diameters and numbers of teeth are meshed

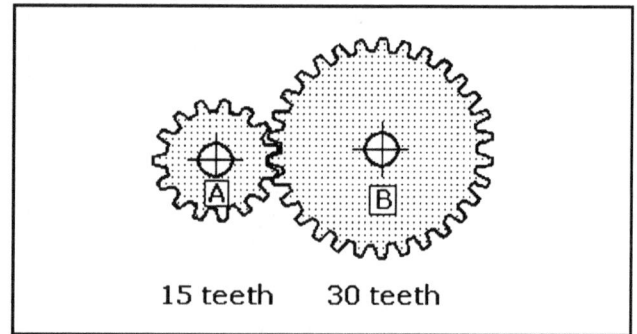

Fig. 23 Gear B has twice as many teeth as gear A, and it turns at half the speed of gear A because gear speed is inversely proportional to the number of teeth on each gear wheel.

together, the number of teeth on each gear determines gear ratio, velocity ratio, distance ratio, and mechanical advantage. In Fig. 23, gear A with 15 teeth is the driving gear and gear B with 30 teeth is the driven gear. The gear ratio GR is determined as:

$$GR = \frac{\text{number of teeth on driven gear B}}{\text{number of teeth on driving gear A}}$$

$$= \frac{30}{15} = \frac{2}{1} \text{ (also written as 2:1)}$$

The number of teeth in both gears determines the rotary distance traveled by each gear and their angular speed or velocity ratio. The angular speeds of gears are inversely proportional to the numbers of their teeth. Because the smaller driving gear A in Fig. 23 will revolve twice as fast as the larger driven gear B, velocity ratio VR is:

$$VR = \frac{\text{velocity of driving gear A}}{\text{velocity of driven gear B}} = \frac{2}{1} \text{ (also written as 2:1)}$$

In this example load is represented by driven gear B with 30 teeth and the effort is represented by driving gear A with 15 teeth. The distance moved by the load is twice that of the effort. Using the general formula for mechanical advantage MA:

$$MA = \frac{load}{effort} = \frac{30}{15} = 2$$

Simple Gear Trains

A gear train made up of multiple gears can have several drivers and several driven gears. If the train contains an odd number of gears, the output gear will rotate in the same direction as the input gear, but if the train contains an even number of gears, the output gear will rotate opposite that of the input gear. The number of teeth on the intermediate gears does not affect the overall velocity ratio, which is governed purely by the number of teeth on the first and last gear.

In simple gear trains, high or low gear ratios can only be obtained by combining large and small gears. In the simplest basic gearing involving two gears, the driven shaft and gear revolves in a direction opposite that of the driving shaft and gear. If it is desired that the two gears and shafts rotate in the same direction, a third *idler gear* must be inserted between the driving gear and the driven gear. The idler revolves in a direction opposite that of the driving gear.

A simple gear train containing an idler is shown in Fig. 24. Driven idler gear B with 20 teeth will revolve 4 times as fast counterclockwise as driving gear A with 80 teeth turning clockwise. However, gear C, also with 80 teeth, will only revolve one turn clockwise for every four revolutions of idler gear B, making the velocities of both gears A and C equal except that gear C turns in the same direction as gear A. In general, the velocity ratio of the first and last gears in a train of simple gears is not changed by the number of gears inserted between them.

Fig. 24 Gear train: When gear A turns once clockwise, gear B turns four times counter clockwise, and gear wheel C turns once clockwise. Gear B reverses the direction of gear C so that both gears A and C turn in the same direction with no change in the speed of gear C.

Compound Gear Trains

More complex compound gear trains can achieve high and low gear ratios in a restricted space by coupling large and small gears on the same axle. In this way gear ratios of adjacent gears can be multiplied through the gear train. Figure 25 shows a set of compound gears with the two gears B and D mounted on the middle shaft. Both rotate at the same speed because they are fastened together. If gear A (80 teeth) rotates at 100 rpm clockwise, gear B (20 teeth) turns at 400 rpm counterclockwise because of its velocity ratio of 1 to 4. Because gear D (60 teeth) also turns at 400 rpm and its velocity ratio is 1 to 3 with respect to gear C

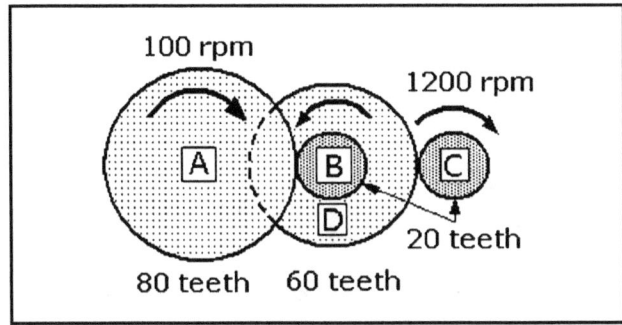

Fig. 25 Compound gears: Two gears B and D are mounted on a central shaft and they turn at the same speed. If gear A rotates at 100 rpm clockwise, gears B and D turn counterclockwise at 400 rpm, and gear C, driven by gear D, turns clockwise at 1200 rpm.

(20 teeth), gear C will turn at 1200 rpm clockwise. The velocity ratio of a compound gear train can be calculated by multiplying the velocity ratios for all pairs of meshing gears. For example, if the driving gear has 45 teeth and the driven gear has 15 teeth, the velocity ratio is $^{15}/_{45} = ^{1}/_{3}$.

Gear Classification

All gears can be classified as either external gears or internal or annual gears:

- *External gears* have teeth on the outside surface of the disk or wheel.
- *Internal* or *annual gears* have teeth on the inside surface of a ring or cylinder.

Spur gears are cylindrical external gears with teeth that are cut straight across the edge of the disk or wheel parallel to the axis of rotation. The spur gears shown in Fig. 26a are the simplest gears. They normally translate rotating motion between two parallel shafts. An *internal* or *annual* gear, as shown in Fig. 26b, is a variation of the spur gear except that its teeth are cut on the inside of a ring or flanged wheel rather than on the outside. Internal gears usually drive or are driven by a pinion. The disadvantage of a simple spur gear is its tendency to produce thrust that can misalign other meshing gears along their respective shafts, thus reducing the face widths of the meshing gears and reducing their mating surfaces.

Rack gears, as the one shown in Fig. 26c, have teeth that lie in the same plane rather than being distributed around a wheel. This gear configuration provides straight-line rather than rotary motion. A rack gear functions like a gear with an infinite radius.

Pinions are small gears with a relatively small number of teeth which can be mated with rack gears.

Rack and pinion gears, shown in Fig. 26c, convert rotary motion to linear motion; when mated together they can transform the rotation of a pinion into reciprocating motion, or vice versa. In some systems, the pinion rotates in a fixed position and engages the rack which is free to move; the combination is found in the steering mechanisms of vehicles. Alternatively, the rack is fixed while the pinion rotates as it moves up and down the rack: Funicular railways are based on this drive mechanism; the driving pinion on the rail car engages the rack positioned between the two rails and propels the car up the incline.

Bevel gears, as shown in Fig. 26d, have straight teeth cut into conical circumferences which mate on axes that intersect, typically at right angles between the input and output shafts. This class of gears includes the most common straight and spiral bevel gears as well as miter and hypoid gears.

Fig. 26 Gear types: Eight common types of gears and gear pairs are shown here.

Straight bevel gears are the simplest bevel gears. Their straight teeth produce instantaneous line contact when they mate. These gears provide moderate torque transmission, but they are not as smooth running or quiet as spiral bevel gears because the straight teeth engage with full-line contact. They permit medium load capacity.

Spiral bevel gears have curved oblique teeth. The spiral angle of curvature with respect to the gear axis permits substantial tooth overlap. Consequently, the teeth engage gradually and at least two teeth are in contact at the same time. These gears have lower tooth loading than straight bevel gears and they can turn up to 8 times faster. They permit high load capacity.

Miter gears are mating bevel gears with equal numbers of teeth used between rotating input and output shafts with axes that are 90° apart.

Hypoid gears are helical bevel gears used when the axes of the two shafts are perpendicular but do not intersect. They are commonly used to connect driveshafts to rear axles of automobiles, and are often incorrectly called *spiral gearing*.

Helical gears are external cylindrical gears with their teeth cut at an angle rather than parallel to the axis. A simple helical gear, as shown in Fig. 26e, has teeth that are offset by an angle with respect to the axis of the shaft so that they spiral around the shaft in a helical manner. Their offset teeth make them capable of smoother and quieter action than spur gears, and they are capable of driving heavy loads because the teeth mesh at an acute angle rather than at 90°. When helical gear axes are parallel they are called parallel helical gears, and when they are at right angles they are called helical gears. Herringbone and worm gears are based on helical gear geometry.

Herringbone or double helical gears, as shown in Fig. 26f, are helical gears with V-shaped right-hand and left-hand helix angles side by side across the face of the gear. This geometry neutralizes axial thrust from helical teeth.

Worm gears, also called *screw gears*, are other variations of helical gearing. A worm gear has a long, thin cylindrical form with one or more continuous helical teeth that mesh with a helical gear. The teeth of the worm gear slide across the teeth of the driven gear rather than exerting a direct rolling pressure as do the teeth of helical gears. Worm gears are widely used to transmit rotation, at significantly lower speeds, from one shaft to another at a 90° angle.

Face gears have straight tooth surfaces, but their axes lie in planes perpendicular to shaft axes. They are designed to mate with instantaneous point contact. These gears are used in right-angle drives, but they have low load capacities.

Practical Gear Configurations

Isometric drawing Fig. 27 shows a *special planetary gear configuration*. The external driver spur gear (lower right) drives the outer ring spur gear (center) which, in turn, drives three internal planet spur gears; they transfer torque to the driven gear (lower left). Simultaneously, the central planet spur gear produces a summing motion in the pinion gear (upper right) which engages a rack with a roller follower contacting a radial disk cam (middle right).

Fig. 27 A special planetary-gear mechanism: The principal of relative motion of mating gears illustrated here can be applied to spur gears in a planetary system. The motion of the central planet gear produces the motion of a summing gear.

Isometric drawing Fig. 28 shows a *unidirectional drive*. The output shaft B rotates in the same direction at all times, regardless of the rotation of the input shaft A. The angular velocity of output shaft B is directly proportional to the angular velocity of input shaft A. The spur gear C on shaft A has a face width that is twice as wide as the faces on spur gears F and D, which are mounted on output shaft B. Spur gear C meshes with idler E and with spur gear D. Idler E meshes with the spur gears C and F. Output shaft B carries two free-wheel disks, G and H, which are oriented unidirectionally.

When input shaft A rotates clockwise (bold arrow), spur gear D rotates counterclockwise and it idles around free-wheel disk H. Simultaneously, idler E, which is also rotating counterclockwise, causes spur gear F to turn clockwise and engage the rollers on free-wheel disk G. Thus, shaft B is made to rotate clockwise. On the other hand, if the input shaft A turns counterclockwise

Fig. 28 The output shaft of this unidirectional drive always rotates in the same direction regardless of the direction of rotation of the input shaft.

(dotted arrow), spur gear F will idle while spur gear D engages free-wheel disk H, which drives shaft B so that it continues to rotate clockwise.

Gear Tooth Geometry

The geometry of gear teeth, as shown in Fig. 29, is determined by pitch, depth, and pressure angle.

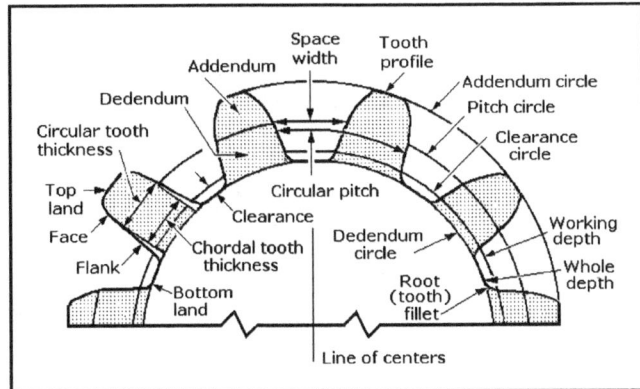

Fig. 29 Gear-tooth geometry.

Gear Terminology

addendum: The radial distance between the *top land* and the *pitch circle*. This distance is measured in inches or millimeters.

addendum circle: The circle defining the outer diameter of the gear.

circular pitch: The distance along the pitch circle from a point on one tooth to a corresponding point on an adjacent tooth. It is also the sum of the *tooth thickness* and the *space width*. This distance is measured in inches or millimeters.

clearance: The radial distance between the *bottom land* and the *clearance circle*. This distance is measured in inches or millimeters.

contact ratio: The ratio of the number of teeth in contact to the number of teeth not in contact.

dedendum: The radial distance between the *pitch circle* and the *dedendum circle*. This distance is measured in inches or millimeters.

dedendum circle: The theoretical circle through the *bottom lands* of a gear.

depth: A number standardized in terms of pitch. Full-depth teeth have a *working depth* of 2/P. If the teeth have equal *addenda* (as in standard interchangeable gears), the addendum is 1/P. Full-depth gear teeth have a larger contact ratio than stub teeth, and their working depth is about 20 percent more than stub gear teeth. Gears with a small number of teeth might require *undercutting* to prevent one interfering with another during engagement.

diametral pitch (P): The ratio of the number of teeth to the *pitch diameter*. A measure of the coarseness of a gear, it is the index of tooth size when U.S. units are used, expressed as teeth per inch.

pitch: A standard pitch is typically a whole number when measured as a *diametral pitch (P)*. *Coarse pitch gears* have teeth larger than a diametral pitch of 20 (typically 0.5 to 19.99). *Fine-pitch gears* usually have teeth of diametral pitch greater than 20. The usual maximum fineness is 120 diametral pitch, but involute-tooth gears can be made with diametral pitches as fine as 200, and cycloidal tooth gears can be made with diametral pitches to 350.

pitch circle: A theoretical circle upon which all calculations are based.

pitch diameter: The diameter of the *pitch circle*, the imaginary circle that rolls without slipping with the pitch circle of the mating gear, measured in inches or millimeters.

pressure angle: The angle between the *tooth profile* and a line perpendicular to the *pitch circle*, usually at the point where the pitch circle and the tooth profile intersect. Standard angles are 20° and 25°. It affects the force that tends to separate mating gears. A high pressure angle decreases the *contact ratio*, but it permits the teeth to have higher capacity and it allows gears to have fewer teeth without *undercutting*.

Gear Dynamics Terminology

backlash: The amount by which the width of a tooth space exceeds the thickness of the engaging tooth measured on the pitch circle. It is the shortest distance between the noncontacting surfaces of adjacent teeth.

gear efficiency: The ratio of output power to input power taking into consideration power losses in the gears and bearings and from windage and the churning of the gear lubricant.

gear power: A gear's load and speed capacity. It is determined by gear dimensions and type. Helical and helical-type gears have capacities to approximately 30,000 hp, spiral bevel gears to about 5000 hp, and worm gears to about 750 hp.

gear ratio: The number of teeth in the larger gear of a pair divided by the number of teeth in the *pinion* gear (the smaller gear of a pair). It is also the ratio of the speed of the pinion to the speed of the gear. In reduction gears, the ratio of input speed to output speed.

gear speed: A value determined by a specific pitchline velocity. It can be increased by improving the accuracy of the gear teeth and the balance of all rotating parts.

undercutting: The recessing in the bases of gear tooth flanks to improve clearance.

PULLEYS AND BELTS

Pulleys and belts transfer rotating motion from one shaft to another. Essentially, pulleys are gears without teeth that depend on the frictional forces of connecting belts, chains, ropes, or cables to transfer torque. If both pulleys have the same diameter, they will rotate at the same speed. However, if one pulley is larger than the other, mechanical advantage and velocity ratio are gained. As with gears, the velocities of pulleys are inversely proportional to their diameters. A large drive pulley driving a smaller driven pulley by means of a belt or chain is shown in Fig. 30. The smaller pulley rotates faster than the larger pulley in the same direction as shown in Fig. 30a. If the belt is crossed, as shown in Fig. 30b, the smaller pulley also rotates faster than the larger pulley, but its rotation is in the opposite direction.

A familiar example of belt and pulley drive can be seen in automotive cooling fan drives. A smooth pulley connected to the engine crankshaft transfers torque to a second smooth pulley coupled to the cooling fan with a reinforced rubber endless belt. Before reliable direct-drive industrial electric motors were developed, a wide variety of industrial machines equipped with smooth pulleys of various diameters were driven by endless leather belts from an overhead driveshaft. Speed changes were achieved by switching the belt to pulleys of different diameters on the same

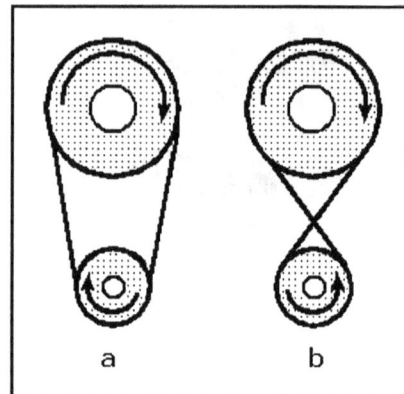

Fig. 30 Belts on pulleys: With a continuous belt both pulleys rotate in the same direction (a), but with a crossed belt both pulleys rotate in opposite directions (b).

machine. The machines included lathes and milling machines, circular saws in sawmills, looms in textile plants, and grinding wheels in grain mills. The source of power could have been a water wheel, windmill, or a steam engine.

SPROCKETS AND CHAINS

Sprockets and chains offer another method for transferring rotating motion from one shaft to another where the friction of a drive belt would be insufficient to transfer power. The speed relationships between sprockets of different diameters coupled by chains are the same as those between pulleys of different diameters coupled by belts, as shown in Fig. 30. Therefore, if the chains are crossed, the sprockets will rotate in different directions. Bicycles

have sprocket and chain drives. The teeth on the sprockets mesh with the links on the chains. Powered winches on large ships act as sprockets because they have teeth that mate with the links of heavy chain for raising anchors. Another example can be seen in tracked equipment including bulldozers, cranes, and military tanks. The flexible treads have teeth that mate with teeth on driving sprockets that propel these machines.

CAM MECHANISMS

A *cam* is a mechanical component capable of transmitting motion to a follower by direct contact. In a cam mechanism, the cam is the driver and the driven member is called the *follower*. The follower can remain stationary, translate, oscillate, or rotate. The general form of a plane cam mechanism is illustrated in the kinematic diagram Fig. 31. It consists of two shaped members A and B with smooth, round, or elongated contact surfaces connected to a third body C. Either body A or body B can be the driver, while the other body is the follower. These shaped bodies can be replaced by an equivalent mechanism. Points 1 and 2 are pin-jointed at the centers of curvature of the contacting surfaces. If any change is made in the relative positions of bodies A and B, points 1 and 2 are shifted, and the links of the equivalent mechanisms have different lengths.

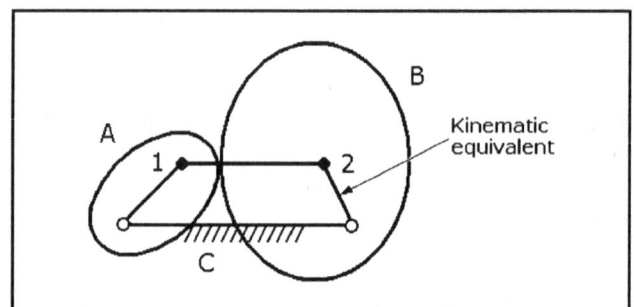

Fig. 31 Basic cam mechanism and its kinematic equivalent. Points 1 and 2 are centers of curvature of the contact point.

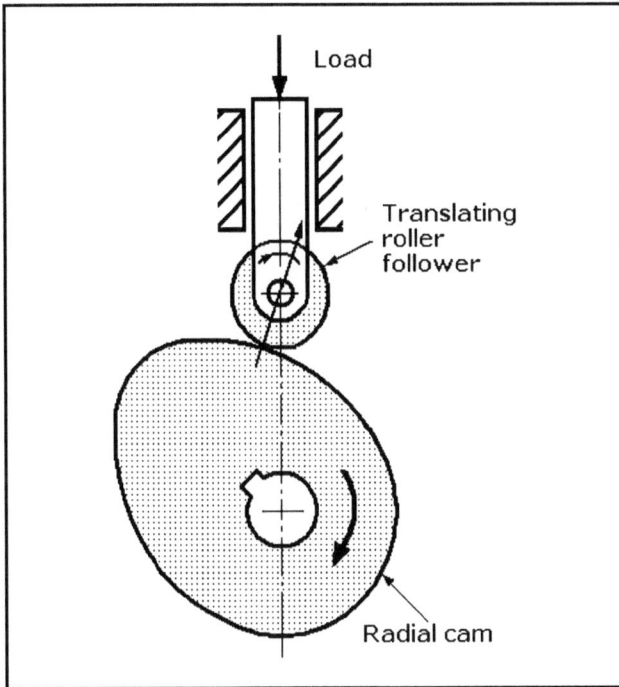

Fig. 32 Radial open cam with a translating roller follower. The roller is kept in contact with the cam by the mass of the load.

A widely used open radial-cam mechanism is shown in Fig. 32. The roller follower is the most common follower used in these mechanisms because it can transfer power efficiently between the cam and follower by reducing friction and minimizing wear between them. The arrangement shown here is called a *gravity constraint cam*; it is simple and effective and can be used with rotating disk or end cams if the weight of the follower system is enough to keep it in constant contact with the cam profile. However, in most practical cam mechanisms, the cam and follower are constrained at all operating speeds by preloaded compression springs. Cams can be designed by three methods:

- Shaping the cam body to some known curve, such as a spiral, parabola, or circular arc
- Designing the cam mathematically to determine follower motion and then plotting the tabulated data to form the cam
- Drawing the cam profile freehand using various drafting curves

The third method is acceptable only if the cam motion is intended for low speeds that will permit the use of a smooth, "bumpless" curve. In situations where higher loads, mass, speed, or elasticity of the members are encountered, a detailed study must be made of both the dynamic aspects of the cam curve and the accuracy of cam fabrication.

Many different kinds of machines include cams, particularly those that operate automatically such as printing presses, textile looms, gear-cutters, and screw machines. Cams open and close the valves in internal combustion engines, index cutting tools on machine tools, and operate switches and relays in electrical control equipment. Cams can be made in an infinite variety of shapes from metal or hard plastic. Some of the most important cams will be considered here. The possible applications of mechanical cams are still unlimited despite the introduction of *electronic cams* that mimic mechanical cam functions with appropriate computer software.

Classification of Cam Mechanisms

Cam mechanisms can be classified by their input/output motions, the configuration and arrangement of the follower, and the shape of the cam. Cams can also be classified by the kinds of motions made by the follower and the characteristics of the cam profile. The possible kinds of input/output motions of cam mechanisms with the most common disk cams are shown in Figs. 33a to e; they are examples of rotating disk cams with translating followers. By contrast, Fig. 33f shows a follower arm with a roller that swings or oscillates in a circular arc with respect to the follower hinge as the cam rotates. The follower configurations in Figs. 33a to d are named according to their characteristics: a *knife-edge;* b, e, and f *roller;* c *flat-faced;* and d *spherical-faced.* The face of the flat follower can also be oblique with respect to the cam. The follower is an element that moves either up and down or side to side as it follows the contour of the cam.

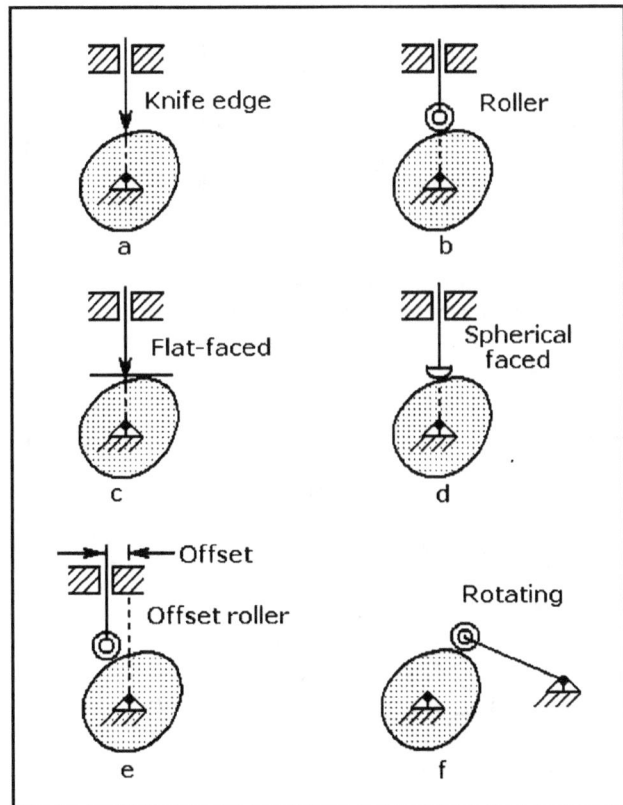

Fig. 33 Cam configurations: Six different configurations of radial open cams and their followers.

There are two basic types of follower: *in-line* and *offset.* The centerline of the in-line follower passes through the centerline of the camshaft. Figures 33a to d show five followers that move in a plane perpendicular to the axis of rotation of the camshaft. By contrast, the centerline of the offset follower, as illustrated in Fig. 33e, does not pass through the centerline of the camshaft. The amount of offset is the horizontal distance between the two centerlines. Follower offset reduces the side thrust introduced by the roller follower. Figure 33f illustrates a translating or swing-arm rotating follower that must be constrained to maintain contact with the cam profile.

The most common rotating disk or plate cams can be made in a variety of shapes including offset round, egg-shaped, oval, and cardioid or heart-shaped. Most cams are mounted on a rotating shaft. The cam and follower must be constrained at all operating speeds to keep them in close contact throughout its cycle if a cam mechanism is to function correctly. Followers are typically spring-loaded to maintain constant contact with the shaped surface of the cam, but gravity constraint is still an option.

If it is anticipated that a cam mechanism will be subjected to severe shock and vibration, a *grooved disk cam,* as shown in Fig. 34, can be used. The cam contour is milled into the face of a disk so that the roller of the cam follower will be confined and continuously constrained within the side walls of the groove throughout the cam cycle. The groove confines the follower roller during the entire cam rotation. Alternatively, the groove can be milled on the outer circumference of a cylinder or barrel to form a *cylindrical* or *barrel cam,* as shown in Fig. 35. The follower of this cam can translate or oscillate. A similar groove can also be milled around the conical exterior surface of a *grooved conical cam.*

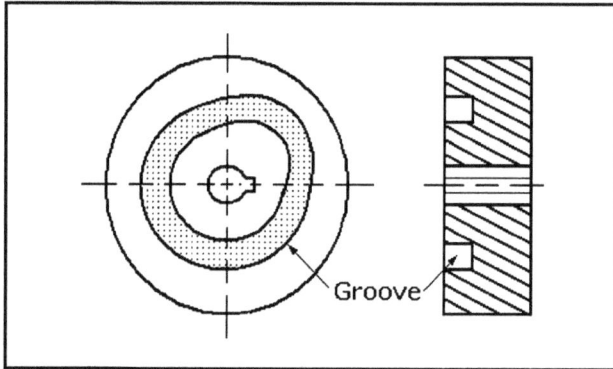

Fig. 34 Grooved cam made by milling a contoured cam groove into a metal or plastic disk. A roller follower is held within the grooved contour by its depth, eliminating the need for spring-loading.

Fig. 35 Cylindrical or barrel cam: A roller follower tracks the groove precisely because of the deep contoured groove milled around the circumference of the rotating cylinder.

By contrast, the barrel-shaped *end cam,* shown in Fig. 36, has a contour milled on one end. This cam is usually rotated, and its follower can also either translate or oscillate, but the follower system must be carefully controlled to exercise the required constraint because the follower roller is not confined by a groove. Another distinct form of cam is the *translating cam,* as shown in Fig. 37. It is typically mounted on a bed or carrier that moves back and forth in a linear reciprocal motion under a stationary vertical translating follower, usually with a roller. However, the cam can also be mounted so that it remains stationary while a follower system moves in a linear reciprocal motion over the limited range of the cam.

The unusual dual-rotary cam configuration shown in Fig. 38 is a *constant-diameter cam*; it consists of two identical disk cams

Fig. 36 End cam: A roller follower tracks a cam contour machined at the end of this rotating cylindrical cam.

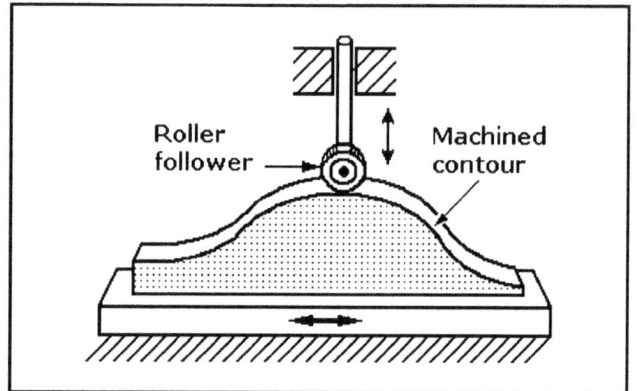

Fig. 37 Translating cam: A roller follower either tracks the reciprocating motion of the cam profile or is driven back and forth over a stationary cam profile.

Fig. 38 Constant-diameter cam: Two identical cams, 1 and 2, are separated on the same shaft and offset at an angle that provides a virtual constant diameter. Cam 1 with roller follower 1 is the functioning cam, and cam 2 with roller follower 2 constrains cam 1 to smooth its motion.

with followers mounted a fixed distance apart on a common shaft, but the cams are offset so that if superimposed their contours form a virtual circle of constant diameter. Cam 1 is the functional cam while cam 2 acts as a constraint, effectively canceling out the irregular motion that occurs with a single rotary cam and follower.

The motions of the followers of all of these cam mechanisms can be altered to obtain a different sequence by changing the contour of the cam profile. The timing of the sequence of disk and cylinder cams can be changed by altering the rotational speed of their camshafts. The timing of the sequence of the translation cam can be changed by altering the rate of reciprocal motion of the bed on which it is mounted on its follower system. The rotation of the follower roller does not influence the motion of any of the cam mechanisms.

Cam Terminology

Figure 39 illustrates the nomenclature for a radial open disk cam with a *roller follower* on a plate cam.

base circle: The circle with the shortest radius from the cam center to any part of the cam profile.

cam profile: The outer surface of a disk cam as it was machined.

follower travel: For a *roller follower* of a disk cam it is the vertical distance of follower travel measured at the center point of the roller as it travels from the *base circle* to the *cam profile*.

motion events: When a cam rotates through one cycle, the follower goes through rises, dwells, and returns. A *rise* is the motion of the follower away from the cam center; a *dwell* occurs when the follower is resting; and a *return* is the motion of the follower toward the cam center.

pitch curve: For a *roller follower* of a disk cam it is the path generated by the center point of the roller as the follower is rotated around a stationary plate cam.

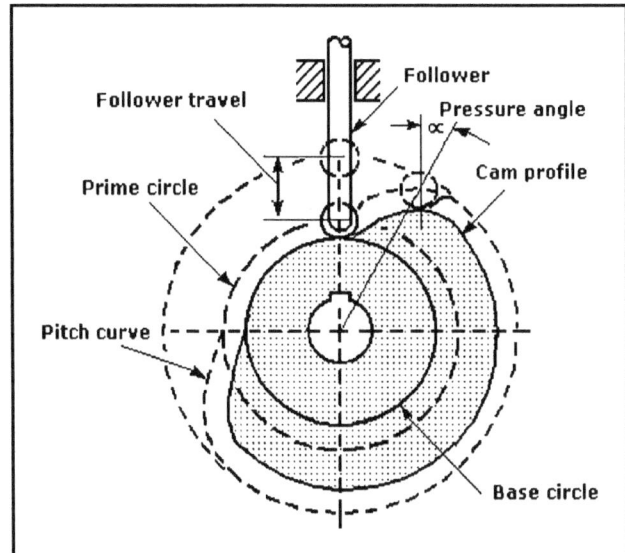

Fig. 39 Cam nomenclature: This diagram identifies the industry-accepted technical terms for cam features.

pressure angle: For a *roller follower* of a disk cam it is the angle at any point between the normal to the pitch curve and the instantaneous direction of follower motion. This angle is important in cam design because it indicates the steepness of the cam profile.

prime circle (reference circle): For a *roller follower* of a disk cam it is the circle with the shortest radius from the cam center to the pitch curve.

stroke or throw: The longest distance or widest angle through which the follower moves or rotates.

working curve: The working surface of a cam that contacts the follower. For a *roller follower* of a plate cam it is the path traced by the center of the roller around the cam profile.

CLUTCH MECHANISMS

A *clutch* is defined as a coupling that connects and disconnects the driving and driven parts of a machine; an example is an engine and a transmission. Clutches typically contain a driving shaft and a driven shaft, and they are classed as either externally or internally controlled. *Externally controlled clutches* can be controlled either by friction surfaces or components that engage or mesh positively. *Internally controlled clutches* are controlled by internal mechanisms or devices; they are further classified as *overload*, *overriding*, and *centrifugal*. There are many different schemes for a driving shaft to engage a driven shaft.

Externally Controlled Friction Clutches

Friction-Plate Clutch. This clutch, shown in Fig. 40, has a control arm, which when actuated, advances a sliding plate on the driving shaft to engage a mating rotating friction plate on the same shaft; this motion engages associated gearing that drives the driven shaft. When reversed, the control arm disengages the sliding plate. The friction surface can be on either plate, but is typically only on one.

Cone Clutch. A clutch operating on the same principle as the friction-plate clutch except that the control arm advances a cone on the driving shaft to engage a mating rotating friction cone on the same shaft; this motion also engages any associated gearing that drives the driven shaft. The friction surface can be on either cone but is typically only on the sliding cone.

Expanding Shoe Clutch. This clutch is similar to the friction-plate clutch except that the control arm engages linkage that forces several friction shoes radially outward so they engage the inner surface of a drum on or geared to the driven shaft.

Externally Controlled Positive Clutches

Jaw Clutch. This clutch is similar to the plate clutch except that the control arm advances a sliding jaw on the driving shaft to make positive engagement with a mating jaw on the driven shaft.

Other examples of externally controlled positive clutches are the *planetary transmission clutch* consisting essentially of a sun gear keyed to a driveshaft, two planet gears, and an outer driven

Fig. 40 Friction plate clutch: When the left sliding plate on the driving shaft is clamped by the control arm against the right friction plate idling on the driving shaft, friction transfers the power of the driving shaft to the friction plate. Gear teeth on the friction plate mesh with a gear mounted on the driven shaft to complete the transfer of power to the driven mechanism. Clutch torque depends on the axial force exerted by the control arm.

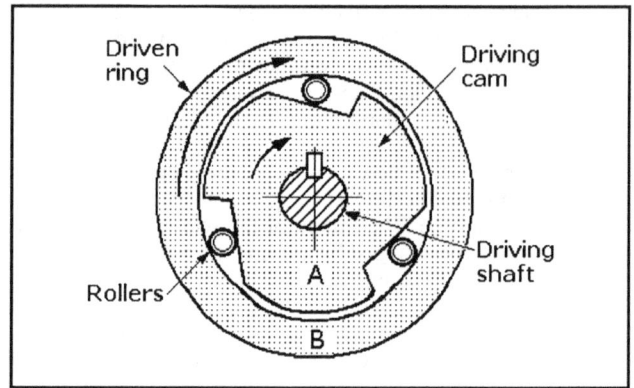

Fig. 41 Overrunning clutch: As driving cam A revolves clockwise, the rollers in the wedge-shaped gaps between cam A and outer ring B are forced by friction into those wedges and are held there; this locks ring B to cam A and drives it clockwise. However, if ring B is turned counterclockwise, or is made to revolve clockwise faster than cam A, the rollers are freed by friction, the clutch slips, and no torque is transmitted.

ring gear. The *pawl and ratchet clutch* consists essentially of a pawl-controlled driving ratchet keyed to a driven gear.

Internally Controlled Clutches

Internally controlled clutches can be controlled by springs, torque, or centrifugal force. The *spring and ball radial-detent clutch,* for example, disengages when torque becomes excessive, allowing the driving gear to continue rotating while the driveshaft stops rotating. The *wrapped-spring clutch* consists of two separate rotating hubs joined by a coil spring. When driven in the right direction, the spring tightens around the hubs increasing the friction grip. However, if driven in the opposite direction the spring relaxes, allowing the clutch to slip.

The *expanding-shoe centrifugal clutch* is similar to the externally controlled *expanding shoe clutch* except that the friction shoes are pulled in by springs until the driving shaft attains a preset speed. At that speed centrifugal force drives the shoes radially outward so that they contact the drum. As the driveshaft rotates faster, pressure between the shoes and drum increases, thus increasing clutch torque.

The *overrunning or overriding clutch,* as shown in Fig. 41, is a specialized form of a cam mechanism, also called a *cam and roller clutch*. The inner driving cam A has wedge-shaped notches on its outer rim that hold rollers between the outer surface of A and the inner cylindrical surfaces of outer driven ring B. When driving cam A is turning clockwise, frictional forces wedge the rollers tightly into the notches to lock outer driven ring B in position so it also turns in a clockwise direction. However, if driven ring B is reversed or runs faster clockwise than driving cam A (when it is either moving or immobile) the rollers are set free, the clutch will slip and no torque is transmitted. Some versions of this clutch include springs between the cam faces and the rollers to ensure faster clutching action if driven ring B attempts to drive driving cam A by overcoming residual friction. A version of this clutch is the basic free-wheel mechanism that drives the rear axle of a bicycle.

Some low-cost, light-duty overrunning clutches for one-direction-only torque transmission intersperse cardioid-shaped pellets called *sprags* with cylindrical rollers. This design permits cylindrical internal drivers to replace cammed drivers. The sprags bind in the concentric space between the inner driver and the outer driven ring if the ring attempts to drive the driver. The torque rating of the clutch depends on the number of sprags installed. For acceptable performance a minimum of three sprags, equally spaced around the circumference of the races, is usually necessary.

GLOSSARY OF COMMON MECHANICAL TERMS

acceleration: The time rate of change of velocity of a body. It is always produced by force acting on a body. Acceleration is measured as feet per second per second (ft/s^2) or meters per second per second (m/s^2).

component forces: The individual forces that are the equivalent of the resultant.

concurrent forces: Forces whose lines of action or directions pass through a common point or meet at a common point.

crank: A side link that revolves relative to the frame.

crank-rocker mechanism: A four-bar linkage characterized by the ability of the shorter side link to revolve through 360° while the opposing link rocks or oscillates.

couple: Two equal and opposite parallel forces that act at diametrically opposite points on a body to cause it to rotate around a point or an axis through its center.

displacement: Distance measured from a fixed reference point in a specified direction; it is a vector quantity; units are measured in inches, feet, miles, centimeters, meters, and kilometers.

double-crank mechanism: A four-bar linkage characterized by the ability of both of its side links to oscillate while the shortest link (opposite the foundation link) can revolve through 360°.

dynamics: The study of the forces that act on bodies not in equilibrium, both balanced and unbalanced; it accounts for the masses and accelerations of the parts as well as the external forces acting on the mechanisms. It is a combination of *kinetics* and *kinematics*.

efficiency of machines: The ratio of a machine's output divided by its input is typically expressed as a percent. There are energy or power losses in all moving machinery caused primarily by friction. This causes inefficiency, so a machine's output is always less than its input; both output and input must be expressed in the same units of power or energy. This ratio, always a fraction, is multiplied by 100 to obtain a percent. It can also be determined by dividing the machine's mechanical advantage by its velocity ratio and multiplying that ratio by 100 to get a percent.

energy: A physical quantity present in three-dimensional space in which forces can act on a body or particle to bring about physical change; it is the capacity for doing work. Energy can take many forms, including mechanical, electrical, electromagnetic, chemical, thermal, solar, and nuclear. Energy and work are related and measured in the same units: foot-pounds, ergs, or joules; it cannot be destroyed, but it can be wasted.

- *Kinetic energy* is the kind of energy a body has when it is in motion. Examples are a rolling soccer ball, a speeding automobile, or a flying airplane.
- *Potential energy* is the kind of energy that a body has because of its position or state. Examples are a concrete block poised at the edge of a building, a shipping container suspended above ground by a crane, or a roadside bomb.

equilibrium: In mechanics, a condition of balance or static equilibrium between opposing forces. An example is when there are equal forces at both ends of a seesaw resting on a *fulcrum*.

force: Strength or energy acting on a body to push or pull it; it is required to produce acceleration. Except for gravitation, one body cannot exert a force on another body unless the two are in contact. The Earth exerts a force of attraction on bodies, whether they are in contact or not. Force is measured in poundals (lb-ft/s^2) or newtons (kg-m/s^2).

fulcrum: A pivot point or edge about which objects are free to rotate.

kinematic chain: A combination of links and pairs without a fixed link.

kinematics: The study of the motions of bodies without considering how the variables of force and mass influence the motion. It is described as the geometry of motion.

kinetics: The study of the effects of external forces including gravity upon the motions of physical bodies.

lever: A simple machine that uses opposing torque around a fulcrum to perform work.

linear motion: Motion in a straight line. An example is when a car is driving on a straight road.

link: A rigid body with pins or fasteners at its ends to connect it to other rigid bodies so it can transmit a force or motion. All machines contain at least one link, either in a fixed position relative to the Earth or capable of moving the machine and the link during the motion; this link is the *frame* or *fixed link* of the machine.

linkages: Mechanical assemblies consisting of two or more levers connected to produce a desired motion. They can also be mechanisms consisting of rigid bodies and lower *pairs*.

machine: An assembly of mechanisms or parts or mechanisms capable of transmitting force, motion, and energy from a power source; the objective of a machine is to overcome some form of resistance to accomplish a desired result. There are two functions of machines: (1) the transmission of relative motion and (2) the transmission of force; both require that the machine be strong and rigid. While both machines and mechanisms are combinations of rigid bodies capable of definite relative motions, machines transform energy, but mechanisms do not. A *simple machine* is an elementary mechanism. Examples are the lever, wheel and axle, pulley, inclined plane, wedge, and screw.

machinery: A term generally meaning various combinations of machines and mechanisms.

mass: The quantity of matter in a body indicating its inertia. Mass also initiates gravitational attraction. It is measured in ounces, pounds, tons, grams, and kilograms.

mechanical advantage: The ratio of the load (or force W) divided by the effort (or force F) exerted by an operator. If friction is considered in determining mechanical advantage, or it has been determined by the actual testing, the ratio W/F is the mechanical advantage MA. However, if the machine is assumed to operate without friction, the ratio W/F is the theoretical mechanical advantage TA. Mechanical advantage and velocity ratio are related.

mechanics: A branch of physics concerned with the motions of objects and their response to forces. Descriptions of mechanics begin with definitions of such quantities as acceleration, displacement, force, mass, time, and velocity.

mechanism: In mechanics, it refers to two or more rigid or resistant bodies connected together by joints so they exhibit definite relative motions with respect to one another. Mechanisms are divided into two classes:

- *Planar*: Two-dimensional mechanisms whose relative motions are in one plane or parallel planes.
- *Spatial*: Three-dimensional mechanisms whose relative motions are not all in the same or parallel planes.

moment of force or torque: The product of the force acting to produce a turning effect and the perpendicular distance of its line of action from the point or axis of rotation. The perpendicular distance is called the *moment* arm or the *lever arm torque*. It is measured in pound-inches (lb-in.), pound-feet (lb-ft), or newton-meters (N-m).

moment of inertia: A physical quantity giving a measure of the *rotational inertia* of a body about a specified axis of rotation; it depends on the mass, size, and shape of the body.

nonconcurrent forces: Forces whose lines of action do not meet at a common point.

noncoplanar forces: Forces that do not act in the same plane.

oscillating motion: Repetitive forward and backward circular motion such as that of a clock pendulum.

pair: A joint between the surfaces of two rigid bodies that keeps them in contact and relatively movable. It might be as simple as a pin, bolt, or hinge between two links or as complex as a universal joint between two links. There are two kinds of pairs in mechanisms classified by the type of contact between the two bodies of the pair: *lower pairs* and *higher pairs*.

- Lower pairs are *surface-contact pairs* classed either as *revolute* or *prismatic*. Examples: a hinged door is a revolute pair and a sash window is a prismatic pair.
- Higher pairs include *point, line, or curve pairs*. Examples: paired rollers, cams and followers, and meshing gear teeth.

power: The time rate of doing work. It is measured in foot-pounds per second (ft-lb/s), foot-pounds per minute (ft-lb/min), horsepower, watts, kilowatts, newton-meters/s, ergs/s, and joules/s.

reciprocating motion: Repetitive back and forth linear motion as that of a piston in an internal combustion engine.

resultant: In a system of forces, it is the single force equivalent of the entire system. When the resultant of a system of forces is zero, the system is in equilibrium.

rotary motion: Circular motion as in the turning of a bicycle wheel.

skeleton outline: A simplified geometrical line drawing showing the fundamentals of a simple machine devoid of the actual details of its construction. It gives all of the geometrical information needed for determining the relative motions of the main links. The relative motions of these links might be complete circles, semicircles, or arcs, or even straight lines.

statics: The study of bodies in equilibrium, either at rest or in uniform motion.

torque: An alternative name for *moment of force*.

velocity: The time rate of change with respect to distance. It is measured in feet per second (ft/s), feet per minute (ft/min), meters per second (m/s), or meters per minute (m/min).

velocity ratio: A ratio of the distance movement of the effort divided by the distance of movement of the load per second for a machine. This ratio has no units.

weight: The force on a body due to the gravitational attraction of the Earth; weight W = mass $n \times$ acceleration g due to the Earth's gravity; mass of a body is constant but g, and therefore W vary slightly over the Earth's surface.

work: The product of force and distance: the distance an object moves in the direction of force. Work is not done if the force exerted on a body fails to move that body. Work, like energy, is measured in units of ergs, joules, or foot-pounds.

CHAPTER 2
MOTION CONTROL
SYSTEMS

MOTION CONTROL SYSTEMS OVERVIEW

Introduction

A modern motion control system typically consists of a motion controller, a motor drive or amplifier, an electric motor, and feedback sensors. The system might also contain other components such as one or more belt-, ballscrew-, or leadscrew-driven linear guides or axis stages. A motion controller today can be a stand-alone programmable controller, a personal computer containing a motion control card, or a programmable logic controller (PLC).

Fig. 1 This multiaxis X-Y-Z motion platform is an example of a motion control system.

All of the components of a motion control system must work together seamlessly to perform their assigned functions. Their selection must be based on both engineering and economic considerations. Figure 1 illustrates a typical multiaxis X-Y-Z motion platform that includes the three linear axes required to move a load, tool, or end effector precisely through three degrees of freedom. With additional mechanical or electromechanical components on each axis, rotation about the three axes can provide up to six degrees of freedom, as shown in Fig. 2.

Motion control systems today can be found in such diverse applications as materials handling equipment, machine tool

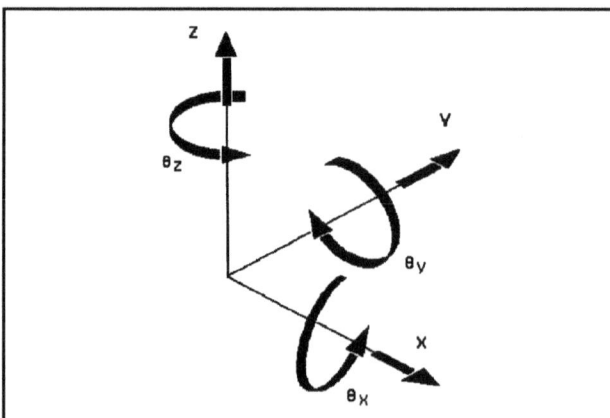

Fig. 2 The right-handed coordinate system showing six degrees of freedom.

centers, chemical and pharmaceutical process lines, inspection stations, robots, and injection molding machines.

Merits of Electric Systems

Most motion control systems today are powered by electric motors rather than hydraulic or pneumatic motors or actuators because of the many benefits they offer:

- More precise load or tool positioning, resulting in fewer product or process defects and lower material costs
- Quicker changeovers for higher flexibility and easier product customizing
- Increased throughput for higher efficiency and capacity
- Simpler system design for easier installation, programming, and training
- Lower downtime and maintenance costs
- Cleaner, quieter operation without oil or air leakage

Electric-powered motion control systems do not require pumps or air compressors, and they do not have hoses or piping that can leak hydraulic fluids or air. This discussion of motion control is limited to electric-powered systems.

Motion Control Classification

Motion control systems can be classified as *open-loop* or *closed-loop*. An open-loop system does not require that measurements of any output variables be made to produce error-correcting signals; by contrast, a closed-loop system requires one or more feedback sensors that measure and respond to errors in output variables.

Closed-Loop System

A *closed-loop motion control system*, as shown in block diagram Fig. 3, has one or more feedback loops that continuously compare the system's response with input commands or settings to correct errors in motor and/or load speed, load position, or motor torque. Feedback sensors provide the electronic signals for correcting deviations from the desired input commands. Closed-loop systems are also called servosystems.

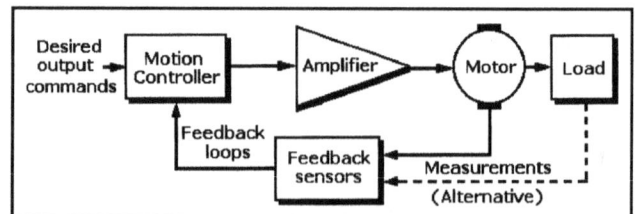

Fig. 3 Block diagram of a basic closed-loop control system.

Each motor in a servosystem requires its own feedback sensors, typically encoders, resolvers, or tachometers, that close loops around the motor and load. Variations in velocity, position, and torque are typically caused by variations in load conditions, but changes in ambient temperature and humidity can also affect load conditions.

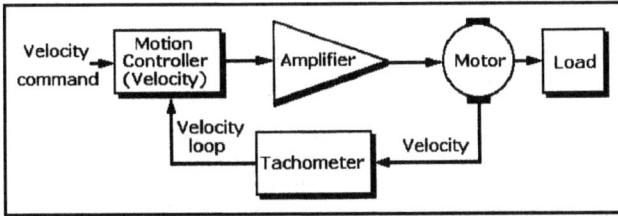

Fig. 4 Block diagram of a velocity-control system.

A *velocity-control loop,* as shown in block diagram Fig. 4, typically contains a tachometer that is able to detect changes in motor speed. This sensor produces error signals that are proportional to the positive or negative deviations of motor speed from its preset value. These signals are sent to the motion controller so that it can compute a corrective signal for the amplifier to keep motor speed within those preset limits despite load changes.

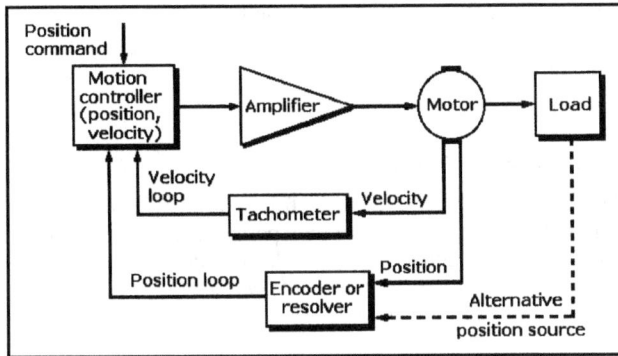

Fig. 5 Block diagram of a position-control system.

A *position-control loop,* as shown in block diagram Fig. 5, typically contains either an encoder or resolver capable of direct or indirect measurements of load position. These sensors generate error signals that are sent to the motion controller, which produces a corrective signal for the amplifier. The output of the amplifier causes the motor to speed up or slow down to correct the position of the load. Most position-control closed-loop systems also include a velocity-control loop.

The *ballscrew slide mechanism,* shown in Fig. 6, is an example of a mechanical system that carries a load whose position must be controlled in a closed-loop servosystem because it is not

Fig. 6 Ballscrew-driven single-axis slide mechanism without position feedback sensors.

Fig. 7 Examples of position feedback sensors installed on a ballscrew-driven slide mechanism: (a) rotary encoder, (b) linear encoder, and (c) laser interferometer.

equipped with position sensors. Three examples of feedback sensors mounted on the ballscrew mechanism that can provide position feedback are shown in Fig. 7: (a) is a rotary optical encoder mounted on the motor housing with its shaft coupled to the motor shaft; (b) is an optical linear encoder with its graduated scale mounted on the base of the mechanism; and (c) is the less commonly used but more accurate and expensive laser interferometer.

A *torque-control loop* contains electronic circuitry that measures the input current applied to the motor and compares it with a value proportional to the torque required to perform the desired task. An error signal from the circuit is sent to the motion controller, which computes a corrective signal for the motor amplifier to keep motor current, and hence torque, constant. Torque-control loops are widely used in machine tools where the load can change due to variations in the density of the material being machined or the sharpness of the cutting tools.

Trapezoidal Velocity Profile

If a motion control system is to achieve smooth, high-speed motion without overstressing the servomotor, the motion controller must command the motor amplifier to ramp up motor velocity gradually until it reaches the desired speed and then ramp it down gradually until it stops after the task is complete. This keeps motor acceleration and deceleration within limits.

The trapezoidal profile, shown in Fig. 8, is widely used because it accelerates motor velocity along a positive linear "upramp" until the desired constant velocity is reached. When the motor is shut down from the constant velocity setting, the profile decelerates velocity along a negative "down ramp" until

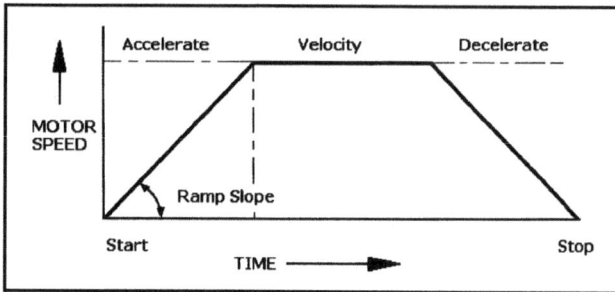

Fig. 8 Servomotors are accelerated to constant velocity and decelerated along a trapezoidal profile to assure efficient operation.

the motor stops. Amplifier current and output voltage reach maximum values during acceleration, then step down to lower values during constant velocity and switch to negative values during deceleration.

Closed-Loop Control Techniques

The simplest form of feedback is *proportional control*, but there are also *derivative* and *integral control* techniques, which compensate for certain steady-state errors that cannot be eliminated from proportional control. All three of these techniques can be combined to form *proportional-integral-derivative (PID) control*.

- In *proportional control* the signal that drives the motor or actuator is directly proportional to the linear difference between the input command for the desired output and the measured actual output.
- In *integral control* the signal driving the motor equals the *time integral* of the difference between the input command and the measured actual output.
- In *derivative control* the signal that drives the motor is proportional to the *time derivative* of the difference between the input command and the measured actual output.
- In *proportional-integral-derivative (PID) control* the signal that drives the motor equals the weighted sum of the difference, the time integral of the difference, and the time derivative of the difference between the input command and the measured actual output.

Open-Loop Motion Control Systems

A typical *open-loop motion control system* includes a stepper motor with a programmable indexer or pulse generator and motor driver, as shown in Fig. 9. This system does not need feedback sensors because load position and velocity are controlled by the predetermined number and direction of input digital pulses sent to the motor driver from the controller. Because load position is not continuously sampled by a feedback sensor (as in a closed-loop servosystem), load positioning accuracy is lower and position errors (commonly called step errors) accumulate over time. For these reasons open-loop systems are most often specified in applications where the load remains constant, load motion is simple, and low positioning speed is acceptable.

Fig. 9 Block diagram of an open-loop motion control system.

Kinds of Controlled Motion

There are five different kinds of motion control: *point-to-point, sequencing, speed, torque,* and *incremental.*

- In *point-to-point motion control* the load is moved between a sequence of numerically defined positions where it is stopped before it is moved to the next position. This is done at a constant speed, with both velocity and distance monitored by the motion controller. Point-to-point positioning can be performed in single-axis or multiaxis systems with servomotors in closed loops or stepping motors in open loops. X-Y tables and milling machines position their loads by multiaxis point-to-point control.
- *Sequencing control* is the control of such functions as opening and closing valves in a preset sequence or starting and stopping a conveyor belt at specified stations in a specific order.
- *Speed control* is the control of the velocity of the motor or actuator in a system.
- *Torque control* is the control of motor or actuator current so that torque remains constant despite load changes.
- *Incremental motion control* is the simultaneous control of two or more variables such as load location, motor speed, or torque.

Motion Interpolation

When a load under control must follow a specific path to get from its starting point to its stopping point, the movements of the axes must be coordinated or interpolated. There are three kinds of interpolation: *linear, circular,* and *contouring.*

Linear interpolation is the ability of a motion control system having two or more axes to move the load from one point to another in a straight line. The motion controller must determine the speed of each axis so that it can coordinate their movements. True linear interpolation requires that the motion controller modify axis acceleration, but some controllers approximate true linear interpolation with programmed acceleration profiles. The path can lie in one plane or be three dimensional.

Circular interpolation is the ability of a motion control system having two or more axes to move the load around a circular trajectory. It requires that the motion controller modify load acceleration while it is in transit. Again the circle can lie in one plane or be three dimensional.

Contouring is the path followed by the load, tool, or end-effector under the coordinated control of two or more axes. It requires that the motion controller change the speeds on different axes so that their trajectories pass through a set of predefined points. Load speed is determined along the trajectory, and it can be constant except during starting and stopping.

Computer-Aided Emulation

Several important types of programmed computer-aided motion control can emulate mechanical motion and eliminate the need for actual gears or cams. *Electronic gearing* is the control by software of one or more axes to impart motion to a load, tool, or end effector that simulates the speed changes that can be performed by actual gears. *Electronic camming* is the control by software of one or more axes to impart a motion to a load, tool, or end effector that simulates the motion changes that are typically performed by actual cams.

Mechanical Components

The mechanical components in a motion control system can be more influential in the design of the system than the electronic circuitry used to control it. Product flow and throughput, human operator requirements, and maintenance issues help to determine

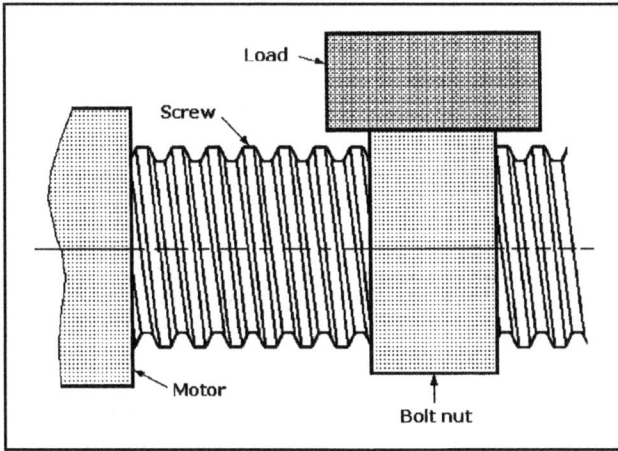

Fig. 10 Leadscrew drive: As the leadscrew rotates, the load is translated in the axial direction of the screw.

the mechanics, which in turn influence the motion controller and software requirements.

Mechanical actuators convert a motor's rotary motion into linear motion. Mechanical methods for accomplishing this include the use of leadscrews, shown in Fig. 10, ballscrews, shown in Fig. 11, worm-drive gearing, shown in Fig. 12, and belt, cable, or chain drives. Method selection is based on the relative costs of the alternatives and consideration for the possible effects of backlash. All actuators have finite levels of torsional and axial stiffness that can affect the system's frequency response characteristics.

Linear guides or stages constrain a translating load to a single degree of freedom. The linear stage supports the mass of the load

Fig. 11 Ballscrew drive: Ballscrews use recirculating balls to reduce friction and gain higher efficiency than conventional leadscrews.

Fig. 12 Worm-drive systems can provide high speed and high torque.

Fig. 13 Ballscrew-driven single-axis slide mechanism translates rotary motion into linear motion.

to be actuated and assures smooth, straight-line motion while minimizing friction. A common example of a linear stage is a ballscrew-driven single-axis stage, illustrated in Fig. 13. The motor turns the ballscrew, and its rotary motion is translated into the linear motion that moves the carriage and load by the stage's bolt nut. The bearing ways act as linear guides. As shown in Fig. 7, these stages can be equipped with sensors such as a rotary or linear encoder or a laser interferometer for feedback.

A ballscrew-driven single-axis stage with a rotary encoder coupled to the motor shaft provides an indirect measurement. This method ignores the tolerance, wear, and compliance in the mechanical components between the carriage and the position encoder that can cause deviations between the desired and true positions. Consequently, this feedback method limits position accuracy to ballscrew accuracy, typically ±5 to 10 µm per 300 mm.

Other kinds of single-axis stages include those containing antifriction rolling elements such as recirculating and nonrecirculating balls or rollers, sliding (friction contact) units, air-bearing units, hydrostatic units, and magnetic levitation (Maglev) units.

A single-axis air-bearing guide or stage is shown in Fig. 14. Some models being offered are 3.9 ft (1.2 m) long and include a carriage for mounting loads. When driven by a linear servomotor the loads can reach velocities of 9.8 ft/s (3 m/s). As shown in Fig. 7, these stages can be equipped with feedback devices such as cost-effective linear encoders or ultrahigh-resolution laser interferometers. The resolution of this type of stage with a noncontact linear encoder can be as fine as 20 nm and accuracy can be ±1 µm. However, these values can be increased to 0.3 nm resolution and submicron accuracy if a laser interferometer is installed.

The pitch, roll, and yaw of air-bearing stages can affect their resolution and accuracy. Some manufacturers claim ±1 arc-s per

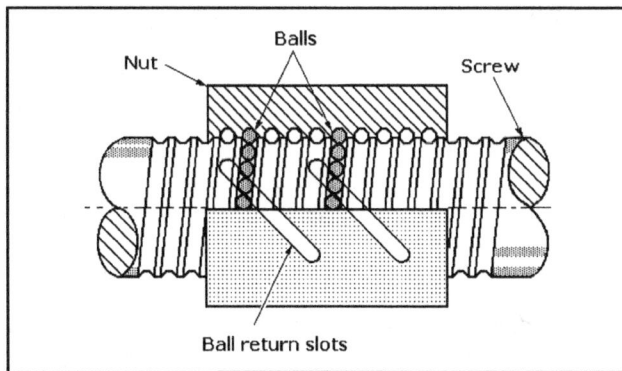

Fig. 14 This single-axis linear guide for load positioning is supported by air bearings as it moves along a granite base.

100 mm as the limits for each of these characteristics. Large air-bearing surfaces provide excellent stiffness and permit large load-carrying capability.

The important attributes of all these stages are their dynamic and static friction, rigidity, stiffness, straightness, flatness, smoothness, and load capacity. Also considered is the amount of work needed to prepare the host machine's mounting surface for their installation.

The structure on which the motion control system is mounted directly affects the system's performance. A properly designed base or host machine will be highly damped and act as a compliant barrier to isolate the motion system from its environment and minimize the impact of external disturbances. The structure must be stiff enough and sufficiently damped to avoid resonance problems. A high static mass to reciprocating mass ratio can also prevent the motion control system from exciting its host structure to harmful resonance.

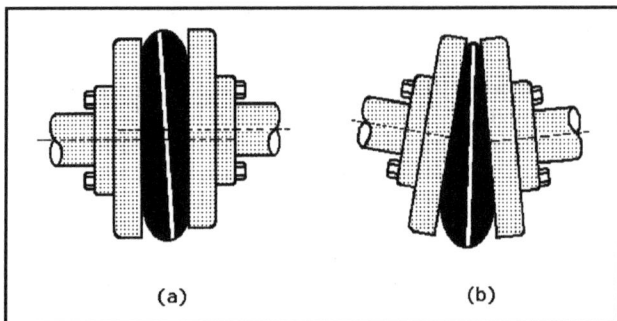

Fig. 15 Flexible shaft couplings adjust for and accommodate parallel misalignment (a) and angular misalignment between rotating shafts (b).

Any components that move will affect a system's response by changing the amount of inertia, damping, friction, stiffness, or resonance. For example, a flexible shaft coupling, as shown in Fig. 15, will compensate for minor parallel (a) and angular (b) misalignment between rotating shafts. Flexible couplings are available in other configurations such as bellows and helixes, as shown in Fig. 16. The bellows configuration (a) is acceptable for light-duty applications where misalignments can be as great as 9° angular or $1/4$ in. parallel. By contrast, helical couplings (b) prevent backlash at constant velocity with some misalignment, and they can also be run at high speed.

Other moving mechanical components include cable carriers that retain moving cables, end stops that restrict travel, shock absorbers to dissipate energy during a collision, and way covers to keep out dust and dirt.

Fig. 16 Bellows couplings (a) are acceptable for light-duty applications. Misalignments can be 9° angular or $1/4$ in. parallel. Helical couplings (b) prevent backlash and can operate at constant velocity with misalignment and be run at high speed.

Electronic System Components

The motion controller is the "brain" of the motion control system and performs all of the required computations for motion path planning, servo-loop closure, and sequence execution. It is essentially a computer dedicated to motion control that has been programmed by the end user for the performance of assigned tasks. The motion controller produces a low-power motor command signal in either a digital or analog format for the motor driver or amplifier.

Significant technical developments have led to the increased acceptance of programmable motion controllers over the past 5 to 10 years. These include the rapid decrease in the cost of microprocessors as well as dramatic increases in their computing power. Added to that are the decreasing cost of more advanced semiconductor and disk memories. During the past 5 to 10 years, the capability of these systems to improve product quality, increase throughput, and provide just-in-time delivery has improved significantly.

The motion controller is the most critical component in the system because of its dependence on software. By contrast, the selection of most motors, drivers, feedback sensors, and associated mechanisms is less critical because they can usually be changed during the design phase or even later in the field with less impact on the characteristics of the intended system. However, making field changes can be costly in terms of lost productivity.

The decision to install any of the three kinds of motion controllers should be based on their ability to control both the number and types of motors required for the application as well as the availability of the software that will provide the optimum performance for the specific application. Also to be considered are the system's multitasking capabilities, the number of input/output (I/O) ports required, and the need for such features as linear and circular interpolation and electronic gearing and camming.

In general, a motion controller receives a set of operator instructions from a host or operator interface and it responds with corresponding command signals for the motor driver or drivers that control the motor or motors driving the load.

Motor Selection

The most popular motors for motion control systems are stepping or stepper motors and permanent-magnet (PM) DC brush-type and brushless DC servomotors. Stepper motors are selected for systems because they can run open-loop without feedback sensors. These motors are indexed or partially rotated by digital pulses that turn their rotors a fixed fraction or a revolution where they will be clamped securely by their inherent holding torque. Stepper motors are cost-effective and reliable choices for many applications that do not require the rapid acceleration, high speed, and position accuracy of a servomotor.

However, a feedback loop can improve the positioning accuracy of a stepper motor without incurring the higher costs of a complete servosystem. Some stepper motor motion controllers can accommodate a closed loop.

Brush and brushless PM DC servomotors are usually selected for applications that require more precise positioning. Both of these motors can reach higher speeds and offer smoother low-speed operation with finer position resolution than stepper motors, but both require one or more feedback sensors in closed loops, adding to system cost and complexity.

Brush-type permanent-magnet (PM) DC servomotors have wound armatures or rotors that rotate within the magnetic field produced by a PM stator. As the rotor turns, current is applied sequentially to the appropriate armature windings by a mechanical commutator consisting of two or more brushes sliding on a ring of insulated copper segments. These motors are quite mature, and modern versions can provide very high performance for very low cost.

There are variations of the brush-type DC servomotor with its iron-core rotor that permit more rapid acceleration and deceleration because of their low-inertia, lightweight cup- or disk-type armatures. The disk-type armature of the pancake-frame motor, for example, has its mass concentrated close to the motor's faceplate permitting a short, flat cylindrical housing. This configuration makes the motor suitable for faceplate mounting in restricted space, a feature particularly useful in industrial robots or other applications where space does not permit the installation of brackets for mounting a motor with a longer length dimension.

The brush-type DC motor with a cup-type armature also offers lower weight and inertia than conventional DC servomotors. However, the tradeoff in the use of these motors is the restriction on their duty cycles because the epoxy-encapsulated armatures are unable to dissipate heat buildup as easily as iron-core armatures and are therefore subject to damage or destruction if overheated.

However, any servomotor with brush commutation can be unsuitable for some applications due to the electromagnetic interference (EMI) caused by brush arcing or the possibility that the arcing can ignite nearby flammable fluids, airborne dust, or vapor, posing a fire or explosion hazard. The EMI generated can adversely affect nearby electronic circuitry. In addition, motor brushes wear down and leave a gritty residue that can contaminate nearby sensitive instruments or precisely ground surfaces. Thus, brush-type motors must be cleaned constantly to prevent the spread of the residue from the motor. Also, brushes must be replaced periodically, causing unproductive downtime.

Brushless DC PM motors overcome these problems and offer the benefits of electronic rather than mechanical commutation. Built as inside-out DC motors, typical brushless motors have PM rotors and wound stator coils. Commutation is performed by internal noncontact Hall-effect devices (HEDs) positioned within the stator windings. The HEDs are wired to power transistor switching circuitry, which is mounted externally in separate modules for some motors but is mounted internally on circuit cards in other motors. Alternatively, commutation can be performed by a commutating encoder or by commutation software resident in the motion controller or motor drive.

Brushless DC motors exhibit low rotor inertia and lower winding thermal resistance than brush-type motors because their high-efficiency magnets permit the use of shorter rotors with smaller diameters. Moreover, because they are not burdened with sliding brush-type mechanical contacts, they can run at higher speeds (50,000 rpm or greater), provide higher continuous torque, and accelerate faster than brush-type motors. Nevertheless, brushless motors still cost more than comparably rated brush-type motors (although that price gap continues to narrow) and their installation adds to overall motion control system cost and complexity. Table 1 summarizes some of the outstanding characteristics of stepper, PM brush, and PM brushless DC motors.

Table 1 Stepping and Permanent-Magnet DC Servomotors Compared.

	Stepping	PM Brush	PM Brushless
Cost	Low	Medium	High
Smoothness	Low to medium	Good to excellent	Good to excellent
Speed range	0–1500 rmp (typical)	0–6000 rpm	0–10,000 rpm
Torque	High- (falls off with speed)	Medium	High
Required feedback	None	Position or velocity	Commutation and position or velocity
Maintenance	None	Yes	None
Cleanliness	Excellent	Brush dust	Excellent

The linear motor, another drive alternative, can move the load directly, eliminating the need for intermediate motion translation mechanism. These motors can accelerate rapidly and position loads accurately at high speed because they have no moving parts in contact with each other. Essentially rotary motors that have been sliced open and unrolled, they have many of the characteristics of conventional motors. They can replace conventional rotary motors driving leadscrew-, ballscrew-, or belt-driven single-axis stages, but they cannot be coupled to gears that could change their drive characteristics. If increased performance is required from a linear motor, the existing motor must be replaced with a larger one.

Linear motors must operate in closed feedback loops, and they typically require more costly feedback sensors than rotary motors. In addition, space must be allowed for the free movement of the motor's power cable as it tracks back and forth along a linear path. Moreover, their applications are also limited because of their inability to dissipate heat as readily as rotary motors with metal frames and cooling fins, and the exposed magnetic fields of some models can attract loose ferrous objects, creating a safety hazard.

Motor Drivers (Amplifiers)

Motor drivers or amplifiers must be capable of driving their associated motors—stepper, brush, brushless, or linear. A drive circuit for a stepper motor can be fairly simple because it needs only several power transistors to sequentially energize the motor phases according to the number of digital step pulses received from the motion controller. However, more advanced stepping motor drivers can control phase current to permit "microstepping," a technique that allows the motor to position the load more precisely.

Servodrive amplifiers for brush and brushless motors typically receive analog voltages of ±10-VDC signals from the motion controller. These signals correspond to current or voltage commands. When amplified, the signals control both the direction and magnitude of the current in the motor windings. Two types of amplifiers are generally used in closed-loop servosystems: *linear* and *pulse-width modulated* (PWM).

Pulse-width modulated amplifiers predominate because they are more efficient than linear amplifiers and can provide up to 100 W. The transistors in PWM amplifiers (as in PWM power supplies) are optimized for switchmode operation, and they are capable of switching amplifier output voltage at frequencies up to 20 kHz. When the power transistors are switched on (on state), they saturate, but when they are off, no current is drawn. This operating mode reduces transistor power dissipation and boosts amplifier efficiency. Because of their higher operating frequencies, the magnetic components in PWM amplifiers can be smaller and lighter than those in linear amplifiers. Thus, the entire drive module can be packaged in a smaller, lighter case.

By contrast, the power transistors in linear amplifiers are continuously in the on state although output power requirements can be varied. This operating mode wastes power, resulting in lower amplifier efficiency while subjecting the power transistors to thermal stress. However, linear amplifiers permit smoother motor operation, a requirement for some sensitive motion control systems. In addition linear amplifiers are better at driving low-inductance motors. Moreover, these amplifiers generate less EMI than PWM amplifiers, so they do not require the same degree of filtering. By contrast, linear amplifiers typically have lower maximum power ratings than PWM amplifiers.

Feedback Sensors

Position feedback is the most common requirement in closed-loop motion control systems, and the most popular sensor for providing this information is the rotary optical encoder. The axial shafts of these encoders are mechanically coupled to the drive-shafts of the motor. They generate either sine waves or pulses that can be counted by the motion controller to determine the motor or load position and direction of travel at any time to permit precise positioning. Analog encoders produce sine waves that must be conditioned by external circuitry for counting, but digital encoders include circuitry for translating sine waves into pulses.

Absolute rotary optical encoders produce binary words for the motion controller that provide precise position information. If they are stopped accidentally due to power failure, these encoders preserve the binary word because the last position of the encoder code wheel acts as a memory.

Linear optical encoders, by contrast, produce pulses that are proportional to the actual linear distance of load movement. They work on the same principles as the rotary encoders, but the graduations are engraved on a stationary glass or metal scale while the read head moves along the scale.

Tachometers are generators that provide analog signals that are directly proportional to motor shaft speed. They are mechanically coupled to the motor shaft and can be located within the motor frame. After tachometer output is converted to a digital format by the motion controller, a feedback signal is generated for the driver to keep motor speed within preset limits.

Other common feedback sensors include resolvers, linear variable differential transformers (LVDTs), Inductosyns, and potentiometers. Less common are the more accurate laser interferometers. Feedback sensor selection is based on an evaluation of the sensor's accuracy, repeatability, ruggedness, temperature limits, size, weight, mounting requirements, and cost, with the relative importance of each determined by the application.

Installation and Operation of the System

The design and implementation of a cost-effective motion-control system require a high degree of expertise on the part of the person or persons responsible for system integration. It is rare that a diverse group of components can be removed from their boxes, installed, and interconnected to form an instantly effective system. Each servosystem (and many stepper systems) must be tuned (stabilized) to the load and environmental conditions. However, installation and development time can be minimized if the customer's requirements are accurately defined, optimum components are selected, and the tuning and debugging tools are applied correctly. Moreover, operators must be properly trained in formal classes or, at the very least, must have a clear understanding of the information in the manufacturers' technical manuals gained by careful reading.

GLOSSARY OF MOTION CONTROL TERMS

Abbe error: A linear error caused by a combination of an underlying angular error along the line of motion and a dimensional offset between the position of the object being measured and the accuracy-determining element such as a leadscrew or encoder.

acceleration: The change in velocity per unit time.

accuracy: (1) absolute accuracy: The motion control system output compared with the commanded input. It is actually a measurement of inaccuracy and it is typically measured in millimeters. (2) motion accuracy: The maximum expected difference between the actual and the intended position of an object or load for a given input. Its value depends on the method used for measuring the actual position. (3) on-axis accuracy: The uncertainty of load position after all linear errors are eliminated. These include such factors as inaccuracy of leadscrew pitch, the angular deviation effect at the measuring point, and thermal expansion of materials.

backlash: The maximum magnitude of an input that produces no measurable output when the direction of motion is reversed. It can result from insufficient preloading or poor meshing of gear teeth in a gear-coupled drivetrain.

error: (1) The difference between the actual result of an input command and the ideal or theoretical result. (2) following error: The instantaneous difference between the actual position as reported by the position feedback loop and the ideal position, as commanded by the controller. (3) steady-state error: The difference between the actual and commanded position after all corrections have been applied by the controller.

hysteresis: The difference in the absolute position of the load for a commanded input when motion is from opposite directions.

inertia: The measure of a load's resistance to changes in velocity or speed. It is a function of the load's mass and shape. The torque required to accelerate or decelerate the load is proportional to inertia.

overshoot: The amount of overcorrection in an underdamped control system.

play: The uncontrolled movement due to the looseness of mechanical parts. It is typically caused by wear, overloading the system, or improper system operation.

precision: See *repeatability*.

repeatability: The ability of a motion control system to return repeatedly to the commanded position. It is influenced by the presence of *backlash* and *hysteresis*. Consequently, *bidirectional repeatability*, a more precise specification, is the ability of the system to achieve the commanded position repeatedly regardless of the direction from which the intended position is approached. It is synonymous with *precision*. However, accuracy and precision are not the same.

resolution: The smallest position increment that the motion control system can detect. It is typically considered to be display or encoder resolution because it is not necessarily the smallest motion the system is capable of delivering reliably.

runout: The deviation between ideal linear (straight-line) motion and the actual measured motion.

sensitivity: The minimum input capable of producing output motion. It is also the ratio of the output motion to the input drive. This term should not be used in place of resolution.

settling time: The time elapsed between the entry of a command to a system and the instant the system first reaches the commanded position and maintains that position within the specified error value.

velocity: The change in distance per unit time. Velocity is a *vector* and speed is a *scalar*, but the terms can be used interchangeably.

MECHANICAL COMPONENTS FORM SPECIALIZED MOTION-CONTROL SYSTEMS

Many different kinds of mechanical components are listed in manufacturers' catalogs for speeding the design and assembly of motion control systems. These drawings illustrate what, where, and how one manufacturer's components were used to build specialized systems.

Fig. 1 Punch Press: Catalog pillow blocks and rail assemblies were installed in this system for reducing the deflection of a punch press plate loader to minimize scrap and improve its cycle speed.

Fig. 2 Microcomputer-Controlled X-Y Table: Catalog pillow blocks, rail guides, and ballscrew assemblies were installed in this rigid system that positions workpieces accurately for precise milling and drilling on a vertical milling machine.

Fig. 3 Pick and Place X-Y System: Catalog support and pillow blocks, ballscrew assemblies, races, and guides were in the assembly of this X-Y system that transfers workpieces between two separate machining stations.

Fig. 4 X-Y Inspection System: Catalog pillow and shaft-support blocks, ballscrew assemblies, and a preassembled motion system were used to build this system, which accurately positions an inspection probe over small electronic components.

SERVOMOTORS, STEPPER MOTORS, AND ACTUATORS FOR MOTION CONTROL

Many different kinds of electric motors have been adapted for use in motion control systems because of their linear characteristics. These include both conventional rotary and linear alternating current (AC) and direct current (DC) motors. These motors can be further classified into those that must be operated in closed-loop servosystems and those that can be operated open-loop.

The most popular servomotors are permanent magnet (PM) rotary DC servomotors that have been adapted from conventional PM DC motors. These servomotors are typically classified as brush-type and brushless. The brush-type PM DC servomotors include those with wound rotors and those with lighter weight, lower inertia cup- and disk coil-type armatures. Brushless servomotors have PM rotors and wound stators.

Some motion control systems are driven by two-part linear servomotors that move along tracks or ways. They are popular in applications where errors introduced by mechanical coupling between the rotary motors and the load can introduce unwanted errors in positioning. Linear motors require closed loops for their operation, and provision must be made to accommodate the back-and-forth movement of the attached data and power cable.

Stepper or stepping motors are generally used in less demanding motion control systems, where positioning the load by stepper motors is not critical for the application. Increased position accuracy can be obtained by enclosing the motors in control loops.

Permanent-Magnet DC Servomotors

Permanent-magnet (PM) field DC rotary motors have proven to be reliable drives for motion control applications where high efficiency, high starting torque, and linear speed–torque curves are desirable characteristics. While they share many of the characteristics of conventional rotary series, shunt, and compound-wound brush-type DC motors, PM DC servomotors increased in popularity with the introduction of stronger ceramic and rare-earth magnets made from such materials as neodymium–iron–boron and the fact that these motors can be driven easily by microprocessor-based controllers.

The replacement of a wound field with permanent magnets eliminates both the need for separate field excitation and the electrical losses that occur in those field windings. Because there are both brush-type and brushless DC servomotors, the term *DC motor* implies that it is brush-type or requires mechanical commutation unless it is modified by the term *brushless*. Permanent-magnet DC brush-type servomotors can also have armatures formed as laminated coils in disk or cup shapes. They are lightweight, low-inertia armatures that permit the motors to accelerate faster than the heavier conventional wound armatures.

The increased field strength of the ceramic and rare-earth magnets permitted the construction of DC motors that are both smaller and lighter than earlier generation comparably rated DC motors with alnico (aluminum–nickel–cobalt or AlNiCo) magnets. Moreover, integrated circuitry and microprocessors have increased the reliability and cost-effectiveness of digital motion controllers and motor drivers or amplifiers while permitting them to be packaged in smaller and lighter cases, thus reducing the size and weight of complete, integrated motion-control systems.

Brush-Type PM DC Servomotors

The design feature that distinguishes the brush-type PM DC servomotor, as shown in Fig. 1, from other brush-type DC motors is

Fig. 1 Cutaway view of a fractional horsepower permanent-magnet DC servomotor.

the use of a permanent-magnet field to replace the wound field. As previously stated, this eliminates both the need for separate field excitation and the electrical losses that typically occur in field windings.

Permanent-magnet DC motors, like all other mechanically commutated DC motors, are energized through brushes and a multisegment commutator. While all DC motors operate on the same principles, only PM DC motors have the linear speed–torque curves shown in Fig. 2, making them ideal for closed-loop and variable-speed servomotor applications. These linear characteristics conveniently describe the full range of motor performance.

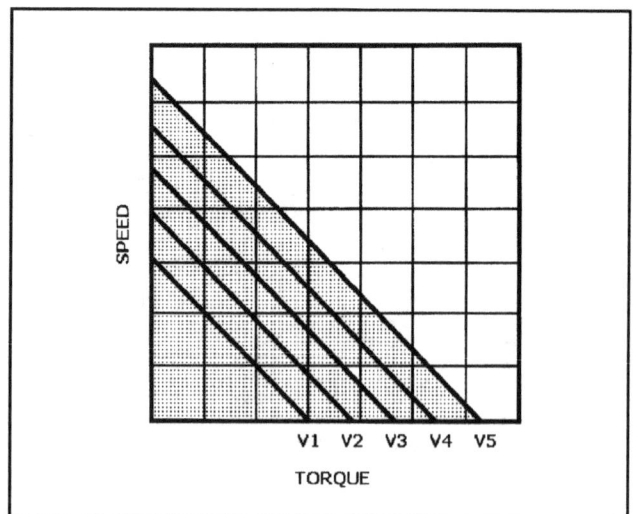

Fig. 2 A typical family of speed/torque curves for a permanent-magnet DC servomotor at different voltage inputs, with voltage increasing from left to right (V1 to V5).

It can be seen that both speed and torque increase linearly with applied voltage, indicated in the diagram as increasing from V1 to V5.

The stators of brush-type PM DC motors are magnetic pole pairs. When the motor is powered, the opposite polarities of the energized windings and the stator magnets attract, and the rotor rotates to align itself with the stator. Just as the rotor reaches alignment, the brushes move across the commutator segments and energize the next winding. This sequence continues as long as power is applied, keeping the rotor in continuous motion. The commutator is staggered from the rotor poles, and the number of its segments is directly proportional to the number of windings. If the connections of a PM DC motor are reversed, the motor will change direction, but it might not operate as efficiently in the reversed direction.

Disk-Type PM DC Motors

The disk-type motor shown in the exploded view in Fig. 3 has a disk-shaped armature with stamped and laminated windings. This nonferrous laminated disk is made as a copper stamping bonded between epoxy–glass insulated layers and fastened to an axial shaft. The stator field can either be a ring of many individual ceramic magnet cylinders, as shown, or a ring-type ceramic magnet attached to the dish-shaped end bell, which completes the magnetic circuit. The spring-loaded brushes ride directly on stamped commutator bars.

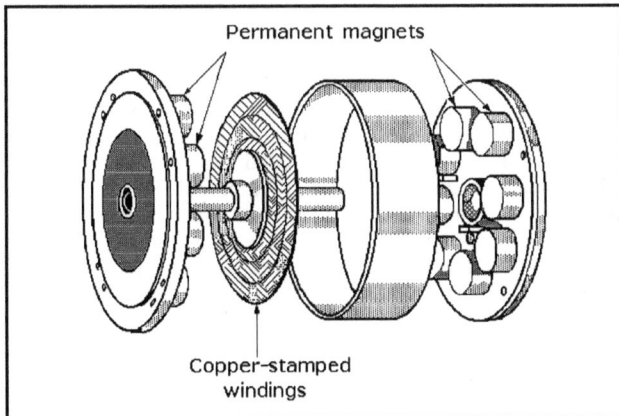

Fig. 3 Exploded view of a permanent-magnet DC servomotor with a disk-type armature.

These motors are also called *pancake motors* because they are housed in cases with thin, flat form factors whose diameters exceed their lengths, suggesting pancakes. Earlier generations of these motors were called *printed-circuit motors* because the armature disks were made by a printed-circuit fabrication process that has been superseded. The flat motor case concentrates the motor's center of mass close to the mounting plate, permitting it to be easily surface mounted. This eliminates the awkward motor overhang and the need for supporting braces if a conventional motor frame is to be surface mounted. Their disk-type motor form factor has made these motors popular as axis drivers for industrial robots where space is limited.

The principal disadvantage of the disk-type motor is the relatively fragile construction of its armature and its inability to dissipate heat as rapidly as iron-core wound rotors. Consequently, these motors are usually limited to applications where the motor can be run under controlled conditions and a shorter duty cycle allows enough time for armature heat buildup to be dissipated.

Cup- or Shell-Type PM DC Motors

Cup- or shell-type PM DC motors offer low inertia and low inductance as well as high acceleration characteristics, making them useful in many servo applications. They have hollow cylindrical armatures made as aluminum or copper coils bonded by polymer resin and fiberglass to form a rigid "ironless cup," which is fastened to an axial shaft. A cutaway view of this class of servomotor is illustrated in Fig. 4.

Fig. 4 Cutaway view of a permanent-magnet DC servomotor with a cup-type armature.

Because the armature has no iron core, it, like the disk motor, has extremely low inertia and a very high torque-to-inertia ratio. This permits the motor to accelerate rapidly for the quick response required in many motion-control applications. The armature rotates in an air gap within very high magnetic flux density. The magnetic field from the stationary magnets is completed through the cup-type armature and a stationary ferrous cylindrical core connected to the motor frame. The shaft rotates within the core, which extends into the rotating cup. Spring brushes commutate these motors.

Another version of a cup-type PM DC motor is shown in the exploded view in Fig. 5. The cup-type armature is rigidly fastened to the shaft by a disk at the right end of the winding, and the magnetic field is also returned through a ferrous metal housing. The brush assembly of this motor is built into its end cap or flange, shown at the far right.

The principal disadvantage of this motor is also the inability of its bonded armature to dissipate internal heat buildup rapidly because of its low thermal conductivity. Without proper cooling and sensitive control circuitry, the armature could be heated to destructive temperatures in seconds.

Fig. 5 Exploded view of a fractional horsepower brush-type DC servomotor.

Brushless PM DC Motors

Brushless DC motors exhibit the same linear speed–torque characteristics as the brush-type PM DC motors, but they are electronically commutated. The construction of these motors, as shown in Fig. 6, differs from that of a typical brush-type DC motor in that they are "inside-out." In other words, they have permanent magnet rotors instead of stators, and the stators rather than the rotors are wound. Although this geometry is required for brushless DC motors, some manufacturers have adapted this design for brush-type DC motors.

Fig. 6 Cutaway view of a brushless DC motor.

The mechanical brush and bar commutator of the brushless DC motor is replaced by electronic sensors, typically Hall-effect devices (HEDs). They are located within the stator windings and wired to solid-state transistor switching circuitry located either on circuit cards mounted within the motor housings or in external packages. Generally, only fractional horsepower brushless motors have switching circuitry within their housings.

The cylindrical magnet rotors of brushless DC motors are magnetized laterally to form opposing north and south poles across the rotor's diameter. These rotors are typically made from neodymium–iron–boron or samarium–cobalt rare-earth magnetic materials, which offer higher flux densities than alnico magnets. These materials permit motors offering higher performance to be packaged in the same frame sizes as earlier motor designs or those with the same ratings to be packaged in smaller frames than the earlier designs. Moreover, rare-earth or ceramic magnet rotors can be made with smaller diameters than those earlier models with alnico magnets, thus reducing their inertia.

A simplified diagram of a DC brushless motor control with one HED for the electronic commutator is shown in Fig. 7. The HED is a Hall-effect sensor integrated with an amplifier in a silicon chip. This IC is capable of sensing the polarity of the rotor's magnetic field and then sending appropriate signals to power transistors T1 and T2 to cause the motor's rotor to rotate continuously. This is accomplished as follows:

(1) With the rotor motionless, the HED detects the rotor's north magnetic pole, causing it to generate a signal that turns on transistor T2. This causes current to flow, energizing winding W2 to form a south-seeking electromagnetic rotor pole. This pole then attracts the rotor's north pole to drive the rotor in a counterclockwise (CCW) direction.

(2) The inertia of the rotor causes it to rotate past its neutral position so that the HED can then sense the rotor's south magnetic pole. It then switches on transistor T1, causing current to flow in winding W1, thus forming a north-seeking stator pole

Fig. 7 Simplified diagram of Hall-effect device (HED) commutation of a brushless DC motor.

that attracts the rotor's south pole, causing it to continue to rotate in the CCW direction.

The transistors conduct in the proper sequence to ensure that the excitation in the stator windings W2 and W1 always leads the PM rotor field to produce the torque necessary to keep the rotor in constant rotation. The windings are energized in a pattern that rotates around the stator.

There are usually two or three HEDs in practical brushless motors that are spaced apart by 90 or 120° around the motor's rotor. They send the signals to the motion controller that actually triggers the power transistors, which drive the armature windings at a specified motor current and voltage level.

The brushless motor in the exploded view Fig. 8 illustrates a design for a miniature brushless DC motor that includes Hall-effect commutation. The stator is formed as an ironless sleeve of copper coils bonded together in polymer resin and fiberglass to form a rigid structure similar to cup-type rotors. However, it is fastened inside the steel laminations within the motor housing.

Fig. 8 Exploded view of a brushless DC motor with Hall-effect device (HED) commutation.

This method of construction permits a range of values for starting current and specific speed (rpm/V) depending on wire gauge and the number of turns. Various terminal resistances can be obtained, permitting the user to select the optimum motor for a specific application. The Hall-effect sensors and a small magnet disk that is magnetized widthwise are mounted on a disk-shaped partition within the motor housing.

Position Sensing in Brushless Motors

Both magnetic sensors and resolvers can sense rotor position in brushless motors. The diagram in Fig. 9 shows how three magnetic sensors can sense rotor position in a three-phase electronically commutated brushless DC motor. In this example, the magnetic sensors are located inside the end bell of the motor. This inexpensive version is adequate for simple controls.

Fig. 9 A magnetic sensor as a rotor position indicator: stationary brushless motor winding (1), permanent-magnet motor rotor (2), three-phase electronically commutated field (3), three magnetic sensors (4), and the electronic circuit board (5).

In the alternate design shown in Fig. 10, a resolver on the end cap of the motor is used to sense rotor position when greater positioning accuracy is required. The high-resolution signals from the resolver can be used to generate sinusoidal motor currents within the motor controller. The currents through the three motor windings are position independent and respectively 120° phase shifted.

Fig. 10 A resolver as a rotor position indicator: stationary motor winding (1), permanent-magnet motor rotor (2), three-phase electronically commutated field (3), three magnetic sensors (4), and the electronic circuit board (5).

Brushless Motor Advantages

Brushless DC motors have at least four distinct advantages over brush-type DC motors that are attributable to the replacement of mechanical commutation by electronic commutation.

- There is no need to replace brushes or remove the gritty residue caused by brush wear from the motor.
- Without brushes to cause electrical arcing, brushless motors do not present fire or explosion hazards in an environment where flammable or explosive vapors, dust, or liquids are present.
- Electromagnetic interference (EMI) is minimized by replacing mechanical commutation, the source of unwanted radio frequencies, with electronic commutation.
- Brushless motors can run faster and more efficiently with electronic commutation. Speeds of up to 50,000 rpm can be achieved versus the upper limit of about 5000 rpm for brush-type DC motors.

Brushless DC Motor Disadvantages

There are at least four disadvantages of brushless DC servomotors.

- Brushless PM DC servomotors cannot be reversed by simply reversing the polarity of the power source. The order in which the current is fed to the field coil must be reversed.
- Brushless DC servomotors cost more than comparably rated brush-type DC servomotors.
- Additional system wiring is required to power the electronic commutation circuitry.
- The motion controller and driver electronics needed to operate a brushless DC servomotor are more complex and expensive than those required for a conventional DC servomotor.

Consequently, the selection of a brushless motor is generally justified on the basis of specific application requirements or its hazardous operating environment.

Characteristics of Brushless Rotary Servomotors

It is difficult to generalize about the characteristics of DC rotary servomotors because of the wide range of products available commercially. However, they typically offer continuous torque ratings of 0.62 lb-ft (0.84 N-m) to 5.0 lb-ft (6.8 N-m), peak torque ratings of 1.9 lb-ft (2.6 N-m) to 14 lb-ft (19 N-m), and continuous power ratings of 0.73 hp (0.54 kW) to 2.76 hp (2.06 kW). Maximum speeds can vary from 1400 to 7500 rpm, and the weight of these motors can be from 5.0 lb (2.3 kg) to 23 lb (10 kg). Feedback typically can be either by resolver or encoder.

Linear Servomotors

A linear motor is essentially a rotary motor that has been opened out into a flat plane, but it operates on the same principles. A permanent-magnet DC linear motor is similar to a permanent-magnet rotary motor, and an AC induction squirrel cage motor is similar to an induction linear motor. The same electromagnetic force that produces torque in a rotary motor also produces torque in a linear motor. Linear motors use the same controls and programmable position controllers as rotary motors.

Before the invention of linear motors, the only way to produce linear motion was to use pneumatic or hydraulic cylinders, or to translate rotary motion to linear motion with ballscrews or belts and pulleys.

A linear motor consists of two mechanical assemblies: *coil* and *magnet*, as shown in Fig. 11. Current flowing in a winding in a magnetic flux field produces a force. The copper windings

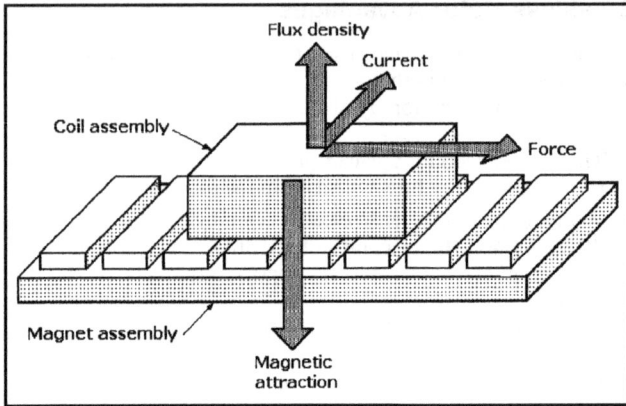

Fig. 11 Operating principles of a linear servomotor.

conduct current (I), and the assembly generates magnetic flux density (B). When the current and flux density interact, a force (F) is generated in the direction shown in Fig. 11, where $F = I \times B$.

Even a small motor will run efficiently, and large forces can be created if a large number of turns are wound in the coil and the magnets are powerful rare-earth magnets. The windings are phased 120 electrical degrees apart, and they must be continually switched or commutated to sustain motion.

Only brushless linear motors for closed-loop servomotor applications are discussed here. Two types of these motors are available commercially—*steel-core* (also called *iron-core*) and *epoxy-core* (also called *ironless*). Each of these linear servomotors has characteristics and features that are optimal in different applications.

The coils of steel-core motors are wound on silicon steel to maximize the generated force available with a single-sided magnet assembly or way. Figure 12 shows a steel-core brushless linear motor. The steel in these motors focuses the magnetic flux to produce very high force density. The magnet assembly consists of rare-earth bar magnets mounted on the upper surface of a steel baseplate arranged to have alternating polarities (i.e., N, S, N, S).

Fig. 12 A linear iron-core linear servomotor consists of a magnetic way and a mating coil assembly.

The steel in the cores is attracted to the permanent magnets in a direction that is perpendicular (normal) to the operating motor force. The magnetic flux density within the air gap of linear motors is typically several thousand gauss. A constant magnetic force is present whether or not the motor is energized. The normal force of the magnetic attraction can be up to 10 times the continuous force rating of the motor. This flux rapidly diminishes

to a few gauss as the measuring point is moved a few centimeters away from the magnets.

Cogging is a form of magnetic "detenting" that occurs in both linear and rotary motors when the motor coil's steel laminations cross the alternating poles of the motor's magnets. Because it can occur in steel-core motors, manufacturers include features that minimize cogging. The high thrust forces attainable with steel-core linear motors permit them to accelerate and move heavy masses while maintaining stiffness during machining or process operations.

The features of epoxy-core or ironless-core motors differ from those of the steel-core motors. For example, their coil assemblies are wound and encapsulated within epoxy to form a thin plate that is inserted in the air gap between the two permanent-magnet strips fastened inside the magnet assembly, as shown in Fig. 13. Because the coil assemblies do not contain steel cores, epoxy-core motors are lighter than steel-core motors and less subject to cogging.

Fig. 13 A linear ironless servomotor consists of an ironless magnetic way and an ironless coil assembly.

The strip magnets are separated to form the air gap into which the coil assembly is inserted. This design maximizes the generated thrust force and also provides a flux return path for the magnetic circuit. Consequently, very little magnetic flux exists outside the motor, thus minimizing residual magnetic attraction.

Epoxy-core motors provide exceptionally smooth motion, making them suitable for applications requiring very low bearing friction and high acceleration of light loads. They also permit constant velocity to be maintained, even at very low speeds.

Linear servomotors can achieve accuracies of 0.1 μm. Normal accelerations are 2 to 3 g, but some motors can reach 15 g. Velocities are limited by the encoder data rate and the amplifier voltage. Normal peak velocities are from 0.04 in./s (1 mm/s) to about 6.6 ft/s (2 m/s), but the velocity of some models can exceed 26 ft/s (8 m/s).

Ironless linear motors can have continuous force ratings from about 5 to 55 lbf (22 to 245 N) and peak force ratings from about 25 to 180 lbf (110 to 800 N). By contrast, iron-core linear motors are available with continuous force ratings of about 30 to 1100 lbf (130 to 4900 N) and peak force ratings of about 60 to 1800 lbf (270 to 8000 N).

Commutation

The linear motor windings that are phased 120° apart must be continually switched or commutated to sustain motion. There are two ways to commutate linear motors: *sinusoidal* and *Hall-effect device (HED)*, or *trapezoidal*. The highest motor efficiency is achieved with sinusoidal commutation, while HED commutation is about 10 to 15 percent less efficient.

In sinusoidal commutation, the linear encoder that provides position feedback in the servosystem is also used to commutate the motor. A process called "phase finding" is required when the motor is turned on, and the motor phases are then incrementally advanced with each encoder pulse. This produces extremely smooth motion. In HED commutation a circuit board containing Hall-effect ICs is embedded in the coil assembly. The HED sensors detect the polarity change in the magnet track and switch the motor phases every 60°.

Sinusoidal commutation is more efficient than HED commutation because the coil windings in motors designed for this commutation method are configured to provide a sinusoidally shaped back EMF waveform. As a result, the motors produce a constant force output when the driving voltage on each phase matches the characteristic back EMF waveform.

Installation of Linear Motors

In a typical linear motor application the coil assembly is attached to the moving member of the host machine and the magnet assembly is mounted on the nonmoving base or frame. These motors can be mounted vertically, but if they are, they typically require a counterbalance system to prevent the load from dropping if power temporarily fails or is routinely shut off. The counterbalance system, typically formed from pulleys and weights, springs, or air cylinders, supports the load against the force of gravity.

If power is lost, servo control is interrupted. Stages in motion tend to stay in motion while those at rest tend to stay at rest. The stopping time and distance depend on the stage's initial velocity and system friction. The motor's back EMF can provide dynamic braking, and friction brakes can be used to attenuate motion rapidly. However, positive stops and travel limits can be built into the motion stage to prevent damage in situations where power or feedback might be lost or the controller or servo driver fail.

Linear servomotors are supplied to the customer in kit form for mounting on the host machine. The host machine structure must include bearings capable of supporting the mass of the motor parts while maintaining the specified air gap between the assemblies and also resisting the normal force of any residual magnetic attraction.

Linear servomotors must be used in closed loop positioning systems because they do not include built-in means for position sensing. Feedback is typically supplied by such sensors as linear encoders, laser interferometers, LVDTs, or linear Inductosyns.

Advantages of Linear vs. Rotary Servomotors

The advantages of linear servomotors over rotary servomotors include:

- *High stiffness:* The linear motor is connected directly to the moving load, so there is no backlash and practically no compliance between the motor and the load. The load moves instantly in response to motor motion.
- *Mechanical simplicity:* The coil assembly is the only moving part of the motor, and its magnet assembly is rigidly mounted to a stationary structure on the host machine. Some linear motor manufacturers offer modular magnetic assemblies in various modular lengths. This permits the user to form a track of any desired length by stacking the modules end to end, allowing virtually unlimited travel. The force produced by the motor is applied directly to the load without any couplings, bearings, or other conversion mechanisms. The only alignments required are for the air gaps, which typically are from 0.039 in. (1 mm) to 0.020 in. (0.5 mm).

- *High accelerations and velocities:* Because there is no physical contact between the coil and magnet assemblies, high accelerations and velocities are possible. Large motors are capable of accelerations of 3 to 5 *g*, but smaller motors are capable of more than 10 *g*.
- *High velocities:* Velocities are limited by feedback encoder data rate and amplifier bus voltage. Normal peak velocities are up to 6.6 ft/s (2 m/s), although some models can reach 26 ft/s (8 m/s). This compares with typical linear speeds of ballscrew transmissions, which are commonly limited to 20 to 30 in./s (0.5 to 0.7 m/s) because of resonances and wear.
- *High accuracy and repeatability:* Linear motors with position feedback encoders can achieve positioning accuracies of ±1 encoder cycle or submicrometer dimensions, limited only by encoder feedback resolution.
- *No backlash or wear:* With no contact between moving parts, linear motors do not wear out. This minimizes maintenance and makes them suitable for applications where long life and long-term peak performance are required.
- *System size reduction:* With the coil assembly attached to the load, no additional space is required. By contrast, rotary motors typically require ballscrews, rack-and-pinion gearing, or timing belt drives.
- *Clean room compatibility:* Linear motors can be used in clean rooms because they do not need lubrication and do not produce carbon brush grit.

Coil Assembly Heat Dissipation

Heat control is more critical in linear motors than in rotary motors because they do not have the metal frames or cases that can act as large heat-dissipating surfaces. Some rotary motors also have radiating fins on their frames that serve as heat sinks to augment the heat dissipation capability of the frames. Linear motors must rely on a combination of high motor efficiency and good thermal conduction from the windings to a heat-conductive, electrically isolated mass. For example, an aluminum attachment bar placed in close contact with the windings can aid in heat dissipation. Moreover, the carriage plate to which the coil assembly is attached must have effective heat-sinking capability.

Stepper Motors

A *stepper* or *stepping motor* is an AC motor whose shaft is indexed through part of a revolution or *step angle* for each DC pulse sent to it. Trains of pulses provide input current to the motor in increments that can "step" the motor through 360°, and the actual angular rotation of the shaft is directly related to the number of pulses introduced. The position of the load can be determined with reasonable accuracy by counting the pulses entered.

The stepper motors suitable for most open-loop motion control applications have wound stator fields (electromagnetic coils) and iron or permanent magnet (PM) rotors. Unlike PM DC servomotors with mechanical brush-type commutators, stepper motors depend on external controllers to provide the switching pulses for commutation. Stepper motor operation is based on the same electromagnetic principles of attraction and repulsion as other motors, but their commutation provides only the torque required to turn their rotors.

Pulses from the external motor controller determine the amplitude and direction of current flow in the stator's field windings, and they can turn the motor's rotor either clockwise or counterclockwise, stop and start it quickly, and hold it securely at desired positions. Rotational shaft speed depends on the frequency of the pulses. Because controllers can step most motors at audio frequencies, their rotors can turn rapidly.

Between the application of pulses when the rotor is at rest, its armature will not drift from its stationary position because of the stepper motor's inherent holding ability or *detent torque*. These motors generate very little heat while at rest, making them suitable for many different instrument drive-motor applications in which power is limited.

The three basic kinds of stepper motors are *permanent magnet, variable reluctance,* and *hybrid.* The same controller circuit can drive both hybrid and PM stepping motors.

Permanent-Magnet (PM) Stepper Motors

Permanent-magnet stepper motors have smooth armatures and include a permanent magnet core that is magnetized widthwise or perpendicular to its rotation axis. These motors usually have two independent windings, with or without center taps. The most common step angles for PM motors are 45 and 90°, but motors with step angles as fine as 1.8° per step as well as 7.5, 15, and 30° per step are generally available. Armature rotation occurs when the stator poles are alternately energized and deenergized to create torque. A 90° stepper has four poles and a 45° stepper has eight poles, and these poles must be energized in sequence. Permanent-magnet steppers step at relatively low rates, but they can produce high torques and they offer very good damping characteristics.

Variable Reluctance Stepper Motors

Variable reluctance (VR) stepper motors have multitooth armatures with each tooth effectively an individual magnet. At rest these magnets align themselves in a natural detent position to provide larger holding torque than can be obtained with a comparably rated PM stepper. Typical VR motor step angles are 15 and 30° per step. The 30° angle is obtained with a 4-tooth rotor and a 6-pole stator, and the 15° angle is achieved with an 8-tooth rotor and a 12-pole stator. These motors typically have three windings with a common return, but they are also available with four or five windings. To obtain continuous rotation, power must be applied to the windings in a coordinated sequence of alternately deenergizing and energizing the poles.

If just one winding of either a PM or VR stepper motor is energized, the rotor (under no load) will snap to a fixed angle and hold that angle until external torque exceeds the holding torque of the motor. At that point, the rotor will turn, but it will still try to hold its new position at each successive equilibrium point.

Hybrid Stepper Motors

The hybrid stepper motor combines the best features of VR and PM stepper motors. A cutaway view of a typical industrial-grade hybrid stepper motor with a multitoothed armature is shown in Fig. 14. The armature is built in two sections, with the teeth in the second section offset from those in the first section. These motors also have multitoothed stator poles that are not visible in the figure. Hybrid stepper motors can achieve high stepping rates, and they offer high detent torque and excellent dynamic and static torque.

Hybrid steppers typically have two windings on each stator pole so that each pole can become either magnetic north or south, depending on current flow. A cross-sectional view of a hybrid stepper motor illustrating the multitoothed poles with dual windings per pole and the multitoothed rotor is illustrated in Fig. 15. The shaft is represented by the central circle in the diagram.

The most popular hybrid steppers have 3- and 5-phase wiring, and step angles of 1.8 and 3.6° per step. These motors can provide more torque from a given frame size than other stepper types because either all or all but one of the motor windings are energized at every point in the drive cycle. Some 5-phase motors have high resolutions of 0.72° per step (500 steps per revolution).

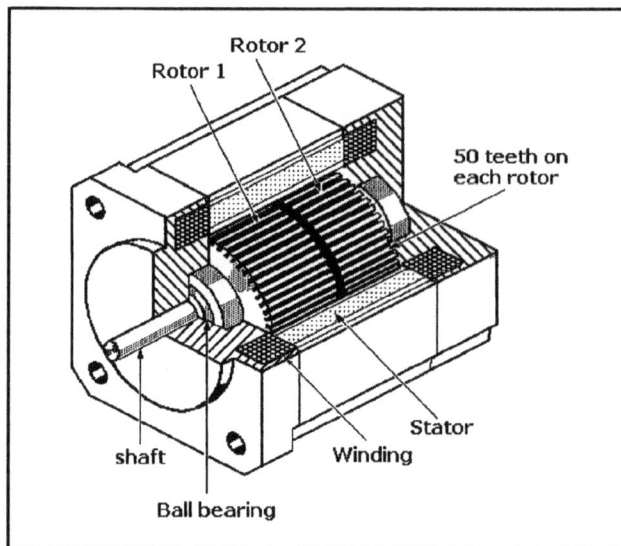

Fig. 14 Cutaway view of a 5-phase hybrid stepping motor. A permanent magnet is within the rotor assembly, and the rotor segments are offset from each other by 3.5°.

Fig. 15 Cross section of a hybrid stepping motor showing the segments of the magnetic-core rotor and stator poles with its wiring diagram.

With a compatible controller, most PM and hybrid motors can be run in half-steps, and some controllers are designed to provide smaller fractional steps, or *microsteps*. Hybrid stepper motors capable of a wide range of torque values are available commercially. This range is achieved by scaling length and diameter dimensions. Hybrid stepper motors are available in NEMA size 17 to 42 frames, and output power can be as high as 1000 W peak.

Stepper Motor Applications

Many different technical and economic factors must be considered in selecting a hybrid stepper motor. For example, the ability of the stepper motor to repeat the positioning of its multitoothed

Fig. 16 This linear actuator can be powered by either an AC or DC motor. It contains ballscrew, reduction gear, clutch, and brake assemblies.

rotor depends on its geometry. A disadvantage of the hybrid stepper motor operating open-loop is that, if overtorqued, its position "memory" is lost and the system must be reinitialized. Stepper motors can perform precise positioning in simple open-loop control systems if they operate at low acceleration rates with static loads. However, if higher acceleration values are required for driving variable loads, the stepper motor must be operated in a closed loop with a position sensor.

DC and AC Motor Linear Actuators

Actuators for motion control systems are available in many different forms, including both linear and rotary versions. One popular configuration is that of a Thomson Saginaw PPA, shown in section view in Fig. 16. It consists of an AC or DC motor mounted parallel to either a ballscrew or Acme screw assembly through a reduction gear assembly with a slip clutch and integral brake assembly. Linear actuators of this type can perform a wide range of commercial, industrial, and institutional applications.

One version designed for mobile applications can be powered by a 12-, 24-, or 36-VDC permanent-magnet motor. These motors are capable of performing such tasks as positioning antenna reflectors, opening and closing security gates, handling materials, and raising and lowering scissors-type lift tables, machine hoods, and light-duty jib crane arms.

Other linear actuators are designed for use in fixed locations where either 120- or 220-VAC line power is available. They can have either AC or DC motors. Those with 120-VAC motors can be equipped with optional electric brakes that virtually eliminate coasting, thus permitting point-to-point travel along the stroke.

Where variable speed is desired and 120-VAC power is available, a linear actuator with a 90-VDC motor can be equipped with a solid-state rectifier/speed controller. Closed-loop feedback provides speed regulation down to one-tenth of the maximum travel rate. This feedback system can maintain its selected travel rate despite load changes.

Thomson Saginaw also offers its linear actuators with either Hall-effect or potentiometer sensors for applications where it is necessary or desirable to control actuator positioning. With Hall-effect sensing, six pulses are generated with each turn of the output shaft during which the stroke travels approximately $\frac{1}{32}$ in. (0.033 in. or 0.84 mm). These pulses can be counted by a separate control unit and added or subtracted from the stored

pulse count in the unit's memory. The actuator can be stopped at any 0.033-in. increment of travel along the stroke selected by programming. A limit switch can be used together with this sensor.

If a 10-turn, 10,000-ohm potentiometer is used as a sensor, it can be driven by the output shaft through a spur gear. The gear ratio is established to change the resistance from 0 to 10,000 ohms over the length of the actuator stroke. A separate control unit measures the resistance (or voltage) across the potentiometer, which varies continuously and linearly with stroke travel. The actuator can be stopped at any position along its stroke.

Stepper-Motor Based Linear Actuators

Linear actuators are available with axial integral threaded shafts and bolt nuts that convert rotary motion to linear motion. Powered by fractional horsepower permanent-magnet stepper motors, these linear actuators are capable of positioning light loads. Digital pulses fed to the actuator cause the threaded shaft to rotate, advancing or retracting it so that a load coupled to the shaft can be moved backward or forward. The bidirectional digital linear actuator shown in Fig. 17 can provide linear resolution as fine as 0.001 in. per pulse. Travel per step is determined by the pitch of the leadscrew and step angle of the motor. The maximum linear force for the model shown is 75 oz.

Fig. 17 This light-duty linear actuator based on a permanent-magnet stepping motor has a shaft that advances or retracts.

SERVOSYSTEM FEEDBACK SENSORS

A servosystem feedback sensor in a motion control system transforms a physical variable into an electrical signal for use by the motion controller. Common feedback sensors are encoders, resolvers, and linear variable differential transformers (LVDTs) for motion and position feedback, and tachometers for velocity feedback. Less common but also in use as feedback devices are potentiometers, linear velocity transducers (LVTs), angular displacement transducers (ADTs), laser interferometers, and potentiometers. Generally speaking, the closer the feedback sensor is to the variable being controlled, the more accurate it will be in assisting the system to correct velocity and position errors.

For example, direct measurement of the linear position of the carriage carrying the load or tool on a single-axis linear guide will provide more accurate feedback than an indirect measurement determined from the angular position of the guide's leadscrew and knowledge of the drivetrain geometry between the sensor and the carriage. Thus, direct position measurement avoids drivetrain errors caused by backlash, hysteresis, and leadscrew wear that can adversely affect indirect measurement.

Rotary Encoders

Rotary encoders, also called *rotary shaft encoders* or *rotary shaft-angle* encoders, are electromechanical transducers that convert shaft rotation into output pulses, which can be counted to measure shaft revolutions or shaft angle. They provide rate and positioning information in servo feedback loops. A rotary encoder can sense a number of discrete positions per revolution. The number is called *points per revolution* and is analogous to the *steps per revolution* of a stepper motor. The speed of an encoder is in units of counts per second. Rotary encoders can measure the motor-shaft or leadscrew angle to report position indirectly, but they can also measure the response of rotating machines directly.

The most popular rotary encoders are *incremental optical shaft-angle encoders* and the *absolute optical shaft-angle encoders*. There are also *direct contact* or *brush-type* and *magnetic rotary encoders*, but they are not as widely used in motion control systems.

Commercial rotary encoders are available as standard or catalog units, or they can be custom made for unusual applications or survival in extreme environments. Standard rotary encoders are packaged in cylindrical cases with diameters from 1.5 to 3.5 in. Resolutions range from 50 cycles per shaft revolution to 2,304,000 counts per revolution. A variation of the conventional configuration, the *hollow-shaft encoder,* eliminates problems associated with the installation and shaft runout of conventional models. Models with hollow shafts are available for mounting on shafts with diameters of 0.04 to 1.6 in. (1 to 40 mm).

Incremental Encoders

The basic parts of an incremental optical shaft-angle encoder are shown in Fig. 1. A glass or plastic code disk mounted on the encoder shaft rotates between an internal light source, typically a light-emitting diode (LED), on one side and a mask and matching photodetector assembly on the other side. The incremental code disk contains a pattern of equally spaced opaque and transparent segments or spokes that radiate out from its center as shown. The electronic signals that are generated by the encoder's electronics board are fed into a motion controller that calculates position and

Fig. 1 Basic elements of an incremental optical rotary encoder.

velocity information for feedback purposes. An exploded view of an industrial-grade incremental encoder is shown in Fig. 2.

Glass code disks containing finer graduations capable of 11- to more than 16-bit resolution are used in high-resolution encoders, and plastic (Mylar) disks capable of 8- to 10-bit resolution are used in the more rugged encoders that are subject to shock and vibration.

Fig. 2 Exploded view of an incremental optical rotary encoder showing the stationary mask between the code wheel and the photodetector assembly.

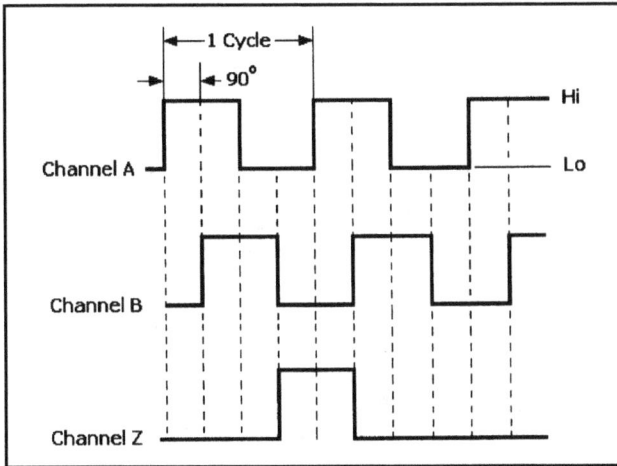

Fig. 3 Channels A and B provide bidirectional position sensing. If channel A leads channel B, the direction is clockwise; if channel B leads channel A, the direction is counterclockwise. Channel Z provides a zero reference for determining the number of disk rotations.

The quadrature encoder is the most common type of incremental encoder. Light from the LED passing through the rotating code disk and mask is "chopped" before it strikes the photodetector assembly. The output signals from the assembly are converted into two channels of square pulses (A and B) as shown in Fig. 3. The number of square pulses in each channel is equal to the number of code disk segments that pass the photodetectors as the disk rotates, but the waveforms are 90° out of phase. If, for example, the pulses in channel A lead those in channel B, the disk is rotating in a clockwise direction, but if the pulses in channel A lag those in channel B lead, the disk is rotating counterclockwise. By monitoring both the number of pulses and the relative phases of signals A and B, both position and direction of rotation can be determined.

Many incremental quadrature encoders also include a third output Z channel to obtain a zero reference or index signal that occurs once per revolution. This channel can be gated to the A and B quadrature channels and used to trigger certain events accurately within the system. The signal can also be used to align the encoder shaft to a mechanical reference.

Absolute Encoders

An *absolute shaft-angle optical encoder* contains multiple light sources and photodetectors, and a code disk with up to 20 tracks of segmented patterns arranged as annular rings, as shown in Fig. 4. The code disk provides a binary output that uniquely defines each shaft angle, thus providing an absolute measurement. This type of encoder is organized in essentially the same way as the incremental encoder shown in Fig. 2, but the code disk rotates between linear arrays of LEDs and photodetectors arranged radially, and a LED opposes a photodetector for each track or annular ring.

The arc lengths of the opaque and transparent sectors decrease with respect to the radial distance from the shaft. These disks, also made of glass or plastic, produce either the natural binary or Gray code. Shaft position accuracy is proportional to the number of annular rings or tracks on the disk. When the code disk rotates, light passing through each track or annular ring generates a continuous stream of signals from the detector array. The electronics board converts that output into a binary word. The value of the output code word is read radially from the most significant bit (MSB) on the inner ring of the disk to the least significant bit (LSB) on the outer ring of the disk.

Fig. 4 Binary-code disk for an absolute optical rotary encoder. Opaque sectors represent a binary value of 1, and the transparent sectors represent binary 0. This four-bit binary-code disk can count from 1 to 15.

The principal reason for selecting an absolute encoder over an incremental encoder is that its code disk retains the last angular position of the encoder shaft whenever it stops moving, whether the system is shut down deliberately or as a result of power failure. This means that the last readout is preserved, an important feature for many applications.

Linear Encoders

Linear encoders can make direct accurate measurements of unidirectional and reciprocating motions of mechanisms with high resolution and repeatability. Figure 5 illustrates the basic parts of an optical linear encoder. A movable scanning unit contains the light source, lens, graduated glass scanning reticule, and an array of photocells. The scale, typically made as a strip of glass with opaque graduations, is bonded to a supporting structure on the host machine.

A beam of light from the light source passes through the lens, four windows of the scanning reticule, and the glass scale to the array of photocells. When the scanning unit moves, the scale modulates the light beam so that the photocells generate sinusoidal signals.

The four windows in the scanning reticule are each 90° apart in phase. The encoder combines the phase-shifted signal to produce

Fig. 5 Optical linear encoders direct light through a moving glass scale with accurately etched graduations to photocells on the opposite side for conversion to a distance value.

two symmetrical sinusoidal outputs that are phase shifted by 90°. A fifth pattern on the scanning reticule has a random graduation that, when aligned with an identical reference mark on the scale, generates a reference signal.

A fine-scale pitch provides high resolution. The spacing between the scanning reticule and the fixed scale must be narrow and constant to eliminate undesirable diffraction effects of the scale grating. The complete scanning unit is mounted on a carriage that moves on ball bearings along the glass scale. The scanning unit is connected to the host machine slide by a coupling that compensates for any alignment errors between the scale and the machine guideways.

External electronic circuitry interpolates the sinusoidal signals from the encoder head to subdivide the line spacing on the scale so that it can measure even smaller motion increments. The practical maximum length of linear encoder scales is about 10 ft (3 m), but commercial catalog models are typically limited to about 6 ft (2 m). If longer distances are to be measured, the encoder scale is made of steel tape with reflective graduations that are sensed by an appropriate photoelectric scanning unit.

Linear encoders can make direct measurements that overcome the inaccuracies inherent in mechanical stages due to backlash, hysteresis, and leadscrew error. However, the scale's susceptibility to damage from metallic chips, grit oil, and other contaminants, together with its relatively large space requirements, limits applications for these encoders.

Commercial linear encoders are available as standard catalog models, or they can be custom made for specific applications or extreme environmental conditions. There are both fully enclosed and open linear encoders with travel distances from 2 in. to 6 ft (50 mm to 1.8 m). Some commercial models are available with resolutions down to 0.07 μm, and others can operate at speeds of up to 16.7 ft/s (5 m/s).

Magnetic Encoders

Magnetic encoders can be made by placing a transversely polarized permanent magnet in close proximity to a Hall-effect device sensor. Figure 6 shows a magnet mounted on a motor shaft in close proximity to a two-channel HED array which detects changes in magnetic flux density as the magnet rotates. The output signals from the sensors are transmitted to the motion controller. The encoder output, either a square wave or a quasi sine wave (depending on the type of magnetic sensing device) can be used to count revolutions per minute (rpm) or determine motor shaft accurately. The phase shift between channels A and B permits them to be compared by the motion controller to determine the direction of motor shaft rotation.

Fig. 6 Basic parts of a magnetic encoder.

Resolvers

A resolver is essentially a rotary transformer that can provide position feedback in a servosystem as an alternative to an encoder. Resolvers resemble small AC motors, as shown in Fig. 7, and

Fig. 7 Exploded view of a brushless resolver frame (a), and rotor and bearings (b). The coil on the rotor couples speed data inductively to the frame for processing.

generate an electrical signal for each revolution of their shaft. Resolvers that sense position in closed-loop motion control applications have one winding on the rotor and a pair of windings on the stator, oriented at 90°. The stator is made by winding copper wire in a stack of iron laminations fastened to the housing, and the rotor is made by winding copper wire in a stack of laminations mounted on the resolver's shaft.

Figure 8 is an electrical schematic for a brushless resolver showing the single rotor winding and the two stator windings 90° apart. In a servosystem, the resolver's rotor is mechanically coupled to the drive motor and load. When a rotor winding is excited by an AC reference signal, it produces an AC voltage output that varies in amplitude according to the sine and cosine of shaft position. If the phase shift between the applied signal to the rotor and the induced signal appearing on the stator coil is measured, that angle is an analog of rotor position. The absolute position of the load being driven can be determined by the ratio of the sine output amplitude to the cosine output amplitude as the resolver shaft turns through one revolution. (A single-speed resolver produces one sine and one cosine wave as the output for each revolution.)

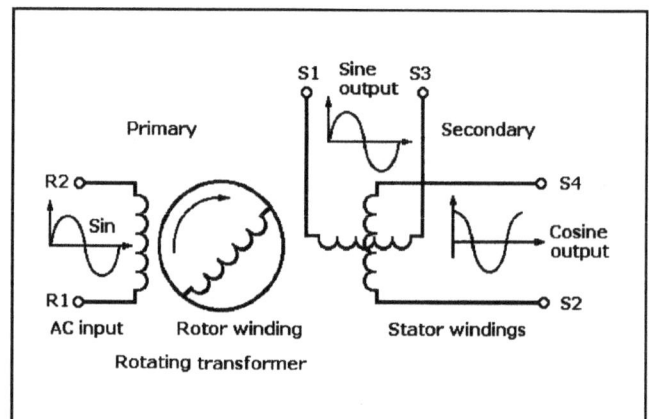

Fig. 8 Schematic for a resolver shows how rotor position is transformed into sine and cosine outputs that measure rotor position.

Fig. 9 Section view of a resolver and tachometer in the same frame as the servomotor.

Connections to the rotor of some resolvers can be made by brushes and slip rings, but resolvers for motion control applications are typically brushless. A rotating transformer on the rotor couples the signal to the rotor inductively. Because brushless resolvers have no slip rings or brushes, they are more rugged than encoders and have operating lives that are up to ten times those of brush-type resolvers. Bearing failure is the most likely cause of resolver failure. The absence of brushes in these resolvers makes them insensitive to vibration and contaminants. Typical brushless resolvers have diameters from 0.8 to 3.7 in. Rotor shafts are typically threaded and splined.

Most brushless resolvers can operate over a 2- to 40-volt range, and their windings are excited by an AC reference voltage at frequencies from 400 to 10,000 Hz. The magnitude of the voltage induced in any stator winding is proportional to the cosine of the angle, q, between the rotor coil axis and the stator coil axis. The voltage induced across any pair of stator terminals will be the vector sum of the voltages across the two connected coils. Accuracies of ±1 arc-minute can be achieved.

In feedback loop applications, the stator's sinusoidal output signals are transmitted to a resolver-to-digital converter (RDC), a specialized analog-to-digital converter (ADC) that converts the signals to a digital representation of the actual angle required as an input to the motion controller.

Tachometers

A tachometer is a DC generator that can provide velocity feedback for a servosystem. The tachometer's output voltage is directly proportional to the rotational speed of the armature shaft that drives it. In a typical servosystem application, it is mechanically coupled to the DC motor and feeds its output voltage back to the controller and amplifier to control drive motor and load speed. A cross-sectional drawing of a tachometer built into the same housing as the DC motor and a resolver is shown in Fig. 9. Encoders or resolvers are part of separate loops that provide position feedback.

As the tachometer's armature coils rotate through the stator's magnetic field, lines of force are cut so that an electromotive force is induced in each of its coils. This emf is directly proportional to the rate at which the magnetic lines of force are cut as well as being directly proportional to the velocity of the motor's drive shaft. The direction of the emf is determined by Fleming's generator rule.

The AC generated by the armature coil is converted to DC by the tachometer's commutator, and its value is directly proportional to shaft rotation speed while its polarity depends on the direction of shaft rotation.

There are two basic types of DC tachometer: *shunt wound* and *permanent magnet* (PM), but PM tachometers are more widely used in servosystems today. There are also moving-coil tachometers which, like motors, have no iron in their armatures. The armature windings are wound from fine copper wire and bonded with glass fibers and polyester resins into a rigid cup, which is bonded to its coaxial shaft. Because this armature contains no iron, it has lower inertia than conventional copper and iron armatures, and it exhibits low inductance. As a result, the moving-coil tachometer is more responsive to speed changes and provides a DC output with very low ripple amplitudes.

Tachometers are available as stand-alone machines. They can be rigidly mounted to the servomotor housings, and their shafts can be mechanically coupled to the servomotor's shafts. If the DC servomotor is either a brushless or moving-coil motor, the stand-alone tachometer will typically be brushless and, although they are housed separately, a common armature shaft will be shared.

A brush-type DC motor with feedback furnished by a brush-type tachometer is shown in Fig. 10. Both tachometer and motor rotor coils are mounted on a common shaft. This arrangement provides a high resonance frequency. Moreover, the need for separate tachometer bearings is eliminated.

Fig. 10 The rotors of the DC motor and tachometer share a common shaft.

Fig. 11 This coil-type DC motor obtains velocity feedback from a tachometer whose rotor coil is mounted on a common shaft and position feedback from a two-channel photoelectric encoder whose code disk is also mounted on the same shaft.

In applications where precise positioning is required in addition to speed regulation, an incremental encoder can be added on the same shaft, as shown in Fig. 11.

Linear Variable Differential Transformers (LVDTs)

A linear variable differential transformer (LVDT) is a sensing transformer consisting of a primary winding, two adjacent secondary windings, and a ferromagnetic core that can be moved axially within the windings, as shown in the cutaway view in Fig. 12. LVDTs are capable of measuring position, acceleration, force, or pressure, depending on how they are installed. In motion control systems, LVDTs provide position feedback by measuring the variation in mutual inductance between their primary and secondary windings caused by the linear movement of the ferromagnetic core.

Fig. 12 Cutaway view of a linear variable displacement transformer (LVDT).

The core is attached to a spring-loaded sensing shaft. When depressed, the shaft moves the core axially within the windings, coupling the excitation voltage in the primary (middle) winding P1 to the two adjacent secondary windings S1 and S2.

Figure 13 is a schematic diagram of a LVDT. When the core is centered between S1 and S2, the voltages induced in S1 and S2 have equal amplitudes and are 180° out of phase. With a series-opposed connection, as shown, the net voltage across the secondaries is zero because both voltages cancel. This is called the *null position* of the core.

Fig. 13 Schematic for a linear variable differential transformer (LVDT) showing how the movable core interacts with the primary and secondary windings.

However, if the core is moved to the left, secondary winding S1 is more strongly coupled to primary winding P1 than secondary winding S2, and an output sine wave in phase with the primary voltage is induced. Similarly, if the core is moved to the right and winding S2 is more strongly coupled to primary winding P1, an output sine wave that is 180° out-of-phase with the primary voltage is induced. The amplitudes of the output sine waves of the LVDT vary symmetrically with core displacement, either to the left or right of the null position.

Linear variable differential transformers require signal conditioning circuitry that includes a stable sine wave oscillator to excite the primary winding P1, a demodulator to convert secondary AC voltage signals to DC, a low-pass filter, and an amplifier to buffer the DC output signal. The amplitude of the resulting DC voltage output is proportional to the magnitude of core displacement, either to the left or right of the null position. The phase of the DC voltage indicates the position of the core relative to the null (left or right). A LVDT containing an integral oscillator/demodulator is a DC-to-DC LVDT, also known as a DCDT.

Linear variable differential transformers can make linear displacement (position) measurements as precise as 0.005 in. (0.127 mm). Output voltage linearity is an important LVDT characteristic, and it can be plotted as a straight line within a specified range. Linearity is the characteristic that largely determines the LVDT's absolute accuracy.

Linear Velocity Transducers (LVTs)

A linear velocity transducer (LVT) consists of a magnet positioned axially within two wire coils. When the magnet is moved through the coils, it induces a voltage within the coils in accordance with the Faraday and Lenz laws. The output voltage from the coils is directly proportional to the magnet's field strength and axial velocity over its working range.

When the magnet is functioning as a transducer, both of its ends are within the two adjacent coils, and when it is moved axially, its north pole will induce a voltage in one coil and its south pole will induce a voltage in the other coil. The two coils can be connected in series or parallel, depending on the application. In both configurations, the DC output voltage from the coils is proportional to magnet velocity. (A single coil would only produce zero voltage because the voltage generated by the north pole would be canceled by the voltage generated by the south pole.)

The characteristics of the LVT depend on how the two coils are connected. If they are connected in series opposition, the output is added and maximum sensitivity is obtained. Also, noise generated in one coil will be canceled by the noise generated in the other coil. However, if the coils are connected in parallel, both sensitivity and source impedance are reduced. Reduced sensitivity improves high-frequency response for measuring high velocities, and the lower output impedance improves the LVT's compatibility with its signal-conditioning electronics.

Angular Displacement Transducers (ADTs)

An angular displacement transducer (ADT) is an air-core variable differential capacitor that can sense angular displacement. As shown in exploded view in Fig. 14, it has a movable metal rotor sandwiched between a single stator plate and segmented stator plates. When a high-frequency AC signal from an oscillator is placed across the plates, it is modulated by the change in capacitance value due to the position of the rotor with respect to the segmented stator plates. The angular displacement of the rotor can then be determined accurately from the demodulated AC signal.

Fig. 14 Exploded view of an angular displacement transducer (ADT) based on a differential variable capacitor.

The base is the mounting platform for the transducer assembly. It contains the axial ball bearing that supports the shaft to which the rotor is fastened. The base also supports the transmitting board, which contains a metal surface that forms the lower plate of the differential capacitor. The semicircular metal rotor mounted on the shaft is the variable plate or rotor of the capacitor. Positioned above the rotor is the receiving board containing two separate semicircular metal sectors on its lower surface. The board acts as the receiver for the AC signal that has been modulated by the capacitance difference between the plates caused by rotor rotation.

An electronics circuit board mounted on top of the assembly contains the oscillator, demodulator, and filtering circuitry. The ADT is powered by DC, and its output is a DC signal that is proportional to angular displacement. The cup-shaped housing encloses the entire assembly, and the base forms a secure cap.

DC voltage is applied to the input terminals of the ADT to power the oscillator, which generates a 400- to 500-kHz voltage that is applied across the transmitting and receiving stator plates. The receiving plates are at virtual ground, and the rotor is at true ground. The capacitance value between the transmitting and receiving plates remains constant, but the capacitance between the separate receiving plates varies with rotor position.

A null point is obtained when the rotor is positioned under equal areas of the receiving stator plates. In that position, the capacitance between the transmitting stator plate and the receiving stator plates will be equal, and there will be no output voltage. However, as the rotor moves clockwise or counterclockwise, the capacitance between the transmitting plate and one of the receiving plates will be greater than it is between the other receiving plate. As a result, after demodulation, the differential output DC voltage will be proportional to the angular distance the rotor moved from the null point.

Inductosyns

The Inductosyn is a proprietary AC sensor that generates position feedback signals that are similar to those from a resolver. There are rotary and linear Inductosyns. Much smaller than a resolver, a rotary Inductosyn is an assembly of a scale and slider on insulating substrates in a loop. When the scale is energized with AC, the voltage couples into the two slider windings and induces voltages proportional to the sine and cosine of the slider spacing within a cyclic pitch.

An Inductosyn-to-digital (I/D) converter, similar to a resolver-to-digital (R/D) converter, is needed to convert these signals into a digital format. A typical rotary Inductosyn with 360 cyclic pitches per rotation can resolve a total of 1,474,560 sectors for each resolution. This corresponds to an angular rotation of less than 0.9 arc-s. This angular information in a digital format is sent to the motion controller.

Laser Interferometers

Laser interferometers provide the most accurate position feedback for servosystems. They offer very high resolution (to 1.24 nm), noncontact measurement, a high update rate, and intrinsic accuracies of up to 0.02 ppm. They can be used in servosystems either as passive position readouts or as active feedback sensors in a position servo loop. The laser beam path can be precisely aligned to coincide with the load or a specific point being measured, eliminating or greatly reducing Abbe error.

A single-axis system based on the Michaelson interferometer is illustrated in Fig. 15. It consists of a helium–neon laser, a polarizing beam splitter with a stationary retroreflector, a moving retroreflector that can be mounted on the object whose position is to be measured, and a photodetector, typically a photodiode.

Fig. 15 Diagram of a laser interferometer for position feedback that combines high resolution with noncontact sensing, high update rates, and accuracies of 0.02 ppm.

Light from the laser is directed toward the polarizing beam splitter, which contains a partially reflecting mirror. Part of the laser beam goes straight through the polarizing beam splitter, and part of the laser beam is reflected. The part that goes straight through the beam splitter reaches the moving reflectometer, which reflects it back to the beam splitter, that passes it on to the photodetector. The part of the beam that is reflected by the beam splitter reaches the stationary retroreflector, a fixed distance away. The retroreflector reflects it back to the beam splitter before it is also reflected into the photodetector.

As a result, the two reflected laser beams strike the photodetector, which converts the combination of the two light beams into an electrical signal. Because of the way laser light beams interact, the output of the detector depends on a *difference* in the distances traveled by the two laser beams. Because both light beams travel the same distance from the laser to the beam splitter and from the beam splitter to the photodetector, these distances are not involved in position measurement. The laser interferometer measurement depends only on the difference in distance between the round-trip laser beam travel from the beam splitter to the moving retroreflector and the fixed round-trip distance of laser beam travel from the beam splitter to the stationary retroreflector.

If these two distances are exactly the same, the two light beams will recombine in phase at the photodetector, which will produce a high electrical output. This event can be viewed on a video display as a bright *light fringe*. However, if the difference between the distances is as short as one-quarter of the laser's wavelength, the light beams will combine out-of-phase, interfering with each other so that there will be no electrical output from the photodetector and no video output on the display, a condition called a *dark fringe*.

As the moving retroreflector mounted on the load moves farther away from the beam splitter, the laser beam path length will increase and a pattern of light and dark fringes will repeat uniformly. This will result in electrical signals that can be counted and converted to a distance measurement to provide an accurate position of the load. The spacing between the light and dark fringes and the resulting electrical pulse rate is determined by the wavelength of the light from the laser. For example, the wavelength of the light beam emitted by a helium–neon (He–Ne) laser, widely used in laser interferometers, is 0.63 μm, or about 0.000025 in.

Thus, the accuracy of load position measurement depends primarily on the known stabilized wavelength of the laser beam. However, that accuracy can be degraded by changes in humidity and temperature as well as airborne contaminants such as smoke or dust in the air between the beam splitter and the moving retroreflector.

Precision Multiturn Potentiometers

The rotary precision multiturn potentiometer shown in the cutaway in Fig. 16 is a simple, low-cost feedback instrument.

Fig. 16 A precision potentiometer is a low-cost, reliable feedback sensor for servosystems.

Originally developed for use in analog computers, precision potentiometers can provide absolute position data in analog form as a resistance value or voltage. Precise and resettable voltages correspond to each setting of the rotary control shaft. If a potentiometer is used in a servosystem, the analog data will usually be converted to digital data by an integrated circuit analog-to-digital converter (ADC). Accuracies of 0.05 percent can be obtained from an instrument-quality precision multiturn potentiometer, and resolutions can exceed 0.005° if the output signal is converted with a 16-bit ADC.

Precision multiturn potentiometers have wirewound or hybrid resistive elements. Hybrid elements are wirewound elements coated with resistive plastic to improve their resolution. To obtain an output from a potentiometer, a conductive wiper must be in contact with the resistive element. During its service life, wear on the resistive element caused by the wiper can degrade the precision of the precision potentiometer.

SOLENOIDS AND THEIR APPLICATIONS

Solenoids: An Economical Choice for Linear or Rotary Motion

A solenoid is an electromechanical device that converts electrical energy into linear or rotary mechanical motion. All solenoids include a coil for conducting current and generating a magnetic field, an iron or steel shell or case to complete the magnetic circuit, and a plunger or armature for translating motion. Solenoids can be actuated by either direct current (DC) or rectified alternating current (AC).

Solenoids are built with conductive paths that transmit maximum magnetic flux density with minimum electrical energy input. The mechanical action performed by the solenoid depends on the design of the plunger in a linear solenoid or the armature in a rotary solenoid. Linear solenoid plungers are either spring-loaded or use external methods to restrain axial movement caused by the magnetic flux when the coil is energized and restore it to its initial position when the current is switched off.

Fig. 1 The pull-in and push-out functions of a solenoid are shown. End A of the plunger pushes out when the solenoid is energized while the clevis-end B pulls in.

Cutaway drawing in Fig. 1 illustrates how pull-in and push-out actions are performed by a linear solenoid. When the coil is energized, the plunger pulls in against the spring, and this motion can be translated into either a "pull-in" or a "push-out" response. All solenoids are basically pull-in-type actuators, but the location of the plunger extension with respect to the coil and spring determines its function. For example, the plunger extension on the left end (end A) provides "push-out" motion against the load, while a plunger extension on the right end terminated by a clevis (end B) provides "pull-in" motion. Commercial solenoids perform only one of these functions. Figure 2 is a cross-sectional view of a typical pull-in commercial linear solenoid.

Rotary solenoids operate on the same principle as linear solenoids except that the axial movement of the armature is converted into rotary movement by various mechanical devices. One of these is the use of internal lands or ball bearings and slots or races that convert a pull-in stroke to rotary or twisting motion.

Fig. 2 Cross-section view of a commercial linear pull-type solenoid with a clevis. The conical end of the plunger increases its efficiency. The solenoid is mounted with its threaded bushing and nut.

Motion control and process automation systems use many different kinds of solenoids to provide motions ranging from simply turning an event on or off to the performance of extremely complex sequencing. When there are requirements for linear or rotary motion, solenoids should be considered because of their relatively small size and low cost when compared with alternatives such as motors or actuators. Solenoids are easy to install and use, and they are both versatile and reliable.

Technical Considerations

Important factors to consider when selecting solenoids are their rated torque/force, duty cycles, estimated working lives, performance curves, ambient temperature range, and temperature rise. The solenoid must have a magnetic return path capable of transmitting the maximum amount of magnetic flux density with minimum energy input. Magnetic flux lines are transmitted to the plunger or armature through the bobbin and air gap back through the iron or steel shell. A ferrous metal path is more efficient than air, but the air gap is needed to permit plunger or armature movement. The force or torque of a solenoid is inversely proportional to the square of the distance between pole faces. By optimizing the ferrous path area, the shape of the plunger or armature, and the magnetic circuit material, the output torque/force can be increased.

The torque/force characteristic is an important solenoid specification. In most applications the force can be a minimum at the start of the plunger or armature stroke but must increase at a rapid rate to reach the maximum value before the plunger or armature reaches the backstop.

The magnetizing force of the solenoid is proportional to the number of copper wire turns in its coil, the magnitude of the current, and the permeance of the magnetic circuit. The pull force required by the load must not be greater than the force developed by the solenoid during any portion of its required stroke, or the plunger or armature will not pull in completely. As a result, the load will not be moved the required distance.

Heat buildup in a solenoid is a function of power and the length of time the power is applied. The permissible temperature rise limits the magnitude of the input power. If constant voltage is applied, heat buildup can degrade the efficiency of the coil by effectively reducing its number of ampere turns. This, in turn, reduces flux density and torque/force output. If the temperature of the coil is permitted to rise above the temperature rating of its insulation, performance will suffer and the solenoid could fail prematurely. Ambient temperature in excess of the specified limits will limit the solenoid cooling expected by convection and conduction.

Heat can be dissipated by cooling the solenoid with forced air from a fan or blower, mounting the solenoid on a heat sink, or circulating a liquid coolant through a heat sink. Alternatively, a larger solenoid than the one actually needed could be used.

The heating of the solenoid is affected by the duty cycle, which is specified from 10 percent to 100 percent, and is directly proportional to solenoid *on* time. The highest starting and ending torque are obtained with the lowest duty cycle and *on* time. Duty cycle is defined as the ratio of *on* time to the sum of *on* time and *off* time. For example, if a solenoid is energized for 30 s and then turned off for 90 s, its duty cycle is $^{30}/_{120} = ^{1}/_{4}$, or 25 percent.

The amount of work performed by a solenoid is directly related to its size. A large solenoid can develop more force at a given stroke than a small one with the same coil current because it has more turns of wire in its coil.

Open-Frame Solenoids

Open-frame solenoids are the simplest and least expensive models. They have open steel frames, exposed coils, and movable plungers centered in their coils. Their simple design permits them to be made inexpensively in high-volume production runs so that they can be sold at low cost. The two forms of open-frame solenoid are the *C-frame solenoid* and the *box-frame solenoid*. They are usually specified for applications where very long life and precise positioning are not critical requirements.

C-Frame Solenoids

C-frame solenoids are low-cost commercial solenoids intended for light-duty applications. The frames are typically laminated steel formed in the shape of the letter C to complete the magnetic circuit through the core, but they leave the coil windings without a complete protective cover. The plungers are typically made as laminated steel bars. However, the coils are usually potted to resist airborne and liquid contaminants. These solenoids can be found in appliances, printers, coin dispensers, security door locks, cameras, and vending machines. They can be powered with either AC or DC current. Nevertheless, C-frame solenoids can have operational lives of millions of cycles, and some standard catalog models are capable of strokes up to 0.5 in. (13 mm).

Box-Frame Solenoids

Box-frame solenoids have steel frames that enclose their coils on two sides, improving their mechanical strength. The coils are wound on phenolic bobbins, and the plungers are typically made from solid bar stock. The frames of some box-type solenoids are made from stacks of thin insulated sheets of steel to control eddy currents as well as keep stray circulating currents confined in solenoids powered by AC. Box-frame solenoids are specified for higher-end applications such as tape decks, industrial controls, tape recorders, and business machines because they offer mechanical and electrical performance that is superior to those of C-frame solenoids. Standard catalog commercial box-frame solenoids can be powered by AC or DC current, and can have strokes that exceed 0.5 in. (13 mm).

Tubular Solenoids

The coils of *tubular solenoids* have coils that are completely enclosed in cylindrical metal cases that provide improved magnetic circuit return and better protection against accidental damage or liquid spillage. These DC solenoids offer the highest volumetric efficiency of any commercial solenoids, and they are specified for industrial and military/aerospace equipment where

the space permitted for their installation is restricted. These solenoids are specified for printers, computer disk and tape drives, and military weapons systems; both pull-in and push-out styles are available. Some commercial tubular linear solenoids in this class have strokes up to 1.5 in. (38 mm), and some can provide 30 lbf (14 kgf) from a unit less than 2.25 in. (57 mm) long. Linear solenoids find applications in vending machines, photocopy machines, door locks, pumps, coin-changing mechanisms, and film processors.

Rotary Solenoids

Rotary solenoid operation is based on the same electromagnetic principles as linear solenoids except that their input electrical energy is converted to rotary or twisting rather than linear motion. Rotary actuators should be considered if controlled speed is a requirement in a rotary stroke application. One style of rotary solenoid is shown in the exploded view in Fig. 3. It includes an armature-plate assembly that rotates when it is pulled into the housing by magnetic flux from the coil. Axial stroke is the linear distance that the armature travels to the center of the coil as the solenoid is energized. The three ball bearings travel to the lower ends of the races in which they are positioned.

The operation of this rotary solenoid is shown in Fig. 4. The rotary solenoid armature is supported by three ball bearings that

Fig. 3 Exploded view of a rotary solenoid showing its principal components.

Fig. 4 Cutaway views of a rotary solenoid de-energized (a) and energized (b). When energized, the solenoid armature pulls in, causing the three ball bearings to roll into the deeper ends of the lateral slots on the faceplate, translating linear to rotary motion.

travel around and down the three inclined ball races. The de-energized state is shown in (a). When power is applied, a linear electromagnetic force pulls in the armature and twists the armature plate, as shown in (b). Rotation continues until the balls have traveled to the deep ends of the races, completing the conversion of linear to rotary motion.

This type of rotary solenoid has a steel case that surrounds and protects the coil, and the coil is wound so that the maximum amount of copper wire is located in the allowed space. The steel housing provides the high permeability path and low residual flux needed for the efficient conversion of electrical energy to mechanical motion.

Rotary solenoids can provide well over 100 lb-in. (115 kgf-cm) of torque from a unit less than 2.25 in. (57 mm) long. Rotary solenoids are found in counters, circuit breakers, electronic component pick-and-place machines, ATM machines, machine tools, ticket-dispensing machines, and photocopiers.

Rotary Actuators

The rotary actuator shown in Fig. 5 operates on the principle of attraction and repulsion of opposite and like magnetic poles as a motor. In this case the electromagnetic flux from the actuator's solenoid interacts with the permanent magnetic field of a neodymium–iron disk magnet attached to the armature but free to rotate.

The patented Ultimag rotary actuator from the Ledex product group of TRW, Vandalia, Ohio, was developed to meet the need for a bidirectional actuator with a limited working stroke of less than 360° but capable of offering higher speed and torque than a rotary solenoid. This fast, short-stroke actuator is finding applications in industrial, office automation, and medical equipment as well as automotive applications.

The PM armature has twice as many poles (magnetized sectors) as the stator. When the actuator is not energized, as shown in (a), the armature poles each share half of a stator pole, causing the shaft to seek and hold mid-stroke.

When power is applied to the stator coil, as shown in (b), its associated poles are polarized north above the PM disk and south

Fig. 5 This bidirectional rotary actuator has a permanent magnet disk mounted on its armature that interacts with the solenoid poles. When the solenoid is deenergized (a), the armature seeks and holds a neutral position, but when the solenoid is energized, the armature rotates in the direction shown. If the input voltage is reversed, armature rotation is reversed (c).

beneath it. The resulting flux interaction attracts half of the armature's PM poles while repelling the other half. This causes the shaft to rotate in the direction shown.

When the stator voltage is reversed, its poles are reversed so that the north pole is above the PM disk and south pole is below it. Consequently, the opposite poles of the actuator armature are attracted and repelled, causing the armature to reverse its direction of rotation.

According to the manufacturer, Ultimag rotary actuators are rated for speeds over 100 Hz and peak torques over 100 oz-in. Typical actuators offer a 45° stroke, but the design permits a maximum stroke of 160°. These actuators can be operated in an *on/off* mode or proportionally, and they can be operated either open- or closed-loop. Gears, belts, and pulleys can amplify the stroke, but this results in reducing actuator torque.

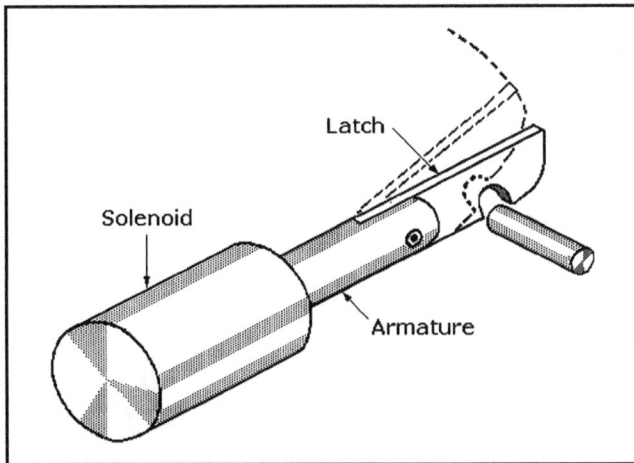

Latching: Linear solenoid push-out or pull-in motion can be used in a wide variety of latching applications such as locking vault doors, safe deposit boxes, secure files, computers, and machine tools, depending on how the movable latch is designed.

Pinchoff of Flexible Tubing: This push-out linear solenoid with an attached blade can control or pinch off liquid flowing in flexible tubing when energized by a remote operator. This arrangement can eliminate valves or other devices that could leak or admit contaminants. It can be used in medical, chemical, and scientific laboratories where fluid flow must be accurately regulated.

Parts or Material Diversion: This diverter arrangement consists of a rotary solenoid with a gate attached to its armature. The gate can swing to either of two alternate positions under push button or automatic control to regulate the flow of parts or materials moving on belts or by gravity feed.

Parts Rejection: A push-out linear solenoid can rapidly expel or reject parts that are moving past it into a bin when triggered. An electronic video or proximity sensing system is required to energize the solenoid at the right time.

Rotary Positioning: A linear push-out solenoid is paired with a multistation drum containing objects that are indexed by a linear solenoid or actuator. This arrangement would permit the automatic assembly of parts to those objects or the application of adhesives to them as the drum is indexed.

Ratcheting Mechanism: A pull-in solenoid with a rack mounted on its plunger becomes a ratcheting mechanism capable of turning a gear for the precise positioning of objects under operator or automated control.

CHAPTER 3
STATIONARY AND MOBILE ROBOTS

INTRODUCTION TO ROBOTS

Robots are favored subjects in science fiction books and movies, and they are widely covered in a variety of scientific and engineering publications. There has been a lot of discussion lately about the rapid advances being made in the ability of robots to emulate human behavior and thought processes. One recent popular article asked the question: "Can robots be trusted?" Some people fear that robot intelligence could exceed that of humans and they could take over the control of human lives, a threat that first appeared in science fiction. But experts agree that even the most advanced robots today only have the mental abilities of a retarded cockroach. However, others see a more insidious threat—that robots will advance to the level that we will want to adopt them as workers, companions, and even pets.

The truth about robots today is far more prosaic. Most robots are being manufactured for industry, and most of them are being purchased by automobile manufacturers around the world. There are also other growing markets for robots. One of the most important is for military robots capable of operating on the ground, in the air, and underwater. A prime motivation for them here is to use robots as proxies to save the lives of soldiers, sailors, and airmen deployed in wars taking place in unfamiliar locations. Some can be used for reconnaissance and others can be put to work detecting and destroying improvised explosive devices (IEDs) as well as land and sea mines. Moreover, flying robots, usually called drones, can make surprise pinpoint strikes against enemies hiding in otherwise inaccessible locations.

Yet another growing market is for commercial and consumer service robots. Today, some are able to perform such tasks as shelving and retrieving merchandise in warehouses, delivering supplies in hospitals, sweeping and cleaning floors, and even mowing lawns. In addition, one must not overlook the role of robots in scientific exploration in space, on or around distant moons and planets, and under the sea.

The term robot is now so commonly used in our language and by the media that most people are unaware of its true definition. The term "robotic" is now frequently seen in advertising promoting any machine with movable parts that look like human arms and hands, whether they are controlled by humans or are automated. They could be toys, appliances, industrial machines, or medical apparatus.

The Robot Defined

The definition for a robot in this chapter might differ from others that have been published, but it is more comprehensive. A robot is: *An electronically programmed, multitasking machine capable of carrying out a range of functions, typically, but not exclusively by autonomous means.* True robots can be reprogrammed electronically to perform other duties by means of signals sent through a connecting cable. Although a true robot's tools can be changed either automatically or manually, no new internal mechanical parts or electronic circuitry is needed to be installed.

This definition rules out automatic machines custom-designed and built to perform the same task repetitively because those tasks can only be changed by replacing internal electrical or mechanical components. It also rules out player pianos and

numerically controlled machines, formerly considered robots because they were directed by punched cards or paper tapes that programmed them. Some appliances and instruments with factory-programmed microcontrollers are not considered to be robots because their functions cannot be altered without changing their built-in microcontrollers.

On the other hand, some machines that are true robots do not match the popular conception of what a robot should look like or how it should act. But these machines can be reprogrammed by software to perform different mundane tasks such as cutting, fastening, folding, gluing, or stacking products in factories or warehouses around the clock. Or they could also be performing routine laboratory tests, all under computer control.

Laboratory devices with manipulator arms for handling toxic or radioactive materials isolated in chambers are not robots because an operator's hands control the manipulator. These devices are more accurately called *teleoperators* because they are not designed to act autonomously. Similarly, a manipulator arm on a deep-diving submersible is a teleoperator because it is also controlled by an operator's hand motions to retrieve biological or archaeological specimens from the ocean floor. Radio- or wire-controlled model vehicles (called "bots"), boats, or aircraft are not robots unless they contain some software programmed functions that are vital for their operation.

Modern robots are controlled by software stored in central processing units (CPUs) with externally programmable memory such as that installed in all desktop and laptop computers. The microprocessors and related peripheral components of industrial robots are typically located in control consoles separate from the robot. These consoles also contain power supplies, keyboards, and disk drives as well as sensing circuitry for providing feedback. Industrial robots are programmed to include commands that shut it down if a person accidentally intrudes into its operating "envelope," and also prevents it from damaging nearby property.

Industrial robots typically include a handheld control pendant on a cable that permits the operator to turn the robot on and off and make periodic software changes or updates in the stored program to improve the robot's performance. Some pendants can also be used to "teach" the robot to perform tasks such as painting, welding, or materials handling. An expert skilled in those activities moves the robot wrist manually through all the motions necessary to perform those tasks efficiently while the robot movements are recorded in memory so they can be played back to perform the task precisely as taught.

Stationary Autonomous Industrial Robots

Modern industrial robots are *autonomous* and usually *stationary*. This means that once programmed with software or "taught" by a human operator, these robots will repeatedly perform their assigned tasks without human intervention from the same positions. However, some industrial robots are designed and built to be programmed so they move over short distances on tracks or rails to accomplish their assigned tasks. They are called *movable* rather than *mobile* robots.

Industrial robots can work tirelessly around the clock, and they do not take breaks or slow down because of boredom or fatigue. Most are assigned to such dangerous tasks as welding, grinding, moving heavy loads rapidly from place to place, repeatedly emptying and stacking parts from incoming pallets, or transferring parts between machines in coordinated work cells. They also do spray painting rapidly in booths or tunnels where the paint spray would be toxic if inhaled by a human painter. Some industrial robots are designed to be versatile while others are optimized to perform a single task faster, more efficiently, and more economically than the versatile models. As a result of having no extra components that are not in continuous use, these specialized robots can cost and weigh less and usually take up less floor space.

The leading specifications of industrial robots are: (1) number of axes, (2) maximum payloads or handling capacities at the wrist (in pounds or kilograms), (3) arm reach (in feet or meters), (4) repeatability (in plus or minus millimeters), and (5) weight in tons or kilograms. Illustrations and specifications data on four current industrial robot families are presented in this chapter.

Some Robot History

The Czech playwright Karel Capek (1890–1938) was the first person to use the word "robot." It was derived from the Czech word for forced labor or serf. Capek was reportedly a candidate for the Nobel Prize several times for his work, and he was a very influential and prolific writer and playwright. The use of the word robot was introduced to the public in Capek's play *R.U.R. (Rossum's Universal Robots)*, which opened in Prague in January 1921. In *R.U.R.*, Capek discusses a paradise where, at first, these machines bring many benefits to the people, but in the end, he declared, they will bring an equal amount of misery in the form of unemployment and social unrest. He was not exactly a fan of robots.

The Worldwide Robot Market

The Robotics Industry of America (RIA) reports that materials handling remains the largest application area for new robot orders, accounting for some 60 percent of the units sold in North America in the first quarter of 2010. RIA estimates that some 190,000 robots are now at work in U. S. factories. This places the United States second only to Japan in its overall use of robots. The RIA estimated that in 2010 more than one million robots were in use worldwide. It also reported that countries with expanding industrialization and population, such as China and India, are rapidly expanding their purchases of robots.

In 2009 the automotive industry, the largest customer for robots in North America, experienced economic distress. This is one of the reasons why 2009 was a very difficult year for the robotics industry. The RIA reported that robotics companies around the world experienced declines as they dealt with the impacts of an economic recession that has put the brakes on capital equipment purchases in many industries as well as the automotive industry. However, in 2010 there were increases in orders for robots from the life science, pharmaceutical, and biomedical industries, and smaller increases in orders from the food and consumer goods sector.

INDUSTRIAL ROBOTS

Industrial robots are defined by the characteristics of their control systems; manipulator or arm geometry; modes of operation; and their end effectors or the tools mounted on a robot's wrist. Industrial robots can be classified by their programming modes which correlate with their performance capabilities: *limited* versus *unlimited sequence control*. These terms refer to the paths that can be taken by the end effector as it is stepped through its programmed motions. Four classes are recognized: limited sequence control and three forms of unlimited sequence control: *point-to-point, continuous-path*, and *controlled-path*.

Another distinction between industrial robots is in the way they are controlled: either *servoed or nonservoed*. A servoed robot includes a closed-loop which provides feedback and enables it to have one of the three forms of unlimited sequence control. This is achieved if the closed loop contains a velocity sensor, a position sensor, or both. By contrast, a nonservoed robot has open-loop control, meaning that it has no feedback and is therefore a limited sequence robot.

Industrial robots can be powered by electric motors or hydraulic or pneumatic actuators. Electric motors are now the most popular drives for industrial robots because they are the least complicated and most efficient power sources. Hydraulic drives have been installed on industrial robots, but this technology has lost favor, particularly for robots that must work within a controlled and populated environment. Hydraulic drives are noisy and subject to oil leakage, which presents a fire hazard in an enclosed space. Moreover, hydraulic drives are more maintenance prone than electric drives.

Nevertheless, hydraulic-drive robots can handle loads of 500 lb or more, and they can be used safely outdoors or in uncontrolled spaces. They are also used in situations where volatile gases or substances are present; these hazards rule out electric motors because a fire or explosion could be caused by electric arcing within the electric motor. Some limited-duty benchtop robots are powered by pneumatic actuators, but they are typically simple two- or three-axis robots. On the other hand, pneumatic power is now widely used to operate end effectors such as "hands" or grippers mounted on the wrists of electric-drive robots. An example is a wrist assembly that includes two rotary pneumatic actuators capable of moving a gripper around two axes, roll and yaw.

The term *degrees-of-freedom* (DOF) as applied to a robot indicates the number of its axes, an important indicator of a robot's capability. Limited sequence robots typically have only

Fig. 1 Components of a floor-standing, six-degree-of-freedom industrial robot.

two or three axes while unlimited sequence robots typically have five or six axes because they are intended to perform more complex tasks. However, the basic robot manipulator arm might have only three axes: arm sweep (base rotation), shoulder swivel (reach), and elbow extension (elevation), but a wrist can provide as many as three additional axes—pitch, roll, and yaw.

The heavy-duty, floor-standing robot shown in Fig. 1 has six principal axes, each driven by an electric motor. The console contains a digital computer that has been programmed with an operating system and applications software so that it can perform the robot's assigned tasks. The operator or programmer can control the movements of the robot arm or manipulator with push buttons on the control console so that it can be run manually through its complete program sequence. During programming, adjustments can be made in the program to prevent any part of the robot from colliding with nearby objects.

Some industrial robots are equipped with training pendants—handheld control boxes that are connected to the computer control console by cable. The pendant typically contains a push-button panel and a color graphic liquid-crystal display (LCD). It permits an operator or programmer to "teach" the robot by leading the wrist and end effector manually through the complete assigned task. The movements of each of the axes in the path sequence are stored in memory so that the robot will play back the routines precisely when commanded to do so.

Some floor-standing industrial robots are built so that they can be mounted upside down, vertically, or at an angle to gain better access to their intended work areas. The inverted robots are typically suspended from structural frames. Those frames might have rails on them to permit the robot to travel over limited distances while engaged in work such as welding long seams or painting long objects. Similarly, the robots might be positioned in a fixed position on a wall or they could be mounted on a vertical rail if vertical movement is required. They could also be mounted on rails set at an angle with respect to the floor for angular excursions.

Industrial Robot Advantages

An industrial robot can be programmed to perform a wider range of tasks than a dedicated automatic machine such as a numerically controlled machining center, even if that machine can accept a wide selection of different tools for doing different jobs. While industrial robots are considered to be multipurpose machines, most manufacturers design robots with certain characteristics that favor specific applications such as welding, painting, loading products or parts in cartons, or performing assembly work.

Manufacturers' literature and specifications sheets list these specialties, but their designations do not mean that the robots are limited to those functions; they will function satisfactorily in the performance of other applications where similar characteristics are required. Those characteristics that make a robot well suited for specific applications are determined by their specifications; these include such factors as size, weight, payload, reach, repeatability, range, speed, and the cost of operation and maintenance.

The decision to purchase a robot should be based on the evaluation of a long checklist of requirements for justifying the purchase. First and foremost of these decisions is the customer's conclusion that a lower cost dedicated machine cannot do the work more cost-effectively than a robot. Other factors to be evaluated, both technical and economic, are

- Ability of the owner to integrate the robot with existing manufacturing facilities
- Cost of training or retraining operators and programmers for the robot
- Cost of writing new applications software to direct the process to be automated
- Estimation of the overhead cost and time lost during downtime while a human operator changes tools between jobs or performs routine robot maintenance

52

The full benefits of an industrial robot cannot be realized unless it is properly integrated with the other conventional machines, conveyers, and materials handling equipment that form a coordinated work cell. Early robot purchasers learned a costly lesson when they found that isolated robots could not pay for themselves because they were not integrated into the normal workflow of the factory, so they were abandoned. Carefully engineered work cells assure that there is a coordinated and timely flow of work.

Industrial robots have been most cost effective in situations where they perform arduous, repetitive tasks, especially in hostile environments where human operators are exposed to life-threatening environmental conditions. These locations include environments where

- Temperatures or humidity are excessive
- Noxious or toxic fumes can damage the lungs
- Welding arcs can damage unprotected eyes
- Molten metal spray or open flame can burn unprotected skin
- High-voltage sources present a constant electrocution hazard

Nevertheless, robots have frequently proven themselves in work situations where none of these factors were present because they were able to demonstrate more consistent and higher quality workmanship than could be performed by skilled and experienced workers. Examples are found in welding, painting, and repetitive assembly work, even in conditioned indoor environments such as automotive assembly lines and appliance factories.

Industrial robots are now found at work in a wide range of industries from machine tool, automotive, aircraft, and ship-building to consumer appliance manufacturing. In addition, many robotic machines that are not easily recognizable perform such nimble tasks as pick-and-place assembly of electronic components on circuit boards. In addition, robots capable of moving rapidly along the length and height of extensive shelving in automated warehouses are storing and retrieving various objects and packages under remote computer control.

Industrial Robot Characteristics

The important specifications to consider in a robot purchase decision are payload, reach, repeatability, interference radius, motion range and speed, payload capacity, and weight. Reach is measured in inches or millimeters, and motion range is determined by the robot's three-dimensional (3D) semispherical work envelope. This is the locus of points that can be reached by the robot's workpoint when all of their axes are in their extreme positions. Motion speed is measured for each axis in degrees per second. The robot must be able to reach all the parts or tools needed to perform its task, so the working range typically determines the size and weight of the robot required.

Robot axis motion speed is typically in the range from 100°/s to 300°/s. High rates of acceleration and deceleration are favored. Payloads are most important if the robot is to do a significant amount of lifting. These are measured in pounds and kilograms. Some production industrial robots are able to handle maximum loads up to 880 lb or 400 kg, but most requirements are far lower—generally less than 50 lb. A large floor-standing robot can weigh as much as 2 tons.

Stiffness is another important robot specification. This term means that the robot arm must be rigid enough in all of its possible positions to perform its assigned tasks without flexing or shifting under load. If the robot has sufficient stiffness it can perform repetitive tasks uniformly without deviating from its programmed dimensional tolerances. This characteristic is specified as *repeatability*, which correlates with stiffness and is measured in inches or millimeters of deviation.

Industrial Robot Geometry

There are four principal stationary robot geometries: (1) articulated, revolute, or jointed arm; (2) polar-coordinate or gun turret; (3) Cartesian; and (4) cylindrical. A low-shoulder articulated robot is shown in Fig. 2 and a high-shoulder articulated robot is shown in Fig. 3. The articulated robot geometry is the most commonly used configuration today for floor-standing industrial robots, but there are many variations. The polar-coordinate geometry

Fig. 2 Low-shoulder, articulated, revolute, or jointed geometry robot.

Fig. 3 High-shoulder, articulated, revolute, or jointed geometry robot.

Fig. 4 Polar-coordinate or gun-turret geometry robot.

robot is illustrated in Fig. 4; the Cartesian-coordinate geometry robot is illustrated in Fig. 5; and the cylindrical-coordinate geometry robot is illustrated in Fig. 6. Among the variations of these basic designs is the vertically jointed-geometry robot shown in Fig. 7.

A robot's wrist at the end of the robot's arm serves as a mounting plate for end effectors or tools. There are two common designs for robot wrists: two-degree-of-freedom (2DOF) and three-degree-of-freedom (3DOF). An example of a 2DOF wrist is shown in Fig. 8; it permits roll around the arm axis and pitch around an axis at right angles to the arm axis. Another version of a 2DOF wrist, illustrated in Fig. 9, has the capability of a second

Fig. 6 Cylindrical-coordinate geometry robot.

Fig. 7 Vertically-jointed robot.

independent roll around the arm axis in addition to the pitch around an axis at right angles to the arm axis. A 3DOF wrist is shown in Fig. 10; in addition to roll and pitch it offers yaw around a third axis perpendicular to both the pitch and roll axes. More degrees-of-freedom can be added by installing end effectors or tools that can move around axes independent of the wrist.

Many different kinds of end effectors are available for robots, but among the most common are pincer- or claw-like two-fingered grippers or hands that can pick up, move, and release objects. Some of these grippers are general purpose, but others have finger gripping surfaces that have been machined to contours that fit precisely around or inside specific objects oriented in preestablished positions. Fingers fashioned to grasp the outside of objects such as cylinders move inward to grasp and lift the object; fingers fashioned to grasp the inside of objects such as pipes or cylinders

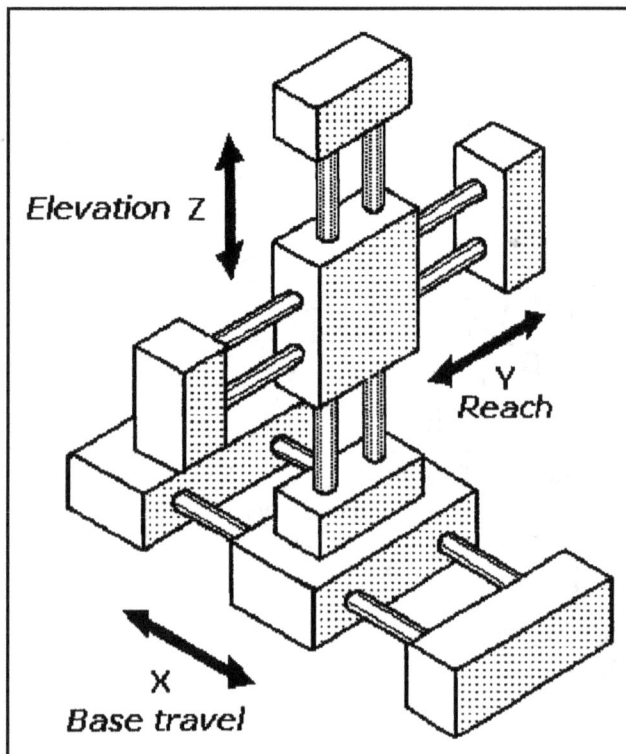

Fig. 5 Cartesian-coordinate geometry robot.

Fig. 8 Two-degree-of-freedom robot wrist can move a tool or end effector attached to its mounting plate around both pitch and roll axes.

Fig. 9 Two-degree-of-freedom robot wrist can move a tool or end effector attached to its mounting plate around pitch and two roll axes.

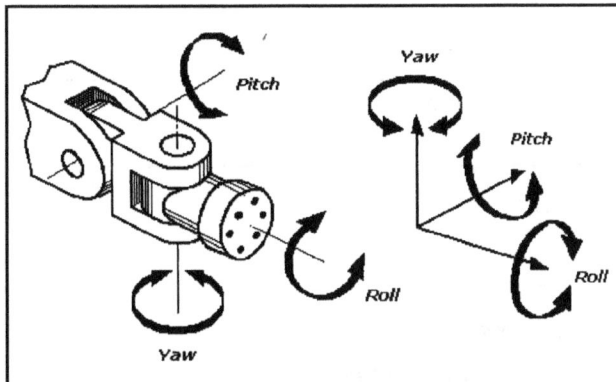

Fig. 10 Three-degree-of-freedom robot wrist can move a tool or end effector attached to its mounting plate around three axes: pitch, roll, and yaw.

Fig. 11 Robotic gripper operated by a reciprocating mechanism. Links open and close the "fingers" permitting them to grasp and release objects. A separate power source (not shown) is required.

Fig. 12 Robotic gripper operated by a rack and pinion mechanism. Rack and pinions open and close the "fingers" permitting them to grasp and release objects. A separate power source (not shown) is required to operate this gripper.

Fig. 13 Robotic gripper operated by a pneumatic or hydraulic piston. Piston opens and closes the "fingers" permitting them to grasp and release objects.

move inside the object and expand outward to grasp inside surfaces and lift the object.

The end effectors shown in Figs. 11 and 12 require independent actuators to power them. These are typically electric motors or pneumatic cylinders with pistons that are mounted between the end effector and the robot's wrist. However, the gripper shown in Fig. 13 includes an actuator that could either be a pneumatic or hydraulic piston for opening or closing the gripper fingers.

More sophisticated and versatile multifingered robotic grippers are now available, but they must be controlled either by software within the host robot or by an independent controller such as a laptop computer. For example, the gripper can have three fingers and an opposing thumb which can curl around objects of varying sizes, shapes, and orientations when actuated by electric motors to establish a firm grasp on the objects. While these grippers are more expensive, they eliminate the need for custom machining fingers to fit objects, and they can pick up randomly positioned objects.

FOUR DIFFERENT ABB INDUSTRIAL ROBOTS

Fig. 14 ABB IRB 2400.

Fig. 15 ABB IRB 6400RF.

Fig. 16 ABB IRB 6640.

Fig. 17 ABB IRB 7600.

The information for the following descriptions of ABB robots was obtained from ABB literature available from ABB Robotics. Four different ABB industrial robots are described with leading characteristics such as load capacity and arm reach.

IRB 2400

ABB's IRB 2400 family (Fig.14) includes three versions all capable of a wide range of tasks from arc welding (various processes) to machine tending and materials handling.

- IRB 2400L has a handling capacity of 7 kg (15 lb) and a reach of 1.80 m (5.91 ft).
- IRB 2400/10 has a handling capacity of 12 kg (26 lb) and a reach of 1.50 m (4.92 ft).
- IRB 2400/16 has a handling capacity of 20 kg (44 lb) and a reach of 1.50 m (4.92 ft).

All versions have six axes, and all can be mounted in an inverted position; some are washable with high-pressure steam to meet increased environmental protection standards.

IRB 6400RF

Two versions of ABB's IRB 6400RF (Fig.15) are powerful and accurate material removal robots.

- IRB 6400RF 2.5 200 has a handling capacity of 200 kg (440 lb) and a reach of 2.50 m (8.20 ft).
- IRB 6400RF 2.8 200 has a handling capacity of 200 kg (440 lb) and a reach of 2.80 m (9.19 ft).

These robots offer a powerful, stiff, and sturdy structure allowing high-speed and high-power material removal capability without compromising path control. They can also perform machining normally carried out by machine tools with the flexibility and cost-efficiency of an industrial robot. Their controllers provide short and precise cycle times with rapid changeovers and consistent high precision. The robots are adapted for demanding foundry environments because they have special paint, sealing and covers and their motors and connectors are protected, enabling them to withstand high-pressure washing with steam. Their mechanically balanced arms are equipped with double bearings. Options for advanced motion control and collision detection greatly reduce the risk of them damaging tools and workpieces.

IRB 6640

ABB's IRB 6640 family (Fig. 16) includes seven versions, each offering different arm lengths and handling capacities. It replaces an earlier successful IRB 6600 family and is based on that family's proven components.

- IRB 6640-180 has a handling capacity of 180 kg (396 lb) and a reach of 2.55 m (8.37 ft).
- IRB 6640-235 has a handling capacity of 235 kg (517 lb) and a reach of 2.55 m (8.37 ft).
- IRB 6640-205 has a handling capacity of 205 kg (451 lb) and a reach of 2.75 m (9.02 ft).
- IRB 6640-185 has a handling capacity of 185 kg (407 lb) and a reach of 2.80 m (9.19 ft).

- IRB 6640-130 has a handling capacity of 130 kg (286 lb) and a reach of 3.20 m (10.50 ft).
- IRB 6640ID-200 has a handling capacity of 200 kg (440 lb) and a reach of 2.55 m (8.37 ft).
- IRB 6640ID-170 has a handling capacity of 170 kg (374 lb) and a reach of 2.75 m (9.02 ft).

Arm reach values over this range are 2.55 to 3.20 m (8.37 to 10.50 ft), so it is important to remember that the longer a robot's reach, the lower its handling capacity. The two ID (internal dressing) robots have their process cables routed inside their upper arms. Because of this feature, the cables follow every motion of the robot arm rather than swinging in irregular patterns; it is especially valued in spot-welding applications.

Upper arm extenders and a selection of different wrist modules permit each robot's work process to be customized. These robots can bend backward completely; a feature that greatly extends their working range while also permitting their adaptation for operation on crowded production floors. Typical applications for these robots are material handling, machine tending, and spot welding. However, each robot can be modified with different features to adapt it to different working environments such as foundries and clean rooms. Passive safety features include load identification, movable mechanical stops, and electronic positioning switches.

IRB 7600

ABB's new IRB 77600 power robot family (Fig. 17) is available in five versions:

- IRB 77600-500 has a handling capacity of 500 kg (1100 lb) and a reach of 2.55 m (8.37 ft).
- IRB 77600-400 has a handling capacity of 400 kg (880 lb) and a reach of 2.55 m (8.37 ft).
- IRB 77600-340 has a handling capacity of 340 kg (750 lb) and a reach of 2.80 m (9.19 ft).
- IRB 77600-325 has a handling capacity of 325 kg (715 lb) and a reach of 3.10 m (10.17 ft).
- IRB 77600-150 has a handling capacity of 150 kg (330 lb) and a reach of 3.50 m (11.48 ft).

The IRB 7600 robots are built with sufficient heavy-lifting capability to satisfy all the industries they serve. Typical examples are rotating car bodies on assembly lines; lifting engines into position; moving heavy objects in foundries; loading and unloading parts in machine cells; and relocating large, heavily loaded pallets.

Robots moving payloads of more than 500 kg (1100 lb) present significant safety concerns for persons working in factories and warehouses if the payload should fall or come in contact with them. In addition to injuring the workers, the robot could be damaged. ABB software installed in the IRB 7600 robots monitors their motion and load as part of a collision detection system which reduces the impact of any unwanted contact between the loaded robot and nearby objects. The system's electronic path stabilization function, combined with an active braking system, keeps the robot on its programmed path despite opposing physical forces. Optional passive safety features available for the robots include load identification, moveable mechanical stops, and safe-position switches.

AUTONOMOUS AND SEMIAUTONOMOUS MOBILE ROBOTS

Mobile robots, in contrast with industrial robots, can be either semiautonomous or autonomous, and they are built in a variety of sizes and shapes and are not always recognizable. They can be wheeled or tracked land vehicles, surface or sub-surface water vehicles, or rotary- or fixed-wing aircraft. These robots can be classified in many different categories: military, law enforcement/public safety, scientific, commercial, and consumer. The consumer subclasses are appliance, educational, and entertainment.

Most mobile robotic vehicles, whether they operate on land, sea, or air, are *semiautonomous* because they are normally controlled by a human operator with commands sent by radio or over flexible wire or cable links. Essential feedback for guiding the robot is returned over those same communication links. The exceptions are submarine vehicles, which carry out their programs autonomously because it is not possible to send commands and receive meaningful feedback while they are under water. But they can perform as semiautonomous robots while being sent to the starting points of their missions and after they complete their missions and surface in response to signals from the mother ship.

Options for Communication and Control

Semiautonomous mobile robot operators typically use controllers (rugged, modified forms of laptop computers) to send signals to the robot that can start, stop, and maneuver it in three dimensions. The controllers have liquid crystal screens that can display real-time video and data from the various robot sensors in different formats on split-screen windows. The operator needs this feedback from the robot's sensors to direct it effectively and be able to override any onboard programmed features. This data can include robot speed, distance traveled, the state of battery charge, and even temperature readings taken on the robot.

Land-based Mobile Robots Can Scout and Retrieve

Land-based robots typically have sensors that permit them to avoid collisions with obstacles such as boulders, walls, trees, or very steep drop-offs in their paths. They are also equipped with closed-circuit TV, night illumination systems, and communication systems. Military and law enforcement mobile robots have, either tracks or wheels and are capable of surveillance, search and rescue, and bomb disposal. Some public service robots have also been equipped with pumps and water tanks for fighting fires in dangerous or inaccessible locations.

Military robots are being manufactured in quantity, and their basic chassis or platforms can be modified for specific missions by the addition of specialized tools, sensors, or weapons. These include video cameras, sonar, lidar (laser-based light detection and ranging), and infrared sensors. In addition, the operating software for militarized controllers can be modified or updated to adapt them to changes in tactical operations in combat zones, and their normal ranges can be extended by installing longer-life rechargeable batteries. They must be able to withstand shock and vibration, extremes in temperature, rain, saltwater spray, and wind-blown sand and dust. They must also be able to function at high altitudes, in jungles, and in deserts.

Video cameras allow the operator to evaluate obstacles in the robot's path and alter its course accordingly. The robot can be maneuvered from behind protective walls, earthen berms, or other substantial barriers. The controller's screen permits the operator to see in real-time where the robot is, how far it has gone, and how far it must go to reach its objective. This is particularly useful at night, in fog, rain, or smoke, and especially if visible light from the robot would make it a target for enemy snipers. Some military robots have been equipped with weapons such as machine guns and grenade launchers that permit their operators to take offensive action or defend themselves in combat situations.

Submersible Mobile Robots Can Search and Explore

There are several kinds of undersea mobile robots called autonomous underwater vehicles (AUVs) or unmanned underwater vehicles (UUVs). They look like naval torpedoes and are intended for marine reconnaissance, surveillance, target acquisition, and other tasks including the detection of undersea mines or obstructions such as fishing nets that could trap the vehicle. They can operate in three dimensions, limited only by the pressures at ocean depths.

As stated earlier, these robots must perform their missions autonomously after they submerge to their specified starting depths because radio waves and acoustic signals are inadequate for sending underwater commands to the robot. However, the principal sensors on these submersible robots are still cameras, navigation systems, side-looking sonar, acoustic Doppler current profilers, and GPS antennas. These robots, like land-based counterparts, are powered by rechargeable batteries.

Once submerged, the AUV begins a repetitive scanning pattern tracing out a specified area of ocean in overlapping parallel lanes. This pattern permits its side-looking sonar returns to record an overall display of the complete search area, which can cover many square miles of the ocean floor. These returns provide shape and dimension information about submerged targets, making them easier to identify. The program in its onboard computer directs the AUV to move its rudder and stabilizers to compensate for changes in current force, water temperature, and salinity that could force it off course. When the scanning is complete, the AUV surfaces and signals its position so that it can be recalled by radio to the mother ship.

Research is now in progress to develop sensors that would enable these vehicles to detect and avoid underwater obstacles such as fishing nets, floating or dangling lines or cables, and kelp that could ensnare them. The sensors could also permit it to escape if it should become entangled. This technology will be particularly useful for search operations in shallow, coastal (littoral) areas where fishing activity is extremely heavy and where mines or other targets are likely to be located.

Robotic Aircraft (Drones) Can Search and Destroy

Aerial robots, popularly known as *drones*, perform air reconnaissance over restricted areas occupied by an enemy or across the borders of nations that refuse to allow the admission of ground observers. Essentially radio-controlled unmanned aircraft, these robots are semiautonomous because they must be "flown" by an operator or pilot on the ground. Some versions are based near targets, and others fly thousands of miles from a home bases to reach their targets; many contain automatic pilots permitting them to fly autonomously and correct for wind that could force them off course. Once over the target area, control is switched back to the ground-based pilot. Air controllers on the ground near target areas can direct these robot aircraft to observe or attack targets of opportunity.

Some robot aircraft are equipped with air-to-ground missiles that can be fired by the pilot at targets identified by the aircraft's onboard cameras or designated by ground-based observers with laser pointers. Unlike ground-based mobile robots, these aircraft require enough instrumentation and guidance equipment to fill a mobile van or large room. The pilots must be qualified to fly conventional fixed-wing aircraft.

Planetary Exploration Robots Can Examine and Report

The best known and most publicized scientific robots are the two NASA Mars Exploration Rovers, *Spirit* and *Opportunity*. Both had been in continuous operation exploring the Mars surface for more than six years. *Spirit* landed on Mars on January 3, 2004 and *Opportunity* landed 21 days later on January 24, 2004. They have investigated hills, craters, and sandy plains in searches for water and life (or at least evidence of it in the past) as well as geological studies of the Red Planet. They both depend on large photovoltaic (PV) solar cell panels to maintain the charges of their onboard batteries for powering their instruments.

Laboratory/Scientific Robots Can Mimic Human Behavior

So many different scientific robots are being developed in government and university laboratories that they are difficult to categorize. They are being built to meet a wide variety of research objectives. Some with humanoid features are used in psychological studies and medical research. Made to look like humans, they have flexible pigmented silicone skin molded to form human-appearing faces and hands. Miniature cameras are hidden behind false eyes, microphones are their ears, and, their voices come from speakers synchronized with jaw movement. They can mimic human abilities such as walking, speaking, and recognizing objects or people, and many have high levels of manual dexterity. Others have proved useful in developing advanced computer programs, more refined sensors, and improved robot components.

Commercial Robots Can Deliver and Retrieve Goods

Many different kinds of robots have been developed to perform routine functions in factories, warehouses, and high-rise buildings, including hospitals. Most would not be recognizable as robots unless they were specifically identified. They cannot easily be categorized by their geometry like industrial robots. Many are custom built to perform, under computer control, such tasks as storing and retrieving merchandise in designated locations on high shelving. Another is a small, flat, briefcase-size wheeled robot that is the prime mover of carts containing supplies for delivery to various locations in hospitals. Because of its established and preplanned movements, it has sensors that permit it to function autonomously in response to commands designating delivery sites. It reaches its destination with the aid of its guidance sensors that permit it to navigate corridors and use elevators.

Consumer Robots Clean Floors and Mow Lawns

The consumer market for robots is offering smart toys, hobby robots and kits for building them, educational robots, home and lawn care appliances, and personal robots (or companions). These are not true robots unless they can be electronically programmed by their owners to perform other activities. Only the hobby robots offer possibilities for reprogramming. Some floor- and carpet-cleaning robots are made in disk shapes with roller-type wheels so they can move in two dimensions. They have onboard sonar or infrared sensors that prevent them from stopping when they encounter walls or the edges of stairways. These obstructions cause them to reverse direction and continue their cleaning in overlapping paths. What are the drawbacks? They cannot clean corners in rooms, and they can injure small children who pick them up thinking they are toys and pets that get in the way of them or attempt to stop them.

Some Robots Entertain or Educate

Some museums and theme parks exhibit humanoid robots portraying historic persons programmed to give short speeches about important events in their lives. They can make appropriate changes in facial expressions and hand and arm gestures. Early versions were mechanical-but now they are programmed electronically so their speeches and routines can be reprogrammed to permit different performances. However, many of the "robots" appearing in moving pictures are actually puppets manipulated behind the scenes by technicians using either invisible wires or internal motors to move them. Their voices are "voice-overs" by professional actors timed to coordinate with the puppet motions.

SEVEN MOBILE AUTONOMOUS AND SEMIAUTONOMOUS ROBOTS (FIG. 18)

Fig. 18 This simplified block diagram of a semiautonomous mobile robot identifies its onboard communications and command and control circuitry (within the dotted outline) and external sensors. They are connected by one- and two-way radio or cable links to the operator's control unit and display (left).

addition, the rear wheels are mounted on a separate arm able to pivot around a separate pin, assuring maximum flexibility. Steering servos permit all six wheels to be turned under computer control. Each wheel has a diameter of 10 in. (25.9 cm) and includes inner springs connecting the rim to the hub to absorb shock and prevent it from being transmitted to sensitive equipment on the rover. In addition, each wheel tread has cleats for providing traction when the rover traverses powdery soil or climbs over rocks.

The retractable instrument mast supports the horizontal rover "head" containing two navigation cameras, two panoramic cameras, and a thermal emission spectrometer behind the cameras. A servomotor rotates the mast through ±180° in azimuth and another tilts the head through +45° and −30°. Three antennas on each rover communicate with Earth: low-gain, high-gain, and UHF are able to send and receive on various frequencies. Both rovers were equipped with the tools needed to search for signs of ancient water and climate on

Two Robots Have Explored Mars for Six Years

The NASA Mars Exploration Rovers *Spirit* and *Opportunity* (Fig. 19) are wheeled, semiautonomous robots. They have large wing-like solar panels that measure 8 by 5 ft (2.4 by 1.5 m). Both rovers have returned vast amounts of scientific information that have advanced our knowledge of the Red Planet. In addition, they have demonstrated far higher levels of reliability than could be expected from any complex electronic system functioning on an alien, cold, windy, and dusty planet millions of miles from Earth.

Both *Spirit* and *Opportunity* were designed to fold up into small, relatively flat packages to survive the long space voyage to Mars. Once there, they withstood bouncing on shock-absorbing air-filled bladders without damage before being commanded to unfold and activate themselves. The wheels and suspension system extended and the instrument mast and all onboard antennas were raised in a predetermined order. Table 1 gives the leading specifications for both Martian rovers.

All six wheels are part of a rocker-bogie mobility system. They are attached to rocker assemblies mounted on a differential axle allowing each wheel assembly to pivot. In

Fig. 19 Two semiautonomous mobile robots, NASA Mars Exploration Rovers *Spirit* and *Opportunity*, landed on Mars in January 2004. Both have explored and communicated successfully from the surface of Mars for more than six years, adding substantially to our knowledge about the Red Planet.

Table 1 Leading Specifications for Mars Exploration Rovers *Opportunity* and *Spirit*

Builder: NASA Jet Propulsion Laboratory	
Rover Chassis	
Length	5.2 ft (1.6 m)
Width	7.5 ft (2.3 m)
Height	4.9 ft (1.5 m)
Weight, overall	408 lb (185 kg)
Weight, wheels/suspension	80 lb (35 kg)
Weight, instruments	11 lb (5 kg)
Communication	Radio frequency
Antennas	Low-gain, high-gain, and UHF
Cameras	Two panoramic (Pancam)
	Two navigational
Spectrometers	Mini-thermal emission (Mini-TES)
	Mossbauer (MB)
	Alpha particle x-ray (APXS)
Special science	Microscopic imager (MI)
	Magnet array
	Rock abrasion tool (RAT)

Table 2 Leading Specifications for the NASA Mars Science Laboratory Rover

Major Contractors: Boeing, Lockheed Martin	
Rover length	9.0 ft (2.7 m)
Rover weight	1982 lb (900 kg)
Instrument weight	176 lb (80 kg)
Speed (max)	300 ft/h (90 m/h)
Obstacle height (max)	30 in. (76 cm)
Generator output (max)	125 W
Electrical output/day	2.5 kW
Proposed payload	
Cameras	
Mast/Cam, MAHLI, MARDI	
Spectrometers	
ChemCam, CheMin	
Alpha-particle x-ray spectrometer (APXS)	
Sample analysis at Mars (SAM)	
Radiation detectors	
Radiation assessment detector (RAD)	
Dynamic albedo of neutrons (DAN)	
Environmental sensors	
Rover environmental monitoring station (REMS)	
Navigation instruments	
Entry descent and landing instruments (MEDLI)	
Hazard avoidance cameras (Hazcams)	
Navigation cameras (Navcams)	

Mars. Because neither robot has a turnoff switch, they may be able to provide useful data for years to come, even if both are disabled and bogged down in the Martian soil.

As of February 2, 2011, NASA Mission Controllers had not heard from *Spirit* since March 22, 2010 when the rover was facing its toughest challenge—trying to survive the harsh Martian winter. However, in September 2010 *Opportunity* photographed its first dust devil (a whirling cloud of dust), a challenging feat in the area where it was working. The last communication with the rover was on January 31, 2011.

This Robot Will Carry on the Work of Spirit and Opportunity

The *Mars Science Laboratory (MSL)*, is a large, boxlike NASA rover about the size of a small automobile. Named *Curiosity*, it has six wheels and weighs nearly 1 ton (900 kg). It is scheduled to head to Mars near the end of 2011 and will land on the Red Planet in August 2012. After making the first-ever precision landing on Mars, it will be expected to take over from the long-lived *Spirit* and *Opportunity* as the lead Mars explorer. Its mission is to determine if Mars ever was, or is still today, an environment able to support microbial life. There it will analyze dozens of samples drilled from rocks or scooped from the ground as it explores, and it will also try to determine the planet's habitability.

The MSL rover (Fig. 20), more than five times as heavy and about twice as long as *Spirit* and *Opportunity*, will carry more than 10 times the weight of their scientific instruments. In addition to an onboard geology lab, *Curiosity* will have a rock-vaporizing laser called ChemCam, which will be able to remove thin layers of material from Martian rocks or soil targets up to 30 ft (9 m) away. It will have a spectrometer to identify the types of atoms excited by the beam and a telescope to capture detailed images of the area illuminated by the beam.

The leading specifications for the MSL rover are in Table 2. It is expected that this rover will be launched by an Atlas V 541 rocket, and it will be designed to operate for at least one Martian year, equivalent to 686 Earth days. The MSL rover will be capable of exploring over far greater distances than any previous Mars rover. The three goals of the MSL rover are to: (1) determine if life ever arose on Mars, (2) characterize the

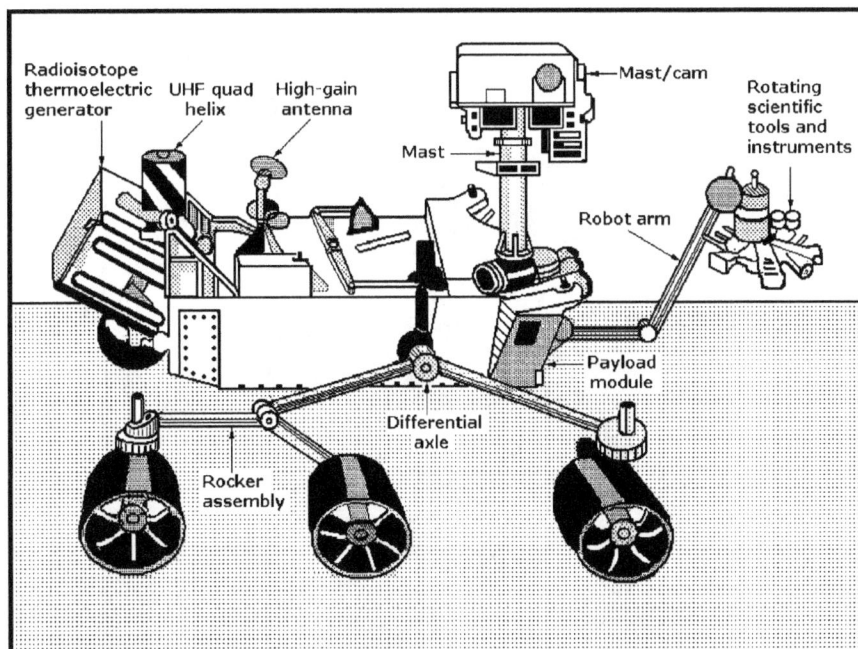

Fig. 20 Mars Science Laboratory (MSL) Rover is a far larger and heavier semiautonomous robot than the two Mars rovers *Spirit* and *Opportunity*, and it will carry many more scientific instruments. When it lands it will be assigned to determine if Mars ever had (or still has) an environment able to support microbial life and if the planet is habitable. Equipped with radioisotope thermoelectric generators with lives of 14 years, it will not depend on solar panels for power.

climate and geology of Mars, and (3) prepare for human exploration. It also has many other detailed scientific objectives.

Curiosity differs significantly in appearance from *Spirit* and *Opportunity* because it will be powered by radioisotope thermoelectric generators (RTGs) so it will not carry solar panels for power. Its radioisotope power system will generate electricity from the natural decay of plutonium 238. The system is designed to produce 125 W of electricity at the start of its Mars mission, but its output is expected to decline to 100 W after 14 years of operation. The MSL rover will generate 2.5 kW/h per day compared to the approximately 0.6 kW/h per day produced by the solar panels of *Spirit* and *Opportunity*.

Heat given off by this decay will be converted into electricity, providing constant power night and day during all Martian seasons. Any waste heat will be distributed through pipes to keep electronic systems warm. This means that no electrical power will need to be diverted for heating system components. The primary mission is planned to last about two Earth years, considerably less then the generator's anticipated minimum 14-year life.

Even if MSL's goals and objectives are reached within two years, more data can be obtained from functioning instruments years later if funds are available to do so. Equipment left on Mars is subject to extreme temperatures [+86° to −197°F (+30 to −127°C)] and coating with Martian dust, but it is not subject to oxidation so it can operate as long as power is available.

This Robot Responds to Civil Emergencies

The TALON Responder (Fig. 21) is a tracked semiautonomous robot designed to respond to civil emergencies. These include locating persons trapped in burning or collapsed buildings and

Table 3 Leading Specifications for Talon Responder

Manufacturer: QinetiQ, North America	
Vehicle	
Width	22.5 in. (57.2 cm)
Length	34 in. (86.4 cm)
Weight	115 to 156 lb (52 to 71 kg)
Speed	To 5.2 mph (8.4 km//h)
Payload capacity	100 lb (45 kg)
Arm lift capacity	20 lb (9 kg) max
Operator control unit (OCU)	
Weight	33 lb (15 kg)
Robot power	Two lead acid, 300 W-h 36 VDC (standard)
OCU power	Nickel-metal hydride 3.6 Ah, 24 VDC
OCU/robot communications	Wireless: digital data/analog video (standard) Fiberoptic cable: 300 m (option)
Cameras (standard)	3 IR illuminated, auto focus color zoom, illumination package
Audio	Two-way audio and loud hailer

participating in their rescue, providing useful information to police during hostage situations, and handling and disposing of suspected or known explosive devices. The leading specifications for this suitcase size robot are given in Table 3.

Responder is a civilian version of the high-performance rugged TALON used by the U.S. military services in Iraq and Afghanistan. Versions of this robot have been equipped with machine guns or other weapons for offensive action as well as self-defense. To help in guiding the motion of its manipulator arm, Responder is equipped with three infrared illuminated color cameras. It also has an automatic focus color zoom camera mounted on a mast that is topped by a bright illumination source. The civilian Responder can, if needed, be upgraded with any of the accessories available for the militarized version.

This robot is directed by an operator with a control unit (OCU), essentially a rugged laptop computer with an attached handheld gamepad-style controller. These OCUs permit Responders to travel as far as 4000 ft (1200 m) from the operator, a range that can keep them safely out of danger from fire, explosive device detonations, or further collapse of damaged structures while viewing the danger zones with its cameras. A fiberoptic cable on a spool can be added as an accessory. The cable can be pulled from the spool in situations where radios could set off a buried explosive device.

Fig. 21 TALON Responder is a civil law-enforcement semiautonomous robot. It can search for and rescue ho stages or victims of natural disasters as well as handle and dispose of explosive devices. A modified version of military robots that have been successfully deployed in Iraq and Afghanistan, this robot can be equipped with optional tools and accessories to adapt it for specific rescue and disposal assignments.

Robot Delivers Hospital Supplies

A two-wheeled autonomous mobile robot the size of a carry-on case is the prime mover of a coupled cart for delivering supplies to stations on all floors of a high-rise hospital. The coupled system (Fig. 22) is called the Automated Robotic Delivery (ARD), system by its manufacturer. The prime-moving robot, called the TUG (Fig. 23), is visible under the front end of its supply and its leading specifications are given in Table 4. ARD systems have been installed in more than 100 hospitals. They can deliver medications and meals to patient rooms and bulk materials to support departments. The system is

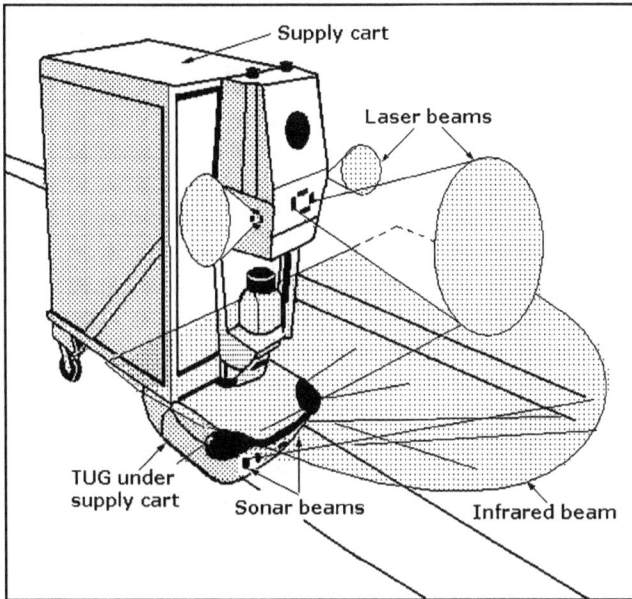

Fig. 22 The TUG, an autonomous robot, is a tractor that is coupled under a supply cart that it moves to distribute supplies. It is guided by laser, infrared, and sonar beams from sensors mounted on its front face and on a column mounted above it. The TUG can travel automatically to designated way stations as it delivers food and medical supplies to all floors of a multistory hospital.

Fig. 23 This cutaway view of the briefcase-sized TUG shows its principal components. It has two wheels, but when coupled to a supply cart (by the pin atop its case) the combination becomes a four-wheel vehicle. The robot is a component in the Automated Robotic Delivery System (ARDS), and it is directed from a command center where it goes to resupply and recharge its batteries at a charging station.

organized to transport materials on either a scheduled or an on-demand basis.

When the TUG's two wheels are coupled to the front end of a two-wheeled cart, a four-wheeled articulated vehicle is formed. The coupling is achieved by a post on top of its housing that fits into a socket under the cart. A selection of two-wheeled supply carts is available to meet different supply requirements. A sensor column attached to the front of the cart, when it is coupled

Table 4 Leading Specifications for the Aethon Tug Robot

Manufacturer: Aethon	
Height	25 in. (64 cm)
Width	20 in. (51 cm)
Length	20 in. (51 cm)
Weight	55 lb (25 kg)
Enclosure	Hi-impact SBS plastic
Drive train	2 Independent 24-VDC motors
Wheels	Aluminum rims: molded urethane treads
Power system	24 VDC: four 12 V lead-acid batteries
Hauling capacity	Up to 500 lb and 250 lb (across a 1-in. elevator gap)
Communications	Wireless radio frequencies: 418 and 900 MHz, 2.4 GHz, pagers, and phones

to the TUG, includes infrared and laser sensors and a sonar sensor is on the front of the TUG. Together, the three sensors permit the combined vehicle to navigate elevators and corridors. The beams of these sensors can be adjusted for the dimensions of the corridor widths on the routes over which the robotic vehicle will travel.

The TUG-cart combination can be summoned by pressing a computer key. This combination, an unescorted vehicle, can reliably make its way through hallways, automatic doors, narrow aisles, and on and off elevators while avoiding contact with passing humans. It accomplishes these feats because its on-board computer is programmed with navigation software giving it a detailed map of the hospital and identifying all other navigation aids along the route. It can reach its destinations without wires or magnets embedded in the corridor walls. Warning tones and lights signal its intention to backup, start, stop, and enter or leave an elevator. A cart-mounted speaker also broadcasts selected voice recordings.

A Military Remotely-Piloted Aircraft Can Observe and Attack the Enemy

The MQ-1 Predator (Fig. 24) is a medium-altitude, long-endurance, remotely piloted vehicle (RPV). It is also an unmanned aerial vehicle (UAV), an aircraft flying without a human crew onboard. Put more simply, it is a semiautonomous flying robot. The term "drone" no longer applies because of its higher level of sophistication. Predators have missions that include reconnaissance, surveillance, and acquisition of critical targets on the ground. An example is the search for terrorist leaders in vehicles or hideouts to be attacked. Predators can provide real-time data on enemy movements, locations, and weather conditions, permitting them to attack targets of opportunity called in by ground controllers.

Predators are key elements in systems consisting of four aircraft and a ground control station with a dedicated satellite link. These systems are supported by operations and maintenance crews for 24-hour, seven-day-a-week missions. Each Predator has a crew of one pilot and one or two sensor operators. They "fly" the aircraft from within the ground control station with either line-of-sight or satellite data links which provide communication with the aircraft thousands of miles away. Predators can take off and land on 5000 by 75 ft (1524 by 23 m) air strips provided they have hard ground surfaces and local line-of-sight communications. The leading specifications of the MQ-1 Predator are given in Table 5.

The MQ-9 Reaper, a later version of the Predator MQ-1 (originally called Predator B), was developed for the United States and several foreign air forces. The first hunter-killer UAV designed for long-endurance, high-altitude surveillance, Reaper

63

Fig. 24 The MQ-1 Predator is a semiautonomous, long-endurance, unmanned aircraft able to make observations from medium altitudes. It provides ongoing, real-time video of vehicular movements and movements of large groups of humans. Predator, one of many different flying robots, is controlled by radio signals from airstrips as far as thousands of miles from its base control station. It can be fitted with missiles to attack and destroy selected enemy targets.

Table 5 Leading Specifications for MQ-1 Predator Drone

Manufacturer: General Atomics Aeronautical Systems, Inc.	
Engine	115 hp Rotax 914F, turbocharged, four-cylinder
Wingspan	48.7 ft (14.8 m)
Wing area	123.3 sq ft (11.5 m^2)
Length	27 ft (8.2 m)
Height	6.9 ft (2.1 m)
Weight empty	1130 lb (5132 kg)
Weight loaded	2250 lb (1020 kg)
Speed cruise	81–103 mi/h (70–90 kn, 130–166 km/h)
Speed maximum	35 m/h (30 kn, 56 km/h)
Range	2302 mi (3705 km)
Ceiling service	25,000 ft (7620 m)
Cameras	2 full-motion video, nose, color
	Variable-aperture TV (daytime)
	Variable-aperture IR (low-light)
Communication	Radio, IFF
Targeting system	Optical IR laser designator with laser illuminator
Missiles	2 laser-guided air-to-ground, armor-piercing

is a larger and more capable aircraft than the Predator. It has a more powerful 950-shaft-horsepower (712 kW) turboprop engine than Predator's 115 hp (86 kW) piston engine. It can carry 15 times more ordnance and cruise at three times the speed of the Predator. Reapers have the same crews and operational support as Predators, but automatic pilots permit them to fly pre-programmed routes to destinations autonomously. And, like Preditors, their weapons are always under the control of the pilot.

Submarine Robot Searches for Underwater Mines and Obstructions

The REMUS 600 (Fig. 25) is a battery-powered, autonomous, submersible robot for undersea surveillance. An AUV, its name is an abbreviation for Remote Environmental Measuring

Units. It looks and moves underwater like a naval torpedo, and is able to operate untethered in the open ocean down to depths of 2000 ft (600 m), giving it great operational range and depth. Its leading specifications are given in Table 6. A scaled-up version of the REMUS 100, it has an operating depth of 350 ft (107 m).

REMUS 600 was developed to search for and locate all types of naval mines—floating, tethered, magnetic, or contact. It can also search for and locate undersea obstructions that could damage or sink ships as well as sunken military ships and equipment. However, it does not have a robotic arm with grippers that would allow it to retrieve underwater objects. It has, however, been adapted to search for objects of scientific interest such as sunken historical wrecks and archaeological artifacts. This AUV is controlled by a rugged specialized laptop computer with a graphic screen. It can provide operational data and charts for the operator, and its proprietary software permits the operator on the mother ship to communicate with and direct it while it is on the surface. The controller can also program its autonomous missions, participate in operator training, and troubleshoot the robot for faults.

While submerged it follows a programmed repetitive-scanning search plan that covers a large area. Three independent fins stabilize the submersible's yaw, pitch, and roll at preset depths. An acoustic Doppler current profiler (ADCP) helps it to navigate underwater by compensating for currents that could set it off from its preset course, and an inertial navigation guidance system that keeps it on course. A side-scan sonar is its primary underwater surveillance instrument, and its output is recorded and stored as video showing objects encountered along its tracking pattern.

Fig. 25 The Remus 600, an autonomous underwater vehicle (AUV), is a submersible mobile robot that can operate untethered anywhere in the open ocean to depths of 600 m (1960 ft). It is equipped with underwater guidance, navigation, and side-scan sonar systems as well as surface-to-home ship communication systems. The Remus 600 can search for and locate naval mines, undersea obstructions, and objects of military or scientific interest on the ocean floor.

Table 6 Leading Specifications for Remus 600 Undersea Vehicle

Manufacturer: Hydroid Inc.	
Diameter	12.75 in. (32.4 cm)
Length	10.7 ft (3.3 m)
Weight	530 lb (240 kg)
Operating depth (max)	1970 ft (600 m)
Endurance	Up to 70 h
Battery	5.2 kWh rechargeable lithium-ion
Propulsion	Direct-drive DC brushless motor
	Open two-blade propeller
Velocity range	Up to 2.6 m/s (5 kn) variable
Controls	Independent fins for yaw, pitch, and roll
Navigation	Inertial guidance, acoustic, GPS
Communication	Acoustic modem, Indium, Wi-Fi
Sensors, standard	Acoustic Doppler current profiler (ADCP), inertial navigation unit, side-scan sonar

Fig. 26 A surgeon seated at this da Vinci system console can perform surgical operations while viewing the surgical site within the patient on a 3D video screen. (Video of the site is obtained from a combination camera and light source called an endoscope.) The surgeon grasps master controls for the surgical instruments located below the display. Hand, wrist, and finger movements are translated into real time instrument movements within the patient, and the endoscope is directed with foot pedals. The instruments, mounted on robotic-type arms, are located on the separate sidecart placed next to the patient.

When its mission is complete, it surfaces and reports its position back to the mother ship by acoustic transponder, GPS, Indium satellite phone, or Wi-Fi. The REMUS 600 can be assembled from various hull modules to form a pressurized hull, and it is easily disassembled for maintenance or shipping. This feature permits it to be reconfigured from a selection of modules to adapt it for mission changes.

This System Offers Less Intrusive Surgery and Faster Recovery

The da Vinci surgical system permits a surgeon to perform operations from a console (Fig. 26) by moving his or her fingers, hands, and wrists to control remote surgical instruments with greater precision than would be possible by performing conventional *laparoscopic* surgery. The site of the operation can be viewed on a video screen from a 3D camera-equipped *endoscope*. According to its manufacturer, Intuitive Surgical, Inc., it provides a minimally invasive technique for performing abdominal operations. Because the instruments are agile and thin, the operations can be performed through small incisions. This lowers patient stress and speeds up recovery time.

The da Vinci system has been called "robotic" because its instruments are controlled by manipulator arms similar to those on true robots, but it is actually a sophisticated *teleoperator* that does not include a programmable computer. All operations are performed entirely by the surgeon. The instruments are mounted on three of the four robotic manipulator arms located on a separate patient sidecart (Fig. 27). It is a movable platform separate from the console and located at the side of the patient. The endoscope, with its high-resolution 3D fiberoptic camera, is mounted on a separate manipulator arm.

Unlike most handheld surgical instruments, many of the da Vinci instruments have jaws that are only 0.5 in. (1.5 cm) long. These instruments include scissors, forceps, and grasping retractors that can perform such tasks as clamping, suturing, and tissue manipulation. They allow the very small and precise incisions needed to carefully dissect diseased tissue.

The camera and lighting arm permits the surgeon to position it exactly where he or she wants it by operating foot pedals at the base of the console. Magnification of up to 10 to 12 × can be achieved with the cameras, which can move to within 2 in. (5 cm) of the tissue being removed during surgery. Two different cameras are also available: straight and 30° oblique. The oblique camera allows the surgeon to see around the corners of organs as well as partially underneath them.

Fig. 27 Patient sidecart. Three or four robotic arms are mounted on this movable platform placed next to the patient. Two or three arms hold the instruments for performing such tasks as clamping, suturing, or tissue manipulation; a fourth arm positions the endoscope. The ends of the small, thin instruments can be moved through seven degrees-of-freedom. The operation, made through small incisions in the patient's body, is carried out completely under the surgeon's control.

GLOSSARY OF ROBOTIC TERMS

actuator: Any transducer that converts electrical, hydraulic, or pneumatic energy into power to perform motions or tasks. Examples are electric motor, air motor, and solenoid.

adaptive control: A method for optimizing performance by continuously and automatically adjusting control variables in response to measured process variables. A robot adaptive control requires two extra features beyond its standard controls: (1) at least one sensor capable of measuring changes in the robot's working conditions and (2) the robot's central processors must be programmed to process sensor information and send signals to correct errors in the robot's operation.

air motor: A device that converts pneumatic pressure and flow into rotary or reciprocating motion.

android: A robot that mimics human appearance and behavior. Other equivalent terms are humanoid or anthropomorphic.

arm: An interconnected set of mechanical levers and powered joints that simulate a human arm and act as a manipulator; it can move an end of arm *wrist* with an attached end effector or tool to any spatial position within its work envelope.

autonomous: A robot capable of carrying out an electronically programmed task or work cycle without human intervention.

axis: A linear direction of travel in any of three dimensions: axes in Cartesian coordinates are labeled X, Y, and Z to orient axis directions with respect to the Earth's surface: X refers to a directional plane or line parallel to the Earth; Y refers to a directional plane or line perpendicular to X and is also parallel to the Earth; and Z refers to a directional plane or line vertical to X and Y and perpendicular to the Earth.

cable drive: A drive that transmits mechanical power by means of flexible cables and pulleys from a motor or actuator to a remote robot joint such as a wrist or ankle. It is also known as a *tendon drive*.

closed loop: A control scheme that compares the output value with the desired input value and sends an error signal when they differ, causing corrective action which restores equality between the values.

collision protection device (CPD): A device attached to a robot wrist that can detect a potential collision between the robot and a foreign object or that a contact has actually been made with one and sends a signal to the robot central processor causing it to stop or divert the motion of the arm before damage can be done. It is also known as a *collision sensor* or a *crash protection device*.

computer-vision system: An electronic system containing a video camera and computer with a vision program that allows a robot to acquire, interpret, and process visual information. The camera is set to view a restricted field into which parts are moved. The vision system can recognize specific parts in various orientations and locations within the camera's field of view and direct the robot to perform specific operations on that part. The system can be programmed to separate individual parts from a mixed group of parts, grasp a part regardless of its orientation for packaging or assembly, measure or inspect parts, and reject faulty or incomplete parts. A two-dimensional (2D) vision system can process 2D images to obtain part identity, position, orientation, and quantity of parts. A 3D vision has the properties of a 2D system but also has depth perception which can be used to avoid assembly errors, detect out-of-place objects, distinguish between similar parts, and correct positioning discrepancies.

continuous-path programming: A robot motion control method that maintains absolute control over the entire motion path of the tool or end effector. Programming can be done by manual teaching or moving the robot wrist sequentially through its work cycle. The robot wrist moves to closely spaced positions according to the program. The end effector performs the assigned tasks while the axes of motion are moving. All axes of motion move simultaneously, each at a different velocity, to trace a smooth continuous 3D path or trajectory. It is recommended in applications where the tool path is critical, such as painting, adhesive placement, or arc welding.

controlled-path programming: A robot motion control method in which all axes move along a straight path between points at a set velocity. Each axis is coordinated so that it accelerates to the specified path velocity and decelerates smoothly and proportionally to provide a predictable, controlled path.

An operator can use a teach pendant to program only the end points of the desired path. This method is used in such applications as parts assembly, welding, materials handling, and machine tending.

degrees-of-freedom (DOF): A value defined by the number of rotational axes through which motion can be obtained by the robot with or without an end effector or tool attached to its wrist. DOF indicates the number of independent ways in which the end effector can be moved.

end effector: Any tool, sensor, or device attached to the robot wrist for performing a task. Examples include grippers or hands, welding torches, paint spray guns, or measuring devices. End effectors are typically powered by pneumatic actuators or electric motors independent of the host robot, and they can add one or more independent degrees of freedom to the robot.

gripper: A mechanical grasping tool or hand attached to a robot's wrist that can pick up and place objects with various shapes and orientations to perform such tasks as assembly, packing, or loading and unloading of parts or materials. Most common grippers have two opposing fingers that are machined to fit specific objects with defined orientations. They can be driven by hydraulic or pneumatic actuators or electric motors powered by supplies that are independent of the host robot. Some versatile grippers with three of more fingers are dexterous enough to grasp objects, regardless of their shape or orientation, when their fingers are directed to close around the object. They are usually computer-controlled by dedicated software either in the host robot's central computer or in an independent notebook computer.

kinematic chain: The combination of rotary and/or translational joints or axes of motion.

kinematic model: A mathematical model used to define the position, velocity, and acceleration of each moving member of a robot without considering its mass and force.

kinematics: A branch of dynamics concerned with aspects of motion between moving elements of a machine or robot separate from considerations of mass and force.

manual teaching: A method for programming a robot by leading the end effector manually through the entire sequence of motions required to perform a task so that all axial positions can be recorded by the operator with a control panel or teach pendant. The position coordinates are stored in the robot's computer memory so that they can be played back automatically to perform the recorded task.

manufacturing cell: A concentrated group of manufacturing equipment typically including one or more industrial robots, a computer-vision system, and ancillary equipment such as parts conveyors, indexing tables, inspection station, end-effector changers, and storage racks dedicated to performing a specific function at one location. All equipment is coordinated and synchronized by a computer to carry out continuous processing. It is also called an *assembly cell* or an *assembly center*.

mobile robot: A robot with its own self-contained means of propulsion: wheels, tracks, propeller, or other mechanism for crawling, climbing, swimming or flying. It contains a central processor and appropriate sensors. It is likely to carry or contain onboard equipment including tools, manipulators, and sensors for navigation and the performance of tasks. These robots participate in but are not limited to Earth or planetary exploration, surveillance, and the disposal of bombs or other hazardous material. Mobile robots can be partly or completely autonomous, but most are directed by a remote operator via two-way wireless, wired, or fiberoptic links which transmit information to and from the robot. Sensors such as video cameras and distance-measuring instruments provide navigational guidance for the remote operator. The links can also be used to reprogram the robot's computer while it is moving. Most mobile robots are more accurately called *telerobots*; they now include unmanned ground vehicles, water and underwater vehicles, and aircraft or drones.

movable robots: Industrial robots mounted on wheels or rollers that can be driven by a power source but are restricted to travel on rails or tracks. They can move in horizontal, vertical, or angled directions while performing assigned tasks such as painting or welding under manual or autonomous control.

open loop: A control technique in which the robot's tasks are performed without error correction; accuracy typically depends on components including a position motion controller and a stepping motor.

payload: The load that can be lifted by a robot, measured in pounds (lb) or kilograms (kg).

payload capacity: The maximum *payload* that can be handled safely by a robot.

point-to-point programming: A robot motion control method in which a series of numerically defined stop points or positions are programmed along a path of motion. The robot moves to a position where it stops, performs an operation, and moves on to the next position. It continues these steps in a sequence performing all operations until the task is completed. This control method is suited for pick-and-place materials handling and other applications where the movements between points need not be controlled.

reach: The maximum distance that can be reached by the *tool point*, a theoretical point beyond the robot's wrist when all axes are extended to their limits. It is measured in inches (in.) or millimeters (mm).

repeatability: The limits of variation or deviation of the robot's tool center point position obtained after repeated cycles under fixed conditions. It is measured as plus or minus millimeters ($\pm$ mm).

resolution: The smallest incremental motion that can be made by a robot; a measure of robot accuracy.

rotary joint: A mechanism consisting of fixed and rotary components which, when attached to the robot wrist, permit the wrist to rotate through 360° without interrupting the supply of compressed air, water, or electricity required for operating various end effectors; utilities are supplied through a slip ring.

rotational inertia: The property of a rotating body that resists changes in angular velocity around its axis of rotation; it is measured as a mass unit multiplied by the square of the unit length—*moment of inertia*, (lb-ft^2 and kg-m^2).

semiautonomous: A robot directed by a human operator that contains electronically programmed sensors or devices capable of performing essential functions such as recording, navigating, and reporting on the robot's conditions that are fed back to the operator for optimum robot operation.

teach pendant: A handheld control box connected by cable to the computer control cabinet that permits the operator to enter programming data defining stop points while "teaching" the robot.

telerobot: A mobile robot with partial autonomous control of its sensors or functions whose movements must be directed by a remote operator via a wireless, wired, or fiberoptic link.

tool center point (TCP): A position between the robot's wrist and that part of the tool or *end effector* that defines the location where tool activity is concentrated; examples are the nozzle of a paint spray gun or the end of an arc-welding electrode.

tool changer: (1) A mechanism with two components (master and tool) for exchanging robotic tooling while it is performing its tasks. The master, attached to the robot wrist, contains half of the coupling; the other half is attached to the end effector or tool to be exchanged. When the two parts of the coupling are mated, they can be locked together automatically by pneumatic or hydraulic pressure. Examples of end effectors that can be exchanged are grippers, welding tools, or deburring machines. (2) A specialized robotic gripper that can pick up, clamp, and release a large variety of end effectors with common arbors or shanks as required. (3) A rotary turret mounted on the robot's wrist containing a selection of tools that can be indexed into a working position as required.

workspace: The locus of all points or envelope that can be reached by the wrist using all of the robot's available axis motions.

wrist: A set of joints, typically rotational, at the end of the robot arm with a mounting plate for attaching end effectors or tools. Wrists can have two or three degrees-of-freedom (DOF): Two DOF wrists can move the tool or end effector around both pitch and roll axes, or pitch and two independent roll axes.

MODIFIED FOUR-LIMBED ROBOT IS A BETTER CLIMBER

Stereoscopic cameras on circular track for omnidirectional vision

End-effector tool options include:
1. Rotary driver
2. CCD video camera
3. Bright LED light
4. Versatile tool collet

Each limb tipped with quick-release mechanism for tool changing

LEMUR IIb is the third generation of Limbed Excursion Mechanical Utility Robots, but it has four legs instead of the six of its predecessor LEMUR II. This simplifies the robot and makes it easier for it to climb inclined surfaces.

Scientists at NASA's Jet Propulsion Laboratory (JPL) have designed and built three different versions of the Limbed Excursion Mechanical Utility Robot, abbreviated as LEMUR. The objective of their development work is to arrive at a design for a robot to be used on space missions that can walk autonomously along a beam toward a mechanical object at a prescribed location and demonstrate its ability to assemble, maintain, and inspect a practical device. All three were test beds for a proposed NASA Autonomous Walking and Inspection Robot (AWIMR), and they are expected to be influential in its design.

The first-generation LEMUR, designated LEMUR I, was described as a "six-legged experimental robot" by NASA scientists and technicians. Then an improved LEMUR II, was built. It was described as a "second-generation six-limbed experimental robot" in the fourth edition of this book. LEMUR I could perform simple mechanical operations by using one or both of its front legs, while LEMUR II could use any of its limbs to perform mechanical operations. Both were equipped with stereoscopic video cameras, image-data-processing circuitry for navigation and mechanical motions, and wireless modems permitting them to be commanded remotely. Both could also transmit images to a host computer.

LEMUR IIb, shown in the figure, is a modified version of the LEMUR II. It is basically the same as LEMUR II but, as can be seen in the figure, it has only four legs or limbs rather than six. LEMUR IIb was equipped with the same cameras, data processing circuitry, and wireless modems as the two earlier versions. As

a consequence, the real difference between them resulted from the elimination of two of its six limbs. This change eliminated the mass of each limb as well as the mass of the robot itself. Each limb of LEMUR II had four degrees-of-freedom (DOFs), but each limb of LEMUR IIb has only three DOFs, a change that also reduced the robot's complexity. This kinematically simplified design made it easier for LEMUR IIb to move on level surfaces and be more adept at climbing inclines.

Despite this decrease in DOFs, the three remaining DOFs are configured to give LEMUR IIb greater dexterity. The leading measurements of LEMUR IIb are:

- Diameter of its hexagonal body   11 in. (28 cm)
- Length of each leg   15 in. (38 cm)
- Span across two limbs extended   41 in. (104 cm)

The reach of the LEMUR IIb is 25 percent longer than that of LEMUR II. The reduction in the number of limbs and DOFs as well as weight in LEMUR IIb decreased the load that must be supported by its limbs and also lowered the robot's center-of-gravity. These benefits were obtained without sacrificing LEMUR IIb's load-carrying capacity

This work was done by Avi Okon, Brett Kennedy, Michael Garrett, and Lee Magnone of Caltech for NASA's Jet Propulsion Laboratory.

SIX-LEGGED ROBOT CRAWLS ON MESH IN LUNAR GRAVITY

Spiderbot is a hand-sized, mobile robot being developed to assemble and repair simple structures and participate in search and rescue during NASA exploratory missions to remote planets. It has six legs with spring-compliant joints and gripping feet. These permit it to propel itself between rungs on a flexible net and walk on flat surfaces in reduced gravity. Its programmed leg motions alternately clamp and unclamp as the robot travels, while it keeps three feet attached to rungs at all times.

Four different research facilities are participating in the design, construction, and testing of hand-sized mobile robots. They resemble spiders but have only six legs, two fewer than spiders, as shown in the figure. Nevertheless, *Spiderbots* are prototypes of a proposed line of relatively inexpensive walking robots. These could be deployed in large numbers to work cooperatively in assembly and repair, and search and rescue activities to support NASA exploratory missions to outer space and remote planets. The Micro-Robot Explorer Project is developing the robots with funds from the Advanced Concepts and Technology Innovations Office of NASA's Jet Propulsion Laboratory (JPL).

The stated objectives of the JPL Micro-Robot Explorer project are:

- Demonstrate a Spiderbot walking within a fixed kinematic envelope on a flexible, deployable mesh in reduced gravity (microgravity).
- Develop the robot, motion-control algorithm, grippers, and grasping algorithms.
- Implement and demonstrate walking behavior with an instrumented Spiderbot.
- Provide a springboard for a future full flight experiment.

Spiderbots are smaller, draw less power, and are more specialized than other legged and wheeled robots that have been developed for NASA. At their present stage of development, they have been designed primarily to demonstrate their ability to crawl on a flexible rectangular net or mesh in reduced gravity and to walk on flat surfaces and assemble simple structures from components brought with them.

All six legs have two spring-compliant joints and gripping actuators that clamp onto mesh rungs and function as feet. Spiderbots are programmed to advance as three legs which are anchored, alternate with three that are not anchored by clamping and unclamping their "feet" to the rungs of the mesh. Each robot will have three "feet" attached to the flexible mesh at all times. These motions will demonstrate that hexapods can crawl on meshes, walk on flat surfaces, and assemble and repair simple structures.

With these characteristics Spiderbots are expected to be able to traverse harsh terrain otherwise inaccessible to wheeled robots. One possibility being studied is that a team of Spiderbots could network together and collaborate in efforts to complete a set of tasks successfully. In addition, the project will explore alternative forms of mobility as well as alternative missions in which Spiderbots can participate.

Spiderbots have been tested during the reduced gravity achieved as an aircraft flies along a parabolic path. During this brief period, the robots were able to demonstrate crawling along the mesh. The results of this testing suggested a possible improvement by adding feedback from sensors on the "feet" of the Spiderbot. It was expected that this addition would indicate the robot's success in gripping the rungs of the mesh as a way to ensure fault-tolerant operation. Caltech, Texas A&M University, International Space University, and Blue Sky Robotics are participating in the Micro-Robot Explorer project.

This work was done by Alberto Behar, Neville Marzwell, Jaret Matthews, and Krandalyn Richardson of Caltech; Jonathan Wall and Michael Poole of Blue Sky Robotics; David Foor of Texas A&M University; and Damian Rogers of ISU (International Space University) for NASA's Jet Propulsion Laboratory.

TWO ROBOTS ANCHOR ANOTHER TRAVERSING STEEP SLOPES

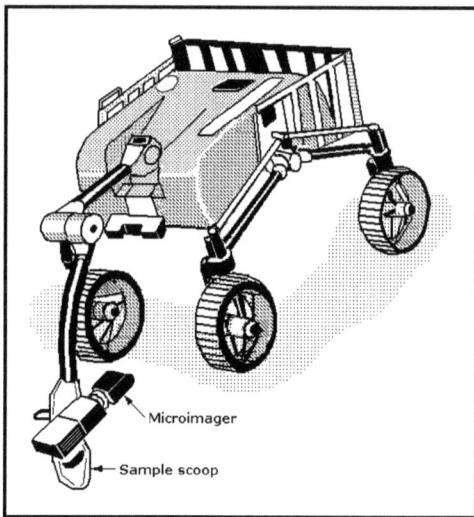

Fig. 1 This Cliffbot is an autonomous robot equipped with instruments for scientific studies so that it can rappel down slopes as steep as 90° angles to explore terrain too steep or dangerous to be explored by humans. It is expected to have applications on Earth as well as the Moon and other planets. The robot will be tethered to and controlled by two other specialized autonomous robots called Anchorbots.

Fig. 2 This computer-controlled winch will be mounted on two Anchorbots (yet to be built) for controlling the descent of a Cliffbot as it rappels down steep slopes.

A system of three autonomous mobile robots that cooperate to permit exploration of steep slopes has been developed at NASA's Jet Propulsion Laboratory. Called TRESSA, the abbreviation for Teamed Robots for Exploration and Science in Steep Areas, the system was developed to permit scientific exploration of steep slopes with angles up to 90°. Originally intended for exploring steep slopes on Mars inaccessible to wheeled robots, TRESSA is seen as having an application here on Earth in rescuing people who are trapped on slopes that are too steep and dangerous for human rescuers to rappel down on ropes.

TRESSA is based on techniques used by extreme mountain climbers that require teamwork and the use of safety tethers. The TRESSA technique has two autonomous robots called *Anchorbots* positioned at the top of the cliff while a third autonomous robot called a *Cliffbot* (see Fig. 1) is lowered down the slope. The Cliffbot drives over the cliff edge supported by tethers payed out from computer-controlled winches on the Anchorbots. They will autonomously control the tension on the tethers to counter the gravitational force on the Cliffbot. The tethers will be let out and reeled in as necessary to keep the wheels of the Cliffbot in constant contact with the cliff face. This will prevent wheel slippage by controlling the speed of descent or ascent. The Cliffbot will be able to drive freely up, down, or across the slope. Winches, as shown in Fig. 2, will be mounted on two four-wheeled Anchorbot rovers that have not yet been built.

The three-robot system requires that the robots be very tightly coupled, and this is achieved by TRESSA software based on two previous NASA software developments: one was for controlling multiple robots and the other was for real-time control of a robot. This software combination makes it possible to keep three robots synchronized and coordinated at all times by using data from all three robots for decision making while simultaneously controlling the wired connections among the robots.

There were two major considerations in the design and operation of TRESSA: tether tension control and detection of faults. In the first, tension is measured by force sensors connected to each other at the Cliffbot. The direction of the tension will also be measured in both azimuth and elevation. The tension controller combines a controller for countering gravitational force, and an optional velocity controller anticipates the motion of the Cliffbot.

The gravity controller estimates the slope angle from the inclination of the tethers. This angle and the weight of the Cliffbot determine the total tension necessary to counteract the weight of the Cliffbot. The total tension required will be allocated into components for each Anchorbot. The difference between this required tension and the tension measured at the Cliffbot creates an error signal that is provided to the gravity controller. The velocity controller computes the tether speed necessary to produce the desired Cliffbot response.

The detection of faults is based on interaction between each robot in the system that monitors its own performance as well as that of the other robots' ability to detect any system faults and prevent any unsafe conditions. At startup, communication links are tested, and if any robot is not in communication, the system will not execute any motion commands. Prior to motion, the Anchorbots will attempt to set tensions in the tethers at optimal levels for counteracting the weight of the Cliffbot. If either Anchorbot fails to reach its optimal tension level within a specified time, it will send a message to the other robots which prevents the command motion from being executed.

If any mechanical error such as a motor stall is detected, the affected robot sends a message that stops any existing motion. Finally, messages are passed among the robots at each time increment (10 Hz) to share sensor information during operations. If messages from any robot cease for more than the allowable time interval, the other robots detect their loss and stop any further motion.

This work was done by Ashley Stroupe, Terrance Huntsberger, Hrand Aghzarian, Paulo Younse, and Michael Garrett for NASA's Jet Propulsion Laboratory.

SIX-LEGGED ROBOT CAN BE STEERED WHILE HOPPING

Fig. 1 This steerable hopping robot has bow-shaped fiberglass springs fastened on all six of its legs for storing the energy needed to power hops. Cords, attached to the ends of each leg, are wound around a motorized spool (not shown). As the spool rotates the cords are tensioned, pulling up all six legs while compressing all leg springs. This results in storing potential energy for powering robot hopping when cord tension is released. Spring compression, provided by cord tension, determines the distances and heights the robot can hop. Timing belt drives keep all legs straight while the robot hops.

Engineers at the Jet Propulsion Laboratory designed, built, and tested a prototype of a steerable six-legged hopping robot for exploring low-gravity environments such as are found on the moon or other planetary satellites. The prototype robot was able to hop vertically 35 cm (14 in.), and at a 30° angle it hopped 30 cm (12 in.) high—a distance of 50 cm (20 in.). The motorized steering of its six legs was demonstrated over a 40° range. Although a gyro was installed on the prototype, gyro-stabilized hopping was demonstrated by a computer-simulated robot model with a controllable flywheel in lunar gravity conditions.

Many of the platforms proposed for planetary and lunar exploration have wheels or moving legs for mobility, but these may not permit efficient travel over rocky or rough terrain. The JPL engineers believed that a stabilized and steerable six-legged robot would be more effective for exploring those landscapes. For example, 83 percent of the lunar surface is densely cratered highlands, and this terrain will be the most challenging for robotic explorers. In those regions, there are slopes of 1 m (3 ft) as steep as 34° as well as slopes of 50 m (164 ft) angled up to 18°.

Each of the six legs of the prototype, as shown in Fig. 1, has two folding ladderlike sections attached to both the upper and lower hexagonal frames of the robot with brackets and pin joints.

Fig. 2 The spool assembly (centrally located on the lower frame) reels in all six leg springs with enough tension to bow them. As the motor winds the spool, an encoder measures the compression of all legs. A hop is triggered when the electromagnetic clutch disengages the spool motor, simultaneously releasing the energy stored in the leg springs. The impact of landing after the first hop stores enough potential energy to continue additional robot hopping.

All of the legs have bowed fiberglass leaf springs fastened between their upper and lower rungs. The springs are 0.5 in. (1.2 cm) wide by 10 in. (25 cm) long. A cord attached to the lowest rung of each leg runs up the centerline of the leg and is wound around a centrally located spool, as shown in Fig. 2.

The cords, when wound on the spool driven by a harmonic motor, apply tension to the springs to bow them to any desired amount of compression, as indicated by the encoder above the spool. To ensure that all legs push off and land in the same direction, a timing belt pulley drive is attached to each leg link, restricting leg extension to a linear motion. The height and distance achieved by the robot hop is directly related to the tension applied to the cords and the resulting spring compression of the leg springs. Also, by varying the dimensions of the springs, different hopping distances and heights can be obtained.

Fig. 3 After six legs are compressed by the cords on the spool, (a) the robot is aimed and ready to hop. In (b), the spool has been released and the robot is in flight, stabilized by an internal gyro. When the robot lands (c) after its first hop, its spring-loaded legs absorb the impact and store enough energy for the next hop while a clutch is released allowing the cords to rewind, locking the legs in compression to prevent the robot from bouncing.

When the robot is ready to hop, as shown in Fig. 3a, all six legs are in compression. The hop occurs when the magnetic clutch disengages the motor from the spool and the spring-loaded legs are released simultaneously. The six legs add stability and reduce the forces that must be transmitted while hopping and landing. An internal gyro was attached under the hexagonal frame of the robot to stabilize it during the hop and reduce the effects of uneven forces between hopping and landing.

Figure 3b shows the robot in flight with the leg springs relaxed. A two degrees-of-freedom steering mechanism with only two motors gives the robot the ability to hop and land at different angles with all six legs pointed in the same direction. The legs swing out from a vertical position. Two pin joints on the lower hexagonal frame control the entire mechanism. The two motor actuators, separated by 120°, contain worm drives that provide enough torque to steer the legs. Because the robot's legs cannot be driven backward, leg orientation is stabilized during hopping and landing. A steering mechanism aims the robot's legs at various angles. (The one in the prototype allowed for a 40° angle in either direction.)

When the robot lands after hopping, as shown in Fig. 3c, the leg springs on its legs compress to absorb the impact and store enough energy for additional hopping. As the legs retract, a constant force spring motor attached to the upper portion of the spool winds up the leg cords to keep them under tension. A unidirectional clutch, in line with the spool and motor drive, allows the spool to quickly overrun the motor when winding the cables, but it locks when the legs try to withdraw the cables. This prevents the robot from bouncing after landing and allows the springs to store the energy from its landing for use in the next hop.

To help stabilize the robot and keep it from tumbling, a motorized gyro was integrated into the base of the robot. Any moments exerted on the robot during hopping are translated into minimal angular displacements from the robot's orientation, removing any concern that the robot could tumble over.

The main structures of the prototype robot were made from aluminum. The cover protecting the mechanical components mounted on the lower hexagonal frame and the six spherical feet were formed from a nylon-based resin by rapid prototyping. The feet are fastened to the lower rungs of the legs by pins allowing them to pivot. Four actuators are in the system: one spring actuator, two steering actuators, and one gyroscope actuator. The testing of the spring actuator, steering actuators, and gyro demonstrated that these components worked successfully.

The work described here was done at the Jet Propulsion Laboratory, California Institute of Technology, under a NASA contract by Paulo Younce and Hrand Aghazarian.

CHAPTER 4
MECHANISMS FOR RENEWABLE POWER GENERATION

OVERVIEW OF RENEWABLE ENERGY SOURCES

As the population of the United States grows, the demand for low-cost electric power expands exponentially. For years the availability of low-cost electricity has been closely correlated with the high American standard of living and productivity. Against this background, the electric utility industry is being asked by the government to install more carbon-dioxide (CO_2)-free renewable energy power plants to reduce this emission to earlier levels because some believe it is the cause of global warming or climate change. How this mandate can be accomplished without increasing the price of electricity, now still largely produced by CO_2-emitting plants, remains to be seen.

The government is calling for a 20 percent reduction of CO_2 emissions from all electricity generating plants by 2020; it is also calling for a 25 percent increase in the number of renewable energy power plants by 2025 and a goal of 80 percent of America's electricity to come from clean energy sources by 2035. It also wants CO_2 emissions to be reduced to 1990 levels by 2025. The two power plant technologies recognized as being capable of producing sufficient renewable, non-hydro, carbon-free, utility-grade electricity today are mechanical: wind turbine and solar thermal farms. Unfortunately, at present only about 4 percent of the nation's electricity is produced by them: 2 percent from wind and 1 percent from solar thermal. However, it is unlikely that enough of these plants can be built to comply with these goals in the time specified without a massive government spending program.

Figure 1, based on information from the U.S. Energy Information Administration (EIA), shows graphically how various fuels are used for generating utility-grade electricity. (This chart has remained virtually unchanged for more than 10 years.) About 70 percent of the nation's electricity is produced by utilities from the combustion of fossil fuels: coal, natural gas, and petroleum (fuel oil). Of this amount, about 45 percent is obtained from coal, 23 percent is obtained from natural gas, and the remaining 1 percent is from fuel-oil. However, coal-fired plants remain the most prominent emitters of CO_2 as well as the other "greenhouse" gases.

Nuclear and hydroelectric plants are now the largest producers of carbon-free power in the United States, about 20 percent from nuclear reactors and less than 7 percent from hydroelectric plants. However, neither of these sources will be significant factors in meeting the government's goals for CO_2 reduction. Nuclear plants (considered to be non-renewable) are not being built fast enough to make a real contribution to those goals. They take years to build, they encounter time-consuming licensing complications, and plans to build more invite public opposition. More hydroelectric plants will not be built because of restrictive environmental regulations and concerns about the availability of fresh water, particularly in the cities of the Southwest.

CO_2 was recently classified as a greenhouse gas and is now included with such toxic gasses as methane (CH_4), nitrous oxide (N_2O), and sulfur dioxide (SO_2), all by-products of fossil-fuel combustion. However, unlike the toxic greenhouse gases, CO_2, the gas in carbonated beverages, is exhaled by all humans and is essential for plant growth. Coal-fired power plants now emit more than 60 percent of the CO_2 from the power industry as well as the largest amounts of the other greenhouse gases. This compares with about 20 percent of CO_2 emitted by natural gas-fired plants and only about 2 percent emitted by fuel-oil-powered plants. However, the electric power industry has cut emissions of SO_2 and nitrogen oxides (NO_x) by 57 percent since 1980.

According to the EIA, utility power plants emit only about 40 percent of the total national CO_2. The other 60 percent is from sources classified as the transportation and buildings and industrial sectors. The transportation sector emits about 30 percent of the CO_2 from such sources as gasoline and diesel-powered vehicles, trains, and aircraft, and the buildings and industrial, sector emits another 25 to 30 percent from such sources as industrial furnaces and heating systems for homes, buildings, and factories. Obviously, it is more difficult to regulate CO_2 emission from these disparate sources than from easily identifiable fossil-fueled plants. However, if all power plants that now emit CO_2 were shut down, millions of tons of CO_2 would still be produced by these other sources, and this would do little to reduce the nation's dependence on imported oil.

Nuclear: The Unlikely Prime Renewable

While nuclear plants are the most expensive power sources to build and their construction can take years, they operate more than 90 percent of the time for as long as 60 years. (This is somewhat more than fossil-fueled power plants and much more than wind farms or solar thermal plants.) All 104 of the operating nuclear plants in 31 states now produce more than 73 percent of all of the nation's carbon-free electrical power. In 2011 there were only two 1100 MW reactors being built in the United States. The Vogtle plants located near Augusta, Georgia, represent the first nuclear reactor construction in the country since the 1970s. (Two existing reactors at the Vogtle plant site became operational in 1987 and 1989.) Plans have been announced for building six more reactors, but they have encountered financial and regulatory obstacles. Because nuclear plant construction has

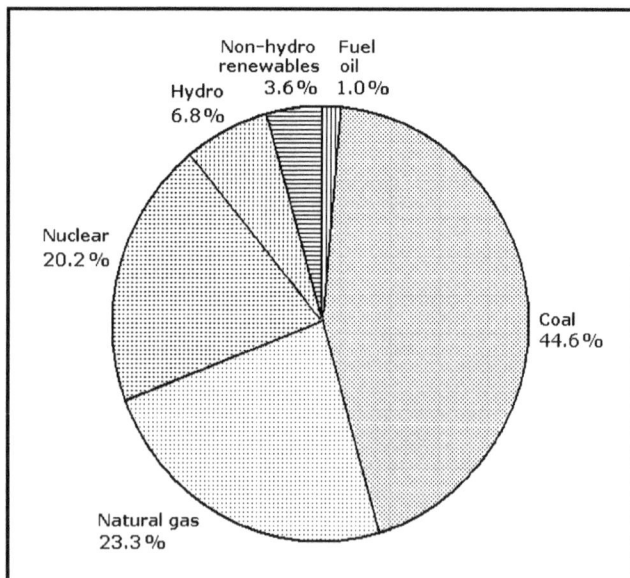

Fig. 1 National fuel mix for electric power generation. Non-hydro renewable power sources include wind, solar, geothermal, and batteries. (*Source: U.S. Department of Energy, Energy Information Administration (EIA) and Edison Electric Institute.*)

not kept up with electricity demand, its contribution to carbon-free generation will decline over the years. It has been estimated that even if the number of nuclear plants doubled within 10 years, the reduction in CO_2 emission would be small.

Alternative Renewable Energy Sources

Conventional power plants (fossil fuel, hydro, and nuclear) can produce from about 300 to 1300 MW of electricity. But these figures apply only to the maximum ratings and overlook inefficiencies and downtime for routine maintenance. Nevertheless, regardless of their actual power output for a given time period, they can operate 24 hours a day for 365 days a year. This can't be said for wind turbine and solar thermal plants.

Photovoltaic (PV) panels are the most familiar form of renewable solar power, but they are not now capable of cost-effective utility-grade power generation. They have the highest cost per kilowatt hour (kWh) of power generation and the lowest efficiency. However, they are now being installed on roofs of homes and buildings for on-site auxiliary power generation where the owners want to reduce their electric bills and gain some independence from power utilities. But it could take years to amortize their installation cost before owners can truly obtain reductions in their electric bills.

While it is unlikely that many of the less efficient coal-fired plants will be shut down, one way to reduce carbon emissions from all coal-fired plants is to liquefy the CO_2 and sequester it in underground caves. However, the cost-effectiveness of none of the liquefaction processes has been proven. Meanwhile, worldwide, more than 40 percent of all electricity is still being generated by coal-fired plants, and another 20 percent is being generated by natural gas-fired plants. They continue to release billions of tons of CO_2 into the atmosphere.

Baseload and Baseload Demand Power Plants

Baseload and *baseload demand* are terms used in the electrical utility industry to refer to contract commitments made by power companies connected to the grid to produce specified amounts of electricity 24 hours per day, 365 days a year. It is necessary that baseloads be maintained to keep the power grid stabilized and meet the utility's obligations to customers. Utility plants are production facilities intended to provide some or all of a given region's energy demand at a constant rate. This means that renewable-energy power plants unable to operate 24 hours a day must have some backup sources such as standby generators or energy storage facilities capable of driving generators. These include batteries, steam boilers, liquefied molten salt vats, and compressed-air chambers.

Windmills: Early Renewable Power Sources

The wind has provided free renewable mechanical power for centuries, making the lives of people easier by relieving them of many labor-intensive tasks. The sails on ships propelled them, eliminating total dependence on oars, and windmills pumped water, ground corn and grain, and sawed lumber. The practical windmill, the true ancestor of the modern wind turbine, first proved itself in low-land European countries facing the North Sea where strong, steady winds prevail and were reliable power sources for centuries.

The Dutch invented the early rotating *hollow-post* mills with rotating four-bladed "sails," made as wood lattices and covered with canvas to increase their mechanical power. When wind struck the sail, it rotated a horizontal shaft geared to a vertical shaft in the center of the post that drove machinery at ground level. The sail and both horizontal and vertical geared shafts were built into a heavy wooden housing that was able to rotate on top of the hollow post to keep the sail facing into the wind. However,

the heavy housing could only be kept facing the changing wind by considerable manual effort with ropes, pulleys, and winches.

This led to the next stage in windmill design, the adoption of round masonry towers, formerly used for storage, as stronger, more stable bases, better suited for mounting rotating machinery. These configurations configured the sail and horizontal and vertical shafts built into a smaller, lighter housing or *cap*. This innovation included bearings and gears making it easier to move the mechanism around the tower's rim. After the available masonry towers had been converted to windmills, it became necessary to build wooden towers.

The result was the now familiar Dutch windmill design with wooden bell-shaped towers having octagonal cross sections, as shown in Fig. 2, rather than the earlier round masonry cross sections. Among its new features were brakes to stop sail rotation for repairs, emergencies, or the end of the workday. The brakes could be applied from the ground by turning a wheel on a winch. Tension applied to the rope attached to levers clamped brake shoes against the horizontal shaft, stopping sail rotation. Also, the rotational speed of the sails could be adapted to changing wind conditions by adding or removing canvas to the blades.

Some windmills were built with a geared wheel inside the cap to permit it to be swiveled manually into the wind from that location. Others had adjustable rotary fans mounted on a tail pole that allowed the wind to swivel the cap automatically like a large weather vane. The vanes on the fan could be opened or closed to adapt its surface area to changes in wind strength as the sail was being automatically swiveled into the wind.

The windmills of Holland pumped water from land being reclaimed and from rain-flooded farmland back into the sea or canals. Dutch windmills also ground corn and grain, drove mechanical saws for cutting logs into lumber, and hoisted ore from mines. Before the era of steam power, it is estimated that there were 8000 windmills in Europe. A typical windmill built before the introduction of steam power was capable of 5 to 10 hp.

Fig. 2 This Dutch windmill was built to mill corn and grain. Its cap assembly can be turned by a geared handwheel to keep the sail pointed into the wind. Canvas can be added or removed from the sail blades to adapt the sail to changing wind conditions, and brake shoes that clamp on the horizontal shaft can be applied from the ground by a brake wheel.

After electricity became the dominant energy source, some large windmills were converted into power stations. Generators connected to gear trains produced electricity making it possible to transmit electricity from the mill over short distances to power nearby homes or factories.

Wind Turbines: Descendents of Windmills

Wind turbine rotors have blades contoured like airplane wings that catch the wind flowing over them to create lift. This lift is translated into a force causing the rotor to turn. When wind conditions are favorable, the rotor shaft turns with enough force to drive an electrical generator. There are three general classes of wind turbines: *utility-scale*, *industrial-scale*, and *residential-scale*.

- *Large-size utility-scale turbines*, rated from 700 kW to 5 MW, generate utility-grade power.
- *Medium-size industrial-scale turbines*, rated from 50 kW to 250 kW, generate power for places beyond the range of utility power lines. (They are often paired with diesel emergency or stand-by generators for backup.)
- *Residential-scale turbines*, rated from 50 W to 50 kW, are for electrification of homes beyond utility power lines or for farms to operate pumps or agricultural equipment.

Most utility-scale wind turbines are horizontal-axis machines with power obtained from three-bladed rotors connected by horizontal shafts and gear trains to an electrical generator. Modern wind turbines adopted many features of the windmill. The power train is located on an enclosed platform or chassis that can be turned to face the wind, and the rotor can be adapted to changes in wind speed by "feathering" its blades (as is done on aircraft propellers). In addition, brakes on the rotor stop it from spinning to permit repairs or routine maintenance and, under extreme wind conditions, they can stop the rotor to protect the turbine from being damaged.

The principal mechanical and electromechanical components of a horizontal-axis wind turbine rated from 1.5 to 3.6 MW are shown in Fig. 3. All major components within the nacelle are connected to form a unified power train. In this design, the three-bladed rotor faces upwind and it has an active yaw-axis drive system to keep it oriented into the wind. The power train consists of a low-speed shaft connecting the rotor to a two- or three-stage gearbox which, in turn, drives the higher-speed coupling required to run the generator. Wind turbine generators are typically asynchronous induction machines that produce from 550 to 1000 VAC. Each utility-scale turbine is equipped with a transformer (not shown) to step up the generator voltage to meet the requirements of the on-site electrical grid connection value, typically 25 to 35 kV.

The nacelle also houses a controller that includes a programmable-computer and both internal and external communications links. The nacelles of turbines in this class typically contain oil and water coolers, and a service crane. All of these components are mounted on the rigid nacelle frame or base which can be rotated on the top of the tower by one or more yaw motors and gearing. The generator's output is transmitted by cables running down the inner wall of the tower to connect to the power grid.

Signals from an anemometer and windvane on top of the nacelle send data reporting actionable changes in wind speed and direction to the controller. These signals are used by the controller to direct the yaw motors that keep the rotor facing into the wind. Turbine performance can be optimized by feathering rotor blade pitch angles to maximize power output under varying wind conditions.

The controller is programmed to turn the turbine on when wind speed reaches the "cut-in" velocity of about 9 mph (4 m/s). It also turns the turbine off if wind speed exceeds the "cut-off" value of 45 to 60 mph (20 to 27 m/s) for more than 10 min to prevent damage to the system. Variability in wind speed causes the turbine to operate at continually changing power levels but electronic circuitry in the latest utility-scale wind turbines regulates the power output to keep it constant despite changes in wind speed. Turbine manufacturers state wind speed for optimal

Fig. 3 Wind turbines with output ratings of 500 kW to 3.6 MW now provide power to the grid. These turbines can operate efficiently in wind speeds of 8 to 55 mph (13 to 86 km/hr). Rotor speeds between 8 and 15 rpm will produce constant power. Three-blade rotors predominate with rotor diameters up to 341 ft (104 m) and tower heights up to 344 ft (105 m).

Table 1 Key Specifications for Two General Electric Wind Turbines

	1.5 MW 1.6-82.5	2.5 MW 2.5-103
Operational Data		
Rated capacity	1.5 MW	2.5 MW
Cut-in wind speed	3.5 m/s (8 mph)	3.0 m/s (7 mph)
Cut-out wind speed	25 m/s (56 mph)	25 m/s (56 mph)
Rated wind speed	8.5 m/s (19 mph)	7.5 m/s (17 mph)
Rotor		
Number of blades	3	3
Rotor diameter	82.5 m (271 ft)	103 m (338 ft)
Swept area	5346 m^2 (57,544 ft^2)	8495 m^2 (91,439 ft^2)
Rotor speed	Variable	8.5-15.3 rpm
Tower		
Hub heights	80 /100 m (262/328 ft)	85/100 m (279/328 ft)
Electrical		
Frequency	50/60 Hz, 690 V	50/60 Hz, 1000 V

operation as rated speed with typical values in the range of 25 to 35 mph (11 to 16 m/s).

Utility-scale turbines, rated from 1.5 to 3.0 MW, have rotor diameters from 270 to 350 ft (82 to 107 m) and tower (hub) heights from 230 to 362 ft (70 to 110 m). The world's largest wind turbine has a 394-ft (120-m) rotor diameter and is capable of generating 5 MW. Larger turbines rated for 6 MW are being tested and 10 MW units are in the planning stage. The key specifications for two General Electric wind turbines are given in Table 1 and the key specifications for two Vestas Wind Systems wind turbines are given in Table 2.

Wind farms can contain hundreds of turbines whose overall power output could be as much as 300 MW. At favorable wind sites, turbines operate at approximately 35 percent of total capacity when averaged over one year. Wind farm turbines are typically networked so they can exchange data, and a master system records their fault and status reports and transmits these to a remote ground control center. The network permits the turbines to be monitored from the center so their onboard controller's computer programs can be overridden for maintenance or in response to emergencies; these could include the probable approach of tornados, hurricanes, or other extreme weather conditions.

Table 2 Key Specifications for Two Vestas Wind Turbines

	V 82-1.65 MW	V90-3.0 MW
Operational Data		
Nominal output	1.65 MW	3.0 MW
Cut-in wind speed	3.5 m/s (8 mph)	3.5 m/s (8 mph)
Cut-out wind speed	20 m/s (45 mph)	25 m/s (56 mph)
Nominal wind speed	13 m/s (29 mph)	15 m/s (34 mph)
Rotor		
Number of blades	3	3
Rotor diameter	82 m (269 ft)	90 m (295 ft)
Swept area	5281 m^2 (56, 844 ft^2)	6362 m^2 (68, 480 ft^2)
Nominal revolutions	14.4 rpm	16.1 rpm
Tower		
Hub heights	70–80 m (230–262 ft)	65–105 m (295–344 ft)
Electrical		
Frequency	50/60 Hz, 690/600 V	50 /60 Hz, 1000 V

Where Are Wind Turbines Located?

Even the most efficient and cost-effective wind turbines, like earlier windmills, must be located where the wind is constant despite seasonal changes and local weather conditions. However, statistically the wind blows only at about 35 percent of strength during the year. Wind farms with nominal rating equal to fossil-fuel or nuclear power plants produce far less power because the strongest wind usually blows at night. Consequently, wind farms must be backed up by auxiliary generators or sources of potential energy to provide the required baseload power. This can be accomplished with standby natural-gas fired turbines, diesel generators, or batteries.

The U.S. Department of Energy considers locations with average annual wind speeds of at least 14 mph (23 km/h) at a 260-ft (80-m) height suitable for wind farm development. In 2010, the states that met these criteria extended south from North Dakota and Minnesota to Texas. The amount of power generated by a typical wind turbine increases exponentially with wind speed. Stated another way, if the wind blows just 15 percent faster, a turbine will produce 50 percent more power. But turbines produce most of their power when the wind blows faster than about 25 mph (40 km/h).

Wind farms located on land require many acres of property, depending on the number of wind turbines planned for the installation. The sites can be on flatlands or in mountainous regions, in deserts or on farmlands, but the choice depends on local wind conditions which must be determined by surveys. However, unlike solar thermal farms, the space between turbines can be used for productive activities such as grazing cattle or growing crops. Nevertheless, all sites that are remote from populated cities where the demand for electric power is high will require transmission line connections to the grid. These can be so long and expensive that a potential site will be discouraged.

Winds are usually stronger and more constant over oceans. If a wind farm is located offshore but near a densely populated city, the transmission line can be shorter than if it were built in a remote part of the country. However, wind turbines capable of withstanding the more demanding marine environment must be more ruggedly built, and they will be more expensive than their land-based equivalents. They must withstand gale-force winds, storms, and the corrosive effects of the saltwater environment with its spray. Also, marine locations make it more difficult for maintenance or repair crews to access them in stormy weather.

The country's first major offshore wind farm, the Cape Wind project, proposed in 2001 for Nantucket Sound, Massachusetts, had been stalled for more than nine years by citizen complaints. It was finally approved for construction by the U.S. government in April 2010. Residents and resort owners argued that the forest of turbines would be an eyesore which could block scenic views of Nantucket Sound and reduce their property values. Moreover, several Indian tribes complained that the turbines would disturb their sacred burial grounds located on the seabed. Cape Wind's CEO predicts that the project will be producing electricity by the end of 2012.

This wind farm will include 130 wind turbines each rated for 3.6 MW that are expected to generate 450 MW. The turbines will be mounted on monopole towers rising 258 ft (79 m) above Nantucket Sound. They will be set in a pattern of rows and spaces covering 25 mi^2 (16,000 acres) between Nantucket Island and Martha's Vineyard. These towers with base diameters of 16 ft (5 m) will be sunk 80 ft (24 m) into the seabed in about 30 ft (9 m) of water. The rotors for the turbines will have diameters of 365 ft (111 m). The closest turbine will be about 5 mi (8 km) from the nearest shore line.

Many objections to wind farms have discouraged their acceptance by the general public. The reasons include the large numbers of towers required to provide enough power to compete with fossil-fueled plants, the loud humming noise emitted by their spinning rotors, their known interference with airport radars, and the fact that migrating birds have been killed in collisions with their spinning rotors.

Despite these drawbacks, the residents of Nolan County, Texas, chose to embrace wind farms for economic reasons. They approved the installation of four of the world's five largest wind farms in or near Sweetwater because their installation provided employment for many people who service or repair the wind turbines. There had been a decades-long decline of jobs in oil drilling, ranching, and growing cotton. The region's wind turbines now produce more than 8300 MW of electricity, enough to power about 2 million homes. It is ironic that wind generation became a major industry in what was once America's petroleum heartland. These projects were successful because of the incentives offered by the state and federal renewable-energy grants and tax credits.

Concentrating Solar Thermal (CST) Systems

Concentrating solar thermal (CST) systems are now the only reliable solar renewable energy systems capable of meeting the requirements for generating utility-grade electricity compatible with the grid. The electricity output from CST systems is called *concentrated solar power* (CSP). CST power generators should be distinguished from solar power generated by solar or *photovoltaic* (PV) cells and panels. Solar cells produce electric current only when single photons hit single electrons, and this physical principle restricts their efficiency to a theoretical maximum of 31 percent.

As a result, the cost per kilowatt hour of photovoltaic power sources is higher than for any of the four widely recognized CST systems that are discussed here. Solar cells are widely used as independent owner-owned power generators for supplementing available utility power in homes, offices, and factories. They also power remote signs and instruments where utility power is either not available or, if it is, making connections to it would be impractical or too expensive. However, PV cells cannot compete economically as sources for utility-scale or grid-quality electrical power. Moreover, their DC output must be converted to AC.

There are four recognized CST system designs for converting the sun's thermal energy into utility-scale or grid-quality electrical power:

1. Parabolic trough mirror solar thermal plants (trough mirror)
2. Power-tower solar thermal plants (power tower)
3. Linear Fresnel reflector solar thermal plants (Fresnel reflector)
4. Parabolic dish Stirling solar thermal plants (dish Stirling)

Large-scale solar thermal farms are designed to collect enough solar heat to drive a turbine-generator capable of providing utility-grade electric power. Designers of these systems assign them projected maximum output ratings in megawatts based on optimum availability of sunlight for a given area. These values can be compared with those stated for fossil-fuel powered plants, but they do not take into account their relative efficiencies. A solar thermal power plant can have efficiency as high as 35 percent.

Output power ratings for CST farms do not account for seasonal changes, clouds, and storms that reduce their power output. In general, the power output from CST farms is far lower than those for nuclear or fossil-fueled plants. The highest anticipated output for a CST plant in the United States will be 250 MW. (Small fossil-fueled power plants produce 300 to 500 MW, while the largest coal-fired plant can produce up to 1300 MW.)

Parabolic Trough Mirror Solar Thermal (CST) Plants

The parabolic trough mirror solar thermal plant (Fig. 4) is one of three CST systems requiring many acres of movable linear

Fig. 4 A trough mirror solar thermal power plant includes a large array of linear trough-shaped mirrors that track the sun as it transits the sky. The mirrors focus solar thermal radiation on large networks of pipes containing oil as a liquid heat transfer medium. The oil in the pipes is sent to a heat exchanger where its temperature is hot enough to convert water in adjacent piping to steam. This steam spins a turbine that drives a generator to produce utility-scale power. The steam is then condensed to water, which is returned to the heat exchanger, keeping heat transfer continuous during daylight hours.

parabolic mirrors to produce grid-compatible power. Each installation has thousands of feet of mirrors shaped as parabolic troughs or flat plates, both arranged in straight parallel rows. The parabolic mirrors are contoured to permit them to focus solar thermal radiation on thousands of feet of black-painted steel pipe located in front of them.

Heat-absorbing mineral oil in the pipes acts as the intermediate heat transfer medium in an extensive closed-loop plumbing system. The pipes, positioned at the focal points of all of the mirrors, gather heat which is then delivered to a common heat exchanger. Solar radiation can keep the circulating oil at 750°F (400°C) over thousands of feet before it reaches the heat exchanger where its heat is transferred to a separate hot-water plumbing system. The water in that system is converted to steam, which spins a turbine to drive an electric generator. After the steam is condensed to water, it is returned to the heat exchanger.

Each of these mirror sections is continuously repositioned on its yaw (vertical) axis by computer control to keep the sunlight focused on the pipe sections during daylight hours. It is not necessary to move the mirrors about their pitch or lateral axes because, as the sun moves east to west with respect to the mirrors, the concentrated solar radiation moves along the lengths of mirrors and pipes, keeping the oil at a high constant temperature. All of the parallel rows of mirrors and pipes are positioned along an east–west axis to obtain maximum available sunlight.

Many parabolic trough mirror solar thermal plants are now operating in California, Arizona, and Nevada, while others are under construction or under development in Arizona, California, and Colorado. An existing example is the Nevada Solar One, a 64-MW plant on 400 acres (0.625 mi^2), that was completed in 2007. A 250-MW plant will be built in California in the Mojave Desert, 100 miles from Los Angeles, on 1765 acres (2.8 mi^2) of desert land. Once it starts operating in 2013, it is expected to generate nearly as much electricity as all existing commercial CST installations in California combined.

Some CST systems are backed up by separate natural gas-fired boilers, which can take over the task of producing steam on cloudy days or after sunset. This means that the turbine and generator can remain functional when adequate sunlight is not available. This backup stabilizes the system and provides supplemental power to meet the baseload demand as required for connection to the grid. A newer plant in Arizona includes tanks of molten salt for storing unused heat, which can also be used to produce steam after sunset or on a cloudy day.

Power-Tower Solar Thermal (CST) Plants

The power-tower solar thermal power plant (Fig. 5) is an alternative design for a CST plant. It uses large arrays of flat rectangular mirrors or *heliostats* to track the sun throughout the day. Each mirror or heliostat is individually moved around both its pitch or lateral axis and its yaw or vertical axis under computer control to keep the sun's thermal radiation focused on a *receiver* or steam boiler positioned on top of a tower designed to be tall enough to receive the reflections from even the most distant heliostats thousands of feet away from it. As in trough mirror systems, all power-tower heliostats are aligned along the east–west axis.

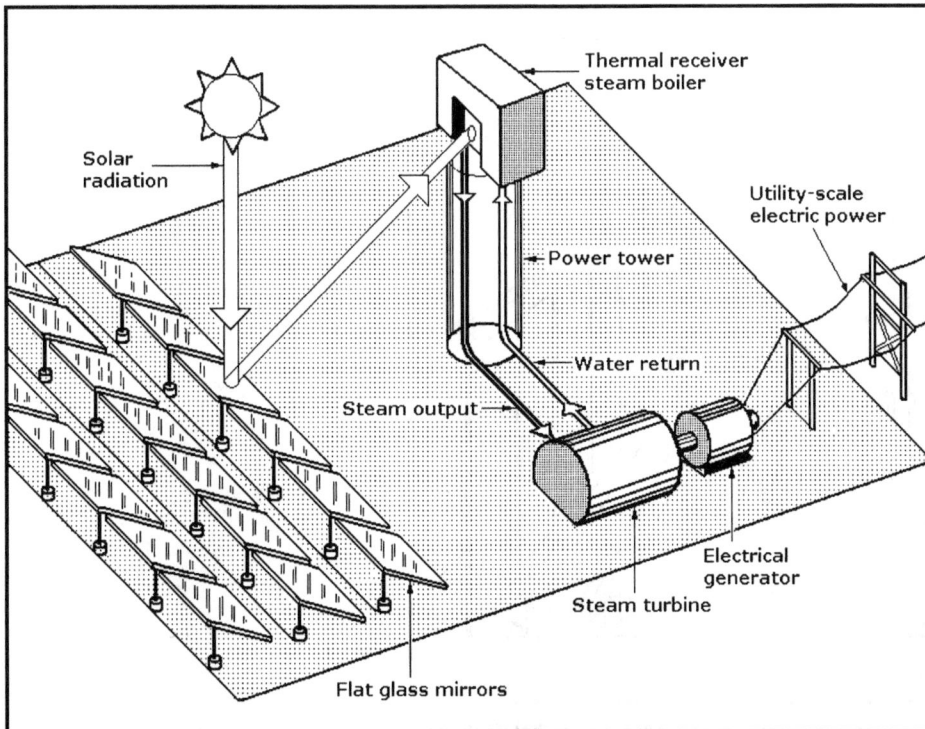

Fig. 5 A tower solar thermal power plant includes large arrays of linear flat mirrors that track the sun as it transits the sky. The mirrors focus solar thermal radiation on a water-filled receiver or boiler on top of a tall tower. The solar heat converts the water to steam, which is piped to the ground where it spins a turbine that drives a generator to produce utility-scale power. The steam is condensed back to water (the only transfer medium in the system) and it is pumped back up to the receiver to keep the process operating continuously during daylight hours.

Water circulating in pipes within the receiver atop the tower is converted directly into steam by the high temperature of the solar heat concentrated on it from the thousands of heliostats. This steam is piped down to the ground and sent thousands of feet to a common turbine which, in turn, drives a generator capable of producing utility-grade, grid-compatible electric power. The steam from the turbine is condensed back to water, which is then pumped back up to the receiver on the tower to maintain a continuous flow of steam to keep the process operational during daylight hours.

The mirrors or heliostats for this technology are smaller than those used in other CST installations; they measure only about 10 ft² (1 m²). Their lower profile and smaller size make them easier to install and maintain. A large power-tower project being built near Lancaster, California, is expected to be completed in 2012. It includes two 160-ft (50-m) high towers, each topped by a thermal receiver. Each thermal receiver can accept the intense reflected solar heat from two of the 6000 heliostat fields positioned on each side of its tower. The project's 24,000 heliostats and two towers are expected to provide enough steam to produce 245 MW of power from the generator.

The power-tower system is considered to be more cost-effective than the parabolic trough system because water is the only heat transfer medium being used. This means that the steam produced by the solar heat directly spins the turbine for driving the generator. The power-tower is a simpler system than the parabolic trough system where heat from oil at about 750°F (400°C) must be exchanged with water to produce the steam necessary to spin the turbine and drive the generator. Some power-tower systems are paired with natural gas-powered boilers, which can take over the task of making enough steam on overcast days or after the sun has set to sustain the baseload.

Plants combining renewable energy generation with a fossil-fueled plant are called *hybrid systems*.

Linear Fresnel Reflector Thermal (CST) Plants

A linear Fresnel reflector (LFR) plant (Fig. 6) is a CST facility consisting of an array of long, narrow mirrors, typically curved, to reflect solar radiation onto steam-filled mild steel pipes mounted on 50-ft (15-m) high posts above the mirrors. These pipes, called linear receivers or absorbers, are heated by solar radiation reflected from the mirrors below them and convey the hot water and steam mixtures to a heat exchanger where the steam drives a turbine generator to produce utility-grade electric power as in the other CST systems.

Steam at a temperature of 608 to 680°F (320 to 360°C) is required for large-scale installations. Both rows of mirrors and linear receiver pipes are equal in length—2000 ft (600 m). The slightly curved mirrors, laminated with a composite metal backing, will be arranged in three 666-ft (200-m) segments. Each LFR mirror segment will be rotated by a single motor and gearbox around its pitch or lateral axis. As in other CST systems, mirrors and receiver pipes will be oriented along the east–west direction. Some LFR systems have small parabolic mirrors positioned over the linear receivers to increase the heat concentration on them.

Some LFR systems have their linear receivers contained within Dewar vacuum bottle jackets. The latest versions of LFR, compact linear Fresnel reflector (CLFR) systems, improve on the original design. The acreage required for a large installation of heliostats can be reduced because they can be positioned in denser concentrations when one or more additional receivers are added. The first LFR systems were designed

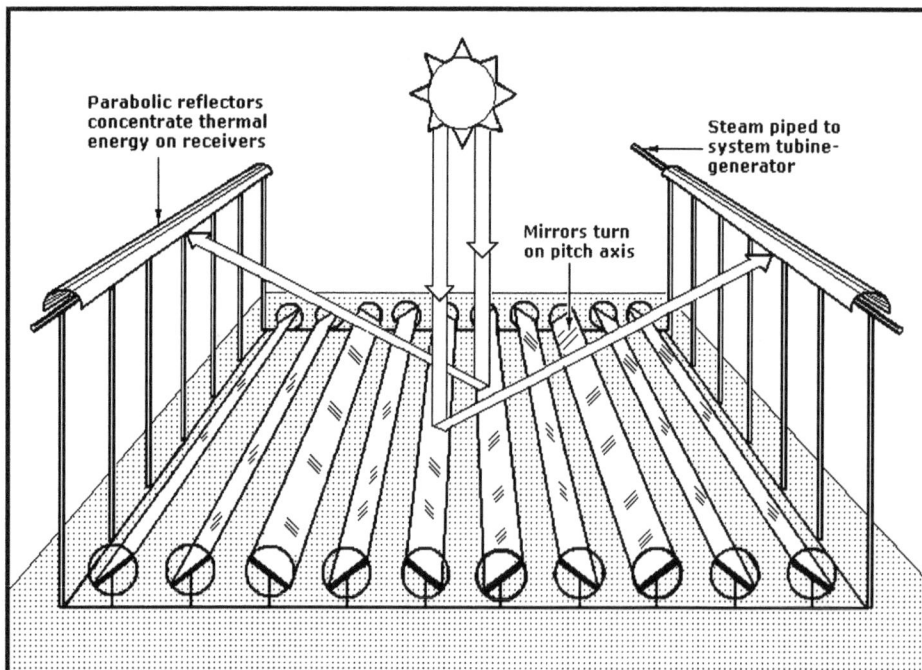

Fig. 6 A Fresnel solar thermal power plant includes large arrays of linear flat mirrors that track the sun as it transits the sky. The mirrors focus thermal radiation on water-filled pipes or receivers mounted on posts 40 to 50 ft (12 to 15 m) above the mirrors. (The mirrors can be switched to focus solar thermal energy to more than one receiver.) The water is converted to steam and, as in other solar thermal systems, it spins a turbine to drive a generator that produces utility-scale power. The piping, in a continuous loop, converts water from the turbine back to steam to keep the process operating continuously during daylight hours.

to reduce construction and operational costs because each linear receiver pipe can obtain reflected solar radiation from several mirrors simultaneously.

As in the power tower systems, LFR systems use only fresh water as the heat transfer medium, eliminating the cost of an intermediate heat transfer fluid. Because their mirror mounting frames do not have to support the receiver pipes, they can be lighter, simpler, and closer to the ground. This permits the mirrors to be cleaned manually, saving on the cost of automated washing equipment. Mirrors controlled by software can continuously switch their focus to different linear receivers to optimize efficiency.

LFR systems are expected to compete with the trough mirror and power-tower systems, but at present their progress lags behind that of other CST technologies. Prototypes have been built in Australia, Belgium, and Germany, but most projects are still in the proposal or planning stage. Consequently, it is difficult to compare the cost-effectiveness and reliability of these CST technologies at this time.

Parabolic Dish Stirling Solar Thermal (CST) Plants

A parabolic dish Stirling solar thermal plant will (when complete) consist of acres of standalone power units, each capable of producing utility-grade electric power. A unit (Fig. 7) consists of a 37-ft (11-m) wide, round parabolic mirror with a perpendicular boom that supports a Stirling engine-generator at its far end. Both dish and engine-generator are mounted atop a vertical pedestal set in the ground. The power-generating unit can be pivoted along both its pitch and yaw axes by computer-controlled motors that enable it to track the sun continuously during the day.

The sun's thermal radiation is continuously focused by the reflective dish onto the Stirling engine-generator. This intense heat expands hydrogen gas within the cylinders of the Stirling engine causing its pistons to turn a crankshaft which, in turn, drives an electric generator. An array of these power generating units must cover hundreds of acres of land to produce enough power to satisfy the grid.

A prototype dish system consisting of four standalone units has been installed at Sandia National Laboratories' Solar Thermal Test Facility near Albuquerque, New Mexico. Upgraded versions of six earlier models, they have been producing up to 150 kW of electrical power. The latest units have round parabolic dishes that are smaller and lighter than the previous rectangular dishes. Made from 40 parabolic glass and metal segments, the segments holding the mirrors are stamped from sheet steel on presses usually used to form contoured auto body parts. A round tubular skeletal frame supports the segments to make a rigid parabolic dish.

The solar Stirling engine is a closed-cycle, four-cylinder reciprocating machine that can convert the sun's heat directly into mechanical power for driving the generator. Unlike the more familiar automotive internal combustion engine, the Stirling engine does not require fuel, and it runs at lower temperatures because no internal combustion is involved. The engine is a sealed system filled with hydrogen gas, which acts as the heat transfer medium. As the gas within the engine heats and cools, its pressure rises and falls to drive the reciprocating

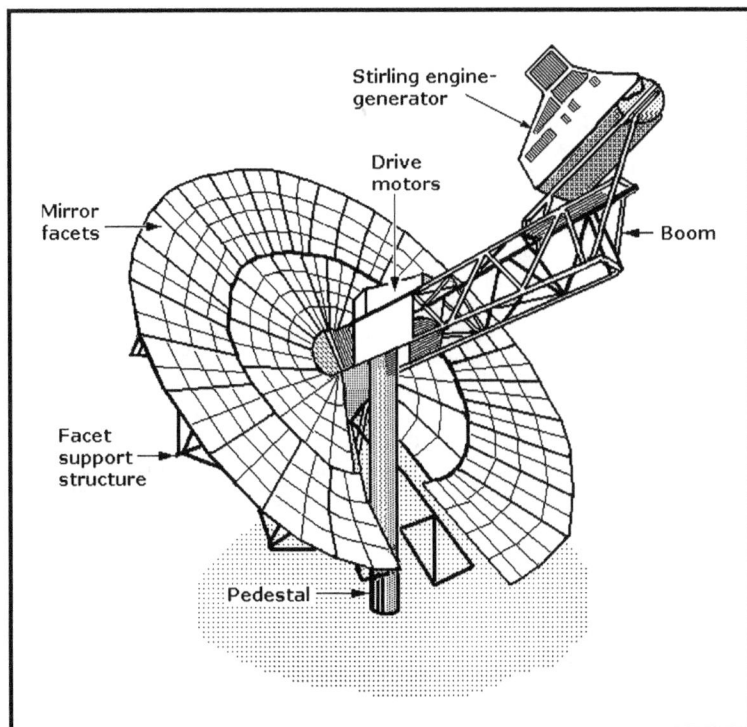

Fig. 7 A dish Stirling solar thermal system consists of a large number of independent solar thermal power generating units. Each unit has a parabolic or radial dish mirror that automatically tracks the sun and focuses its thermal radiation directly on a Stirling engine. This engine converts the intense solar heat to mechanical power that drives a generator capable of producing 25 kW of electrical power. The combined output of many of these units produces utility-scale power.

pistons of the engine; these turn the crankshaft to drive the generator. A liquid-cooling system exhausts the engine's waste heat to the atmosphere.

The higher efficiency of a parabolic dish unit results from the sun's radiation being focused directly on the Sterling engine's hydrogen tubes rather than the long liquid-filled pipes typical of other CST systems. This permits temperatures outside the engine to reach 1450°F (775°C) (compared with the 750°F [400°C] for CST liquid-filled pipes). In addition, the Stirling engine has a flat efficiency curve so it produces near maximum efficiency even when the sun is obscured or is low in the sky.

Arrays of dish Stirling units can occupy less space than the acres of heliostats. Moreover, as self-contained sources, they can start producing 1 MW of grid power when 40 or more units of a planned project are installed, hastening their return on investment. Also, if one Stirling engine-generator fails and shuts down, the total power output of the system is not seriously impacted. Another not so obvious feature is that, unlike other CST plants, the Stirling engine systems do not need a water supply, a distinct advantage if it is located in a hot desert region where water is scarce.

HOW A STIRLING ENGINE WORKS

The Stirling engine was invented by Robert Stirling in Scotland in 1816 to compete against the steam engine, but for a number of reasons, principally its slow starting, it never succeeded. Over the intervening years, the Stirling engine has demonstrated thermodynamic principles in science classes or has become an interesting subject for construction by hobbyists and engine enthusiasts. However, in recent years, its favorable characteristics, high efficiency, and the ability to run with heat from many sources gave it a new lease on life. It is now a key component in dish solar thermal units that drive a generator capable of producing utility-scale electricity.

The Stirling engine is a heat engine that works in repetitive cycles of heat expansion and cooling a constant volume of air or gas. According to the general laws of gas, pressure on a constant volume of gas is proportional to temperature, directly with an increase and inversely with a decrease. However, increasing pressure on the constant volume results in increasing its heat, while decreasing it results in cooling. This is how the Stirling engine converts heat energy into mechanical work.

Stirling engines are similar to the steam engine in that all heat flows in and out through the engine's walls. Steam engines use water in its gaseous state (steam) as the working gas. By contrast, Stirling engines are sealed systems with constant volumes of gas that always remain in their gaseous state. Both engines are *external combustion engines* that differ from *internal combustion engines*, which obtain their hot working gas from the combustion of hydrocarbon fuels within their cylinders.

Stirling engines for solar thermal systems have multiple cylinders for increased power, and hydrogen is the working gas in their sealed systems. The heat required to run the engine is obtained from a large parabolic mirror that concentrates intense solar heat energy on the engine while the cooling required is accomplished by radiators. Slow starting, a common complaint against Stirling engines is not a problem in this application. Unlike other solar thermal systems, each parabolic dish unit produces its own electrical power, which is merged to achieve the electrical grid requirements.

Two of the common configurations of the Sterling engine are the *two-piston alpha* and the *displacer-piston beta* versions. The alpha engine has two separate power pistons and the beta engine, as shown in the upper part of Fig. 8, has a single power piston in the same cylinder with a shaft coaxial with the displacer piston. The displacer has an outside diameter that is smaller than the inside diameter of the cylinder. This permits the working gas to pass freely around it while it shuttles from the hot end of the cylinder to the cooling radiator. It does not draw any power from the expanding gas.

As shown in the lower part of Fig. 8, after the working gas is pushed to the hot end of the cylinder (Figs. 8a and b) the gas expands and pushes the power piston to its limit. When the displacer pushes the gas to the cold end of the cylinder (Fig. 8c), it contracts and the momentum of the machine is enhanced by a flywheel (Fig. 8d) which pushes the power piston the other way to compress the cooled gas. The beta design avoids any problems that could be caused by the leakage of hot moving seals encountered in the alpha engines. (Note: the diagrams do not show internal heat exchangers or a regenerator, which would be located in the gas path around the displacer piston).

Fig. 8 The complete cycle of a beta-type Stirling engine.

a The power piston has compressed the gas and the displacer piston moved so most gas is pushed up against the heat source.

b The hot gas has gained pressure and it pushes the power piston to the extreme limit of the power stroke.

c The displacer piston moves to its limit pushing the gas to the cold end of the cylinder.

d The cooled gas is compressed by the momentum of flywheel, which requires less force because the gas pressure dropped.

The Outlook for CST Renewable Energy

The construction of a solar thermal farm calls for decisions to be made in engineering, economics, climate, environment, and distance to locations of high demand for electricity. A choice must be made between the four basic technologies: parabolic trough mirror, power-tower, linear Fresnel, and parabolic dish Stirling. This calls for an evaluation of their advantages and disadvantages based on efficiency, ease of construction, land use, and cost-effectiveness. At this time, none of these technologies has demonstrated its superiority over the other three.

Power-tower systems are simplest because they all have steam plumbing, but they require tall towers and individual motors to move their heliostats in two dimensions to track the sun's transit. By contrast, parabolic trough systems have more extensive oil and water plumbing but simpler heliostat drive systems that need to be moved in only one dimension. LFR systems require less acreage for a large installation because of their smaller heliostats; these can be in denser concentrations and only need movement in one dimension. Because they use water for heat transfer medium, they also require extensive steam piping. These three systems become inoperative if either a turbine or generator breaks down.

However, dish Stirling farms avoid this drawback because of their individual self-contained power generation units.

All CST plants must be located where there is constant intense sunlight during all seasons and suitable flat land. All but dish Stirling plants require a fresh water supply. This limits all CST systems in the United States to the southwestern states of Arizona, Nevada, New Mexico, and southern California. These states offer flat, hot, non-productive desert land, but these locations are distant from population centers where electricity demand is highest; this means that all will require costly long-distance cable connections to the national grid.

All CST farms require hundreds of acres of land (one planned 250-MW CST plant will occupy 1765 acres) and all will require some form of backup to maintain the baseload. It could be a hydrocarbon-fueled steam generation plant or some form of potential energy storage such as tanks of molten salt kept hot enough to produce steam or compressed air stored in a sealed cave to spin a turbine. The energy from these sources can be tapped day or night when needed.

Harnessing Moving-Water Power

People have used the mechanical power from spinning water wheels to perform such tasks as pumping out flooded fields or grinding grain for centuries. Before the industrial revolution, water wheels powered textile machines as well as machine tools from lathes, milling machines, grinders, and drills. These machines were driven by pulleys or leather belts from the rotating shafts of water wheels.

However, with the invention of the steam engine during the industrial revolution, coal-burning steam engines replaced many of the water wheels in the mill towns adjacent to a river that had formerly performed these tasks. This left hydroelectric dams as the only source of renewable power that was free of toxic smoke, carbon dioxide emissions, and concern over radiation.

Large dams such as Grand Coulee and Boulder were built to exploit the available rivers as energy sources. Despite their merits, environmental considerations and water conservation regulations have put an end to the construction of large dams in the United States. Hydroelectric dams produce less than 7 percent of utility-grade power in this country. However, within recent years other forms of moving water have become contenders for niches in the national efforts to obtain more renewable, carbon-free sources of electricity. Power generated by tides and ocean waves could compete with wind turbines and solar energy in the future

Tidal Electric Power Generation

No utility-grade ocean waves or tidal power generation stations are now operating in the United States. Many experimental prototypes of underwater turbines have been proposed with either horizontal or vertical configurations, but none have resulted in working facilities. However, there are plans in both the United States and the United Kingdom for large numbers or "fields" of these machines.

Because tides are more predictable than the wind, they can be relied upon to supply power in two directions at predictable times of day. However, a major drawback of any tidal power station is that it can only generate power when the tide is flowing in or out; in many parts of the world this is for only for 10 hours each day. To overcome this handicap, networks of tidal power stations at different locations can be formed to keep power flowing at those times when one tidal station is out of action.

Among the many schemes for harnessing electrical energy from the motion of tides, there is the construction of a tidal barrage or tidal fence (a smaller tidal barrage). A tidal barrage is essentially a dam that spans a river estuary. When the tide goes in and out, the water flows through tunnels in the barrage. This ebb and flow spins a turbine directly or it pushes air through a pipe to spin a turbine.

However, individual underwater turbines are seen as the best prospects for tidal generation. Existing models look and act like underwater wind turbines. Since water is about 800 times denser than air, the slow-moving ebb and flow of tides can exert much greater forces than wind on turbines. Consequently, a tidal turbine can have rotors with smaller diameter rotors than wind turbines and still produce the same power output. However, there are concerns about tidal generation because the turbines must be located near the mouths of rivers or estuaries where they could interfere with ship navigation, recreational boating, and commercial and recreational fishing while also being threats to migrating marine wildlife.

Ocean-Wave Power Generation

Ocean-wave electric power generators are seen as a more acceptable alternative to tidal electrical power generation because they would be located offshore in the open ocean. Arrays or fields of floating wave-energy buoys would not interfere with shipping and known fishing grounds. Power from these buoys would be transmitted back to a shore station where it would be fed into the grid. Because of the challenging ocean environment, these buoys must be simple, rugged, and reliable enough to withstand the rigors of storms with hurricane force while remaining securely anchored to the sea floor.

An ingenious wave-energy buoy based on a solenoid (Fig. 9) was developed by Dr. Annette von Jouanne, an engineering professor at Oregon State University. This second modification looks like a large yellow life ring that rises and falls with wave motion over a cylindrical spar projecting through the opening in the ring. The spar would be securely anchored to the sea floor, essentially keeping it in a fixed upright position; it has specially designed copper coils wound around its upper end. The bobbing ring-shaped float encloses a tubular ring of magnets. When the ocean waves move the float up and down over the spar in close proximity to the magnets, their field induces a voltage in the coils of the spar. According to Dr. von Jouanne, this direct-drive linear generator is capable of producing 10 kW of electric power.

A field of similar buoys would be located in an array so that the sum of their output voltages would achieve the power level required by the grid. The current from each buoy would be conducted by cable down to the seafloor where it will be merged with the output of other buoys before being sent by a larger cable to a shore station. Ocean scientists say this concept has an advantage over renewable energy schemes like wind and solar power because it would be available 24 hours a day. They also say that ocean waves have an energy density as much as 50 times greater than wind.

Some of the downsides of arrays of wave-energy buoys are the same as those for tidal turbines. Even if located miles off shore away from shipping channels, those areas would have to be cordoned off and equipped with hazard markers to prevent collisions with fishing vessels and small boats not confined to shipping lanes. Moreover, little is known about how the buoys of this design would withstand long-term battering by heavy storms or how often marine growth would impede their operation. Studies are needed to determine their impact on marine life, whale migrations, and commercial fishing.

Another Possible Mechanical Hydropower Solution

A new concept for generating electric power from ocean currents, tides, and waves would depend on pressurized water or hydraulic fluid from underwater turbine-pumps. In this scheme, the pressurized liquid would be piped above the water to drive generators on land or offshore platforms. As proposed by Caltech researchers, the main advantage this scheme would have over existing or proposed hydropower systems is that all electrical generation and transmission facilities are kept out of the water. This would eliminate the need for periodic removal of

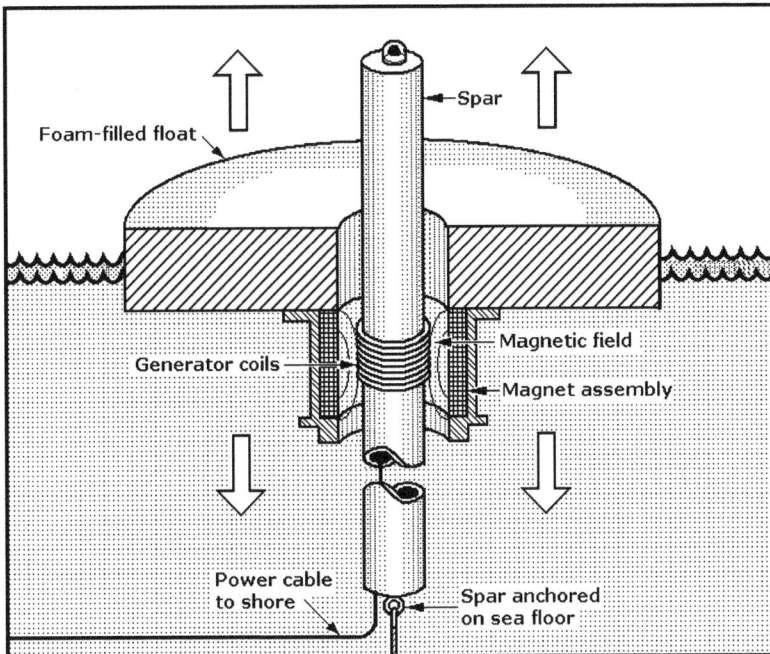

Fig. 9 A wave-energy buoy consists of two components: a float including a tubular permanent magnet and an axial spar containing a copper generator coil. Wave action causes the float to bob up and down over an essentially motionless spar anchored to the sea floor. In this linear direct-drive system, the magnet's field interacts with the coil to induce a voltage capable of producing 10 kW. A cluster of these buoys can produce utility-scale power, which is transmitted by cable to a shore station.

accumulated marine growth from cables and the threat of possible seawater leakage that could damage or destroy an electric generator.

The submerged turbine-pumps (similar in appearance to wind turbines) would capture energy from moving water and drive geared pumps capable of pressurizing fluids to 3000 lb/in.^{2} (20.7 mpA). The pressurized fluids would then be pumped through 3-ft (1-m) diameter pipes to shore or ocean platform-based turbine-generators. The fluids would spin the generators to produce utility-grade electric power.

Hydraulic fluid, if used, could be recirculated between the submerged units and the above-water turbine-generators in a closed-flow system. However, if seawater is used, it could be taken from the ocean by submerged turbine-pumps and discharged back into the ocean above water at the power-generation station. However, pressurized liquids could also be used to fill elevated above-water tanks or other pumped-storage facilities for release to drive turbine-generators during high-load periods.

A large number of submerged turbine-pump units would be needed to carry out this concept. Also, the units could be positioned so that pressurized output flow from separate pipes could be combined into a wider pipe for more efficient delivery to the power-generation or pumped-storage facilities. This concept, however, would have to deal with the same problems associated with locating underwater turbine-generators on the seafloor and avoiding interference with navigation, fishing, and marine life.

The Relative Costs of Renewable Energy

The EIA estimates that a kilowatt-hour (kWh) of electricity from a photovoltaic solar plant entering service in 2016 will cost 40 cents (in 2009 dollars). That is three to five times the projected cost of electricity from the non-renewables, coal, natural gas, and nuclear (uranium) plants. Wind power, by this measure, will cost 12 cents per kWh. Subsidies from the U.S. government are expected to cover 30 percent of capital costs of renewable energy projects, bringing them closer to the prevailing prices per kilowatt-hour of the non-renewables.

An important consideration in plans to introduce large numbers of renewable energy power plants into this country is the present inadequacy of the national power grid. It will be difficult to accommodate many of them because of their remote locations. The lack of adequate high-voltage transmission lines in the United States and the cost of extending them could inhibit the growth of renewable power. According to a Joint Coordinated System Plan from the Midwest Independent System Operator (MISO), the nation will need 15,000 mi (27,000 km) of new transmission lines costing $80 billion. These transmission lines are likely to be rated for 765,000 volts.

According to the EIA, the United States will require the addition of 250 GW of new generation capacity between 2009 and 2035. Of this new capacity, 44 percent or 110 GW will be coal-based. But to comply with state-level renewable energy requirements, the nation will add only 38 GW of renewable energy capacity.

Large fossil-fuel plants remain attractive in all parts of the world in countries such as China and India due to their low initial capital costs; consequently, they are still being built there. Nevertheless, while non-renewable power plant prices are rising, those for renewable plants are falling. This, however, does not mean that the cost of replacing all non-renewable carbon-emitting plants will be significantly less. Moreover, it also does not mean that this replacement could be accomplished in the next 50 years—even if there is the political will to do so.

GLOSSARY OF WIND TURBINE TERMS

anemometer: A standard meteorological instrument mounted on the top rear of a wind turbine nacelle to measure wind speed. Its data output is sent to the controller for changing rotor blade settings in response to changes in wind speed. It is usually paired with a wind vane for measuring wind direction.

availability factor: A measure of the reliability of a wind turbine (or other power plant). The percentage of time a plant is ready to generate electrical power. Manufacturers claim that utility-scale wind turbines can have availability exceeding 90 percent.

blades: The parts of the rotor assembly with airfoil contours (similar to aircraft propeller blades). Lengths can exceed 150 ft (45 m), so rotor diameters can exceed 300 ft (90 m). Utility-scale turbines rated 10 kW and higher typically have three blades, but smaller-scale units typically have two. Air flow over the blades creates lift force causing rotor rotation. Blades are being made of glass-fiber epoxy or carbon-fiber epoxy.

brakes: Mechanisms for stopping rotor rotation. They are applied if wind speed exceeds the "cut-out" value, or for maintenance or in emergencies. Some turbines have two brakes, one on the low-speed rotor shaft and the other for the high-speed generator shaft.

capacity factor (CF): A measure of the productivity of a wind turbine (or other power plants including fossil-fueled, nuclear, and solar thermal) stated as:

$$\frac{\text{Actual power produced (set time period)}}{\text{Maximum power that could be produced (same time period)}}$$

CF for a typical fossil-fueled plant: 40–80 percent; CF for utility-scale wind turbines: 25–40 percent.

controller: A computer-based control center for wind turbines that is programmed to direct turbine functions. It can start rotor rotation at "cut-in" wind speed and shut it down at "cut-out" wind speed to prevent damage to the turbine. (Strong wind can bend the blades back far enough to strike the tower.) The controller receives data from anemometers, wind vanes, and commands radioed from ground-based wind-farm control stations that can override routine programs in response to extreme wind conditions.

cut-in wind speed: The minimum wind speed at which useful power can be generated by the turbine. This value is from 8 to 16 mph (13 to 26 km/h)

cut-out wind speed: The maximum wind speed for safe wind turbine operation is about 50 mph (80 km/h).

gear box: A set of gears that increases the low speed of the rotor shaft (30 to 60 rpm) to the high speed required to drive the generator shaft (1000 to 1800 rpm) so it can produce electricity. (Some wind turbines with generators capable of producing electricity at lower generator shaft speeds have eliminated gear boxes, reducing both weight and cost.)

generator: Typically a commercial induction generator that can produce alternating current (AC) at frequencies of either 50 or 60 Hz.

hub height: The distance from the ground to the center line of the rotor hub, a function of the tower height that can exceed 300 ft (90 m).

nacelle: The streamlined equipment housings mounted on bearings at the top of the tower. It encloses the main bearings, low-speed shaft gear box, brakes generator, controller, transformers, and sometimes a crane. In "upwind" turbines, it is rotated by one or more geared yaw motors that keep the rotor facing into the wind.

pitch: A term for rotating the rotor blades in their hub. The blades are "pitched" when directed into the wind to obtain optimum rotor speed or "feathered" to stop rotor rotation by eliminating "lift" when wind speed is excessive for safe turbine operation or too low to produce electricity.

rotor: The assembly of blades (typically three in utility-scale wind turbines) and hub in a configuration similar to that of a three-bladed aircraft propeller.

tower: The tall hollow column that supports the nacelle that can be made from steel or reinforced concrete depending on the turbine's location. The taller the tower, the more efficient the electrical generation becomes because wind speed increases with elevation.

turbine design: "Upwind" turbines face the wind; "downwind" turbines face 180° from the wind.

utility-scale turbine power range: 700 kW to 5 Mw.

wind vane: A meteorological instrument that measures wind direction and provides a signal to the controller to operate the yaw drives to keep the rotor facing into the wind.

yaw drive: Motor-driven gearing that rotates the nacelle and rotor of "upwind" turbines in response to changes in wind direction. "Downwind" turbines do not have these drives because their nacelles are designed to act as weather vanes. This permits the nacelle to conform automatically to changes in wind direction.

Renewable Energy Resources

Magazine Articles

Discover (June 2009) Rising Power, Future of Energy, pp. 45–48

Popular Mechanics (Nov 2008) Solar's New Dawn, pp. 63–67
(Dec 2009) The New Wildcatters (wind farms), pp. 91–95
(Feb 2010) The Myth of Clean Coal, pp. 50–51
(May 2010) Tech Watch, p. 17
(July 2010) Energyland: The Race to Cheap, Sustainable Power, pp. 70–78p

Scientific American (Mar 2009) Generating Electrical Power, pp. 56–60
(Oct 2009) The Way the Wind Blows, pp. 27–28
(Nov 2009) A Path to Sustainable Energy by 2030, pp. 58–65

Smithsonian (July/Aug 2009) Catching a Wave, pp. 67–71, Shining Example, pp. 114–119
(Aug 2010) Solar's Great Leap Forward, pp. 52–57

Technology Review (Mar/Apr 2009) Green Nuclear, p.10
(Sept/Oct 2009) Can Renewables Become More than a Sideshow? pp. 91–101, Cleaning Coal, p. 10
(Nov/Dec 2009) Natural Gas Changes the Energy Map, pp. 44–50
(Dec 2010) Giant Holes in the Ground, pp. 61–65,
How Not to Make Energy Decisions pp. 79–81

Web Sites

American Wind Energy Association (AWEA)
www.awea.org

Edison Electric Institute (EEI)
www.eei.org

New York State Energy Research and Development Authority (NYSERDA)
www.powernaturally.org

World of Renewables
www.worldofrenewables.com

World of Solar Energy
www.worldofsolarthermal.com

World of Wind Energy
www.worldofwindenergy.com

CHAPTER 5
LINKAGES: DRIVES AND MECHANISMS

FOUR-BAR LINKAGES AND TYPICAL INDUSTRIAL APPLICATIONS

All mechanisms can be broken down into equivalent four-bar linkages. They can be considered to be the basic mechanism and are useful in many mechanical operations.

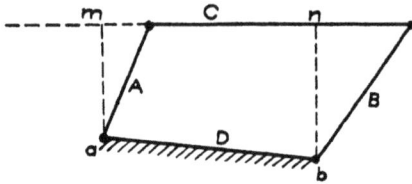

FOUR-BAR LINKAGES—Two cranks, a connecting rod and a line between the fixed centers of the cranks make up the basic four-bar linkage. Cranks can rotate if A is smaller than B or C or D. Link motion can be predicted.

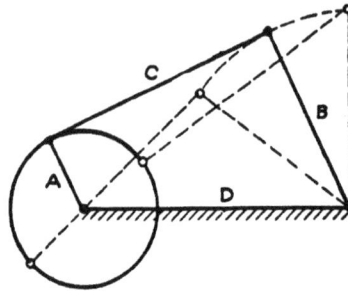

PARALLEL CRANK FOUR-BAR—Both cranks of the parallel crank four-bar linkage always turn at the same angular speed, but they have two positions where the crank can-not be effective.

NON-PARALLEL EQUAL CRANK—The centrodes are formed as gears for passing dead center and they can replace ellipticals.

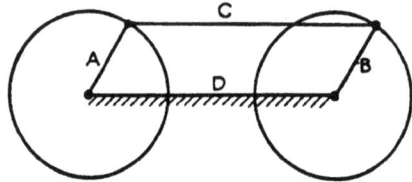

DOUBLE PARALLEL CRANK MECHANISM—This mechanism forms the basis for the universal drafting machine.

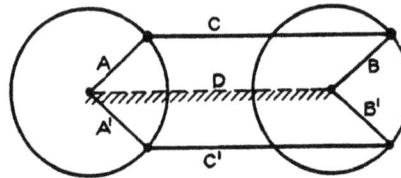

CRANK AND ROCKER—The following relations must hold for its operation: $A + B + C > D$; $A + D + B > C$; $A + C - B < D$, and $C - A + B > D$.

DOUBLE PARALLEL CRANK—This mechanism avoids a dead center position by having two sets of cranks at 90° advancement. The connecting rods are always parallel.

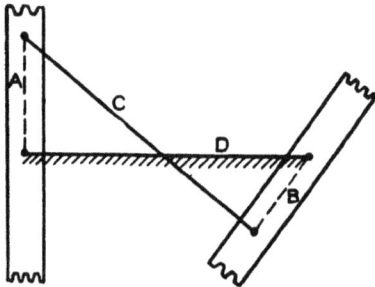

SLOW MOTION LINK—As crank A is rotated upward it imparts motion to crank B. When A reaches its dead center position, the angular velocity of crank B decreases to zero.

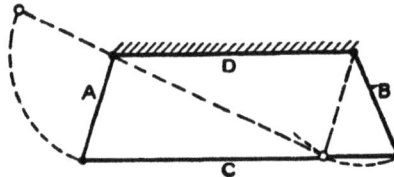

ISOSCELES DRAG LINKS—This "lazy-tong" device is made of several isosceles links; it is used as a movable lamp support.

FOUR-BAR LINK WITH SLIDING MEMBER—One crank is replaced by a circular slot with an effective crank distance of B.

PARALLEL CRANK—Steam control linkage assures equal valve openings.

TRAPAZOIDAL LINKAGE—This linkage is not used for complete rotation but can be used for special control. The inside moves through a larger angle than the outside with normals intersecting on the extension of a rear axle in a car.

WATT'S STRAIGHT-LINE MECHANISM—Point T describes a line perpendicular to the parallel position of the cranks.

STRAIGHT SLIDING LINK—This is the form in which a slide is usually used to replace a link. The line of centers and the crank *B* are both of infinite length.

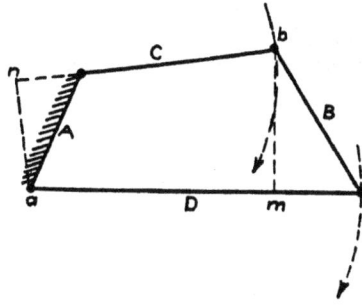

DRAG LINK—This linkage is used as the drive for slotter machines. For complete rotation: $B > A + D - C$ and $B < D + C - A$.

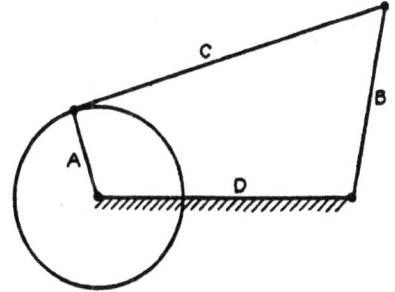

ROTATING CRANK MECHANISM—This linkage is frequently used to change a rotary motion to a swinging movement.

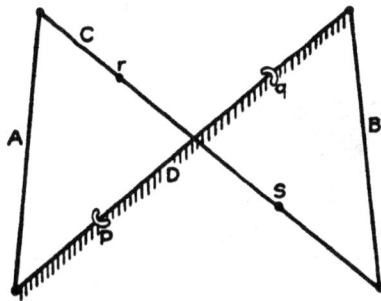

NON-PARALLEL EQUAL CRANK—If crank *A* has a uniform angular speed, *B* will vary.

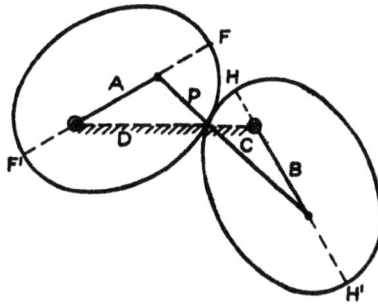

ELLIPTICAL GEARS—They produce the same motion as non-parallel equal cranks.

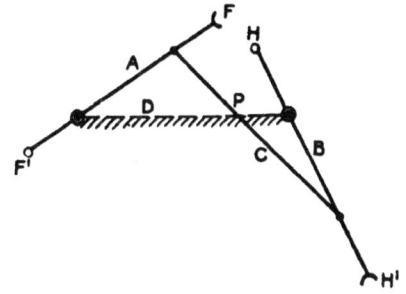

NON-PARALLEL EQUAL CRANK—It is the same as the first example given but with crossover points on its link ends.

TREADLE DRIVE—This four-bar linkage is used in driving grinding wheels and sewing machines.

DOUBLE LEVER MECHANISM—This slewing crane can move a load in a horizontal direction by using the D-shaped portion of the top curve.

PANTOGRAPH—The pantograph is a parallelogram in which lines through *F*, *G*, and *H* must always intersect at a common point.

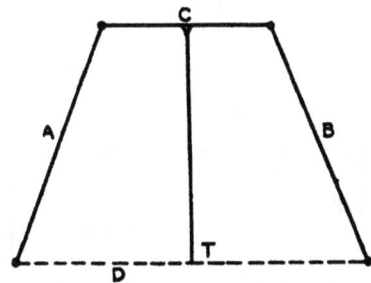

ROBERT'S STRAIGHT-LINE MECHANISM—The lengths of cranks *A* and *B* should not be less than 0.6 *D*; *C* is one half of *D*.

TCHEBICHEFF'S—Links are made in proportion: $AB = CD = 20$, $AD = 16$, $BC = 8$.

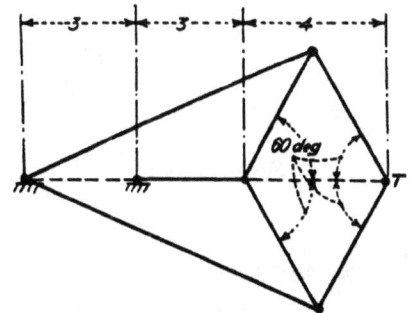

PEAUCELLIER'S CELL—When proportioned as shown, the tracing point *T* forms a straight line perpendicular to the axis.

SEVEN LINKAGES FOR TRANSPORT MECHANISMS

Fig. 1 In this design a rotary action is used. The shafts *D* rotate in unison and also support the main moving member. The shafts are carried in the frame of the machine and can be connected by either a link, a chain and sprocket, or by an intermediate idler gear between two equal gears keyed on the shafts. The rail *A-A* is fixed rigidly on the machine. A pressure or friction plate can hold the material against the top of the rail and prevent any movement during the period of rest.

Transport mechanisms generally move material. The motion, although unidirectional, gives an intermittent advancement to the material being conveyed. The essential characteristic of such a motion is that all points in the main moving members follow similar and equal paths. This is necessary so that the members can be subdivided into sections with projecting parts. The purpose of the projections is to push the articles during the forward motion of the material being transported. The transport returns by a different path from the one it followed in its advancement, and the material is left undisturbed until the next cycle begins. During this period of rest, while the transport is returning to its starting position, various operations can be performed sequentially. The selection

of the particular transport mechanism best suited to any situation depends, to some degree, on the arrangement that can be obtained for driving the materials and the path desired. A slight amount of overtravel is always required so that the projection on the transport can clear the material when it is going into position for the advancing stroke.

The designs illustrated here have been selected from many sources and are typical of the simplest solutions of such problems. The paths, as indicated in these illustrations, can be varied by changes in the cams, levers, and associated parts. Nevertheless, the customary cut-and-try method might still lead to the best solution.

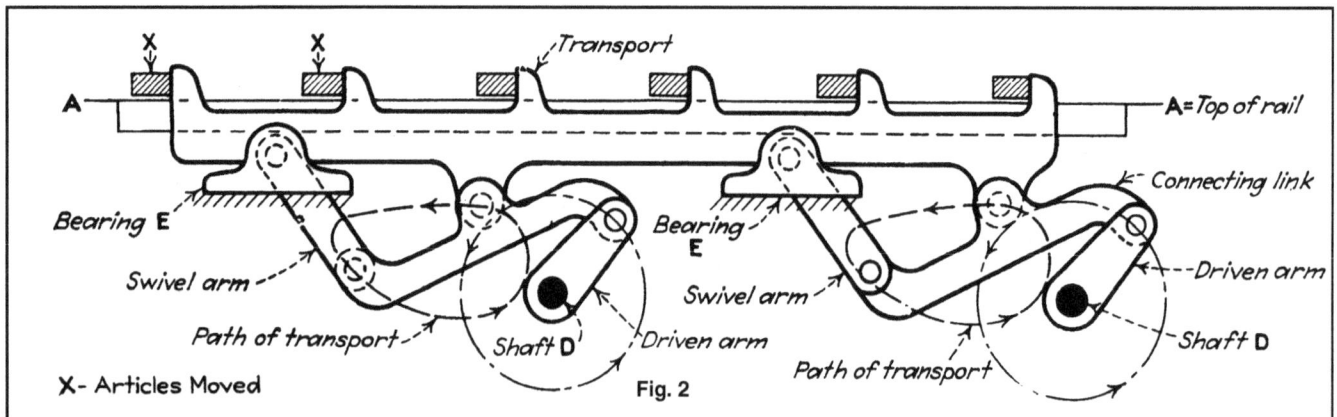

Fig. 2 Here is a simple form of linkage that imparts a somewhat "egg-shaped" motion to the transport. The forward stroke is almost a straight line. The transport is carried on the connecting links. As in the design of Fig. 1, the shafts *D* are driven in unison and are supported in the frame of the machine. Bearings *E* are also supported by the frame of the machine and the rail *A-A* is fixed.

Fig. 3 In another type of action, the forward and return strokes are accomplished by a suitable mechanism, while the raising and lowering is imparted by a friction slide. Thus it can be seen that as the transport supporting slide *B* starts to move to the left, the friction slide *C*, which rests on the friction rail, tends to remain at rest. As a result, the lifting lever starts to turn in a clockwise direction. This motion raises the transport which remains in its raised position against stops until the return stroke starts. At that time the reverse action begins. An adjustment should be provided to compensate for the friction between the slide and its rail. It can readily be seen that this motion imparts a long straight path to the transport.

Fig. 4 This drawing illustrates an action in which the forward motion is imparted by an eccentric while the raising and lowering of the transport is accomplished by a cam. The shafts, *F, E,* and *D,* are positioned by the frame of the machine. Special bell cranks support the transport and are interconnected by a tierod.

Fig. 5 This is another form of transport mechanism based on a link motion. The bearings *C* are supported by the frame as is the driving shaft *D.*

Fig. 6 An arrangement of interconnected gears with equal diameters that will impart a transport motion to a mechanism. The gear and link mechanism imparts both the forward motion and the raising and lowering motions. The gear shafts are supported in the frame of the machine.

Fig. 7 In this transport mechanism, the forward and return strokes are accomplished by the eccentric arms, while the vertical motion is performed by the cams.

FIVE LINKAGES FOR STRAIGHT-LINE MOTION

These linkages convert rotary to straight-line motion without
the need for guides.

Fig. 1 An Evans' linkage has an oscillating drive-arm that should have a maximum operating angle of about 40°. For a relatively short guideway, the reciprocating output stroke is large. Output motion is on a true straight line in true harmonic motion. If an exact straight-line motion is not required, a link can replace the slide. The longer this link, the closer the output motion approaches that of a true straight line. If the link-length equals the output stroke, deviation from straight-line motion is only 0.03 percent of the output stroke.

Fig. 2 A simplified Watt's linkage generates an approximate straight-line motion. If the two arms are of equal length, the tracing point describes a symmetrical figure 8 with an almost straight line throughout the stroke length. The straightest and longest stroke occurs when the connecting-link length is about two-thirds of the stroke, and arm length is 1.5 times the stroke length. Offset should equal half the connecting-link length. If the arms are unequal, one branch of the figure-8 curve is straighter than the other. It is straightest when a/b equals (arm 2)/(arm 1).

Fig. 3 Four-bar linkage produces an approximately straight-line motion. This arrangement provides motion for the stylus on self-registering measuring instruments. A comparatively small drive displacement results in a long, almost-straight line.

Fig. 4 A D-drive is the result when linkage arms are arranged as shown here. The output-link point describes a path that resembles the letter *D*, so there is a straight part of its cycle. This motion is ideal for quick engagement and disengagement before and after a straight driving stroke.

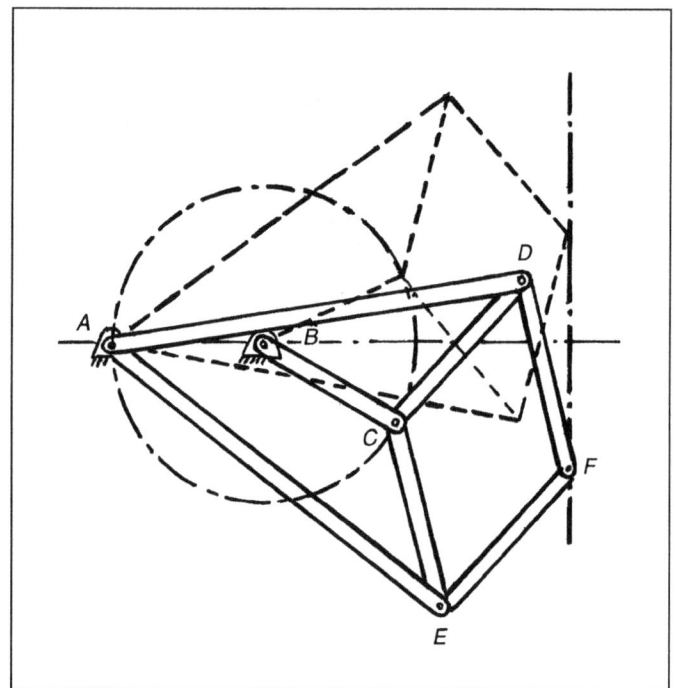

Fig. 5 The "Peaucellier cell" was the first solution to the classical problem of generating a straight line with a linkage. Within the physical limits of the motion, $AC \times AF$ remains constant. The curves described by *C* and *F* are, therefore, inverse; if *C* describes a circle that goes through *A*, then *F* will describe a circle of infinite radius—a straight line, perpendicular to *AB*. The only requirements are that: $AB = BC$; $AD = AE$; and *CD*, *DF*, *FE*, *EC* be equal. The linkage can be used to generate circular arcs of large radius by locating *A* outside the circular path of *C*.

SIX EXPANDING AND CONTRACTING LINKAGES

Parallel bars, telescoping slides, and other devices that can spark answers to many design problems.

Figs. 1 and 2 **Expanding grilles** are often put to work as a safety feature. A single parallelogram (Fig. 1) requires slotted bars; a double parallelogram (Fig. 2) requires none—but the middle grille-bar must be held parallel by some other method.

Fig. 1

Fig. 2

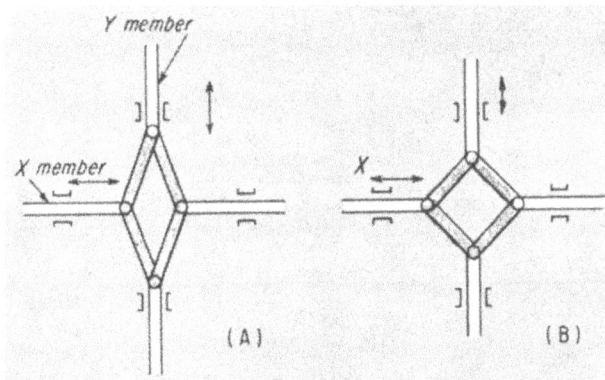

Fig. 3 **Variable motion** can be produced with this arrangement. In (A) position, the Y member is moving faster than the X member. In (B), speeds of both members are instantaneously equal. If the motion is continued in the same direction, the speed of X will become greater.

Fig. 4

Fig. 5

Fig. 6

Figs. 4, 5, and 6 **Multibar** barriers such as shutters and gates (Fig. 4) can take various forms. Slots (Fig. 5) allow for vertical adjustment. The space between bars can be made adjustable (Fig. 6) by connecting the vertical bars with parallel links.

FOUR LINKAGES FOR DIFFERENT MOTIONS

Fig. 1 No linkages or guides are included in this modified hypocyclic drive which is relatively small in relation to the length of its stroke. The sun gear of pitch diameter D is stationary. The drive shaft, which turns the T-shaped arm, is concentric with this gear. The idler and planet gears, with pitch diameters of $D/2$, rotate freely on pivots in the arm extensions. The pitch diameter of the idler has no geometrical significance, although this gear does have an important mechanical function. It reverses the rotation of the planet gear, thus producing true hypocyclic motion with ordinary spur gears only. Such an arrangement occupies only about half as much space as does an equivalent mechanism containing an internal gear. The center distance R is the sum of $D/2$, $D/4$, and an arbitrary distance d, determined by specific applications. Points A and B on the driven link, which is fixed to the planet, describe straight-line paths through a stroke of $4R$. All points between A and B trace ellipses, while the line AB envelopes an astroid.

Fig. 1

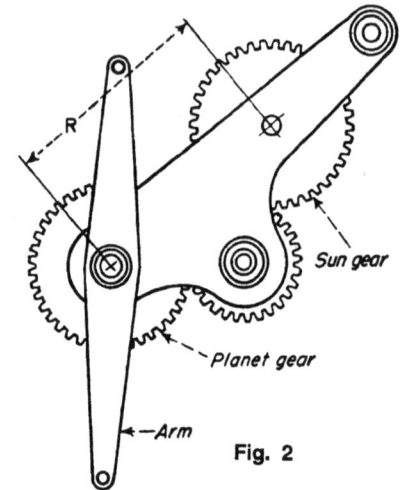

Fig. 2

Fig. 2 A slight modification of the mechanism in Fig. 1 will produce another type of useful motion. If the planet gear has the same diameter as that of the sun gear, the arm will remain parallel to itself throughout the complete cycle. All points on the arm will thereby describe circles of radius R. Here again, the position and diameter of the idler gear have no geometrical importance. This mechanism can be used, for example, to cross-perforate a uniformly moving paper web. The value for R is chosen so that $2\pi R$, or the circumference of the circle described by the needle carrier, equals the desired distance between successive lines of perforations. If the center distance R is made adjustable, the spacing of perforated lines can be varied as desired.

Fig. 3

Fig. 3 To describe a "D" curve, begin at the straight part of path G, and replace the oval arc of C with a circular arc that will set the length of link DC.

Fig. 4

Fig. 4 This mechanism can act as a film-strip hook that will describe a nearly straight line. It will engage and disengage the film perforation in a direction approximately normal to the film. Slight changes in the shape of the guiding slot f permit the shape of the output curve and the velocity diagram to be varied.

NINE LINKAGES FOR ACCELERATING AND DECELERATING LINEAR MOTIONS

When ordinary rotary cams cannot be conveniently applied,
the mechanisms presented here, or adaptations of them, offer
a variety of interesting possibilities for obtaining either
acceleration or deceleration, or both.

Fig. 1

Fig. 1 **A slide block** with a pinion and shaft and a pin for link B reciprocates at a constant rate. The pinion has a crankpin for mounting link D, and it also engages a stationary rack. The pinion can make one complete revolution at each forward stroke of the slide block and another as the slide block returns in the opposite direction. However, if the slide block is not moved through its normal travel range, the pinion turns only a fraction of a revolution. The mechanism can be made variable by making the connection link for F adjustable along the length of the element that connects links B and D. Alternatively, the crankpin for link D can be made adjustable along the radius of the pinion, or both the connection link and the crankpin can be made adjustable.

Fig. 2 **A drive rod,** reciprocating at a constant rate, rocks link BC about a pivot on a stationary block. A toggle between arm B and the stationary block contacts an abutment. Motion of the drive rod through the toggle causes deceleration of driven link B. As the drive rod moves toward the right, the toggle is actuated by encountering the abutment. The slotted link BC slides on its pivot while turning. This lengthens arm B and shortens arm C of link BC. The result is deceleration of the driven link. The toggle is returned by a spring (not shown) on the return stroke, and its effect is to accelerate the driven link on its return stroke.

Fig. 3 **The same direction** of travel for both the drive rod and the drive link is provided by the variation of the Fig. 2 mechanism. Here, acceleration is in the direction of the arrows, and deceleration occurs on the return stroke. The effect of acceleration decreases as the toggle flattens.

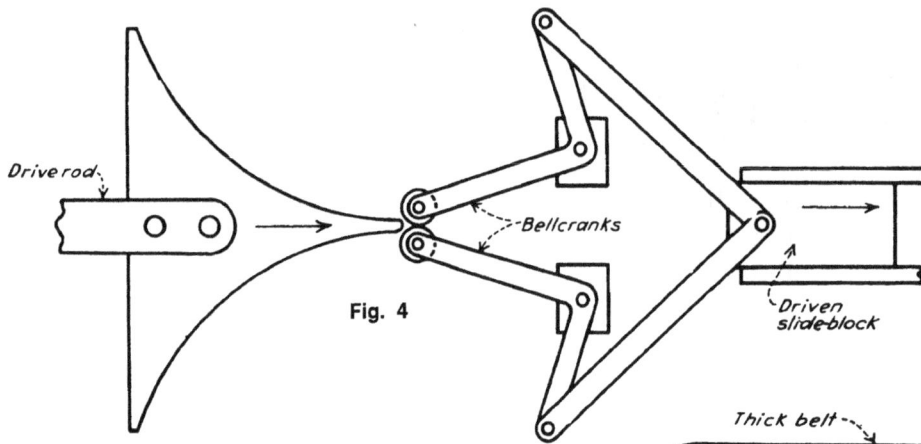

Fig. 4 A bellcrank motion is accelerated as the rollers are spread apart by a curved member on the end of the drive rod, thereby accelerating the motion of the slide block. The driven elements must be returned by spring to close the system.

Fig. 5 A constant-speed shaft winds up a thick belt or similar flexible connecting member, and its effective increase in radius causes the slide block to accelerate. It must be returned by a spring or weight on its reversal.

Fig. 6 An auxiliary block that carries sheaves for a cable which runs between the driving and driven slide block is mounted on two synchronized eccentrics. The motion of the driven block is equal to the length of the cable paid out over the sheaves, resulting from the additive motions of the driving and auxiliary blocks.

Fig. 7 A curved flange on the driving slide block is straddled by rollers that are pivotally mounted in a member connected to the driven slide block. The flange can be curved to give the desired acceleration or deceleration, and the mechanism returns by itself.

Fig. 8 The stepped acceleration of the driven block is accomplished as each of the three reciprocating sheaves progressively engages the cable. When the third acceleration step is reached, the driven slide block moves six times faster than the drive rod.

Fig. 9 A form-turned nut, slotted to travel on a rider, is propelled by reversing its screw shaft, thus moving the concave roller up and down to accelerate or decelerate the slide block.

TWELVE LINKAGES FOR MULTIPLYING SHORT MOTIONS

Shaft rotated by diaphragm thrust on adjustable ball head screw

Pointer Detail

Counterweight

Diaphragm

Pin on shaft turns pointer

Counterweight

Fig. 1 A lever-type transmission in a pressure gage.

Actuating pin on diaphragm

Double cam

Actuating cam

Reset

Lever arm actuates cam on pointer shaft

Pointer shaft

Re-set cam

Fig. 2 A lever and cam drive for a tire gage.

Rocker arm with lever adjustment

Diaphragm stop

Sector gear shaft rotated by lever on rocker shaft

Rocker shaft frame adjustment

Adjustable lever arm on rocker shaft

Fig. 4 A sector gear drive for an aircraft airspeed indicator.

Sector gear turns pointer

Differential pressure lever

Fig. 3 A lever and sector gear in a differential pressure gage.

Pointer

Cord

Glass disk

Cam

Lever

Spring

Diaphragm moves against lever pin

Fig. 5 A lever, cam, and cord transmission in a barometer.

Fig. 6   **A link and chain transmission** for an aircraft rate of climb instrument.

Fig. 7   **A lever system** in an automobile gasoline tank.

Fig. 8   **Interfering magnetic fields** for fluid pressure measurement.

Fig. 9   **A lever system** for measuring atmospheric pressure variations.

Fig. 10   **A lever and chain transmission** for a draft gage.

Fig. 11   **A toggle and cord drive** for a fluid pressure measuring instrument.

Fig. 12   **A spiral feed transmission** for a general purpose analog instrument.

FOUR PARALLEL-LINK MECHANISMS

Eight-bar linkage

Link *AB* in this arrangement will always be parallel to *EF*, and link *CD* will always be parallel to *AB*. Hence *CD* will always be parallel to *EF*. Also, the linkages are so proportioned that point *C* moves in an approximately straight line. The final result is that the output plate will remain horizontal while moving almost straight up and down. The weight permitted this device to function as a disappearing platform in a theater stage.

Double-handed screw mechanism

Turning the adjusting screw spreads or contracts the linkage pairs to raise or lower the table. Six parallel links are shown, but the mechanism can be built with four, eight, or more links.

Tensioning mechanism

A simple parallel-link mechanism that produces tension in webs, wires, tapes, and strip steels. Adjusting the weight varies the drag on the material.

Triple-pivot mechanism

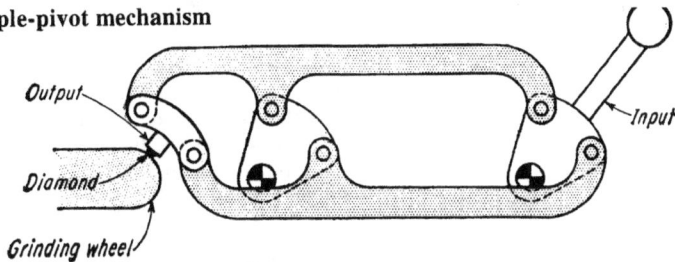

Two triangular plates pivot around fixed points on a machine frame. The output point describes a circular-arc curve. It can round out the cutting surfaces of grinding wheels.

SEVEN STROKE MULTIPLIER LINKAGES

Reciprocating-table drive

Two gears rolling on a stationary bottom rack drive the movable top rack, which is attached to a printing table. When the input crank rotates, the table will move out to a distance of four times the crank length.

Parallel-link feeder

Barrel | Feed bar | Input

One of the cranks is the input, and the other follows to keep the feeding bar horizontal. The feeder can move barrels from station to station.

Parallelogram linkage

Vertical links | Carton feeding | Carton opened

All seven short links are kept in a vertical position while rotating. The center link is the driver. This particular machine feeds and opens cartons, but the mechanism will work in many other applications.

Parallel-link driller

Input crank | Eccentric plate | Cranks | Output shafts

This parallel-link driller powers a group of shafts. The input crank drives the eccentric plate. This, in turn, rotates the output cranks that have the same length at the same speed. Gears would occupy more room between the shafts.

Parallel-plate driver

Input | Connecting rods | Plates | Output

The input and output shafts of this parallel-plate driver rotate with the same angular relationship. The positions of the shafts, however, can vary to suit other requirements without affecting the input-output relationship between the shafts.

Curve-scribing mechanism

Output | P | Frame points | Hood | Automobile body

The output link rotates so that it appears to revolve around a point moving in space (P). This avoids the need for hinges at distant or inaccessible spots. The mechanism is suitable for hinging the hoods of automobiles.

Parallel-link coupling

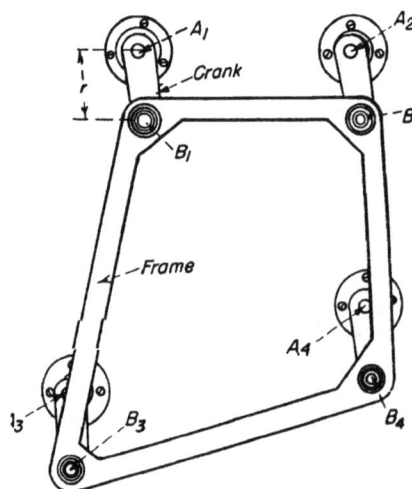
A_1 | Crank | A_2 | B_1 | Frame | A_4 | B_3 | B_4

The absence of backlash makes this parallel-link coupling a precision, low-cost replacement for gear or chain drives that can also rotate parallel shafts. Any number of shafts greater than two can be driven from any one of the shafts, provided two conditions are fulfilled: (1) All cranks must have the same length r; and (2) the two polygons formed by the shafts A and frame pivot centers B must be identical. The main disadvantage of this mechanism is its dynamic unbalance, which limits the speed of rotation. To lessen the effect of the vibrations produced, the frame should be made as light as is consistent with strength requirements for the intended application.

NINE FORCE AND STROKE MULTIPLIER LINKAGES

Wide-angle oscillator

The motion of the input linkage in the diagram is converted into a wide-angle oscillation by the two sprockets and chain. An oscillation of 60° is converted into 180° oscillation.

Gear-sector drive

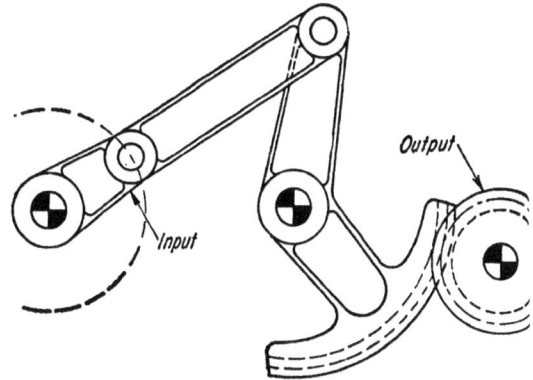

This is actually a four-bar linkage combined with a set of gears. A four-bar linkage usually obtains no more than about 120° of maximum oscillation. The gear segments multiply the oscillation in inverse proportion to the radii of the gears. For the proportions shown, the oscillation is boosted two and one-half times.

Angle-doubling drive

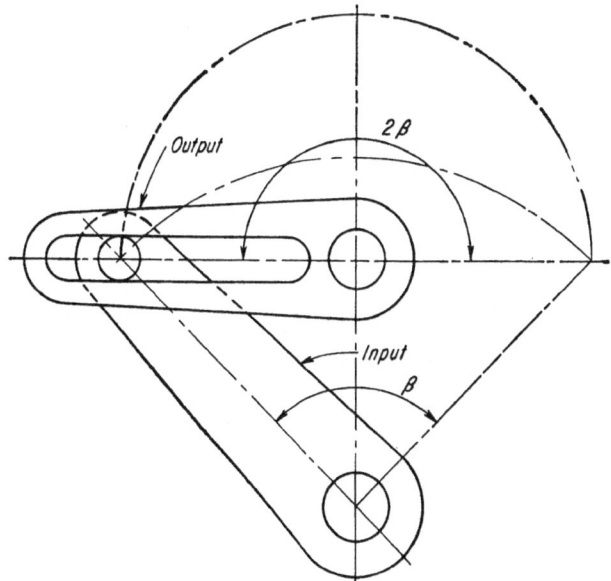

This angle-doubling drive will enlarge the oscillating motion β of one machine member into an output oscillation of 2β. If gears are employed, the direction of rotation cannot be the same unless an idler gear is installed. In that case, the centers of the input and output shafts cannot be too close. Rotating the input link clockwise causes the output to follow in a clockwise direction. For any set of link proportions, the distance between the shafts determines the gain in angle multiplication.

Pulley drive

This pulley drive multiplies the stroke of a hydraulic piston, causing the slider to move rapidly to the right for catapulting objects.

Typewriter drive

This drive multiplies the finger force of a typewriter, producing a strong hammer action at the roller from a light touch. There are three pivot points attached to the frame. The links are arranged so that the type bar can move in free flight after a key has been struck. The mechanism illustrated is actually two four-bar linkages in series. Some typewriters have as many as four four-bar linkages in a series.

Double-toggle puncher

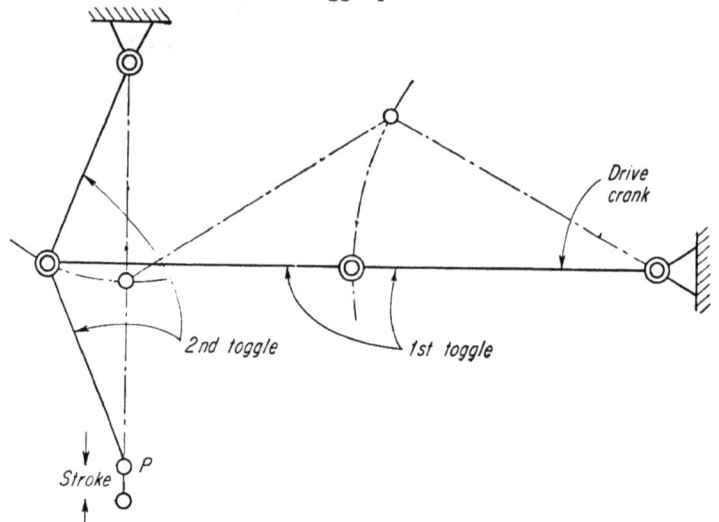

The first toggle of this puncher keeps point P in the raised position although its weight can exert a strong downward force (as in a heavy punch weight). When the drive crank rotates clockwise (e.g., driven by a reciprocating mechanism), the second toggle begins to straighten so as to create a strong punching force.

Gear-rack drive

This drive mechanism converts the motion of an input crank into a much larger rotation of the output (from 30° to 360°). The crank drives the slider and gear rack, which in turn rotates the output gear.

Chain drive

Springs and chains are attached to geared cranks of this drive to operate a sprocket output. Depending on the gear ratio, the output will produce a desired oscillation, e.g., two revolutions of output in each direction for each 360° of input.

Linkage-train drive

Arranging linkages in series on this drive can increase its angle of oscillation. In the version illustrated, the oscillating motion of the L-shaped rocker is the input for the second linkage. The final oscillation is 180°.

EIGHTEEN VARIATIONS OF DIFFERENTIAL LINKAGE

Figure 1 shows the modifications of the differential linkage shown in Fig. 2(A). These are based on the variations in the triple-jointed intermediate link 6. The links are designated as follows: Frame links: links 2, 3 and 4; two-jointed intermediate links: links 5 and 7; three jointed intermediate links: link 6.

Fig. 2

(A) (B)

The input motions to be added are a and b; their sum s is equal to c_1a 1 c_2b, where c_1 and c_2 are scale factors. The links are numbered in the same way as those in Fig. 2(A).

Fig. 1

(A) Three rotary joints.

(B) Two rotary and one sliding joint.

(C) One rotary and two sliding joints.

(D) One rotary and two sliding joints.

(E) Three sliding joints.

(F) Three sliding joints.

(G) One rotary and two cam joints (planetary gear differential)

(H) One rotary and two cam joints (rack and gear).

(I) Two sliding and one cam joint.

(J) One rotary and two cam joints (chain and sprocket).

(K) One rotary and two cam joints.

(L) One rotary and two cam joints.

(M) One rotary and two cam joints (planetary gears differential).

(N) One sliding and two cam joints.

(Q) One rotary, one sliding, and one cam joint.

Front view   End view

(O) Two rotary and one cam joint.

(P) One rotary, one sliding, and one cam joint.

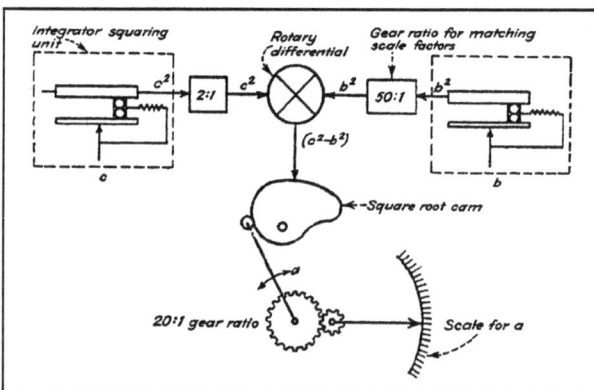

The intergrator method of mechanizing the equation $a = \sqrt{c^2 - b^2}$ is shown in the schematic form. It requires an excessive number of parts.

The cam method of mechanizing $a = \sqrt{c^2 - b^2}$ uses function generators for squaring and a link differential for subtraction. Note the reduction in parts from the integrator method.

FOUR-BAR SPACE MECHANISMS

There are potentially hundreds of them, but only a few have been discovered so far. Here are the best of one class—the four-bar space mechanisms.

R-C-C-C mechanism

P-C-C-C mechanism

H-C-C-C mechanism

R-S-C-R mechanism

R-S-C-P mechanism

R-S-C-H mechanism

P-P-S-C mechanism

P-H-S-C mechanism

H-H-S-C mechanism

Fig. 1 The nine chosen mechanisms.

A virtually unexplored area of mechanism research is the vast domain of three-dimensional linkage, frequently called space mechanism. Only a comparatively few kinds have been investigated or described, and little has been done to classify those that are known. As a result, many engineers do not know much about them, and applications of space mechanisms have not been as widespread as they could be.

Because a space mechanism can exist with a wide variety of connecting joints or "pair" combinations, it can be identified by the type and sequence of its joints. A listing of all of the physically realizable kinematic pairs has been established, based on the number of degrees-of-freedom of a joint. These pairs are all the known ways of connecting two bodies together for every possible freedom of relative motion between them.

The Practical Nine

The next step was to find the combination of pairs and links that would produce practical mechanisms. Based on the "Kutzbach criterion" (the only known mobility criterion—it determines the degree-of-freedom of a mechanism due to the constraints imposed by the pairs), 417 different kinds of space mechanisms have been identified. Detailed examination showed many of these to be mechanically complex and of limited adaptability. But the four-link mechanisms had particular appeal because of their mechanical simplicity. A total of 138 different kinds of four-bar mechanisms have been found. Of these, nine have particular merit (Fig. 1).

Fig. 2   The three mavericks.

These nine four-link mechanisms are the easiest to build because they contain only those joints that have area contact and are self-connecting. In the table, these joints are the five closed, lower pair types:

R = Revolute joint, which permits rotation only

P = Prism joint, which permits sliding motion only

H = Helix or screw type of joint

C = Cylinder joint, which permits both rotation and sliding (hence has two degrees-of-freedom)

S = Sphere joint, which is the common ball joint permitting rotation in any direction (three degrees-of-freedom)

All these mechanisms can produce rotary or sliding output motion from a rotary input—the most common mechanical requirements for which linkage mechanisms are designed.

The type letters of the kinematic pairs in the table identify the mechanism by ordering the letter symbols consecutively around the closed kinematic chain. The first letter identifies the pair connecting the input link and the fixed link; the last letter identifies the output link, or last link, with the fixed link. Thus, a mechanism labeled R-S-C-R is a double-crank mechanism with a spherical pair between the input crank and the coupler, and a cylindrical pair between the coupler and the output crank.

The Mavericks

The Kutzbach criterion is inadequate for the job because it cannot predict the existence of such mechanisms as the Bennett R-R-R-R mechanism, the double-ball joint R-S-S-R mechanism, and the R-C-C-R mechanism (Fig. 2). These "special" mechanisms require special geometric conditions to have a single degree-of-freedom. The R-R-R-R mechanism requires a particular orientation of the revolute axes and a particular ratio of link lengths to function as a single degree-of-freedom space mechanism. The R-S-S-R configuration, when functioning as a single degree-of-freedom mechanism, will have a passive degree-of-freedom of its coupler link. When properly constructed, the configuration R-C-C-R will also have a passive degree-of-freedom of its coupler, and it will function as a single-degree space mechanism.

Of these three special four-link mechanisms, the R-S-S-R mechanism is seen as the outstanding choice. It is the most versatile and practical configuration for meeting double-crank motion requirements.

Classification of kinematic pairs

Degree-of-freedom	Type number*	Type of joint	
		Symbol	Name
1	100	R	Revolute
	010	P	Prism
	001	H	Helix
2	200	T	Torus
	110	C	Cylinder
	101	T_H	Torus-helix
	020	. .	. . . .
	011	. .	. . . .
3	300	S	Sphere
	210	S_S	Sphere-slotted cylinder
	201	S_{SH}	Sphere-slotted helix
	120	P_L	Plane
	021	. .	. . . .
	111	. .	. . . .
4	310	S_G	Sphere-grooved
	301	S_{GH}	Sphere-grooved helix
	220	C_P	Cylinder-plane
	121	. .	. . . .
	211	. .	. . . .
5	320	S_P	Sphere-plane
	221	. .	. . . .
	311		. . . .

*Number of freedoms, given in the order of N_R, N_T, N_H.

SEVEN THREE-DIMENSIONAL LINKAGE DRIVES

The main advantage of three-dimensional drives is their ability to transmit motion between nonparallel shafts. They can also generate other types of helpful motion. This roundup includes descriptions of seven industrial applications for the drives.

Spherical Crank Drive

This type of drive is the basis for most three-dimensional linkages, much as the common four-bar linkage is the basis for the two-dimensional field. Both mechanisms operate on similar principles. (In the accompanying sketches, α is the input angle, and β the output angle. This notation has been used throughout this section.)

In the four-bar linkage, the rotary motion of driving crank 1 is transformed into an oscillating motion of output link 3. If the fixed link is made the shortest of all, then it is a double-crank mechanism; both the driving and driven members make full rotations.

The spherical crank drive, link 1 is the input, and link 3 is the output. The axes of rotation intersect at point O; the lines connecting AB, BC, CD, and DA can be considered to be parts of great circles of a sphere. The length of the link is best represented by angles $\bar{a}$, $\bar{b}$, $\bar{c}$, and $\bar{d}$.

Four-bar linkage

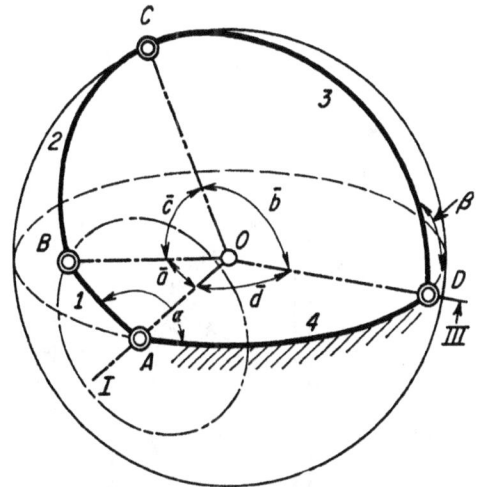

The Spherical Crank

111

Spherical-Slide Oscillator Drive

The two-dimensional slider crank is obtained from a four-bar linkage by making the oscillating arm infinitely long. By making an analogous change in the spherical crank, the spherical slider crank is obtained.

The uniform rotation of input shaft *I* is transferred into a nonuniform oscillating or rotating motion of output shaft *III*. These shafts intersect at an angle δ, corresponding to the frame link *4* of the spherical crank. Angle γ corresponds to the length of link *I*, and axis *II* is at right angle to axis *III*.

The output oscillates when γ is smaller than δ, but it rotates when γ is larger than δ.

The relation between input angle α and output angle β as designated in the skewed Hooke's joint is:

$$\tan \beta = \frac{(\tan \gamma)(\sin \alpha)}{\sin \delta + (\tan \gamma)(\cos \delta)(\cos \alpha)}$$

Top view

Worm driveshaft

Worm wheel

Side view

Skewed Hooke's Joint

Dough-Kneading Mechanism

Skewed Hooke's Joint Drive

This variation of the spherical crank is specified where an almost linear relation is desired between the input and output angles for a large part of the motion cycle.

The equation defining the output in terms of the input can be obtained from the skewed Hooke's joint equation by making δ = 90°. Thus, sin δ = 1, cos δ = 0, and

$$\tan \beta = \tan \gamma \sin \alpha$$

The principle of the skewed Hooke's joint has been applied to the drive of a washing machine (see sketch).

Here, the driveshaft drives the worm wheel *I* which has a crank fashioned at an angle γ. The crank rides between two plates and causes the output shaft *III* to oscillate in accordance with the equation.

The dough-kneading drive is also based on the Hooke's joint, but it follows the path of link *2* to give a wobbling motion that kneads dough in the tank.

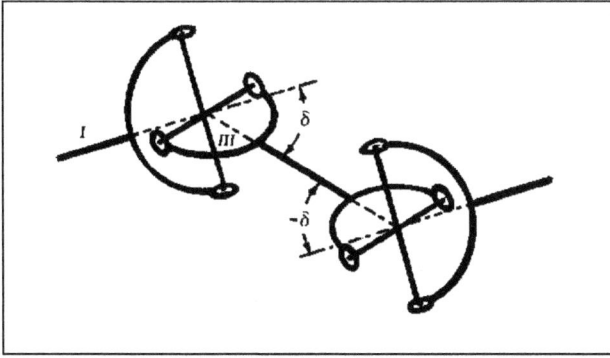

The Universal Joint Drive

The universal joint is a variation of the spherical-slide oscillator, but with angle $\gamma = 90°$. This drive provides a totally rotating output and can be operated as a pair, as shown in the diagram.

The equation relating input with output for a single universal joint, where γ is the angle between the connecting link and shaft I, is:

$$\tan \beta = \tan \alpha \cos \delta$$

The output motion is pulsating (see curve) unless the joints are operated as pairs to provide a uniform motion.

Input-Output Relationship

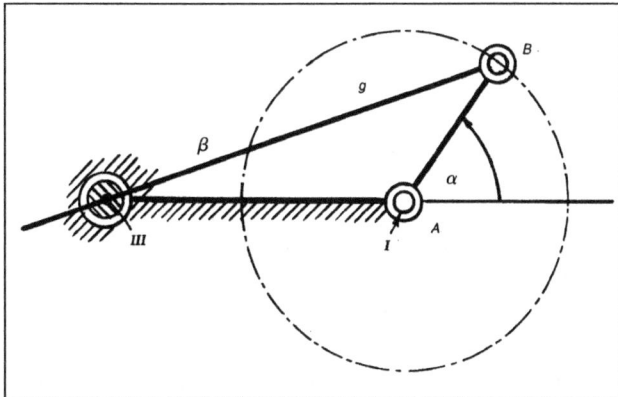

The 3D Crank Slide Drive

The three-dimensional crank slide is a variation of a plane crank slide (see sketch), with a ball point through which link g always slides, while a point B on link g describes a circle. A 3D crank is obtained from this mechanism by shifting output shaft III so that it is not normal to the plane of the circle; another way to accomplish this is to make shafts I and III nonparallel.

A practical variation of the 3D crank slide is the agitator mechanism (see sketch). As input gear I rotates, link g swivels

Agitator Mechanism

around (and also lifts) shaft III. Hence, the vertical link has both an oscillating rotary motion and a sinusoidal harmonic translation in the direction of its axis of rotation. The link performs what is essentially a twisting motion in each cycle.

113

(A) Basic Configuration

(B) Its Inversion

(C) As a 90° Uniform Motion Transmitter

The Space Crank Drive

One of the more recent developments in 3D linkages is the space crank shown in (A). It resembles the spherical crank, but has different output characteristics. The relationship between the input and output displacements is:

$$\cos \beta = (\tan \gamma)(\cos \alpha)(\sin \beta) - \frac{\cos \lambda}{\cos \gamma}$$

The velocity ratio is:

$$\frac{\omega_o}{\omega_i} = \frac{\tan \gamma \sin \alpha}{1 + \tan \gamma \cos \alpha \cot \beta}$$

where ω_0 is the output velocity and ω_i is the constant input velocity.

An inversion of the space crank is shown in (B). It can couple intersecting shafts and permits either shaft to be driven with full rotations. Motion is transmitted up to $37\frac{1}{2}°$ misalignment.

By combining two inversions (C), a method for transmitting an exact motion pattern around a 90° bend is obtained. This unit can also act as a coupler or, if the center link is replaced by a gear, it can drive two output shafts; in addition, it can transmit uniform motion around two bends.

Top view after a 90° input rotation

Steel balls riding within spherical grooves convert a continuous rotary input motion into an output that oscillates the shaft back and forth.

The oscillating motion is powered at right angles. The input shaft, in making full rotations, causes the output shaft to oscillate 120°.

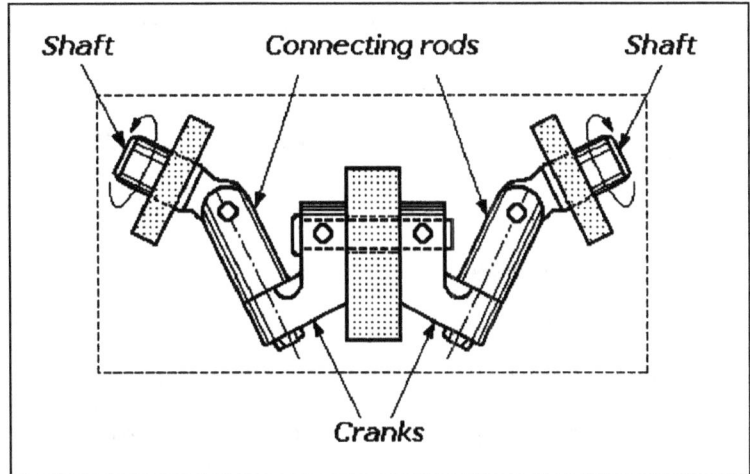

A constant-speed-ratio universal is obtained by placing two "inversions" back-to-back. Motion is transmitted up to a 75° misalignment.

A right-angle limited-stroke drive transmits an exact motion pattern. A multiplicity of fittings can be operated from a common shaft.

The Elliptical Slide Drive

The output motion, β, of a spherical slide oscillator can be duplicated with a two-dimensional "elliptical slide." The mechanism has a link g that slides through a pivot point D and is fastened to a point P moving along an elliptical path. The ellipse can be generated by a Cardan drive, which is a planetary gear system whose planet gear has half the diameter of its internal gear. The center of the planet, point M, describes a circle; any point on its periphery describes a straight line; and any point in between, such as point P, describes an ellipse.

There are special relationships between the dimensions of the 3D spherical slide and the 2D elliptical slide: $\tan \gamma / \sin \delta = a/d$ and $\tan \gamma / \cot \delta = b/d$, where a is the major half-axis, b the minor half-axis of the ellipse, and d is the length of the fixed link DN. The minor axis lies along this link.

If point D is moved within the ellipse, a completely rotating output is obtained, corresponding to the rotating spherical crank slide.

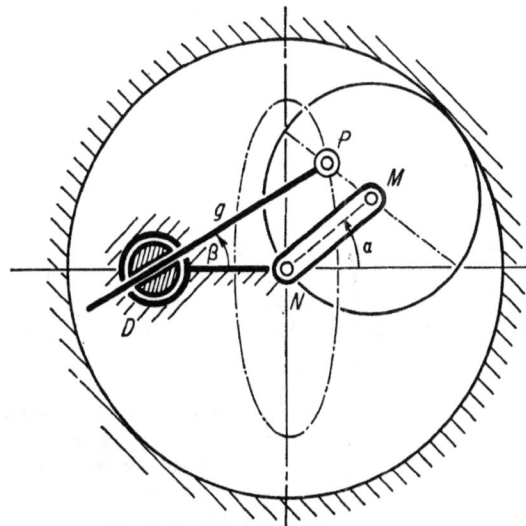

THIRTEEN DIFFERENT TOGGLE LINKAGE APPLICATIONS

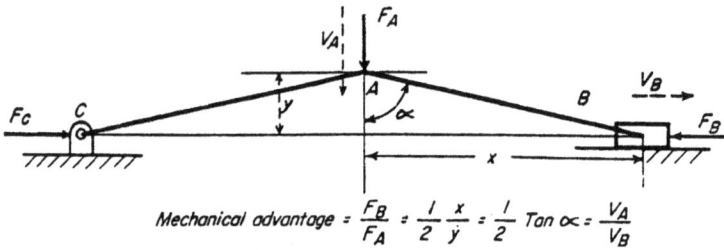

$$\text{Mechanical advantage} = \frac{F_B}{F_A} = \frac{1}{2}\frac{x}{y} = \frac{1}{2}\,\text{Tan}\,\alpha = \frac{V_A}{V_B}$$

Fig. 1   Many mechanical linkages are based on the simple toggle that consists of two links which tend to line up in a straight line at one point in their motion. The mechanical advantage is the velocity ratio of the input point A with respect to the outpoint point B: or V_A/V_B. As the angle is α approaches 90°, the links come into toggle, and the mechanical advantage and velocity ratio both approach infinity. However, frictional effects reduce the forces to much less than infinity, although they are still quite high.

Fig. 2   Forces can be applied through other links, and need not be perpendicular to each other. (A) One toggle link can be attached to another link rather than to a fixed point or slider. (B) Two toggle links can come into toggle by lining up on top of each other rather than as an extension of each other. The resisting force can be a spring.

HIGH MECHANICAL ADVANTAGE

Fig. 3   In punch presses, large forces are needed at the lower end of the work stroke. However, little force is required during the remainder of the stroke. The crank and connecting rod come into toggle at the lower end of the punch stroke, giving a high mechanical advantage at exactly the time it is most needed.

Fig. 4   A cold-heading rivet machine is designed to give each rivet two successive blows. Following the first blow (point 2), the hammer moves upward a short distance (to point 3). Following the second blow (at point 4), the hammer then moves upward a longer distance (to point 1) to provide clearance for moving the workpiece. Both strokes are produced by one revolution of the crank, and at the lowest point of each stroke (points 2 and 4) the links are in toggle.

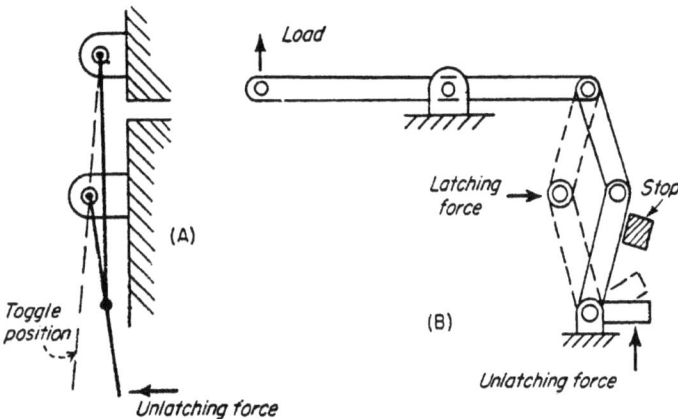

Fig. 5   Locking latches produce a high mechanical advantage when in the toggle portion of the stroke. A simple latch exerts a large force in the locked position (Fig. 5A). For positive locking, the closed position of latch is slightly beyond the toggle position. A small unlatching force opens the linkage (Fig. 5B).

Fig. 6   A stone crusher has two toggle linkages in series to obtain a high mechanical advantage. When the vertical link *I* reaches the top of its stroke, it comes into toggle with the driving crank *II*; at the same time, link *III* comes into toggle and link *IV*. This multiplication results in a very large crushing force.

Fig. 7   A friction ratchet is mounted on a wheel; a light spring keeps the friction shoes in contact with the flange. This device permits clockwise motion of the arm *I*. However, reverse rotation causes friction to force link *II* into toggle with the shoes. This action greatly increases the locking pressure.

Fig. 8

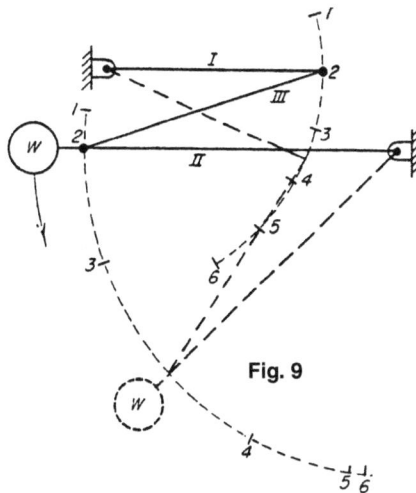

Fig. 9

Fig. 8 Door check linkage gives a high velocity ratio during the stroke. As the door swings closed, connecting link *I* comes into toggle with the shock absorber arm *II*, giving it a large angular velocity. The shock absorber is more effective in retarding motion near the closed position.

Fig. 9 An impact reducer is on some large circuit breakers. Crank *I* rotates at constant velocity while the lower crank moves slowly at the beginning and end of the stroke. It moves rapidly at the mid-stroke when arm *II* and link *III* are in toggle. The accelerated weight absorbs energy and returns it to the system when it slows down.

VARIABLE MECHANICAL ADVANTAGE

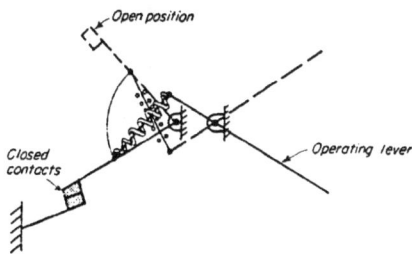

Fig. 10 A toaster switch has an increasing mechanical advantage to aid in compressing a spring. In the closed position, the spring holds the contacts closed and the operating lever in the down position. As the lever is moved upward, the spring is compressed and comes into toggle with both the contact arm and the lever. Little effort is required to move the links through the toggle position; beyond this point, the spring snaps the contacts open. A similar action occurs on closing.

Fig. 11 A toggle press has an increasing mechanical advantage to counteract the resistance of the material being compressed. A rotating handwheel with a differential screw moves nuts *A* and *B* together, and links *I* and *II* are brought into toggle.

Fig. 12_Four-bar linkages can be altered to give a variable velocity ratio (or mechanical advantage). (Fig. 12A) Since the cranks *I* and *II* both come into toggle with the connecting link *III* at the same time, there is no variation in mechanical advantage. (Fig. 12B) increasing the length of link *III* gives an increased mechanical advantage between positions 1 and 2, because crank *I* and connecting link *III* are near toggle. (Fig. 12C) Placing one pivot at the left produces similar effects as in (Fig. 12B). (Fig. 12D) increasing the center distance puts crank *II* and link *III* near toggle at position 1; crank *I* and link *III* approach the toggle position at 4.

Fig. 13 A riveting machine with a reciprocating piston produces a high mechanical advantage with the linkage shown. With a constant piston driving force, the force of the head increases to a maximum value when links *II* and *III* come into toggle.

HINGED LINKS AND TORSION BUSHINGS SOFT-START DRIVES

Centrifugal force automatically draws up the linkage legs, while the torsional resistance of the bushings opposes the deflection forces.

A spidery linkage system combined with a rubber torsion bushing system formed a power-transmission coupling. Developed by a British company, Twiflex Couplings Ltd., Twickenham, England, the device (drawing below) provides ultra-soft starting characteristics. In addition to the torsion system, it also depends on centrifugal force to draw up the linkage legs automatically, thus providing additional soft coupling at high speeds to absorb and isolate any torsional vibrations arising from the prime mover.

The TL coupling has been installed to couple marine main engines to gearbox-propeller systems. Here the coupling reduces propeller vibrations to negligible proportions even at high critical speeds. Other applications are also foreseen, including their use in diesel drives, machine tools, and off-the-road construction equipment. The coupling's range is from 100 hp at 4000 rpm to 20,000 hp at 400 rpm.

Articulating links. The key factor in the TL coupling, an improvement over an earlier Twiflex design, is the circular grouping of hinged linkages connecting the driving and driven coupling flanges. The forked or tangential links have resilient precompressed bonded-rubber bushings at the outer flange attachments, while the other pivots ride on bearings.

When torque is applied to the coupling, the linkages deflect in a positive or negative direction from the neutral position (drawings, below). Deflection is opposed by the torsional resistance of the rubber bushings at the outer pins. When the coupling is rotating, the masses of the linkage give rise to centrifugal forces that further oppose coupling deflection. Therefore, the working position of the linkages depends both on the applied torque and on the speed of the coupling's rotation.

Tests of the coupling's torque/deflection characteristics under load have shown that the torsional stiffness of the coupling increases progressively with speed and with torque when deflected in the positive direction. Although the geometry of the coupling is asymmetrical, the torsional characteristics are similar for both directions of drive in the normal working range. Either half of the coupling can act as the driver for either direction of rotation.

The linkage configuration permits the coupling to be tailored to meet the exact stiffness requirements of individual systems or to provide ultra-low torsional stiffness at values substantially softer than other positive-drive couplings. These characteristics enable the Twiflex coupling to perform several tasks:

- It detunes the fundamental mode of torsional vibration in a power transmission system. The coupling is especially soft at low speeds, which permits complete detuning of the system.
- It decouples the driven machinery from engine-excited torsional vibration. In a typical geared system, the major machine modes driven by the gearboxes are not excited if the ratio of coupling stiffness to transmitted torque is less than about 7:1—a ratio easily provided by the Twiflex coupling.
- It protects the prime mover from impulsive torques generated by driven machinery. Generator short circuits and other causes of impulsive torques are frequently of sufficient duration to cause high response torques in the main shafting.

Using the example of the TL 2307G coupling design—which is suitable for 10,000 hp at 525 rpm—the torsional stiffness at working points is largely determined by coupling geometry and is, therefore, affected to a minor extent by the variations in the properties of the rubber bushings. Moreover, the coupling can provide torsional-stiffness values that are accurate within 5.0 percent.

Articulating links

End view (small flange removed) Coupling in zero deflection position

Section A-A

Articulating links of the new coupling (left) are arranged around the driving flanges. A four-link design (right) can handle torques from a 100-hp prime mover driving at 4000 rpm.

EIGHT LINKAGES FOR BAND CLUTCHES AND BRAKES

Fig. 1 An outside band clutch operated by a roller and cone.

Fig. 2 An outside band clutch made as two half-wrap bands with an intermediate equalizer.

Plan

Fig. 3 An outside band clutch made as two full-wrap bands with an intermediate equalizer.

Side Elevation

Fig. 4 An inside band clutch operated by a yoke having movement along the shaft.

Oil line to brake cylinder

Brake cylinder

Fig. 5 A two-way acting band brake operated hydraulically.

Fig. 6 A hoist-drum band brake operated by a foot pedal.

Dotted lines show brake released

Fig. 7 A band brake with a single toggle action.

Fig. 8 A crawler-drive band brake operated by a ratchet lever.

DESIGN OF CRANK-AND-ROCKER LINKS FOR OPTIMUM FORCE TRANSMISSION

Four-bar linkages can be designed with a minimum of trial and error by a combination of tabular and iteration techniques.

The determination of optimum crank-and-rocker linkages has most effectively been performed on a computer because of the complexity of the equations and calculations involved. Thanks to the work done at Columbia University's Department of Mechanical and Nuclear Engineering, all you need now is a calculator and the computer-generated tables presented here. The computations were done by Mr. Meng-Sang Chew, at the university.

A crank-and-rocker linkage, $ABCD$, is shown in the first figure. The two extreme positions of the rocker are shown schematically in the second figure. Here ψ denotes the rocker swing angle and ϕ denotes the corresponding crank rotation, both measured counterclockwise from the extended dead-center position, AB_1C_1D.

The problem is to find the proportions of the crank-and-rocker linkage for a given rocker swing angle, ψ, a prescribed corresponding crank rotation, ϕ, and optimum force transmission. The latter is usually defined in terms of the transmission angle, m, the angle μ between coupler BC extended and rocker CD.

Considering static forces only, the closer the transmission angle is to 90°, the greater is the ratio of the driving component of the force exerted on the rocker to the component exerting bearing pressure on the rocker. The control of transmission-angle variation becomes especially important at high speeds and in heavy-duty applications.

How to find the optimum. The steps in the determination of crank-and-rocker proportions for a given rocker swing angle, corresponding crank rotation, and optimum transmission are:

- Select (ψ, ϕ) within the following range:

$$0° < \psi < 180°$$
$$(90° + 1/2\ \psi) < \phi < (270° + 1/2\ \psi)$$

- Calculate: $t = \tan 1/2\ \phi$
  $$u = \tan 1/2\ (\phi - \psi)$$
  $$v = \tan 1/2\ \psi$$

The optimum solution for the classic four-bar crank-and-rocker mechanism problem can now be obtained with only the accompanying table and a calculator.

An example in this knee-joint tester designed and built by following the design and calculating procedures outlined in this article.

- Using the table, find the ratio λ_{opt} of coupler to crank length that minimizes the transmission-angle deviation from 90°. The most practical combinations of (ψ, ϕ) are included in the table. If the (ψ, ϕ) combination is not included, or if $\phi = 180°$, go to next steps (a,b,c):

(a) If $\phi = 180°$ and (ψ, ϕ) fall outside the range given in the table, determine the arbitrary intermediate value Q from the equation: $Q^3 + 2Q^2 - t^2Q - (t^2/u^2)(1 + t^2) = 0$ where $(1/u^2 < Q < t^2)$. This is conveniently accomplished by numerical iteration:

Set
$$Q_1 = \frac{1}{2}\left(t^2 + \frac{1}{u^2}\right)$$

Calculate $Q_2, Q_3, \ldots$ from the recursion equation:

$$Q_{i+1} = \frac{2Q_i^2(Q_i + 1) + (t^2/u^2)(1 + t^2)}{Q_i(3Q_i + 4) - t^2}$$

Iterate until the ratio $[(Q_i + 1 - Q_i)/Q_i]$ is sufficiently small, so that you obtain the desired number of significant figures. Then:

$$\lambda_{opt} = t^2/Q$$

(b) If $\phi = 180°$ and the determination of λ_{opt} requires interpolation between two entries in the table, let $Q_1 = t^2 \lambda^2$, where λ corresponds to the nearest entry in the table, and continue as in (a) above to determine Q and λ_{opt}. Usually one or two iterations will suffice.

(c) $\phi = 180°$. In this case, $a^2 + b^2 = c^2 + d^2$; $\psi = 2 \sin^{-1}(b/d)$; and the maximum deviation, Δ, of the transmission angle from 90° is equal to $\sin^{-1}(ab/cd)$.

- Determine linkage proportions as follows:

$$(a')^2 = \frac{u^2 + \lambda_{opt}^2}{1 + u^2}$$

$$(b')^2 = \frac{v^2}{1 + v^2}$$

$$(c')^2 = \frac{\lambda_{opt}^2 v^2}{1 + v^2}$$

$$(d')^2 = \frac{t^2 + \lambda_{opt}^2}{1 + t^2}$$

Then: $a = ka'$; $b = kb'$; $c = kc'$; $d = kd'$ where k is a scale factor, such that the length of any one link, usually the crank, is equal to a design value. The

Optimum values of lambda ratio for given ø and ψ

ø deg	\(\psi\), deg 160	162	164	166	168	170	172	174	176	178
10	2.3532	2.4743	2.6166	2.7873	2.9978	3.2669	3.6284	4.1517	5.0119	6.8642
12	2.3298	2.4491	2.5891	2.7570	2.9636	3.2272	3.5804	4.0899	4.9224	6.6967
14	2.3064	2.4239	2.5617	2.7266	2.9293	3.1874	3.5324	4.0283	4.8342	6.5367
16	2.2831	2.3988	2.5344	2.6964	2.8953	3.1479	3.4848	3.9675	4.7482	6.3853
18	2.2600	2.3740	2.5073	2.6664	2.8615	3.1089	3.4380	3.9080	4.6650	6.2427
20	2.2372	2.3494	2.4805	2.6368	2.8282	3.0704	3.3920	3.8499	4.5848	6.1087
22	2.2145	2.3250	2.4540	2.6076	2.7954	3.0327	3.3470	3.7935	4.5077	5.9826
24	2.1922	2.3010	2.4279	2.5789	2.7631	2.9956	3.3030	3.7388	4.4338	5.8641
26	2.1701	2.2773	2.4022	2.5505	2.7314	2.9594	3.2602	3.6857	4.3628	5.7524
28	2.1483	2.2539	2.3768	2.5227	2.7004	2.9239	3.2185	3.6344	4.2948	5.6469
30	2.1268	2.2309	2.3519	2.4954	2.6699	2.8893	3.1779	3.5847	4.2295	5.5472
32	2.1056	2.2082	2.3273	2.4685	2.6401	2.8554	3.1384	3.5367	4.1668	5.4526
34	2.0846	2.1858	2.3032	2.4421	2.6108	2.8223	3.0999	3.4901	4.1066	5.3628
36	2.0640	2.1637	2.2794	2.4162	2.5821	2.7899	3.0624	3.4449	4.0486	5.2773
38	2.0436	2.1420	2.2560	2.3908	2.5540	2.7583	3.0259	3.4012	3.9927	5.1957
40	2.0234	2.1205	2.2330	2.3657	2.5264	2.7274	2.9903	3.3587	3.9388	5.1177
42	2.0035	2.0994	2.2103	2.3411	2.4994	2.6971	2.9556	3.3175	3.8868	5.0430
44	1.9839	2.0785	2.1879	2.3169	2.4728	2.6675	2.9217	3.2773	3.8364	4.9712
46	1.9644	2.0579	2.1659	2.2931	2.4468	2.6384	2.8886	3.2383	3.7877	4.9023
48	1.9452	2.0375	2.1441	2.2696	2.4211	2.6100	2.8563	3.2003	3.7404	4.8358
50	1.9262	2.0174	2.1227	2.2465	2.3959	2.5820	2.8246	3.1632	3.6945	4.7717

max deviation, Δ, of the transmission angle from 90° is:

$$\sin\Delta = \frac{|(a \pm b)^2 - c^2 - d^2|}{2cd}$$

$0° \leq \Delta \leq 90°$
$+$ sign if $\phi < 180°$
$-$ sign if $\phi < 180°$

An actual example. A simulator for testing artificial knee joints, built by the Department of Orthopedic Surgery, Columbia University, under the direction of Dr. N. Eftekhar, is shown schematically. The drive includes an adjustable crank-and-rocker, ABCD. The rocker swing angle ranges from a maximum of about 48° to a minimum of about one-third of this value. The crank is 4 in. long and rotates at 150 rpm. The swing angle

adjustment is obtained by changing the length of the crank.

Find the proportions of the linkage, assuming optimum-transmission proportions for the maximum rocker swing angle, as this represents the most severe condition. For smaller swing angles, the maximum transmission-angle deviation from 90° will be less.

Crank rotation corresponding to 48° rocker swing is selected at approximately 170°. Using the table, find $\lambda_{opt} = 2.6100$. This gives $a' = 1.5382$, $b' = 0.40674$, $c' = 1.0616$, and $d' = 1.0218$.

For a 4 in. crank, $k = 4/0.40674 = 9.8343$ and $a = 15.127$ in., $b = 4$ in., $c = 10.440$ in., and $d = 10.049$ in., which is very close to the proportions used. The maximum deviation of the transmission angle from 90° is 47.98°.

This procedure applies not only for the transmission optimization of crank-and-rocker linkages, but also for other crank-and-rocker design. For example, if only the rocker swing angle and the corresponding crank rotation are prescribed, the ratio of coupler to crank length is arbitrary, and the equations can be used with any value of λ_2 within the range $(1, u^2t^2)$. The ratio λ can then be tailored to suit a variety of design requirements, such as size, bearing reactions, transmission-angle control, or combinations of these requirements.

The method also was used to design dead-center linkages for aircraft landing-gear retraction systems, and it can be applied to any four-bar linkage designs that meet the requirements discussed here.

Optimum values of lambda ratio for given ϕ and ψ

ϕ deg	ψ, deg 182	184	186	188	190	192	194	196	198	200
10	7.2086	5.3403	4.4560	3.9112	3.5318	3.2478	3.0245	2.8428	2.6911	2.5616
12	7.0369	5.2692	4.4227	3.8969	3.5282	3.2507	3.0317	2.8528	2.7030	2.5748
14	6.8646	5.1881	4.3795	3.8739	3.5174	3.2478	3.0341	2.8589	2.7117	2.5855
16	6.6971	5.1013	4.3287	3.8435	3.5000	3.2392	3.0317	2.8610	2.7171	2.5934
18	6.5371	5.0121	4.2726	3.8071	3.4768	3.2252	3.0245	2.8589	2.7189	2.5982
20	6.3857	4.9226	4.2131	3.7663	3.4487	3.2065	3.0129	2.8528	2.7171	2.5998
22	6.2431	4.8344	4.1518	3.7221	3.4167	3.1837	2.9972	2.8428	2.7117	2.5982
24	6.1090	4.7484	4.0900	3.6759	3.3818	3.1575	2.9780	2.8293	2.7030	2.5934
26	5.9830	4.6652	4.0284	3.6284	3.3447	3.1286	2.9558	2.8127	2.6911	2.5855
28	5.8644	4.5849	3.9676	3.5804	3.3062	3.0976	2.9311	2.7833	2.6763	2.5748
30	5.7527	4.5079	3.9080	3.5324	3.2669	3.0652	2.9045	2.7718	2.6592	2.5616
32	5.6472	4.4339	3.8500	3.4849	3.2272	3.0318	2.8764	2.7484	2.6399	2.5461
34	5.5475	4.3630	3.7936	3.4380	3.1875	2.9979	2.8473	2.7236	2.6190	2.5287
36	5.4529	4.2949	3.7388	3.3920	3.1480	2.9636	2.8175	2.6977	2.5967	2.5097
38	5.3631	4.2296	3.6858	3.3470	3.1089	2.9294	2.7873	2.6711	2.5734	2.4894
40	5.2776	4.1669	3.6345	3.3031	3.0705	2.8953	2.7570	2.6440	2.5492	2.4680
42	5.1960	4.1067	3.5848	3.2602	3.0327	2.8615	2.7266	2.6166	2.5246	2.4459
44	5.1180	4.0487	3.5367	3.2185	2.9956	2.8282	2.6964	2.5891	2.4996	2.4232
46	5.0432	3.9928	3.4901	3.1779	2.9594	2.7954	2.6665	2.5617	2.4744	2.4001
48	4.9715	3.9389	3.4450	3.1384	2.9239	2.7631	2.6369	2.5344	2.4491	2.3767
50	4.9025	3.8869	3.4012	3.0999	2.8893	2.7314	2.6076	2.5073	2.4239	2.3533

DESIGN OF FOUR-BAR LINKAGES FOR ANGULAR MOTION

How to use four-bar linkages to generate continuous or intermittent angular motions required by feeder mechanisms

Fig. 1 **Four-bar linkage** synchronizes two angular movements, ϕ_{12} and ϕ_{13}, with ψ_{12}, and ψ_{14}.

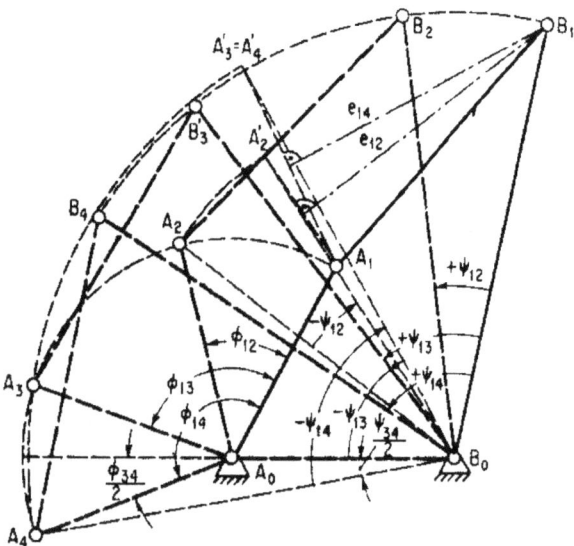

Fig. 2 **Three angular positions**, ϕ_{12}, ϕ_{13}, ϕ_{14}, are synchronized by four-bar linkage here with ψ_{12}, ψ_{13}, and ψ_{14}.

In putting feeder mechanisms to work, it is often necessary to synchronize two sets of angular motions. A four-bar linkage offers one way. For example, in Fig. 1 two angular motions, ϕ_{12} and ϕ_{13}, must be synchronized with two others, ψ_{12} and ψ_{13}, about the given pivot points A_0 and B_0 and the given crank length A_0A. This means that crank length B_0B. must be long enough so that the resulting four-bar linkage will coordinate angular motions ϕ_{12} and ϕ_{13} with ψ_{12} and ψ_{13}. The procedure is:

1. Obtain point A'_2 by revolving A_2 about B_0 through angle $-\psi_{12}$ but in the opposite direction.
2. Obtain point A'_3 similarly by revolving A_3 about B_0 through angle $-\psi_{13}$.
3. Draw lines $A_1A'_2$ and $A_1A'_3$ and the perpendicular bisectors of the lines which intersect at desired point B_1.
4. The quadrilateral $A_0A_1B_1B_0$ represents the four-bar linkage that will produce the required relationship between the angles ϕ_{12}, ϕ_{13}, and ψ_{12}, ψ_{13}.

Three angles with four relative positions can be synchronized in a similar way. Figure 2 shows how to synchronize angles ψ_{12}, ψ_{13}, ψ_{14} with corresponding angles ψ_{12}, ψ_{13}, and ψ_{14}, using freely chosen pivot points A_0 and B_0. In this case, crank length A_0A as well as B_0B is to be determined, and the procedure is:

1. Locate pivot points A_0 and B_0 on a line that bisects angle $A_3A_0A_4$, the length A_0B_0 being arbitrary.
2. Measure off $1/2$ of angle $B_3B_0B_4$ and with this angle draw B_0A_4 which establishes crank length A_0A at intersection of A_0A_4. This also establishes points A_3, A_2 and A_1.
3. With B_0 as center and B_0B_4 as radius mark off angles $-\psi_{14}$, $-\psi_{13}$, $-\psi_{12}$, the negative sign indicating they are in opposite sense to ψ_{14}, ψ_{13} and ψ_{12}. This establishes points A'_2, A'_3 and A'_4, but here A'_3 and A'_4 coincide because of symmetry of A_3 and A_4 about A_0B_0.
4. Draw lines $A_1A'_2$ and $A_1A'_4$, and the perpendicular bisectors of these lines, which intersect at the desired point B_1.
5. The quadrilateral $A_0A_1B_0B_1$ represents the four-bar linkage that will produce the required relationship between the angles ϕ_{12}, ϕ_{13}, ϕ_{14}, and ψ_{12}, ψ_{13}, ψ_{14}.

The illustrations show how these angles must be coordinated within the given space. In Fig. 3A, input angles of the crank must be coordinated with the output angles of the forked escapement. In Fig. 3B, input angles of the crank are coordinated with the output angles of the tilting hopper. In Fig. 3C, the input angles of the crank are coordinated with the output angles of the segment. In Fig. 3D, a box on a conveyor is tilted 90° by an output crank, which is actuated by an input crank through a coupler. Other mechanisms shown can also coordinate the input and output angles; some have dwell periods between the cycles, others give a linear output with dwell periods.

Fig. 3 **Input and output angular movements** of feeder mechanisms are synchronized by a four-bar linkage. (A) The separate feed for spherical or cylindrical parts on a conveyor. (B) Group-sorting of spherical parts by the tilting hopper. (C) A separate feed for spherical or cylindrical parts by gravity. (D) Rectangular parts are turned on a conveyor. (E) Parts are separated by levers, and the conveyor movement is controlled by a trigger at the right. (F) Bar stock is positioned by the angular oscillation of an output lever when the input crank is actuated.

MULTIBAR LINKAGES FOR CURVILINEAR MOTIONS

Four-bar linkages can be combined into six, eight, or more linkages for the feeder mechanisms in cameras, automatic lathes, farm machinery, and torch-cutting machines.

When feeder mechanisms require complex curvilinear motions, it might be necessary to use compound linkages rather than four links. However, four-bar linkages can be synthesized to produce curvilinear motions of various degrees of complexity, and all possibilities for four-bar linkages should be considered before selecting more complex linkages.

For example, a camera film-advancing mechanism, Fig. 1, has a simple four-bar linage with a coupler point d, which generates a curvilinear and straight-line motion a resembling a D. Another more complex curvilinear motion, Fig. 2, is also generated by a coupler point E of a four-bar linkage, which controls an automatic profile cutter. Four-bar linkages can generate many different curvilinear motions, as in Fig. 3. Here the points of the coupler prongs, g_1, g_2, and g_3 on coupler b, and g_4 and g_5 on coupler e, are chosen so that their motions result in the desired progressive feeding of straw into a press.

A similar feeding and elevating device is shown in Fig. 4. The rotating device crank a moves coupler b and swinging lever c, which actuates the guiding arm f through the link e. The bar h carries the prone fingers g_1 through g_7. They generate coupler curves a_1 through a_7.

As another practical example, consider the torch-cutting machine in Fig. 5A designed to cut sheet metal along a curvilinear path a. Here the points A_0 and B_0 are fixed in the machine, and the lever A_0A_1 has an adjustable length to suit the different curvilinear paths a desired.

The length B_1B_1 is also fixed. The challenge is to find the length of the levers A_1B_1 and E_1B_1 in the four-bar linkage to give the desired path a, which is to be traced by the coupler point E on which the cutting torch is mounted.

The graphical solution for this problem, as shown in Fig. 5B, requires the selection of the points A_1 and E_4 so that the distances A_1E_1 to A_8E_8 are equal and the points E_1 to E_8 lie on the desired coupler curved a. In this case, only the points E_4 to E_8 represent the desired profile to be cut. The correct selection of points A_1 and E_1 depends upon making the following triangles congruent:

$$\Delta E_2A_2B_{01} = \Delta E_1A_1B_{02}$$
$$\Delta E_3A_3B_{01} = \Delta E_1A_1B_{03}$$
$$\Delta E_8A_8B_{01} = \Delta E_1A_1B_{08}$$

and so on until $E_8A_8B_{01} = E_1A_1B_{08}$. At the same time, all points A_1 to A_8 must lie on the arc having B_1 as center.

Fig. 1

Fig. 2

Fig. 3

Fig. 4

(A)

(B)

Fig. 5

126

Feeder Mechanisms (*continued*)

Synthesis of an Eight-Bar Linkage

Design a linkage with eight precision points, as shown in Fig. 6. In this mechanism the curvilinear motion of one four-bar linkage is coordinated with the angular oscillation of a second four-bar linkage. The first four-bar linkage consists of $A A_0 B B_0$ with coupler point E which generates γ with eight precision points E_1 through E_8 and drives a second four-bar linkage $H H_0 G G_0$. Coupler point F generates curve δ with precision points F_1 through F_8. The coupler points F_2, F_4, F_6, F_8 are coincident because straight links $G G_0$ and $G H$ are in line with one another in these coupler positions. This is what permits $H H_0$ to oscillate, despite the continuous motion of the coupler point F. The coupler points F_1 coincident with F_5, and F_3 coincident with F_7, have been chosen so that F_1 is the center of a circle k_1 and F_3 is the center of a circle k_3. These circles are tangent to coupler curve γ at E_1, and E_5, E_3, and E_7, and they indicate the limiting positions of the second four-bar linkage $H H_0 G G_0$.

The limiting angular oscillation of $H H_0$, which is one of the requirements of this mechanism, is represented by positions $H_0 H_1$ and $H_0 H_3$. It oscillates four times for each revolution of the input crank $A A_0$, and the positions H_1 to H_8 correspond to input crank positions A_1 to A_8.

The synthesis of a compound linkage with dwell periods and coordinated intermittent motion is shown in Fig. 7. The four-bar linkage $A A_0 B B_0$ generates an approximately triangular curve with coupler point E, which has six precision points E_1 through E_6. A linkage that will do this is not unusual and can be readily proportioned from known methods of four-bar linkage synthesis. However, the linkage incorporates dwell periods that produce coordinated intermittent motion with a second four-bar linkage $F F_0 H B_0$. Here the tangent arcs k_{12}, k_{34} and k_{56} are drawn with EF as the radius from centers F_{12}, F_{34} and F_{56}.

These centers establish the circle with F_0 as the center and pivot point for the second four-bar linkage. Each tangent arc causes a dwell of the link $F F_0$, while $A A_0$ rotates continuously. Thus, the link $F F_0$, with three rest periods in one revolution, can produce intermittent curvilinear motion in the second four-bar linkage $F F_0 H B_0$. In laying out the center, F_0 must be selected so that the angle $E F F_0$ deviates only slightly from 90° because this will minimize the required torque that is to be applied at E. The length of $B_0 H$ can be customized, and the rest periods at H_{34}, H_{12} and H_{56} will correspond to the crank angles ϕ_{34}, ϕ_{12} and ϕ_{56}.

A compound linkage can also produce a 360° oscillating motion with a dwell period, as in Fig. 8. The two four-bar linkages are $A A_0 B B_0$ and $B B_0 F F_0$, and the output coupler curve γ is traversed only through segment E_1, E_2. The oscillating motion is produced by lever $H H_0$, connected to the coupler point by $E H$. The fixed point H_0 is located within the loop of the coupler curve γ. The dwell occurs at point H_3, which is the center of circular arc k tangent to the coupler curve γ during the desired dwell period. In this example, the dwell is made to occur in the middle of the 360° oscillation. The coincident positions H_1 and H_2 indicate the limiting positions of the link $H H_0$, and they correspond to the positions E_1 and E_2 of the coupler point.

Fig. 6

Fig. 7

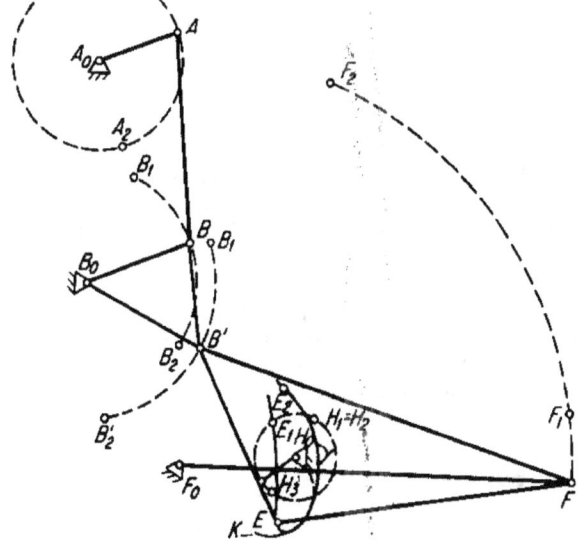

Fig. 8

ROBERTS' LAW HELPS TO DESIGN ALTERNATE FOUR-BAR LINKAGES

The three linkage examples

Fig. 1

Fig. 2   (2)

Fig. 3

Fig. 4

Fig. 5   (5)

When a four-bar linkage has been designed or selected from a catalog to produce a desired coupler curve, it is often found that one of the pivot points is inconveniently located or that the transmission angles are not suitable. (A coupler curve is produced by a point on the connecting rod joining the two cranks of the four-bar linkage). According to *Roberts' Law* there are at least two other four-bar linkages that will generate the same coupler curve. One of these linkages might be more suitable for the application.

Robert's Law states that the two alternate linkages are related to the first by a series of similar triangles. This leads to graphical solutions; three examples are shown. The first involves similar triangles, the second is a more convenient step-by-step method, and the third illustrates the solution of a special case where the coupler point lies along the connecting rod.

Method of Similar Triangles

Four-bar linkage *ABCD* in Fig. 1 uses point *P*, which is actually an extension of the connecting rod *BC*, to produce the desired curve. Point *E* is found by constructing *EP* parallel to *AB*, and *EA* parallel to *PB*. Then triangle *EFP* is constructed similar to triangle *BPC*. This calls for laying out angles *a* and *β*.

Point *H* is found in a similar way, and point *G* is located by drawing *GH* parallel to *FP* and *GF* parallel to *HP*.

The two alternate linkages to *ABCD* are *GFEA* and *GHID*. All use point *P* to produce the desired curve, and given any one of the three, the other two can be determined.

The Step-by-Step Method

With the similar-triangle method just described, slight errors in constructing the proper angles lead to large errors in link dimensions. The construction of angles can be avoided by laying off the link lengths along a straight line.

Thus, linkage *ABCD* in Fig. 2 is laid off as a straight line from *A* to *D* in Fig. 3. Included in the transfers is point *P*. Points *EFGHI* are quickly found by either extending the original lines or constructing parallel lines. Fig. 3, which now has all the correct dimensions of all the links, is placed under a sheet of tracing paper and, with the aid of a compass, links *AB* and *CD* are rotated

Fig. 1 **The method of similar triangles.**
Figs. 2, 3, 4 **A step-by-step method.**
Fig. 5 **This special case** shows the simplicity of applying Roberts' Law.

(see Fig. 4) so that linkage *ABCD* is identical to that in Fig. 2. Links *PEF* and *PHI* are rotated parallel to *AB* and *CD*, respectively. Completion of the parallelogram gives the two alternate linkages, *AEFG* and *GHID*.

Special Case

It is not uncommon for the coupler point *P* to lie on a line through *BC*, as in Fig. 5. Links *EA*, *EP*, and *ID* can be found quickly by constructing the appropriate parallel lines. Point *G* is located by using the proportion: *CB*:*BP* = *DA*:*AG*. Points *H* and *F* are then located by drawing lines parallel to *AB* and *CD*.

DESIGN OF SLIDER-CRANK MECHANISMS

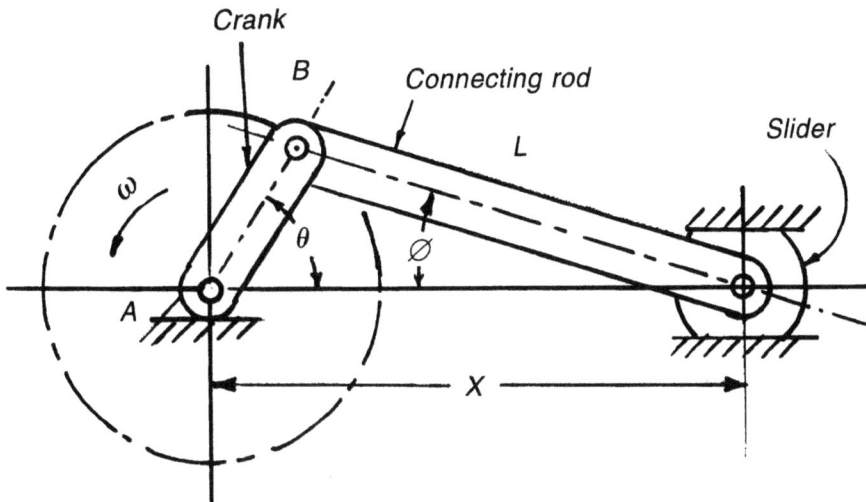

The slider crank, an efficient mechanism for changing reciprocating motion to rotary motion, is widely used in engines, pumps, automatic machinery, and machine tools.

The equations developed here for finding these factors are in a more simplified form than is generally found in text books.

SYMBOLS

L = length of connecting rod
R = crank length; radius of crank circle
x = distance from center of crankshaft A to wrist pin C
x' = slider velocity (linear velocity of point C)
x'' = slider acceleration
θ = crank angle measured from dead center (when slider is fully extended)
ϕ = angular position of connecting rod; $\phi = 0$ when $\theta = 0$
ϕ' = connecting-rod angular velocity = $d\phi/dt$
ϕ'' = connecting-rod angular acceleration = $d^2\phi/dl^2$
ω = constant crank angle velocity

Displacement of slider

$$x = L \cos \phi + R \cos \phi$$

Also:

$$\cos \phi = \left[1 - \left(\frac{R}{L} \right)^2 \sin^2\theta \right]^{1/2}$$

Angular velocity of the connecting rod

$$\phi' = \omega \left[\frac{(R/L) \cos \theta}{[1 - (R/L)^2 \sin^2\theta]^{1/2}} \right]$$

Linear velocity of the piston

$$\frac{x'}{L} = -\omega \left[1 + \frac{\phi'}{\omega} \right] \left(\frac{R}{L} \right) \sin \theta$$

Angular acceleration of the connecting rod

$$\phi'' = \frac{\omega^2 (R/L) \sin \theta [(R/L)^2 - 1]}{[1 - (R/L)^2 \sin^2\theta]^{3/2}}$$

Slider acceleration

$$\frac{x''}{L} = -\omega^2 \left(\frac{R}{L} \right) \left[\cos \theta + \frac{\phi''}{\omega^2} \sin \theta + \frac{\phi'}{\omega} \cos \theta \right]$$

CHAPTER 6
GEARS: DEVICES, DRIVES, AND MECHANISMS

GEARS AND ECCENTRIC DISK PROVIDE QUICK INDEXING

- Index table
- Index gear
- Locking rack
- Connecting rod
- Eccentric disk
- Oscillating planet gear
- Ring gear
- Input gear
- Control gear
- Control cam
- Spring-loaded detent

. . provides choice of indexing modes

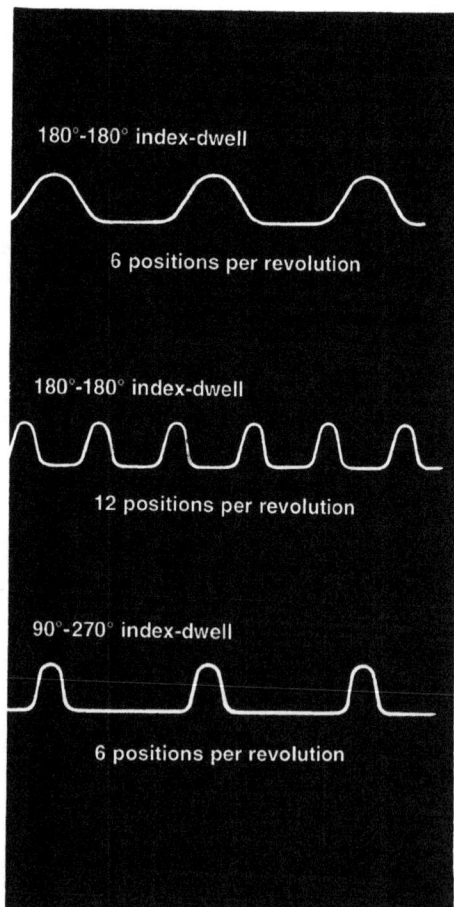

180°-180° index-dwell

6 positions per revolution

180°-180° index-dwell

12 positions per revolution

90°-270° index-dwell

6 positions per revolution

Both stops and dwell are adjustable.

An ingenious intermittent mechanism with its multiple gears, gear racks, and levers provides smoothness and flexibility in converting constant rotary motion into a start-and-stop type of indexing.

It works equally well for high-speed operations, as fast as 2 seconds per cycle, including index and dwell, or for slow-speed assembly functions.

The mechanism minimizes shock loads and offers more versatility than the indexing cams and genevas usually employed to convert rotary motion into start-stop indexing. The number of stations (stops) per revolution of the table can easily be changed, as can the period of dwell during each stop.

Advantages. This flexibility broadens the scope of such automatic machine operations as feeding, sorting, packaging, and weighing that the rotary table can perform. But the design offers other advantages, too:

- Gears instead of cams make the mechanism cheaper to manufacture, because gears are simpler to machine.
- The all-mechanical interlocked system achieves an absolute time relationship between motions.
- Gearing is arranged so that the machine automatically goes into a dwell when it is overloaded, preventing damage during jam-ups.
- Its built-in anti-backlash gear system averts rebound effects, play, and lost motion during stops.

How it works. Input from a single motor drives an eccentric disk and connecting rod. In the position shown in the drawing, the indexing gear and table are locked by the rack—the planet gear rides freely across the index gear without imparting any motion to it. Indexing of the table to its next position begins when the control cam simultaneously releases the locking rack from the index gear and causes the spring control ring gear to pivot into mesh with the planet.

This is a planetary gear system containing a stationary ring gear, a driving planet gear, and a "sun" index gear. As the crank keeps moving to the right, it begins to accelerate the index gear with harmonic motion—a desirable type of motion because of its low acceleration-deceleration characteristics—while it is imparting high-speed transfer to the table.

At the end of 180° rotation of the crank, the control cam pivots the ring-gear segment out of mesh and, simultaneously, engages the locking rack. As the connecting rod is drawn back, the planet gear rotates freely over the index gear, which is locked in place.

The cam control is so synchronized that all toothed elements are in full engagement briefly when the crank arm is in full toggle at both the beginning and end of index. The device can be operated just as easily in the other direction.

Overload protection. The ring gear segment includes a spring-load detent mechanism (simplified in the illustration) that will hold the gearing in full engagement under normal indexing forces. If rotation of the table is blocked at any point in index, the detent spring force is overcome and the ring gear pops out of engagement with the planet gear.

A detent roller (not shown) will then snap into a second detent position, which will keep the ring gear free during the remainder of the index portion of the cycle. After that, the detent will automatically reset itself.

Incomplete indexing is detected by an electrical system that stops the machine at the end of the index cycle.

Easy change of settings. To change indexes for a new job setup, the eccentric is simply replaced with one having a different crank radius, which gives the proper drive stroke for 6, 8, 12, 16, 24, 32, or 96 positions per table rotation.

Because indexing occurs during one-half revolution of the eccentric disk, the input gear must rotate at two or three times per cycle to accomplish indexing of $1/2$, $1/4$, or $1/16$ of the total cycle time (which is the equivalent to index-to-dwell cycles of 180°/180°, 90°/270°, or 60°/300°). To change the cycle time, it is only necessary to mount a difference set of change gears between input gear and control cam gear.

ODD-SHAPED PLANETARY GEARS SMOOTH STOP AND GO

This intermittent-motion mechanism for automatic processing machinery combines gears with lobes; some pitch curves are circular and some are noncircular.

This intermittent-motion mechanism combines circular gears with noncircular gears in a planetary arrangement, as shown in the drawing.

The mechanism was developed by Ferdinand Freudenstein, a professor of mechanical engineering at Columbia University. Continuous rotation applied to the input shaft produces a smooth, stop-and-go unidirectional rotation in the output shaft, even at high speeds.

This jar-free intermittent motion is sought in machines designed for packaging, production, automatic transfer, and processing.

Varying differential. The basis for Freudenstein's invention is the varying differential motion obtained between two sets of gears. One set has lobular pitch circles whose curves are partly circular and partly noncircular.

The circular portions of the pitch curves cooperate with the remainder of the mechanism to provide a dwell time or stationary phase, or phases, for the output member. The non-circular portions act with the remainder of the mechanism to provide a motion phase, or phases, for the output member.

Competing genevas. The main competitors to Freudenstein's "pulsating planetary" mechanism are external genevas and starwheels. These devices have a number of limitations that include:

- Need for a means, separate from the driving pin, for locking the output member during the dwell phase of the motion. Moreover, accurate manufacture and careful design are required to make a smooth transition from rest to motion and vice versa.

- Kinematic characteristics in the geneva that are not favorable for high-speed operation, except when the number of stations (i.e., the number of slots in the output member) is large. For example, there is a sudden change of acceleration of the output member at the beginning and end of each indexing operation.

- Relatively little flexibility in the design of the geneva mechanism. One factor alone (the number of slots

At heart of new planetary (in front view, circular set stacked behind noncircular set), two sets of gears when assembled (side view) resemble conventional unit (schematic).

in the output member) determines the characteristics of the motion. As a result, the ratio of the time of motion to the time of dwell cannot exceed one-half, the output motion cannot be uniform for any finite portion of the indexing cycle, and it is always opposite in sense to the sense of input rotation. The output shaft, moreover, must always be offset from the input shaft.

Many modifications of the standard external geneva have been proposed, including multiple and unequally spaced driving pins, double rollers, and separate entrance and exit slots. These proposals have, however, been only partly successful in overcoming these limitations.

Differential motion. In deriving the operating principle of his mechanism, Freudenstein first considered a conventional epicyclic (planetary) drive in which the input to the cage or arm causes a planet set with gears 2 and 3 to rotate the output "sun," gear 4, while another sun, gear 1, is kept fixed (see drawing).

Letting r_1, r_2, r_3, r_4, equal the pitch radii of the circular 1, 2, 3, 4, then the output ratio, defined as:

$$R = \frac{\text{angular velocity of output gear}}{\text{angular velocity of arm}}$$

is equal to: $\quad R = 1 - \dfrac{r_1 r_3}{r_2 r_4}$

Now, if $r_1 = r_4$ and $r_2 = r_3$, there is no "differential motion" and the output remains stationary. Thus if one gear pair, say 3 and 4, is made partly circular and partly noncircular, then where $r_2 = r_3$ and $r_1 = r_4$ for the circular portion, gear 4 dwells. Where $r_2 \neq r_3$ and $r_1 \neq r_4$ for the noncircular portion, gear 4 has motion. The magnitude of this motion depends on the difference in radii, in accordance with the previous equation. In this manner, gear 4 undergoes an intermittent motion (see graph).

Advantages. The pulsating planetary approach demonstrates some highly useful characteristics for intermittent-motion machines:

- The gear teeth serve to lock the output member during the dwell as well as to drive that member during motion.
- Superior high-speed characteristics are obtainable. The profiles of the pitch curves of the noncircular gears can be tailored to a wide variety of desired kinematic and dynamic characteristics. There need be no sudden terminal acceleration change of the driven member, so the transition from dwell to motion, and vice versa,

will be smooth, with no jarring of machine or payload.

- The ratio of motion to dwell time is adjustable within wide limits. It can even exceed unity, if desired. The number of indexing operations per revolution of the input member also can exceed unity.
- The direction of rotation of the output member can be in the same or opposite sense relative to that of the input member, according to whether the pitch axis P_{34} for the noncircular portions of gears 3 and 4 lies wholly outside or wholly inside the pitch surface of the planetary sun gear 1.
- Rotation of the output member is coaxial with the rotation of the input member.
- The velocity variation during motion is adjustable within wide limits. Uniform output velocity for part of the indexing cycle is obtainable; by

varying the number and shape of the lobes, a variety of other desirable motion characteristics can be obtained.

- The mechanism is compact and has relatively few moving parts, which can be readily dynamically balanced.

Design hints. The design techniques work out surprisingly simply, said Freudenstein. First the designer must select the number of lobes L_3 and L_4 on the gears 3 and 4. In the drawings, $L_3 = 2$ and $L_4 = 3$. Any two lobes on the two gears (i.e., any two lobes of which one is on one gear and the other on the other gear) that are to mesh together must have the same arc length. Thus, every lobe on gear 3 must mesh with every lobe on gear 4, and $T_3/T_4 = L_3/L_4 = 2/3$, where T_3 and T_4 are the numbers of teeth on gears 3 and 4. T_1 and T_2 will denote the numbers of teeth on gears 1 and 2.

Output motion (upper curve) has long dwell periods; velocity curve (center) has smooth transition from zero to peak; acceleration at transition is zero (bottom).

Next, select the ratio S of the time of motion of gear *4* to its dwell time, assuming a uniform rotation of the arm *5*. For the gears shown, $S = 1$. From the geometry,

$$(\theta_{30} + \Delta\theta_{30})L_3 = 360°$$

and

$$S = \Delta\theta_{30}/\theta_{30}$$

Hence

$$\theta_{30}(1 + S)L_3 = 360°$$

For $S = 1$ and $L_3 + 2$,

$$\theta_{30} = 90°$$

and

$$\Delta\theta_3 = 90°$$

Now select a convenient profile for the noncircular portion of gear *3*. One profile (see the profile drawing) that Freudenstein found to have favorable high-speed characteristics for stop-and-go mechanisms is

$$r_3 = R_3$$

$$\left[1 + \frac{\lambda}{2}\left(1 - \cos\frac{2\pi(\theta_3 - \theta_{30})}{\Delta\theta_3}\right)\right]$$

The profile defined by this equation has, among other properties, the characteristic that, at transition from rest to motion and vice versa, gear *4* will have zero acceleration for the uniform rotation of arm *5*.

In the above equation, λ is the quantity which, when multiplied by R_3, gives the maximum or peak value of $r_3 - R_3$, differing by an amount h' from the radius R_3 of the circular portions of the gear. The noncircular portions of each lobe are, moreover, symmetrical about their midpoints, the midpoints of these portions being indicated by m.

To evaluate the quantity λ, Freudenstein worked out the equation:

$$\lambda = \frac{1 - \mu}{\mu} \times$$

$$\frac{[S + \alpha - (1 + \alpha)\mu][\alpha - S - (1 + \alpha)\mu]}{[\alpha - (1 + \alpha)\mu]^2}$$

where $R_3\lambda$ = height of lobe

$$\mu = \frac{R_3}{A} = R_3/(R_3 + R_4)$$

$$\alpha = S + (1 + S)L_3/L_4$$

To evaluate the equation, select a suitable value for μ that is a reasonably simple rational fraction, i.e., a fraction such as $\frac{3}{8}$ whose numerator and denominator are reasonably small integral numbers.

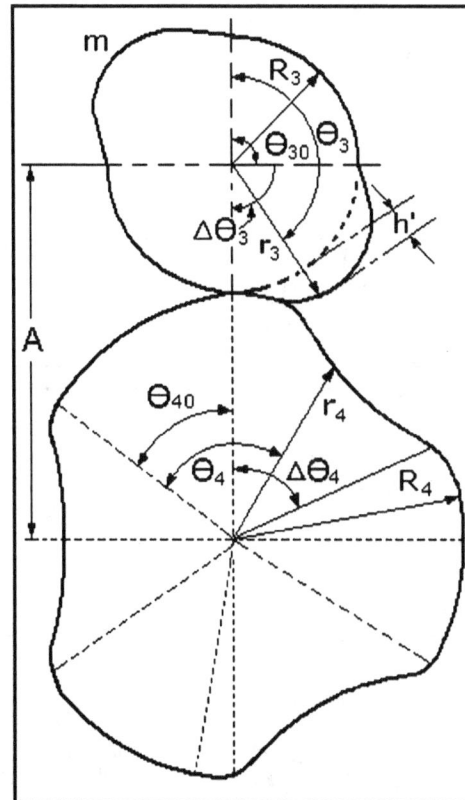

Profiles for noncircular gears are circular arcs blended to special cam curves.

Thus, without a computer or lengthy trial-and-error procedures, the designer can select the configuration that will achieve his objective of smooth intermittent motion.

CYCLOID GEAR MECHANISM
CONTROLS PUMP STROKE

An adjustable ring gear meshes with a planet gear having half of its diameter to provide an infinitely variable stroke in a pump. The adjustment in the ring gear is made by engaging other teeth. In the design below, a yoke replaces the connecting rod.

A metering pump for liquid or gas has an adjustable ring gear that meshes with a special-size planet gear to provide an infinitely variable stroke in the pump. The stroke can be set manually or automatically when driven by a servomotor. Flow control from 180 to 1200 liter/hr. (48 to 317 gal./hr.) is possible while the pump is at a standstill or running.

Straight-line motion is key. The mechanism makes use of a planet gear whose diameter is half that of the ring gear. As the planet is rotated to roll on the inside of the ring, a point on the pitch diameter of the planet will describe a straight line (instead of the usual hypocycloid curve). This line is a diameter of the ring gear. The left end of the connecting rod is pinned to the planet at this point.

The ring gear can be shifted if a second set of gear teeth is machined in its outer surface. This set can then be meshed with a worm gear for control. Shifting the ring gear alters the slope of the straight-line path. The two extreme positions are shown in the diagram. In the position of the mechanism shown, the pin will reciprocate vertically to produce the minimum stroke for the piston. Rotating the ring gear 90° will cause the pin to reciprocate horizontally to produce the maximum piston stroke.

The second diagram illustrates another version that has a yoke instead of a connecting rod. This permits the length of the stroke to be reduced to zero. Also, the length of the pump can be substantially reduced.

GEARS CONVERT ROTARY-TO-LINEAR MOTION

A compact gear system that provides linear motion from a rotating shaft was designed by Allen G. Ford of The Jet Propulsion Laboratory in California. It has a planetary gear system so that the end of an arm attached to the planet gear always moves in a linear path (drawing).

The gear system is set in motion by a motor attached to the base plate. Gear *A*, attached to the motor shaft, turns the case assembly, causing Gear *C* to rotate along Gear *B*, which is fixed.

The arm is the same length as the center distance between Gears *B* and *C*. Lines between the centers of Gear *C*, the end of the arm, and the case axle form an isosceles triangle, the base of which is always along the plane through the center of rotation. So the output motion of the arm attached to Gear *C* will be in a straight line.

When the end of travel is reached, a switch causes the motor to reverse, returning the arm to its original position.

The end of arm moves in a straight line because of the triangle effect (right).

TWIN-MOTOR PLANETARY GEARS OFFER SAFETY AND DUAL-SPEED

Many operators and owners of hoists and cranes fear the possible catastrophic damage that can occur if the driving motor of a unit should fail for any reason. One solution to this problem is to feed the power of two motors of equal rating into a planetary gear drive.

Power supply. Each of the motors is selected to supply half the required output power to the hoisting gear (see diagram). One motor drives the ring gear, which has both external and internal teeth. The second motor drives the sun gear directly.

Both the ring gear and sun gear rotate in the same direction. If both gears rotate at the same speed, the planetary cage, which is coupled to the output, will also revolve at the same speed (and in the same direction). It is as if the entire inner works of the planetary were fused together. There would be no relative motion. Then, if one motor fails, the cage will revolve at half its original speed, and the other motor can still lift with undiminished capacity. The same principle holds true when the ring gear rotates more slowly than the sun gear.

Power flow from two motors combine in a planetary that drives the cable drum.

No need to shift gears. Another advantage is that two working speeds are available as a result of a simple switching arrangement. This makes is unnecessary to shift gears to obtain either speed.

The diagram shows an installation for a steel mill crane.

ELEVEN CYCLOID GEAR MECHANISMS

The appeal of cycloidal mechanisms is that they can be tailored to provide one of these three common motions:

- **Intermittent**—with either short or long dwells
- **Rotary with progressive oscillation**—where the output undergoes a cycloidal motion during which the forward motion is greater than the return motion
- **Rotary**-to-linear with a dwell period

All the cycloidal mechanisms shown here are general. This results in compact positive mechanisms capable of operating at relatively high speeds with little backlash or "slop." These mechanisms can be classified into three groups:

Hypocycloid—the points tracing the cycloidal curves are located on an external gear rolling inside an internal ring gear. This ring gear is usually stationary and fixed to the frame.

Epicycloid—the tracing points are on an external gear that rolls in another external (stationary) gear.

Pericycloid—the tracing points are located on an internal gear that rolls on a stationary external gear.

Basic hypocycloid curves

Double-dwell mechanism

Coupling the output pin to a slotted member produces a prolonged dwell in each of the extreme positions. This is another application of the diamond-type hypocycloidal curve.

The input drives a planet in mesh with a stationary ring gear. Point P_1 on the planet gear describes a diamond-shape curve, point P_2 on the pitch line of the planet describes the familiar cusp curve, and point P_3, which is on an extension rod fixed to the planet gear, describes a loop-type curve. In one application, an end miller located at P_1 machined a diamond-shaped profile.

Long-dwell drive mechanism

Internal drive mechanism

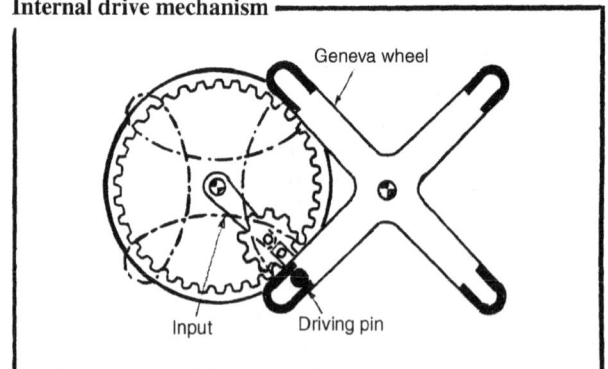

In common with standard, four-station genevas, each rotation of the input of this drive indexes the slotted geneva 90°. A pin fastened to the planet gear causes the drive to describe a rectangular-shaped cycloidal curve. This produces a smoother indexing motion because the driving pin moves on a noncircular path.

A loop-type curve permits the driving pin to enter the slot in a direction that is radially outward from the center. The pin then loops over to index the cross member rapidly. As with other genevas, the output rotates 90° before going into a long dwell period during each 270° rotation of the input element.

Cycloidal parallelogram mechanism

Two identical hypocycloid mechanisms guide the point of the bar along the triangularly shaped path. The mechanisms are useful where space is limited in the area where the curve must be described. These double-cycloid mechanisms can be designed to produce other curve shapes.

Cycloidal short-dwell rotary mechanism

The pitch circle of this planet gear is exactly one-quarter that of the ring gear. A pin on the planet gear will cause the slotted output member to dwell four times during each revolution of the input shaft.

Cycloidal rocker mechanism

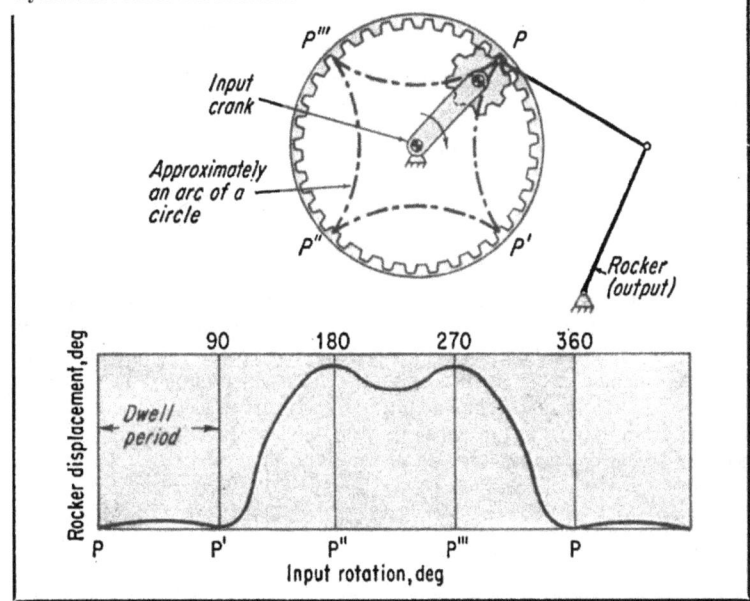

The curvature of the cusp is approximately that of an arc of a circle. Hence the rocker reaches a long dwell at the right extreme position while point *P* moves to *P'*. There is then a quick return from *P'* to *P''*, with a momentary dwell at the end of this phase. The rocker then undergoes a slight oscillation from point *P''* to *P'''*, as shown in the rocker displacement diagram.

Cycloidal reciprocator mechanism

Part of curve P-P' produces a long dwell, but the five-lobe cycloidal curve avoids a marked oscillation at the end of the stroke. There are also two points of instantaneous dwell where the curve is perpendicular to the connecting rod.

By making the pitch diameter of the planet gear equal to half that of the ring gear, every point on the planet gear (such as points P_2 and P_3) will describe elliptical curves which get flatter as the points are selected closer to the pitch circle. Point P_1, at the center of the planet, describes a circle; point P_4, at the pitch circle, describes a straight line. When a cutting tool is placed at P_3, it will cut almost-flat sections from round stock, as when milling flats on a bolt. The other two flats of the bolt can be cut by rotating the bolt or the cutting tool 90°.

Epicycloid reciprocator mechanism

Adjustable harmonic drive mechanism

By making the planet gear diameter half that of the internal gear, a straight-line output curve can be produced by the driving pin which is fastened to the planet gear. The pin engages the slotted member to cause the output to reciprocate back and forth with harmonic (sinusoidal) motion. The position of the fixed ring gear can be changed by adjusting the lever, which in turn rotates the straight-line output curve. When the curve is horizontal, the stroke is at a maximum; when the curve is vertical, the stroke is zero.

Elliptical-motion drive mechanism

Here the sun gear is fixed and the planet is gear driven around it by the input link. There is no internal ring gear as with the hypocycloid mechanisms. Driving pin P on the planet describes the curve shown, which contains two almost-flat portions. If the pin rides in the slotted yoke, a short dwell is produced at both the extreme positions of the output member. The horizontal slots in the yoke ride the end-guides, as shown.

FIVE CARDAN-GEAR MECHANISMS

These gearing arrangements convert rotary into straight-line motion, without the need for slideways.

Fig. 1 Cardan gearing works on the principle that any point on the periphery of a circle rolling on the inside of another circle describes, in general, a hypocyloid. This curve degenerates into a true straight line (diameter of the larger circle) if the diameters of both circles are in the ratio of 1:2. The rotation of the input shaft causes a small gear to roll around the inside of the fixed gear. A pin located on the pitch circle of the small gear describes a straight line. Its linear displacement is proportional to the theoretically true sine or cosine of the angle through which the input shaft is rotated.

Fig. 2 Cardan gearing and a Scotch yoke in combination provide an adjustable stroke. The angular position of the outer gear is adjustable. The adjusted stroke equals the projection of the large diameter, along which the drive pin travels, on the Scotch-yoke's centerline. The yoke motion is simple harmonic.

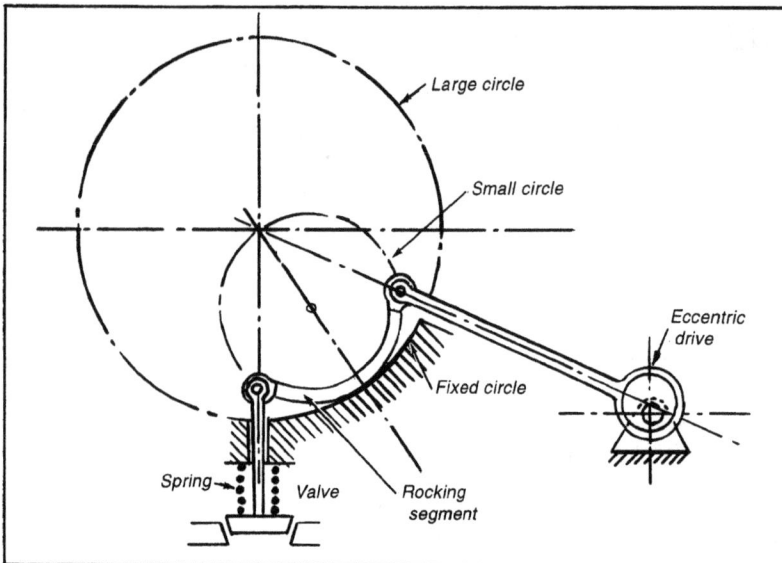

Fig. 3 **A valve drive** demonstrates how the Cardan principle can be applied. A segment of the smaller circle rocks back and forth on a circular segment whose radius is twice as large. The input and output rods are each attached to points on the small circle. Both these points describe straight lines. The guide of the valve rod prevents the rocking member from slipping.

Fig. 4 **A simplified Cardan** mechanism eliminates the need for the relatively expensive internal gear. Here, only spur gears are used, and the basic requirements must be met, i.e., the 1:2 ratio and the proper direction of rotation. The rotation requirement is met by introducing an idler gear of appropriate size. This drive delivers a large stroke for the comparative size of its gears.

Fig. 5 **A rearrangement of gearing** in the simplified Cardan mechanism results in another useful motion. If the fixed sun gear and planet pinion are in the ratio of 1:1, an arm fixed to the planet shaft will stay parallel to itself during rotation, while any point on the arm describes a circle of radius R. When arranged in conjugate pairs, the mechanism can punch holes on moving webs of paper.

CONTROLLED DIFFERENTIAL GEAR DRIVES

By coupling a differential gear assembly to a variable speed drive, a drive's horsepower capacity can be increased at the expense of its speed range. Alternatively, the speed range can be increased at the expense of the horsepower range. Many combinations of these variables are possible. The features of the differential depend on the manufacturer. Some systems have bevel gears, others have planetary gears. Both single and double differentials are employed. Variable-speed drives with differential gears are available with ratings up to 30 hp.

Horsepower-increasing differential (Fig. 1). The differential is coupled so that the output of the motor is fed into one side and the output of the speed variator is fed into the other side. An additional gear pair is employed as shown in Fig. 1

Output speed

$$n_4 = \frac{1}{2}\left(n_1 + \frac{n_2}{R}\right)$$

Output torque

$$T_4 = 2T_3 = 2RT_2$$

Output hp

$$hp = \left(\frac{Rn_1 + n_2}{63,025}\right)T_2$$

hp increase

$$\Delta hp = \left(\frac{Rn_1}{63,025}\right)T_2$$

Speed variation

$$n_{4\,max} - n_{4\,min} = \frac{1}{2R}(n_{2\,max} - n_{2\,min})$$

Speed range increase differential (Fig. 2). This arrangement achieves a wide range of speed with the low limit at zero or in the reverse direction.

Fig. 1

Fig. 2

Transmission assembly

Fig. 3 A variable-speed transmission consists of two sets of worm gears feeding a differential mechanism. The output shaft speed depends on the difference in rpm between the two input worms. When the worm speeds are equal, output is zero. Each worm shaft carries a cone-shaped pulley. These pulley are mounted so that their tapers are in opposite directions. Shifting the position of the drive belt on these pulleys has a compound effect on their output speed.

FLEXIBLE FACE-GEARS ARE EFFICIENT HIGH-RATIO SPEED REDUCERS

A system of flexible face-gearing provides designers with a means for obtaining high-ratio speed reductions in compact trains with concentric input and output shafts.

With this approach, reduction ratios range from 10:1 to 200:1 for single-stage reducers, whereas ratios of millions to one are possible for multi-stage trains. Patents on the flexible face-gear reducers were held by Clarence Slaughter of Grand Rapids, Michigan.

Building blocks. Single-stage gear reducers consist of three basic parts: a flexible face-gear made of plastic or thin metal; a solid, non-flexing face-gear; and a wave former with one or more sliders and rollers to force the flexible gear into mesh with the solid gear at points where the teeth are in phase.

The high-speed input to the system usually drives the wave former. Low-speed output can be derived from either the flexible or the solid face gear; the gear not connected to the output is fixed to the housing.

Teeth make the difference. Motion between the two gears depends on a slight difference in their number of teeth (usually one or two teeth). But drives with gears that have up to a difference of 10 teeth have been devised.

On each revolution of the wave former, there is a relative motion between the two gears that equals the difference in their numbers of teeth. The reduction ratio

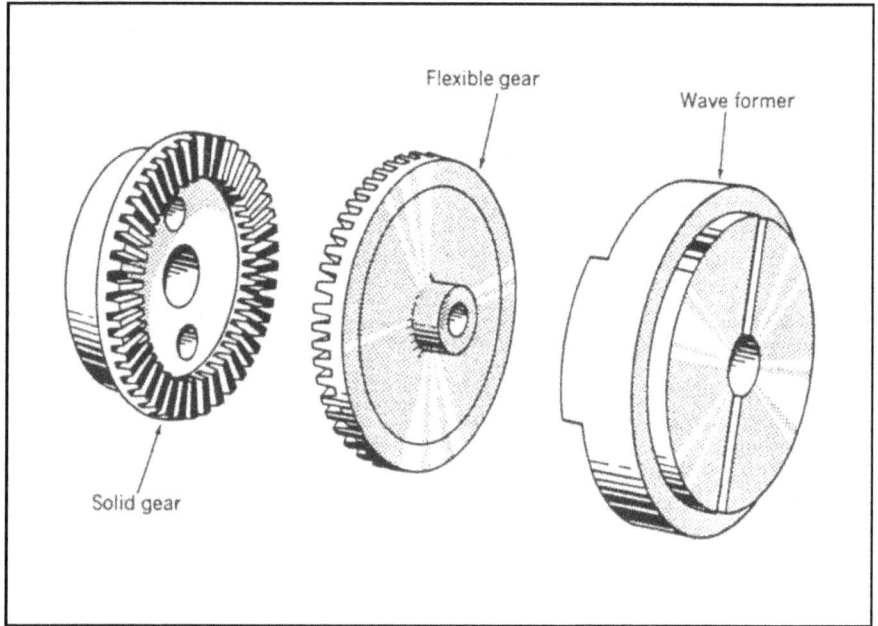

A flexible face-gear is flexed by a rotating wave former into contact with a solid gear at point of mesh. The two gears have slightly different numbers of teeth.

equals the number of teeth in the output gear divided by the difference in their numbers of teeth.

Two-stage and four-stage gear reducers are made by combining flexible and solid gears with multiple rows of teeth

and driving the flexible gears with a common wave former.

Hermetic sealing is accomplished by making the flexible gear serve as a full seal and by taking output rotation from the solid gear.

A two-stage speed reducer is driven by a common-wave former operating against an integral flexible gear for both stages.

A four-stage speed reducer can, theoretically, attain reductions of millions to one. The train is both compact and simple.

ROTARY SEQUENCER GEARS TURN COAXIALLY

Two coaxial rotations, one clockwise and one counterclockwise, are derived from a single clockwise rotation.

A proposed rotary sequencer is assembled from a conventional planetary differential gear-set and a latching mechanism. Its single output and two rotary outputs (one clockwise and one counterclockwise) are coaxial, and the output torque is constant over the entire cycle. Housed in a lightweight, compact, cylindrical package, the sequencer requires no bulky ratchets, friction clutches, or cam-and-track followers. Among its possible applications are sequencing in automated production-line equipment, in home appliances, and in vehicles.

The sequencer is shown in Figure 1. A sun gear connects with four planetary gears that engage a ring gear. With the ring gear held stationary, clockwise rotation of the sun gear causes the entire planetary-gear carrier also to rotate clockwise. If the planetary-gear carrier is held fixed, the ring gear will rotate counterclockwise when the sun gear rotates clockwise.

Figure 2 shows the latch. It consists of a hook (the carrier hook) that is rigidly attached to the planetary-gear carrier, a ring that is rigidly attached to the ring gear, and a latch pivot arm with a pair of latch rollers attached to one end. The other end of the pivot arm rotates about a short shaft that extends from the fixed wall of the housing.

The sequencer cycle starts with the ring latch roller resting in a slot in the ring. This locks the ring and causes the planetary-gear carrier to rotate clockwise with the input shaft (Fig. 2a). When the carrier hook has rotated approximately three-quarters of a complete cycle, it begins to engage the planet-carrier latch roller (Fig. 2b), causing the latch pivot arm to rotate and the ring latch roller to slip out of its slot (Fig. 2c). This frees the ring and ring gear for counterclockwise motion, while locking the carrier. After a short interval of concurrent motion, the planetary-gear output shaft ceases its clockwise motion, and the ring-gear output shaft continues its clockwise motion.

When the ring reaches the position in Fig. 2d, the cycle is complete, and the input shaft is stopped. If required, the input can then be rotated counterclockwise, and the sequence will be reversed until the starting position (Fig. 2a) is reached again.

In a modified version of the sequencer, the latch pivot arm is shortened until its length equals the radii of the rollers. This does away with the short overlap of output rotations when both are in motion. For this design, the carrier motion ceases before the ring begins its rotation.

This work was done by Walter T. Appleberry of Rockwell International Corp. for **Johnson Space Center,** *Houston, Texas.*

Fig. 1 The Rotary Sequencer has a ring-gear output that is coaxial with a planetary-gear output. Clockwise rotation of the input is converted to clockwise rotation of the planetary-gear output followed by counterclockwise rotation of the ring-gear output. The sequence is controlled by the latch action described in Fig. 2.

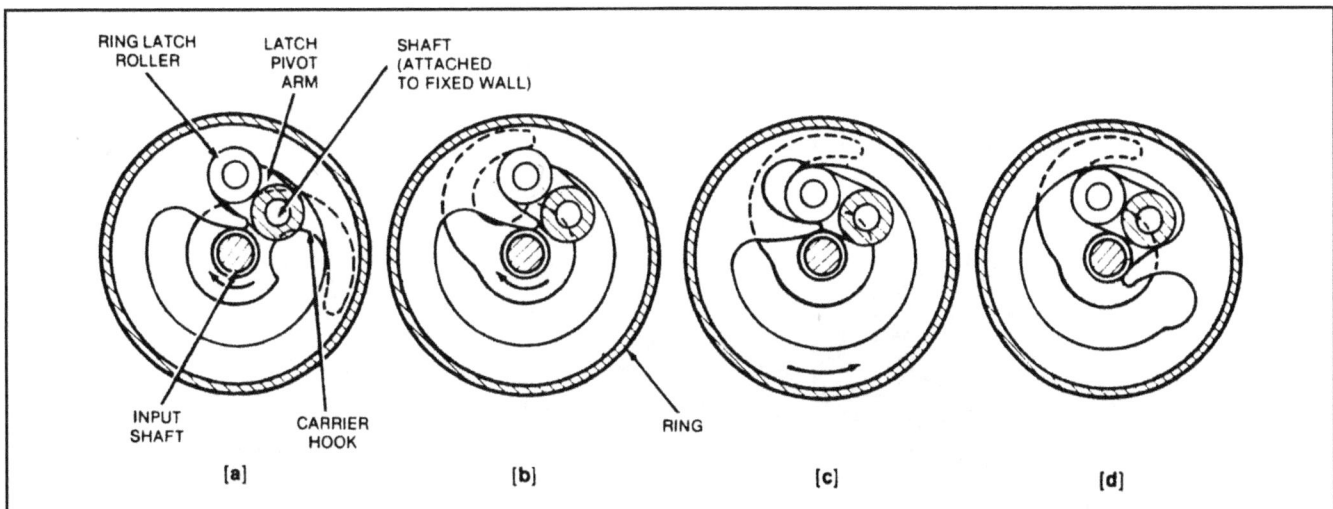

Fig. 2 The Latch Sequence is shown in four steps: (a) The input shaft rotates the carrier clockwise while the ring latch roller holds the ring gear stationary; (b) the carrier hook begins to engage the carrier latch roller; (c) the ring latch roller begins to move out of its slot, and the carrier motion ceases while the ring begins to move; and (d) the sequence has ended with the ring in its final position.

PLANETARY GEAR SYSTEMS

Designers keep finding new and useful planetaries. Forty-eight popular types are given here with their speed-ratio equations.

MISSILE SILO COVER DRIVE

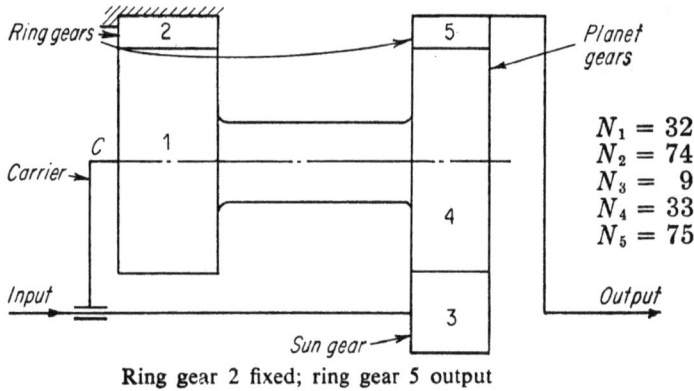

$N_1 = 32$
$N_2 = 74$
$N_3 = 9$
$N_4 = 33$
$N_5 = 75$

Ring gear 2 fixed; ring gear 5 output

Speed-ratio equation

$$R = \frac{1 + \dfrac{N_4 N_2}{N_3 N_1}}{1 - \dfrac{N_4 N_2}{N_5 N_1}} = \frac{1 + \dfrac{(33)(74)}{(9)(32)}}{1 - \dfrac{(33)(74)}{(75)(32)}} = -541\tfrac{2}{3}$$

Symbols

C = carrier (also called "spider")—a non-gear member of gear train whose rotation affects gear ratio

N = number of teeth
R = overall speed reduction ratio
1, 2, 3, etc. = gears in a train (corresponding to labels on schematic diagram)

DOUBLE-ECCENTRIC DRIVE

Input is through double-throw crank (carrier). **Gear 1**

$$R = \frac{1}{1 - \dfrac{N_5 N_3 N_1}{N_6 N_4 N_2}}$$

When $N_1 = 103$, $N_2 = 110$, $N_3 = 109$, $N_4 = 100$, $N_5 = 94$, $N_6 = 96$

$$R = \frac{1}{1 - \dfrac{(94)(109)(103)}{(96)(100)(110)}} = 1505$$

(A)

Coupled ring gear 2, & sun gear 3

2

1   3

Coupled planet carriers

Fixed ring gear 4

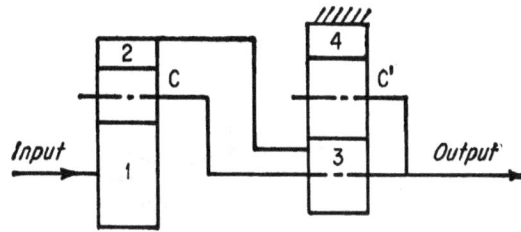

$$R = 1 - \frac{N_2 N_4}{N_1 N_3}$$

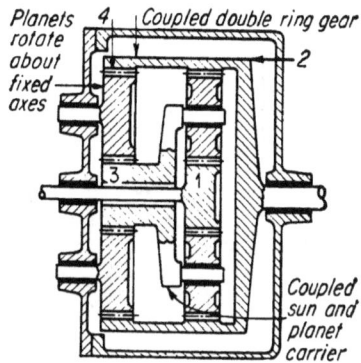

(B)

Planets rotate about fixed axes

4   Coupled double ring gear

2

3   1

Coupled sun and planet carrier

$$R = \left(1 + \frac{N_2}{N_1}\right)\left(-\frac{N_4}{N_3}\right) - \frac{N_2}{N_1}$$

(C)

Planets rotate about fixed axes

Coupled ring 4, and planet carrier

Coupled ring gear 2 and sun 3

3   1

$$R = 1 + \frac{N_2}{N_1}\left(1 + \frac{N_4}{N_3}\right)$$

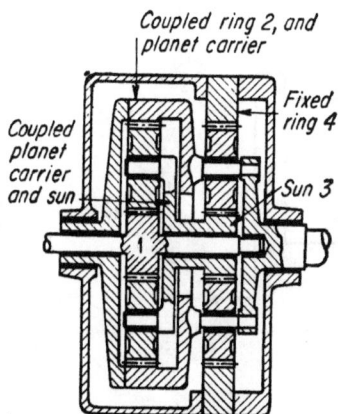

(D)

Coupled ring 2, and planet carrier

Fixed ring 4

Coupled planet carrier and sun

Sun 3

$$R = 1 + \frac{N_4}{N_3}\left(1 + \frac{N_2}{N_1}\right)$$

Output is difference
between speeds of two
parts leading to high
reduction ratios

(A)

Rotating planet carrier

Fixed sun

$N_1 = 20$
$N_2 = 31$
$N_3 = 32$
$N_4 = 19$

$$R = \cfrac{1}{1 - \cfrac{N_3 N_1}{N_4 N_2}} = \cfrac{1}{1 - \cfrac{(32)(20)}{(19)(31)}} = -11.549$$

(B)

Fixed ring gear

Rotating planet carrier

Rotating ring gear

$$R = \cfrac{1}{1 - \cfrac{N_3 N_2}{N_4 N_1}}$$

(C)

Rotating double ring gear

Planets rotate about fixed axes

Rotating planet carrier

$$R = \frac{1 + (N_4/N_6)}{(N_4/N_6) - (N_1/N_3)}$$

(D)

Rotating planet carrier

Fixed annulus

Rotating ring gear

$$R = \cfrac{1 + (N_3/N_1)}{1 - \cfrac{N_4 N_3}{N_5 N_2}}$$

Assembly

Schematic

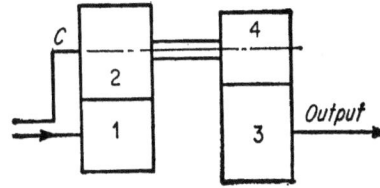

Input member	Fixed member	Output member	Speed-ratio equation
1	C	2	$R = -N_2/N_1$
2	C	1	$R = -N_1/N_2$
1	2	C	$R = 1 + (N_2/N_1)$
2	1	C	$R = 1 + (N_1/N_2)$
C	2	1	$R = \dfrac{1}{1 + (N_2/N_1)}$
C	1	2	$R = \dfrac{1}{1 + (N_1/N_2)}$

Input member	Fixed member	Output member	Speed-ratio equation
1	C	3	$R = \dfrac{N_2 N_3}{N_1 N_4}$
1	3	C	$R = 1 - \dfrac{N_2 N_3}{N_1 N_4}$
3	1	C	$R = 1 - \dfrac{N_1 N_4}{N_2 N_3}$
3	C	1	$R = \dfrac{N_4 N_1}{N_3 N_2}$
C	1	3	$R = 1 \left/ \left(1 - \dfrac{N_1 N_4}{N_2 N_3}\right)\right.$
C	3	1	$R = 1 \left/ \left(1 - \dfrac{N_2 N_3}{N_1 N_4}\right)\right.$

HUMPAGE'S BEVEL GEARS

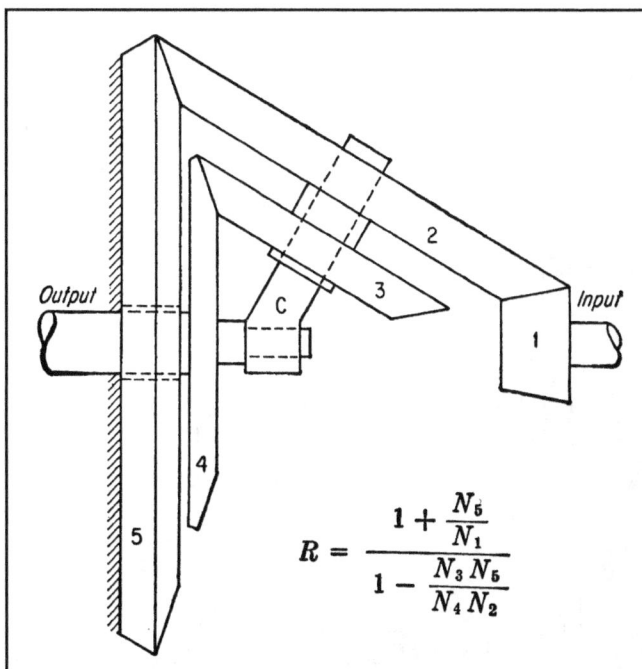

$$R = \frac{1 + \dfrac{N_5}{N_1}}{1 - \dfrac{N_3 N_5}{N_4 N_2}}$$

TWO-SPEED FORDOMATIC (Ford Motor Co.)

$N_1 = 27$
$N_2 = 31$
$N_3 = 90$
$N_4 = 36$
$N_5 = 19$

Low gear—
gear 1 fixed

$$R = 1 + \frac{N_1}{N_4} = 1.75$$

Reverse gear—
gear 3 fixed

$$R = 1 - \frac{N_3}{N_4} = -1.50$$

Note: Power-Glide Transmission is similar to above, but with $N_1 = 23$, $N_2 = 28$, $N_3 = 79$, $N_4 = 28$, $N_5 = 18$. This produces identical ratios in low and reverse.

$$R = 1 + \frac{23}{28} = 1.82 \qquad R = 1 - \frac{79}{28} = -1.82$$

CRUISE-O-MATIC 3-SPEED TRANSMISSION (Ford Motor Co.)

Long planet, $N_3 = 18$
Short planet, $N_2 = 18$
Sun gears, $N_4 = 36$, $N_1 = 30$
Ring gears, $N_5 = 72$

Low gear—Input to 1 C fixed

$$R = \frac{N_5}{N_1} = 2.4$$

Intermediate gear—
Input to 1, gear 4 fixed

$$R = \frac{1 + \dfrac{N_4}{N_1}}{1 + \dfrac{N_4}{N_5}} = 1.467$$

Reverse gear—
Input to 4, C fixed

$$R = \frac{N_5}{N_4} = -2.0$$

HYDRAMATIC 3-SPEED TRANSMISSION (General Motors)

$N_1 = 46$
$N_2 = 82$
$N_3 = 39$
$N_4 = 77$

Low gear—
Input to 3, 4 fixed

$$R = 1 + \frac{N_4}{N_3} = 2.97$$

Intermediate gear— Input to 2, 1 fixed

$$R = 1 + \frac{N_1}{N_2} = 1.56$$

Reverse gear—Input to 3, 2 fixed

$$R = 1 - \frac{N_4 N_2}{N_3 N_1} = -2.52$$

TRIPLE PLANETARY DRIVES

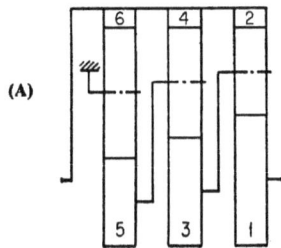

(A)

Input to gear 1, output from gear 6

(A)
$$R = \left(1 + \frac{N_2}{N_1}\right)\left[\left(1 + \frac{N_4}{N_3}\right)\left(-\frac{N_6}{N_5}\right) - \frac{N_4}{N_3}\right] - \frac{N_2}{N_1}$$

(B)
$$R = \left[1 + \frac{N_1}{N_2}\left(1 + \frac{N_4}{N_3}\right)\right]\left(1 + \frac{N_6}{N_5}\right)$$

(C)
$$R = \left[1 + \frac{N_4/N_3}{1 + (N_2/N_1)}\right]\bigg/\left[1 + \frac{N_4/N_3}{1 + (N_6/N_5)}\right]$$

FORD TRACTOR DRIVES

Ring gear 3 coupled to sun gear 1; split output.

$$N_1 = 46$$
$$N_2 = 77$$
$$N_3 = 29$$
$$N_4 = 85$$

(A) $\quad R = 1 + \dfrac{N_3}{N_2}\left(1 + \dfrac{N_4}{N_1}\right)$

(B) $\quad R = \left(\dfrac{1 + (N_1/N_2)}{1 - \dfrac{N_1\,N_3}{N_2\,N_4}}\right)\left(1 + \dfrac{N_6}{N_5}\right)$

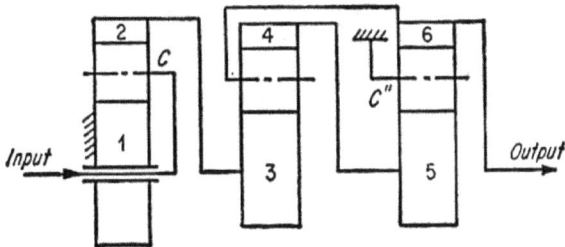

(C) $\quad R = \dfrac{1}{1 + \dfrac{N_1}{N_2}}\left[1 + \dfrac{N_4}{N_3}\left(1 + \dfrac{N_6}{N_5}\right)\right]$

(D) $\quad R = \dfrac{N_3}{N_4}\left(1 - \dfrac{N_4}{N_3} + \dfrac{N_2}{N_1}\right)$

LYCOMING TURBINE DRIVE

$$R = \left(1 + \dfrac{N_3}{N_2}\right)$$
$$\times \left(1 + \dfrac{N_4}{N_1}\right)$$

Input to sun gear 2, output to propeller shaft.
Basically same system as the Ford tractor drive, (gears
are numbered the same way) and will have the same speed-ratio.

COMPOUND SPUR-BEVEL GEAR DRIVE

$$R = \dfrac{1}{-\dfrac{N_1}{N_3}\left(1 + \dfrac{N_7\,N_5}{N_8\,N_6}\right) - \dfrac{N_7\,N_5\,N_2}{N_8\,N_6\,N_4}}$$

151

TWO-GEAR PLANETARY DRIVES

(A)

(B)

The pin in (A) is fixed to the planet which is mounted on the eccentric hub of the input shaft. The ring gear is the output gear. The system in (B) is simplified, but it produces slight pulsations in output.

$$R = \frac{1}{1 - (N_1/N_2)}$$

PLANOCENTRIC DRIVE

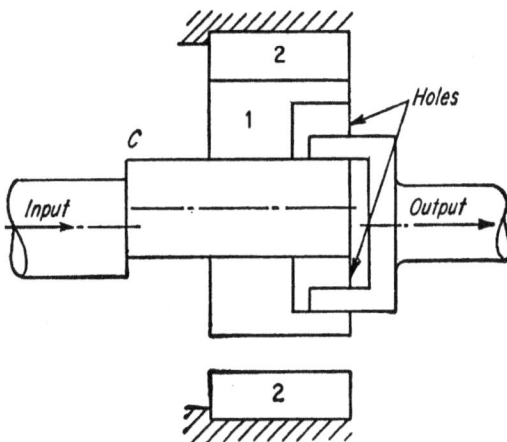

$N_2 = 65$
$N_2 = 64$

The planet gear 1 is eccentrically mounted to the input gear (planet 1 is not rigidly connected to the eccentric). The output is driven by holes.

$$R = \frac{N_1}{N_1 - N_2} = \frac{64}{64 - 65} = -64$$

WOBBLE-GEAR DRIVE

This drive is a close relative of the harmonic drive. The bevel "wobble" gear mesh at only one point on the circumference because of the slight angle of inclination of the driving gear, N_1, which has one tooth more than output gear, N_2. The driving gear, N_1, does not rotate; it yaws and pitches only.

$$R = R_i = \frac{1}{1 - m_{or}}$$

$$R = \frac{1}{1 - \frac{N_1}{N_2}} = \frac{1}{1 - \frac{101}{100}} = -100$$

NONCIRCULAR GEARS ARE BALANCED FOR SPEED

Noncircular gears generally cost more than competitive components such as linkages and cams. But with the development of modern production methods, such as the computer-controlled gear shaper, cost has gone down considerably. Also, in comparison with linkages, noncircular gears are more compact and balanced—and they can be more easily balanced. These are important considerations in high-speed machinery. Furthermore, the gears can produce continuous, unidirectional cyclic motion—a point in their favor when compared with cams. The disadvantage of cams is that they offer only reciprocating motion.

Applications Can Be Classified into Two Groups:

- Where only an over-all change in angular velocity of the driven member is required, as in quick-return drives, intermittent mechanisms in such machines as printing presses, planers, shears, winding machines, and automatic-feed machines.
- Where precise, nonlinear functions must be generated, as in mechanical computing machines for extracting roots of numbers, raising numbers to any power, or generating trigonometric and logarithmic functions.

Noncircular Gears

It is always possible to design a specially shaped gear to roll and mesh properly with a gear of any shape. The sole requirement is that the distance between the two axes must be constant. However, the pitch line of the mating gear might turn out to be an open curve, and the gears can be rotated only for a portion of a revolution—as with two logarithmic-spiral gears (illustrated in Fig. 1).

True elliptical gears can only be made to mesh properly if they are twins, and if they are rotated about their focal points. However, gears resembling ellipses can be generated from a basic ellipse. These "higher-order" ellipses (see Fig. 2) can be meshed in various interesting combinations to rotate about centers *A, B, C,* or *D*. For example, two second-order elliptical gears can be meshed to rotate about their geometric center, however, they will produce two complete speed cycles per revolution. The difference in contour between a basic ellipse and a second-order ellipse is usually very slight. Note also that the fourth-order "ellipses" resemble square gears (this explains why the square gears, sometimes found as ornaments on tie clasps, illustrated in Fig. 3, actually work).

Fig. 1 The logarithmic spiral gears shown in (A), are open-curved. They are usually components in computing devices. The elliptically shaped gears, shown in (B), are closed curved. They are components in automatic machinery. The specially shaped gears, shown in (C), offer a wider range of velocity and acceleration characteristics.

Noncircular Gears (*continued*)

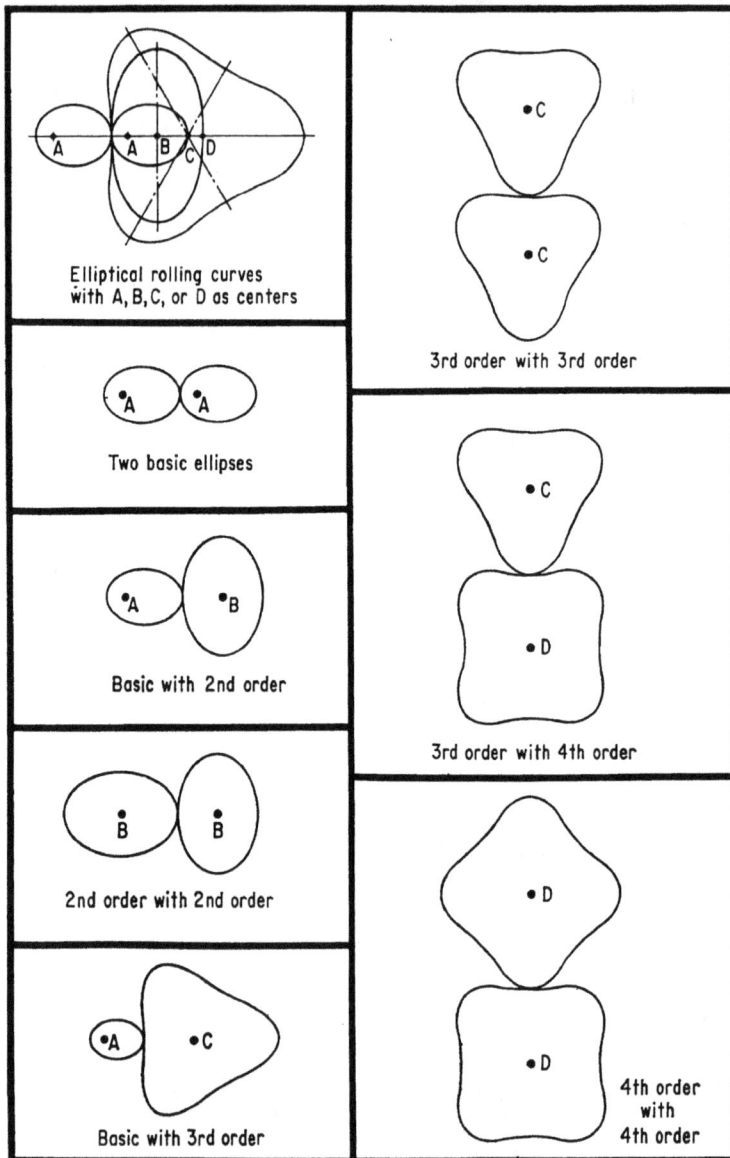

Fig. 2 Basic and high-order elliptical gear combinations.

A circular gear, mounted eccentrically, can roll properly only with specially derived curves (shown in Fig. 4). One of the curves, however, closely resembles an ellipse. For proper mesh, it must have twice as many teeth as the eccentric gear. When the radiis, *r*, and eccentricity, *e*, are known, the major semiaxis of the elliptically shaped gear becomes $2r + e$, and the minor $2r - e$. Note also that one of the gears in this group must have internal teeth to roll with the eccentric gear. Actually, it is possible to generate internal-tooth shapes to rotate with noncircular gears of any shape (but, again, the curves can be of the open type).

Noncircular gears can also be designed to roll with specially shaped racks (shown in Fig. 5). Combinations include an elliptical gear and a sinusoid-like rack. A third-order ellipse is illustrated, but any of the elliptical rolling curves can be used in its place. The main advantage of those curves is that when the ellipse rolls, its axis of rotation moves along a straight line; other combinations include a logarithmic spiral and straight rack. The rack, however, must be inclined to its direction of motion by the angle of the spiral.

Design Equations

Equations for noncircular gears are given here in functional form for three common design requirements. They are valid for any noncircular gear pair. Symbols are defined in the box.

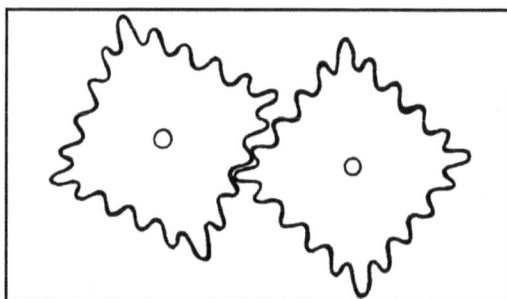

Fig. 3 Square gears seem to defy basic kinematic laws, but they are a takeoff on a pair of fourth-order ellipses.

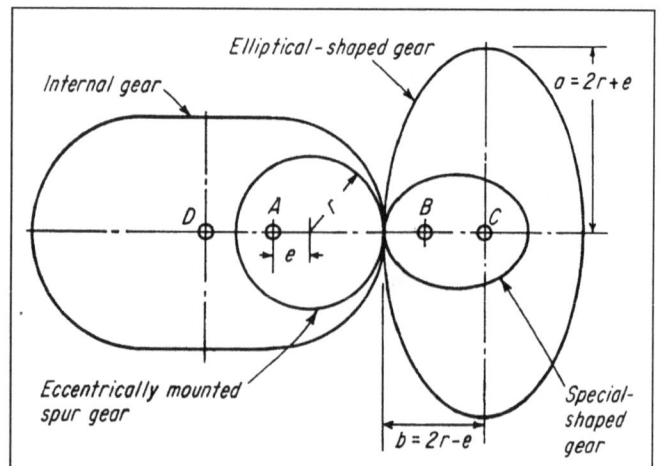

Fig. 4 An eccentric spur gear rotating about point A will mesh properly with any of the three gears shown whose centers are at points B, C, and D.

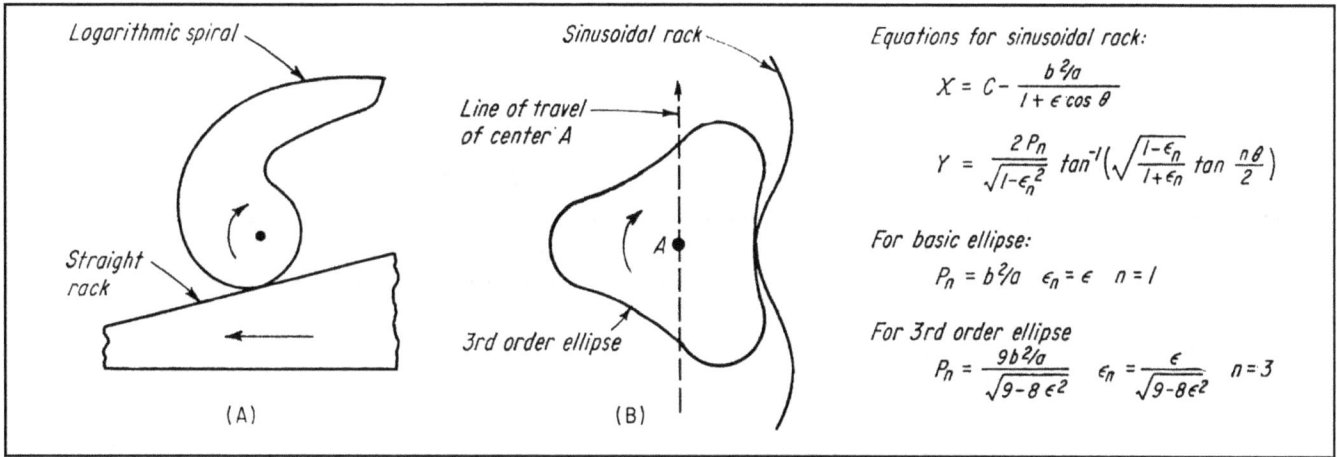

Fig. 5 Rack and gear combinations are possible with noncircular gears. The straight rack for the logarithmic spiral (A) must move obliquely; the center of third-order ellipse (B) follows a straight line.

Equations for sinusoidal rack:

$$X = C - \frac{b^2/a}{1 + \epsilon \cos \theta}$$

$$Y = \frac{2 P_n}{\sqrt{1 - \epsilon_n^2}} \tan^{-1}\left(\sqrt{\frac{1 - \epsilon_n}{1 + \epsilon_n}} \tan \frac{n\theta}{2}\right)$$

For basic ellipse:

$$P_n = b^2/a \quad \epsilon_n = \epsilon \quad n = 1$$

For 3rd order ellipse

$$P_n = \frac{9 b^2/a}{\sqrt{9 - 8\epsilon^2}} \quad \epsilon_n = \frac{\epsilon}{\sqrt{9 - 8\epsilon^2}} \quad n = 3$$

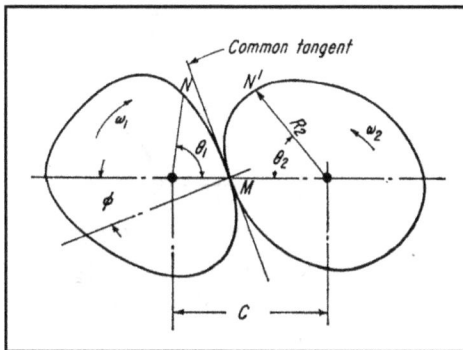

Symbols

a = semi-major axis of ellipse
b = semi-minor axis of ellipse
C = center distance (see above sketch)
ϵ = eccentricity of an ellipse = $\sqrt{1 - (b/a)^2}$
e = eccentricity of an eccentrically mounted spur gear
N = number of teeth
P = diametral pitch
r_c = radius of curvature
R = active pitch radius
S = length of periphery of pitch circle
X, Y = rectangular coordinates
θ = polar angle to R
ϕ = angle of obliquity
ω = angular velocity
$f(\theta), F(\theta), G(\theta)$ = various functions of θ
$f'(\theta), F'(\theta), G'(\theta)$ = first derivatives of functions of θ

Case I Polar equation of one curve and center distance are known; to find the polar equation of the mating gear:

$$R_1 = f(\theta_1)$$

$$R_2 = C - f(\theta_1)$$

$$\theta_2 = -\theta_1 + C \int \frac{d\theta_1}{C - f(\theta_1)}$$

Case II The relationship between angular rotation of the two members and the center distance are known; to find the polar equations of both members:

$$\theta_2 = F(\theta_1)$$

$$R_1 = \frac{CF'(\theta_1)}{1 + F'(\theta_1)}$$

$$R_2 = C - R_1 = \frac{C}{1 + F'(\theta_1)}$$

Case III The relationship between angular velocities of the two members and the center distance are known; to find the polar equations of both members:

$$\omega_2 = \omega_1 G(\theta_1)$$

$$R_1 = \frac{CG(\theta_1)}{1 + G(\theta_1)}$$

$$R_2 = C - R_1$$

$$\theta_1 = \int G(\theta_1) d\theta_1$$

Velocity equations and the characteristics of five types of noncircular gears are listed in the table.

Checking for Closed Curves

Gears can be quickly analyzed to determine whether their pitch curves are open or closed with the following equations:

In case I, if $R = f(\theta) = f(\theta + 2N_\pi)$, the pitch curve is closed.

In case II, if $\theta_1 = F(\theta_2)$ and $F(\theta_0) = 0$, the curve is closed with the equation $F(\theta_0 + 2_\pi/N_1) = 2_\pi/N_2$. It can be satisfied by substituting integers or rational fractions for N_1 and N_2. If fractions must be used to solve this equation, the curve will have double points (intersect itself), which is, or course, an undesirable condition.

In case III, if $\theta_2 = \int G(\theta_1) d\theta_1$, let $G(\theta_1) d \theta_1 = F(\theta_1)$, and use the same method as for case II, with the subscripts reversed.

With some gear sets, the mating gear will be a closed curve only if the correct center distance is employed. This distance can be found from the equation:

$$4\pi = \int_0^{2\pi} \frac{d\theta_1}{C - f(\theta_1)}$$

155

Characteristics of Five Noncircular Gear Systems

Type	Comments	Basic equations	Velocity equations ω_1 = constant
Two ellipses rotating about foci	Gears are identical. Comparatively easy to manufacture. Used for quick-return mechanisms, printing presses, automatic machinery.	$R = \dfrac{b^2}{a\,[\,1 + \epsilon\cos\theta\,]}$ ϵ = eccentricity $\epsilon = \sqrt{1 - \left(\dfrac{b}{a}\right)^2}$ $a = \frac{1}{2}$ major axis $b = \frac{1}{2}$ minor axis	$\omega_2 = \omega_1\left[\dfrac{r^2 + 1 + (r^2-1)\cos\theta_2}{2r}\right]$ where $r = \dfrac{R\max}{R\min}$
2nd order elliptical gears rotating about their geometric centers	Gears are identical. Geometric properties well known. Better balanced than true elliptical gears. Used where two complete speed cycles are required for one revolution.	$R = \dfrac{2ab}{(a+b)-(a-b)\cos 2\theta}$ $C = a + b$ a = maximum radius b = minimum radius	$\omega_2 = \omega_1\left[\dfrac{r^2 + 1 - (r^2-1)\cos 2\theta_2}{2r}\right]$ where $r = \dfrac{a}{b}$
Eccentric circular gear rotating with its conjugate	Standard spur gear can be employed as the eccentric. Mating gear has special shape.	$R_1 = \sqrt{a^2 + e^2 + 2ae\cos\theta_1}$ $\theta_2 = \theta_1 + C\displaystyle\int \dfrac{d\theta_1}{C - R_1}$ $C = R_1 + R_2$	$\dfrac{\omega_2}{\omega_1} = \dfrac{\sqrt{a^2 + e^2 + 2ae\cos\theta_1}}{C - \sqrt{a^2 + e^2 + 2ae\cos\theta_1}}$
Logarithmic spiral gears	Gears can be identical although can be used in combinations to give variety of functions. Must be open gears.	$R_1 = Ae^{k\theta_1}$ $R_2 = C - R_1$ $R_2 = Ae^{k\theta_2}$ $\theta_2 = \dfrac{1}{k}\log\,(C - Ae)^{k\theta_1}$ e = natural log base	$\dfrac{\omega_2}{\omega_1} = \dfrac{Ae^{k\theta_1}}{C - Ae^{k\theta_1}}$ (graph marked 0.693)
Sine-function gears	For producing angular displacement proportional to sine of input angle. Must be open gears.	$\theta_2 = \sin^{-1}(k\theta_1)$ $R_2 = \dfrac{C}{1 + k\cos\theta_1}$ $R_1 = C - R_2$ $R_1 = \dfrac{Ck\cos\theta_1}{1 + k\cos\theta_1}$	$\dfrac{\omega_2}{\omega_1} = k\cos\theta_1$

SHEET-METAL GEARS, SPROCKETS, WORMS, AND RATCHETS FOR LIGHT LOADS

When a specified motion must be transmitted at intervals rather than continuously, and the loads are light, these mechanisms are ideal because of their low cost and adaptability to mass production.

Although not generally considered precision parts, ratchets and gears can be stamped to tolerances of ±0.007 in, and if necessary, shaved to close dimensions.

Fig. 1 The pinion is a sheet metal cup with rectangular holes serving as teeth. The meshing gear is sheet metal, blanked with specially formed teeth. The pinion can be attached to another sheet metal wheel by prongs, as shown, to form a gear train.

Fig. 2 The sheet-metal wheel gear meshes with a wide-face pinion, which is either extruded or machined. The wheel is blanked with teeth of conventional form.

Fig. 3 The pinion mates with round pins on a circular disk made of metal, plastic, or wood. The pins can be attached by staking or with threaded fasteners.

Fig. 4 Two blanked gears, conically formed after blanking, become bevel gears meshing on a parallel axis. Both have specially formed teeth.

Fig. 7 The blanked and formed bevel-type gear meshes with a machined or extruded pinion. Conventional teeth can be used on both the gear and pinion.

Fig. 8 The blanked, cup-shaped wheel meshes with a solid pinion on 90° intersecting axes.

Fig. 5 The horizontal wheel with waves on its out rim replacing teeth, meshes with either one or two sheet-metal pinions. They have specially formed teeth and are mounted on intersecting axes.

Fig. 6 Two bevel-type gears, with specially formed teeth, are mounted on 90° intersecting axes. They can be attached by staking them to hubs.

Fig. 9 Backlash can be eliminated from stamped gears by stacking two identical gears and displacing them by one tooth. The spring then bears one projection on each gear, taking up lost motion.

Fig. 10 A sheet metal cup with indentations replacing worm-wheel teeth, meshes with a standard coarse-thread screw.

Fig. 11 A blanked wheel, with specially formed teeth, meshes with a helical spring mounted on a shaft, which serves as the worm.

Fig. 12 This worm wheel is blanked from sheet metal with specially formed teeth. The worm is a sheet-metal disk that was split and helically formed.

Fig. 13 Blanked ratchets with one-sided teeth are stacked to fit a wide-sheet-metal finger when single thickness is inadequate. The ratchet gears can be spot-welded.

Fig. 14 To avoid stacking, a single ratchet is used with a U-shaped finger, also made of sheet metal.

Fig. 15 This wheel is a punched disk with square-punched holes serving as teeth. The pawl is spring steel.

Fig. 16 This sheet-metal blanked pinion, with specially formed teeth, meshes with windows blanked in a sheet metal cylinder. They form a pinion-and-rack assembly.

Fig. 17 This sprocket, like that in Fig. 13, can be fabricated from separate stampings.

Fig. 18 For a wire chain as shown, the sprocket is made by bending out punched teeth on a drawn cup.

THIRTEEN WAYS GEARS AND CLUTCHES CAN CHANGE SPEED RATIOS

Thirteen ways of arranging gears and clutches to obtain changes in speed ratios

1 Keyed to shaft

2 Not keyed to shaft (rotatable on shaft)

3 Sliding gear keyed to shaft

4 Clutch (shown with clutch keyed to shaft, and gear not keyed to shaft)

Fig. 1 **The schematic symbols** used in the following illustrations to represent gears and clutches.

Fig. 2 **Double-clutch drive.** Two pairs of gears are permanently in mesh. Pair I or II transmits motion to the output shaft depending on the position of the coupling; the other pair idles. The coupling is shown in a neutral position with both gear pairs idle. Herring-bone gears are recommended for quieter running.

Fig. 3 **Sliding-change drive.** Gears are meshed by lateral sliding. Up to three gears can be mounted on a sliding sleeve. Only one pair is in mesh in any operating position. This drive is simpler, cheaper, and more extensively used than the drive of Fig. 2. Chamfering the sides of the teeth eases their engagement.

Fig. 4 **Swivel-gear drive.** Output gears are fastened to the shaft. A handle is pushed down, then shifted laterally to obtain transmission through any output gear. This drive is not suitable for the transmission of large torques because the swivel gear tends to vibrate. Its overall ratio should not exceed 1:3.

Fig. 5 **Slide-key drive.** A spring-loaded slide key rides inside a hollow output shaft. The slide key snaps out of the shaft when it is in position to lock a specific change gear to the output shaft. No central position is shown.

Slide key in hollow shaft

Fig. 6 **This is a combination coupling and slide gears.** It has three ratios: a direct mesh for ratios I and II; a third ratio is transmitted through gears II and III, which couple together.

Fig. 7 **Double-shift drive.** One shift must always be in a neutral position. That might require both levers to be shifted when making a change. However, only two shafts are used to achieve four ratios.

Both shifts operated together

Fig. 8 A triple shaft drive gives four ratios. (A) The output of the first drive serves as the input for the second. The presence of an intermediate shaft eliminates the requirement for ensuring that one shift is always in the neutral position. A wrong shift-lever position cannot cause damage. (B) A space-saving modification; the coupling is on shaft *A* instead of the intermediate shaft (C) Still more space is saved if one gear replaces a pair on the intermediate shaft. Ratios can be calculated to allow this.

Fig. 9 Six ratios are available with two couplings and (A) ten gears, (B) eight gears. Up to six gears can be in permanent mesh. It is not necessary to ensure that one shift is in neutral.

Fig. 10 This eight-ratio drive has two slide gears and a coupling. This arrangement reduces the number of parts and meshes. The position of shifts I and II are interdependent. One shift must be in neutral if the other is in mesh.

Shaft levers I and II must be operated together

Fig. 11 This drive has eight ratios; a coupled gear drive and slide-key drive are in series. Comparatively low strength of the slide key limits the drive to small torque.

GEAR AND CLUTCH SHIFTING MECHANISMS

Shift collar with trunnions

Clutch shifting spool

Nut for end play adjustment

Adjustable nut for assembly

Rectangular bronze shoe

Eccentric crank integral on shaft

E
D

Shifting shoe clamped to shaft

Positioning grooves

Light-duty gears

Screw adjusts clearance between clutch disks

Shifting yoke

Cam

Driving disk

Driven disk

Thrust bearing

Positioning nut

Over running clutch drives disk

Shifting lever attached to this shaft

Gear segment

Shifting fork

Fixed rod

Rack teeth in hub engage gear segment

Pawl disengaging lever

Locking pawl

Screwed to main casting

Gear housing free to turn in bearings

Lever rocks entire gear case

Shifting rod

Gear

Turned grooves position rods

Fork

Screw fastens fork to shift rod

Slot position screw

Spring plunger

E
F
D
C
G
H

Change gears

Tumbler gear

Driven gear

Splined shaft

Shaft cut away for clearance

Shifting Fork

X
X

Finished trunnion

Two-piece clutch yoke

Bell crank shifting fork

Shifting center

Steel shoe (rectangular) attached to fork by inserted pins

Clutch Shoe Clutch

Driving member

Eccentrically turned shifting collar

Adjusting nut

Slot

Handle

Pin

Shifting stud

Shifting collar

Operating pinion

Upper surface of shifting bar has milled teeth

Stationary shaft

Shaft movable longitudinally

Housing

Clutch releasing fork engages flange

Flange free to rotate on bushing

Spacing bushing

Mounting screw

Pin

Gear

161

Labels in top illustrations:
Pivot · *A* · *Clutch groove* · *Shifting bar of flat stock*

Solid yoke · *Flange holds yoke in place* · *Accomodates yoke trunnions* · *Shifting lever* · *Shifting shaft* · *Trunnion*

Lever which actuates locking plunger · *Locking plunger* · *Lug* · *Shift rod* · *Shift lever keyed to rod*

Bearing housing · *Clutch member* · *Grease fitting* · *Shift fork* · *Locks bearing to clutch member* · *Holds bearing housing* · *Shift fork* · *Trunnion*

Locking plunger · *Handle* · *Pinion mounted on rocker arm and driven by D engages change gears* · *Rocker arm* · *Splined shaft drives gears D and E* · *Locking plunger engages these notches* · *Locking plunger engages these notches when driving other idler gears* · *D* · *B* · *E* · *J* · *K*

Spiderless Differential

If you've ever been unable to drive your car out of a ditch because one wheel spun uselessly while the other sat torqueless and immobile, you'll thank the inventors (Seliger and Hegar) of the limited-slip differential shown here.

In straight running, it performs as a drive axle, driven by the driveshaft pinion through the ring gear. The differential action occurs only when one wheel loses traction, travels along a different arc, or otherwise attempts to turn at a speed that is different from that of the other. Then the wedge-type, two-way, over-running clutch (second figure) disengages, freeing the wheel to spin without drag.

Variations. Each clutch has three positions: forward drive, idle, and reverse drive. Thus, there are many combinations of drive-idle, depending on road conditions and turn direction. U.S. Patent 3,124,972 describes a few:

- For left turns, the left wheel is driving, and the right wheel is forced to turn faster, thus over-running and disengaging the clutch. A friction ring built into each clutch assembly does the shifting. Wear is negligible.
- If power should be removed from the driveshaft during the left turn, the friction rings will shift each clutch and cause the left wheel to run free

and the right wheel to drag in full coupling with the car's driveshaft.

- If your car is on the straightaway, under power and one wheel is lifted out of contact with the road, the other immediately transmits full torque. (The conventional spider differential performs in the opposite manner.)

On or off. Note one limitation, however: There is no gradual division of power. A wheel is either clutched in and turning at exactly the same speed as its opposite, or it is clutched out. It is not the same kind of mechanism as the conventional spider differential, which divides the driving load variably at any ratio of speeds.

Drive shaft · *Pinion* · *Left axle* · *Right axle* · *Ring gear* · *Wedge-type clutch*

Two-way over-running clutch disengages the non-driving wheel

TWINWORM GEAR DRIVE OFFERS BIDIRECTIONAL OUTPUT

The term "self-locking" as applied to gear systems denotes a drive that gives the input gear freedom to rotate the output gear in either direction. But the output gear locks with the input when an outside torque attempts to rotate the output in either direction. This characteristic is often sought by designers who want to be sure that loads on the output side of the system cannot affect the position of the gears. Worm gears are one of the few gear systems that can be made self-locking, but at the expense of efficiency. It seldom exceeds 40 percent when the gears are self-locking.

An Israeli engineer, B. Popper, invented a simple dual-worm gear system that not only provided self-locking with over 90 percent efficiency, but exhibited a phenomenon which the inventor called "deceleration-locking."

The "Twinworm" drive has been employed in Israel-designed counters and computers for years with marked success.

The Twinworm drive is simply constructed. Two threaded rods, or "worm" screws, are meshed together. Each worm is wound in a different direction and has a different pitch angle. For proper mesh, the worm axes are not parallel, but slightly skewed. (If both worms had the same pitch angle, a normal, reversible drive would result—similar to helical gears.) But by selecting proper, and different, pitch angles, the drive will exhibit either self-locking, or a combination of self-locking and deceleration-locking characteristics, as desired. Deceleration-locking is a completely new property best described in this way.

When the input gear decelerates (for example, when the power source is shut off, or when an outside force is applied to the output gear in a direction that tends to help the output gear), the entire transmission immediately locks up and comes to an abrupt stop, moderated only by any elastic "stretch" in the system.

Almost any type of thread will work with the new drive—standard threads, 60° screw threads, Acme threads, or any arbitrary shallow-profile thread. Hence, the worms can be manufactured on standard machine-shop equipment.

Jobs for the New Drive

Applications for Twinworm can be divided into two groups:

(1) Those employing self-locking characteristics to prevent the load from affecting the system.

(2) Those employing deceleration-locking characteristics to brake the system to an abrupt stop if the input decelerates.

Self-locking occurs as soon as $\tan \phi_1$ is equal to or smaller than μ, or when

$$\tan \phi_1 = \frac{\mu}{S_1}$$

Angles ϕ_1 and ϕ_2 represent the respective pitch angles of the two worms, and $\phi_2 - \phi_1$ is the angle between the two worm shafts ϕ angle of misalignment. Angle ϕ_1 is quite small (usually in the order of 2° to 5°).

Here, S_1 represents a "safety factor" (selected by the designer). It must be somewhat greater than one to make sure that self-locking is maintained, even if μ should fall below an assumed value. Neither ϕ_2 nor the angle $(\phi_2 - \phi_1)$ affects the self-locking characteristic.

Deceleration-locking occurs as soon as $\tan \phi_2$ is also equal to or smaller than μ, or, if a second safety factor S_2 is employed (where $S_2 > 1$), when

$$\tan \phi_2 = \frac{\mu}{S_2}$$

For the equations to hold true, ϕ_2 must always be made greater than ϕ_1. Also, μ refers to the idealized case where the worm threads are square. If the threads are inclined (as with Acme-threads or V-threads) then a modified value of μ must be employed, where

$$\mu_{modified} = \frac{\mu_{true}}{\cos \theta}$$

A relationship between the input and output forces during rotation is:

$$\frac{p_1}{P_2} = \frac{\sin \phi_1 + \mu \cos \phi_1}{\sin \phi_2 + \mu \cos \phi_2}$$

Efficiency is determined from the equation:

$$\eta = \frac{1 + \mu/\tan \phi_2}{1 + \mu/\tan \phi_1}$$

Twinworm gears
Bevel or helical gears
Input
Output

Twinworm clutch units
Gear meshes
Digit wheels
5 4 3 2 1
Special inputs
Mechanical Counter

Hoists

Load
Rack
Worm
Motor or handle
Racks

Worm
Motor
Ring (base for antenna)
Large Rings

BEVEL AND HYPOID GEAR DESIGN PREVENTS UNDERCUTTING

Lengths and radii of shafts can be chosen to prevent undercutting.
Lewis Research Center, Cleveland, Ohio

A computer-assisted method of analysis of straddle designs for spiral bevel and hypoid gears helps to prevent undercutting of gear shafts during cutting of the gear teeth. Figure 1 illustrates a spiral bevel gear or straddle design, in which the shaft extends from both ends of the toothed surface to provide double bearing support. One major problem in such a design is to choose the length and radius of the shaft at the narrow end (equivalently, the radial coordinate r and axial coordinate u) such that the head cutter that generates the gear teeth does not collide with, and thereby undercut, the shaft.

The analytical method and computer program are based on the equations for the surface traced out by the motion of the head cutter, the equation for the cylindrical surface of the shaft, and the equations that express the relationships among the coordinate systems fixed to the various components of the gear-cutting machine tool and to the gear. The location of a collision between the shaft and the cutter is defined as the vector that simultaneously satisfies the equations for head-cutter-traced and shaft surfaces. The solution of these equations yields the u and r coordinates of the point of collision.

Given input parameters in the form of the basic machine-tool settings for cutting the gear, the computer program finds numerical values of r and u at a representative large number of points along the path of the cutter. These computations yield a family of closed curves (see Fig. 2) that are the loci of collision points. The region below the curves is free of collisions; thus, it contains the values of r and u that can be chosen by the designer to avoid collisions between the shaft and the head cutter.

This work was done by Robert F. Handschuh of the U.S. Army Aviation Systems Command; Faydor L. Litvin, Chihping Kuan, and Jonathan Kieffer of the University of Illinois at Chicago; and Robert Bossler of Lucas Western, Inc., for **Lewis Research Center.**

Fig. 1   A straddle-design spiral bevel gear includes two integral shaft extensions. One of these could terminate near or even beyond the apex of the pitch cone.

Fig. 2   This family of closed curves applies to a typical hypoid gear. It will help in the selection of the length and radius of the shaft at the narrow end. The region below the curves is free of collisions between the head cutter and the shaft.

MACHINING METHOD TO IMPROVE WORM GEAR MESHING

Researchers at the University of Illinois have developed a new method for machining mating face worm and cylindrical worm gears for face worm gear drives that offers better meshing than conventional spiral bevel or hypoid gear drives. In this research, sponsored by NASA's John Glenn Research Laboratory, the proposed gear shapes are based on the concept of generating a face worm gear surface by using a tilted head cutter rather than traditional hob.

The gear-tooth machining equipment would be similar to that used for generating spiral bevel and hypoid gears, but a tilted-head cutter would replace the hob in this method. In addition to being larger than a hob, it could generate face worm and cylindrical worm gear surfaces more precisely and faster than a hob. In this description, the term "head cutter" can mean either a head grinding or cutting tool.

A face worm gear would be generated by a tilted head cutter with blades or grinding surfaces having straight-line profiles, as shown in Fig. 1a. The tilt of the head cutter would prevent interference with teeth adjacent to the groove being cut or ground.

A cylindrical worm able to mesh with the face worm gear would be generated by a tilted head cutter mounted on the cradle of a generating machine, as shown in Fig. 1b. The blades or grinding surfaces of the head cutter would have a parabolic profile and would deviate from the straight-line profiles of the head cutter for the face worm gear. The shortest distance between the cylindrical worm and the cradle would follow a parabolic function during the cycle of meshing in the generating process. This provides a parabolic function of transmission errors to the gear drive

The small mismatch between the profiles of the face worm gear and worm head cutters would make it possible to localize the bearing contact in the worm gear drive. The parabolic function of transmission errors could absorb discontinuous linear functions of transmission errors caused by errors of alignment. This is a significant benefit because these errors are the main sources of noise and vibration in gear drives.

The principal advantage of using tilted head cutters is that cutting speeds are independent of the shape-generation processes. The quality of finished parts are improved by making it possible to choose the cutting speeds for minimizing the high machining temperatures which could deform the gear. Both the profiles of the cutting or grinding surfaces and the position and orientation settings of the head cutter can be derived from the theoretical hob-generated shape. This derivation would be determined by an algorithm that accounts for the tilted-head-cutter geometry and enforces the meshing requirements while eliminating tooth interference.

A computer program for tooth contact analysis was developed for the simulation of meshing and contact of the proposed cylindrical worms and face worm gears. Program test results showed the following:

Fig. 1 The head cutters tilted with respect to the gear blanks improve the machining of face and cylindrical worm gears for more precise meshing in the worm gear drives.

- Desired transmission errors would be low-magnitude parabolic functions.
- Contacts would be localized, and their paths would be longitudinal because they would lie along the gear-tooth surfaces.
- Bearing contact region would be free of severe contact stress areas.
- Contact ratio would be larger than three.

(These results indicated that there would be at least three pairs of teeth in contact at all rotation positions.)

This work was done by Faydor L. Litvin, Alessandro Nava, Qi Fan, and Alfonso Fuentes of the University of Illinois for **NASA's John Glenn Research Laboratory.**

GEARED SPEED REDUCERS OFFER ONE-WAY OUTPUT

This eccentric cam adjusts over a range of high reduction ratios, but unbalance limits it to low speeds. When its direction of input changes, there is no lag in output rotation. The output shaft moves in steps because of a ratchet drive through a pawl which is attached to a U follower.

A traveling gear moves along a worm and transfers drive torque to the other pinion when the input rotation changes direction. To ease the gear engagement, the gear teeth are tapered at their ends. Output rotation is smooth, but there is a lag after direction changes as the gear shifts. The gear cannot be wider than the axial offset between pinions or there will be destructive interference.

Two bevel gears drive through roller clutches. One clutch catches in one direction and the other catches in the opposite direction. There is little or no interruption of smooth output rotation when the input direction changes.

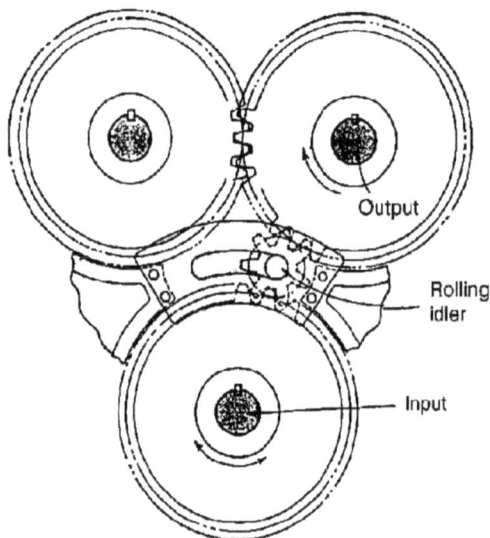

This rolling idler also provides a smooth output and a slight lag after its input direction changes. A small drag on the idler is necessary so that it will transfer smoothly into engagement with the other gear and not remain spinning between the gears.

Roller clutches are on the input gears in this drive. These also give smooth output speed and little output lag as the direction changes.

DESIGN OF GEARED FIVE-BAR MECHANISMS

Geared five-bar mechanisms offer excellent force-transmission characteristics and can produce more complex output motions—including dwells—than conventional four-bar mechanisms.

It is often necessary to design a mechanism that will convert uniform input rotational motion into nonuniform output rotation or reciprocation. Mechanisms designed for such purposes are usually based on four-bar linkages. Those linkages produce a sinusoidal output that can be modified to yield a variety of motions.

Four-bar linkages have their limitations, however. Because they cannot produce dwells of useful duration, the designer might have to include a cam when a dwell is desired, and he might have to accept the inherent speed restrictions and vibration associated with cams. A further limitation of four-bar linkages is that only a few kinds have efficient force-transmission capabilities.

One way to increase the variety of output motions of a four-bar linkage, and obtain longer dwells and better force transmissions, is to add a link. The resulting five-bar linkage would become impractical, however, because it would then have only two degrees of freedom and would, consequently, require two inputs to control the output.

Simply constraining two adjacent links would not solve the problem. The five-bar chain would then function effectively only as a four-bar linkage. If, on the other hand, any two nonadjacent links are constrained so as to remove only one degree of freedom, the five-bar chain becomes a functionally useful mechanism.

Gearing provides solution. There are several ways to constrain two nonadjacent links in a five-bar chain. Some possibilities include the use of gears, slot-and-pin joints, or nonlinear band mechanisms. Of these three possibilities, gearing is the most attractive. Some practical gearing systems (Fig. 1) include paired external gears, planet gears revolving within an external ring gear, and planet gears driving slotted cranks.

In one successful system (Fig. 1A) each of the two external gears has a fixed crank that is connected to a crossbar by a rod. The system has been successful in high-speed machines where it transforms rotary motion into high-impact linear motion. The Stirling engine includes a similar system (Fig. 1B).

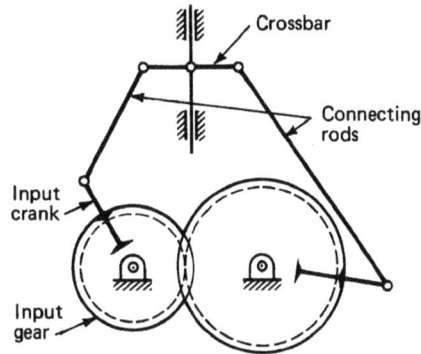

A – Fixed-crank external gear system

B – Stirling engine system

C – Internal planetary gear system

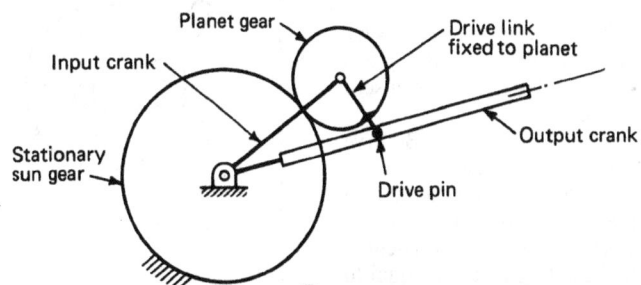

D – External planetary gear system

Fig. 1 **Five-bar mechanism designs** can be based on paired external gears or planetary gears. They convert simple input motions into complex outputs.

In a different system (Fig. 1C) a pin on a planet gear traces an epicyclic, three-lobe curve to drive an output crank back and forth with a long dwell at the extreme right-hand position. A slotted output crank (Fig. 1D) will provide a similar output.

Two professors of mechanical engineering, Daniel H. Suchora of Youngstown State University, Youngstown, Ohio, and Michael Savage of the University of Akron, Akron, Ohio, studied a variation of this mechanism in detail.

Five kinematic inversions of this form (Fig. 2) were established by the two researchers. As an aid in distinguishing between the five, each type is named according to the link which acts as the fixed link. The study showed that the Type 5 mechanism would have the greatest practical value.

In the Type 5 mechanism (Fig. 3A), the gear that is stationary acts as a sun gear. The input shaft at Point E drives the input crank which, in turn, causes the planet gear to revolve around the sun gear. Link a_2, fixed to the planet, then drives the output crank, Link a_4, by means of the connecting link, Link a_3. At any input position, the third and fourth links can be assembled in either of two distinct positions or "phases" (Fig. 3B).

Variety of outputs. The different kinds of output motions that can be obtained from a Type 5 mechanism are based on the different epicyclic curves traced by link joint B. The variables that control the shape of a "B-curve" are the gear ratio GR ($GR = N_2/N_5$), the link ratio a_2/a_1 and the initial position of the gear set, defined by the initial positions of θ_1 and θ_2, designated as θ_{10} and θ_{20}, respectively.

Typical B-curve shapes (Fig. 4) include ovals, cusps, and loops. When the B-curve is oval (Fig. 4B) or semioval (Fig. 4C), the resulting B-curve is similar to the true-circle B-curve produced by a four-bar linkage. The resulting output motion of Link a_4 will be a sinusoidal type of oscillation, similar to that produced by a four-bar linkage.

When the B-curve is cusped (Fig. 4A), dwells are obtained. When the B-curve is looped (Figs. 4D and 4E), a double oscillation is obtained.

In the case of the cusped B-curve (Fig. 4A), dwells are obtained. When the B-curve is looped (Figs. 4D and 4E), a double oscillation is obtained.

In the case of the cusped B-curve (Fig. 4A), by selecting a_2 to be equal to the pitch radius of the planet gear r_2, link joint B becomes located at the pitch circle of the planet gear. The gear ratio in all the cases illustrated is unity ($GR = 1$).

Professors Suchora and Savage analyzed the different output motions produced by the geared five-bar mechanisms by plotting the angular position θ_4 of the

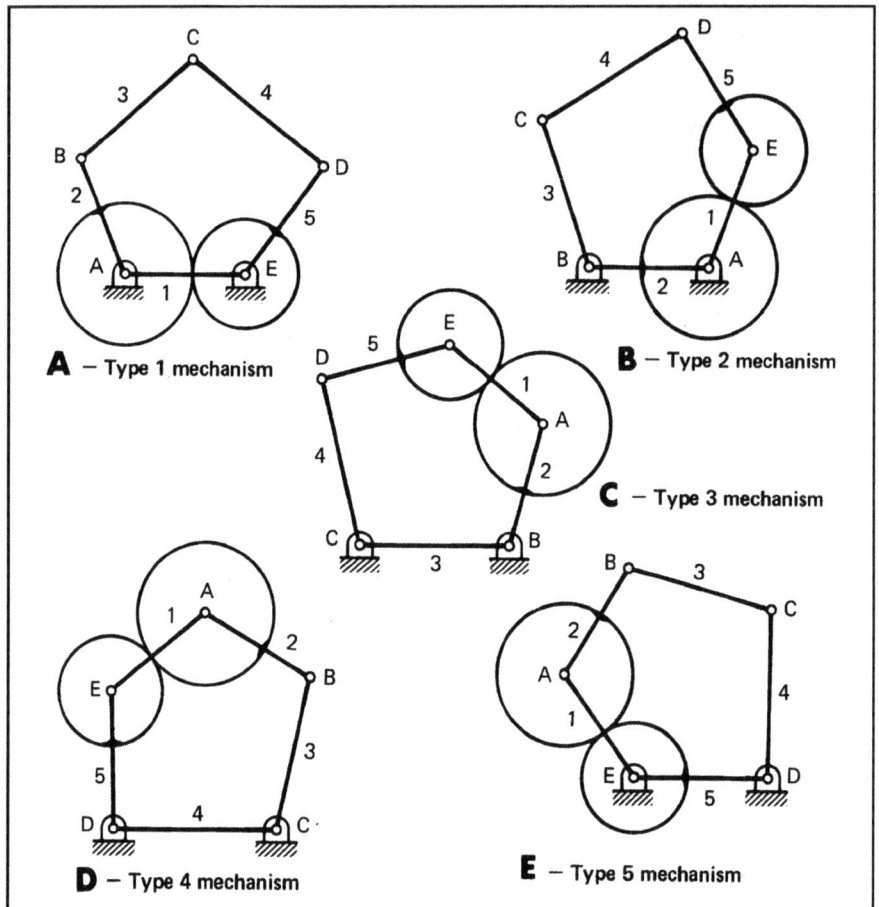

Fig. 2 Five types of geared five-bar mechanisms. A different link acts as the fixed link in each example. Type 5 might be the most useful for machine design.

Fig. 3 A detailed design of a Type 5 mechanism. The input crank causes the planet gear to revolve around the sun gear, which is always stationary.

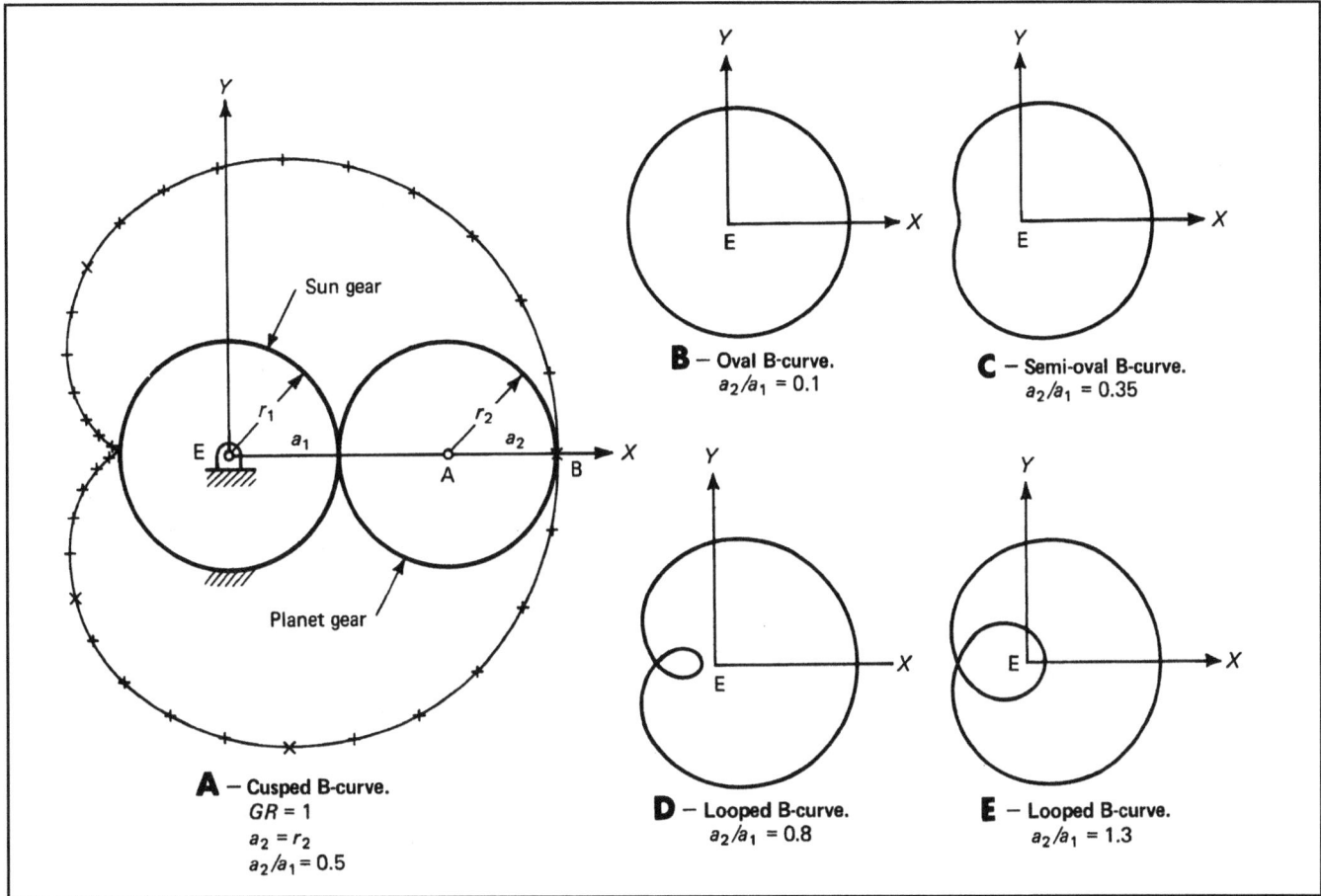

Fig. 4   Typical B-curve shapes obtained from various Type-5 geared five-bar mechanisms. The shape of the epicyclic curve is changed by the link ratio a_2/a_1 and other parameters, as described in the text.

Calculating displacement, velocity, and acceleration

Displacement θ_4 can be found from the following equation:

$$\theta_4 = 2 \tan^{-1} \left(\frac{I \pm \sqrt{I^2 + H^2 - J^2}}{H + J} \right)$$

where $H = a_1 \cos \theta_1 + a_2 \cos \theta_2 - a_5$; $I = a_1 \sin \theta_1 + a_2 \sin \theta_2$; and $J = 1/2a_4 (a^2_3 - a^2_4 - H^2 - I^2)$; and where $\theta_2 = \theta_{20} + (1 + 1/GR) (\theta_1 - \theta_{10})$, where θ_{10} and θ_{20} are the initial values of the angles θ_1 and θ_2, respectively.

For layout purposes, once θ_4 is determined, θ_3 can be found from:

$$\theta_3 = \tan^{-1} \left(\frac{a_4 \sin \theta_4 + I}{a_4 \cos \theta_4 + H} \right)$$

To find velocities θ'_4 and θ'_3, use these equations:

$$\theta'_4 = \frac{a_1 \sin (\theta_3 - \theta_1) + a_2 \sin (\theta_3 - \theta_2) \theta'_2}{a_4 \sin (\theta_4 - \theta_3)}$$

$$\theta'_3 = \frac{a_1 \sin (\theta_1 - \theta_4) + a_2 \sin (\theta_2 - \theta_4) \theta'_2}{a_4 \sin (\theta_4 - \theta_3)}$$

where $\theta'_2 = (1 + 1/GR)$.

Use these equations to determine accelerations θ''_4 and θ''_3:

$$\theta''_4 = \frac{L}{a_3 a_4 \sin (\theta_4 - \theta_3)}$$

$$\theta''_3 = \frac{K}{a_3 a_4 \sin (\theta_4 - \theta_3)}$$

where $K = a_3 a_4 \cos(\theta_3 - \theta_4) \theta'^2_3 + a_4^2 \theta'^2_4 + a_1 a_2 \cos(\theta_1 - \theta_4) + a_2 a_4 \cos(\theta_2 - \theta_4) \theta'^2_2$ and $L = a_3^2 \theta'^2_3 - a_3 a_1 \cos(\theta_3 - \theta_4) \theta'^2_4 - a_1 a_3 \cos(\theta_3 - \theta_1) + a_2 a_3 \cos(\theta_3 - \theta_2) \theta'^2_2$.

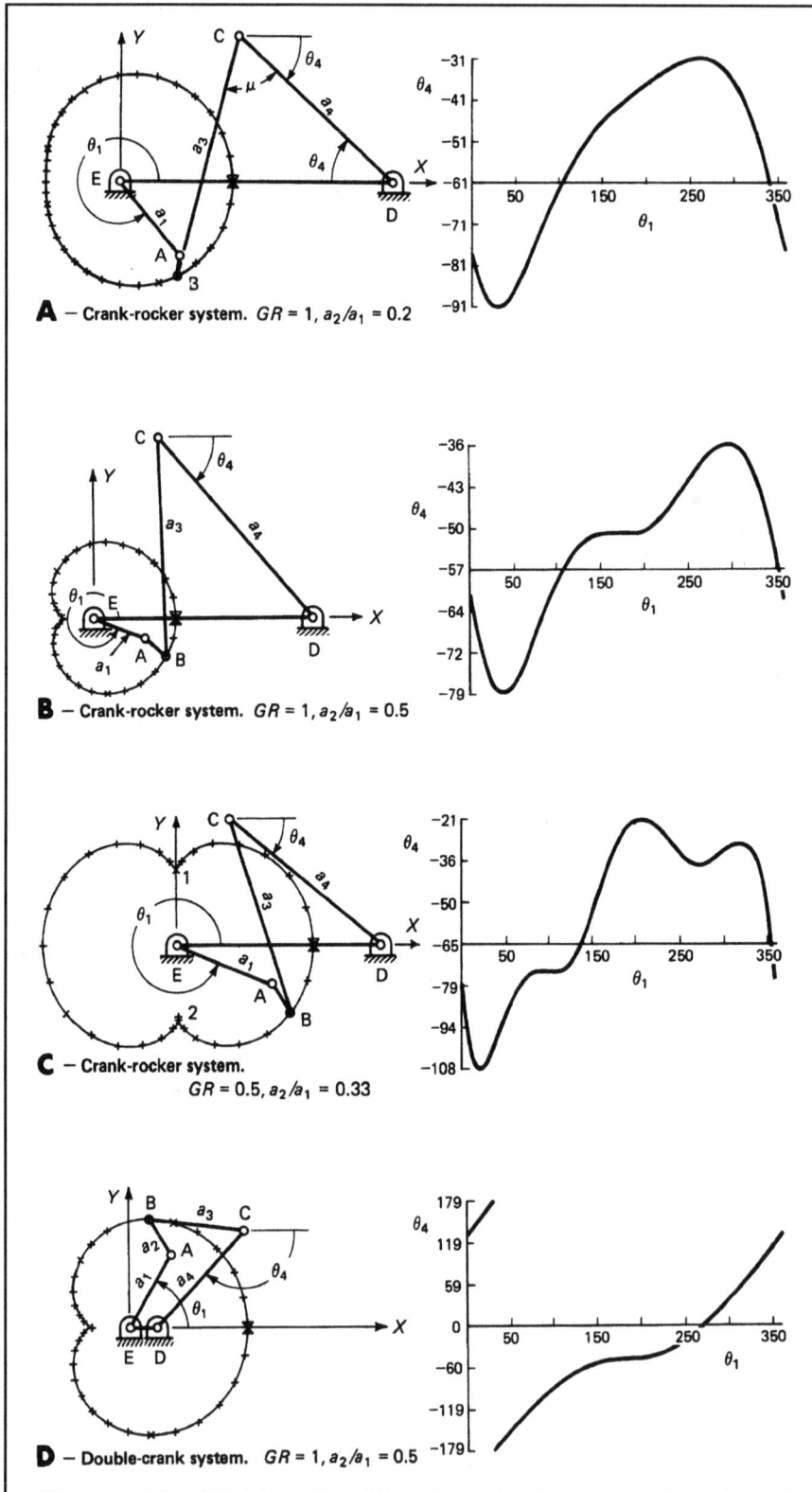

A — Crank-rocker system. $GR = 1$, $a_2/a_1 = 0.2$

B — Crank-rocker system. $GR = 1$, $a_2/a_1 = 0.5$

C — Crank-rocker system.
$GR = 0.5$, $a_2/a_1 = 0.33$

D — Double-crank system. $GR = 1$, $a_2/a_1 = 0.5$

Fig. 5 **A variety of output motions** can be produced by varying the design of five-bar geared mechanisms. Dwells are obtainable with proper design. Force transmission is excellent. In these diagrams, the angular position of the output link is plotted against the angular position of the input link for various five-bar mechanism designs.

output link a_4 against the angular position of the input link θ_1 for a variety of mechanism configurations (Fig. 5).

In three of the four cases illustrated, $GR = 1$, although the gear pairs are not shown. Thus, one input rotation generates the entire path of the B-curve. Each mechanism configuration produces a different output.

One configuration (Fig. 5A) produces an approximately sinusoidal reciprocating output motion that typically has better force-transmission capabilities than equivalent four-bar outputs. The transmission angle μ should be within 45 to 135° during the entire rotation for best results.

Another configuration (Fig. 5B) produces a horizontal or almost-horizontal portion of the output curve. The output link, a_4, is virtually stationary during this period of input rotation—from about 150 to 200° of input rotation θ_1 in the case illustrated. Dwells of longer duration can be designed.

By changing the gear ratio to 0.5 (Fig. 5C), a complex motion is obtained; two intermediate dwells occur at cusps 1 and 2 in the path of the B-curve. One dwell, from $\theta_1 = 80$ to 110°, is of good quality. The dwell from 240 to 330° is actually a small oscillation.

Dwell quality is affected by the location of Point D with respect to the cusp, and by the lengths of links a_3 and a_4. It is possible to design this form of mechanism so it will produce two usable dwells per rotation of input.

In a double-crank version of the geared five-bar mechanism (Fig. 5D), the output link makes full rotations. The output motion is approximately linear, with a usable intermediate dwell caused by the cusp in the path of the B-curve.

From this discussion, it's apparent that the Type 5 geared mechanism with $GR = 1$ offers many useful motions for machine designers. Professors Suchora and Savage have derived the necessary displacement, velocity, and acceleration equations (see the "Calculating displacement, velocity, and acceleration" box).

EQUATIONS FOR DESIGNING GEARED CYCLOID MECHANISMS

Fig. 1   Symbols and equations for epicycloid drives. ——

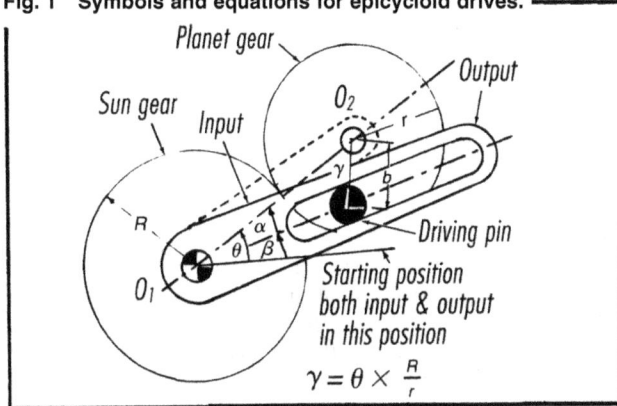

$$\gamma = \theta \times \frac{R}{r}$$

The equations for angular displacement, velocity, and acceleration for a basic epicyclic drive are given below.

$$\tan\beta = \frac{\sin\theta - \left(\dfrac{b}{R-r}\right)\left(\sin\dfrac{R-r}{r}\theta\right)}{\cos\theta + \left(\dfrac{b}{R-r}\right)\left(\cos\dfrac{R-r}{r}\theta\right)} \quad (4)$$

$$V = \omega\frac{1 - \left(\dfrac{R-r}{r}\right)\left(\dfrac{b^2}{(R-r)^2}\right) + \left(\dfrac{2r-R}{r}\right)\left(\dfrac{b}{R-r}\right)\left(\cos\dfrac{R}{r}\theta\right)}{1 + \dfrac{b^2}{(R-r)^2} + \left(\dfrac{2b}{R-r}\right)\left(\cos\dfrac{R}{r}\theta\right)}$$

$$(5)$$

$$A = \omega^2\frac{\left(1 - \dfrac{b^2}{(R-r)^2}\right)\left(\dfrac{b}{R-r}\right)\left(\dfrac{R^2}{r^2}\right)\left(\sin\dfrac{R}{r}\theta\right)}{\left[1 + \dfrac{b^2}{(R-r)^2} + \left(\dfrac{2b}{R-r}\right)\left(\cos\dfrac{R}{r}\theta\right)\right]^2} \quad (6)$$

Angular displacement

$$\tan\beta = \frac{(R+r)\sin\theta - b\sin(\theta+\gamma)}{(R+r)\cos\theta - b\cos(\theta+\gamma)} \quad (1)$$

Angular velocity

$$V = \omega\frac{1 + \dfrac{b^2}{r(R+r)} - \left(\dfrac{2r+R}{r}\right)\left(\dfrac{b}{R+r}\right)\left(\cos\dfrac{R}{r}\theta\right)}{1 + \left(\dfrac{b^2}{R+r}\right)^2 - \left(\dfrac{2b}{R+r}\right)\left(\cos\dfrac{R}{r}\theta\right)} \quad (2)$$

Angular acceleration

$$V = \omega^2\frac{\left(1 - \dfrac{b^2}{(R+r)^2}\right)\left(\dfrac{R^2}{r^2}\right)\left(\dfrac{b}{R+r}\right)\left(\sin\dfrac{R}{r}\theta\right)}{\left[1 + \dfrac{b^2}{(R+r)^2} - \left(\dfrac{2b}{R+r}\right)\left(\cos\dfrac{R}{r}\theta\right)\right]^2} \quad (3)$$

Fig. 2   Symbols for geared cycloid mechanisms. ——

Symbols ——

A = angular acceleration of output, degrees per second^2
b = radius of driving pin from center of planet gear
r = pitch radius of planet gear
R = pitch radius of fixed sun gear
V = angular velocity of output, degrees per second
β = angular displacement of output, degree
$\gamma = \theta R/r$
θ = input displacement, degree
ω = angular velocity of input, degrees per second

Fig. 3 A gear rolling on a gear flattens curves.

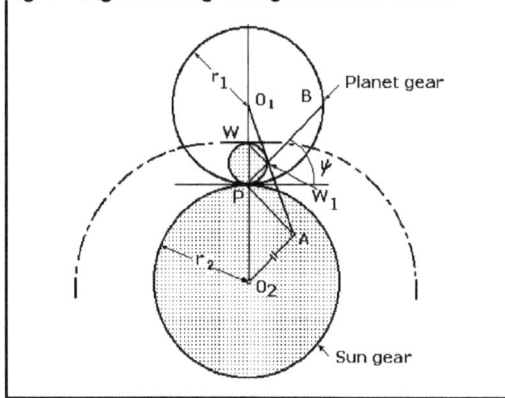

It is frequently desirable to find points on the planet gear that will describe approximately straight lines for portions of the output curves. These points will yield dwell mechanisms. Construction is as follows (see Fig. 3):

1. Draw an arbitrary line PB.
2. Draw its parallel O_2A.
3. Draw its perpendicular PA at P. Locate point A.
4. Draw O_1A. Locate W_1.
5. Draw perpendicular to PW_1 at W_1 to locate W.
6. Draw a circular with PW as the diameter.

Fig. 4 A gear rolling on a rack describes vee curves.

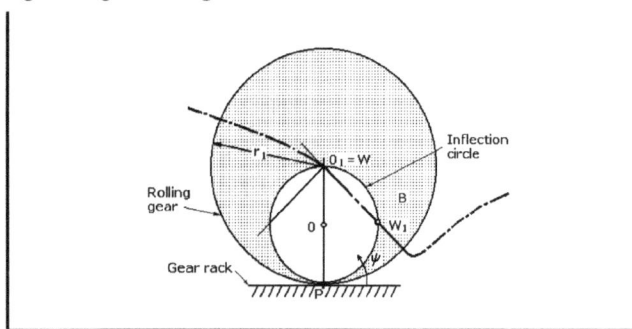

All points on this circle describe curves with portions that are approximately straight. This circle is also called the **inflection circle** because all points describe curves that have a point of inflection at the position illustrated. (The curve passing through point W is shown in Fig. 4.)

This is a special case. Draw a circle with a diameter half that of the gear (diameter O_1P). This is the inflection circle. Any point, such as point W_1, will describe a curve that is almost straight in the vicinity selected. Tangents to the curves will always pass through the center of the gear, O_1 (see Fig. 5).

To find the inflection circle for a gear rolling inside a gear:

1. Draw arbitrary line PB from the contact point P.
2. Draw its parallel O_2A, and its perpendicular PA. Locate A.
3. Draw line AO_1 through the center of the rolling gear. Locate W_1.
4. Draw a perpendicular through W_1. Obtain W. Line WP is the diameter of the inflection circle. Point W_1, which is an arbitrary point on the circle, will trace a curve of repeated almost-straight lines, as shown.

Fig. 5 A gear rolling inside a gear describes a zig-zag.

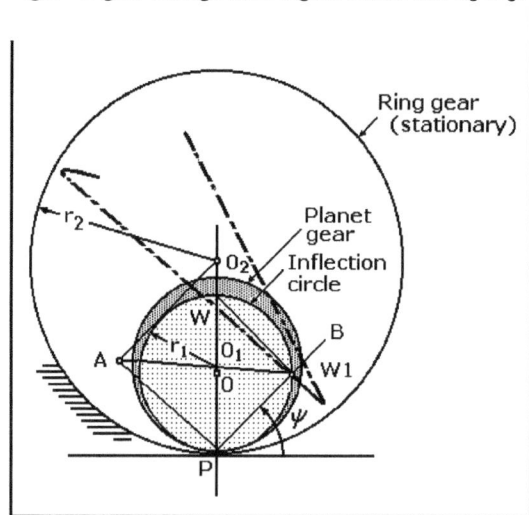

Fig. 6   The center of curvature: a gear rolling on gear.

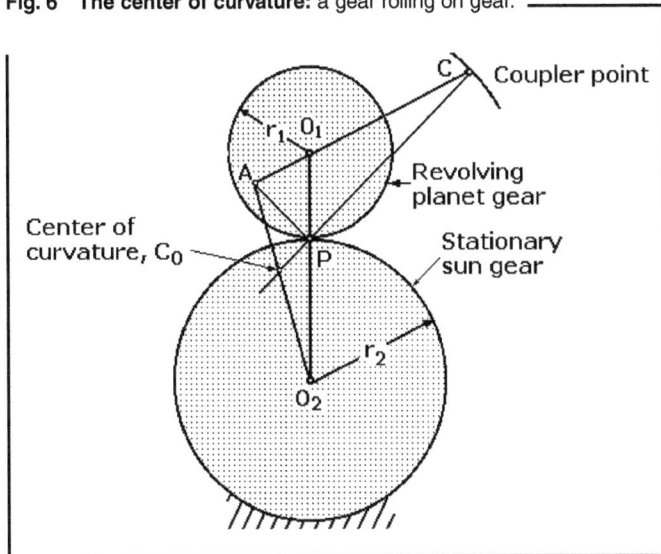

Fig. 7   The center of curvature: a gear rolling on a rack

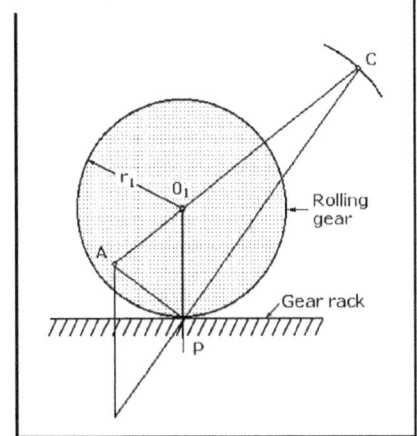

By locating the centers of curvature at various points, one can determine the length of the rocking or reciprocating arm to provide long dwells (see Fig. 6).

1. Draw a line through points C and P.
2. Draw a line through points C and O_1.
3. Draw a perpendicular to CP at P. This locates point A.
4. Draw line AO_2, to locate C_0, the center of curvature.

Construction is similar to Fig. 6.

1. Draw an extension of line CP.
2. Draw a perpendicular at P to locate A.
3. Draw a perpendicular from A to the straight surface to locate C.

Fig. 9   Analytical solutions.

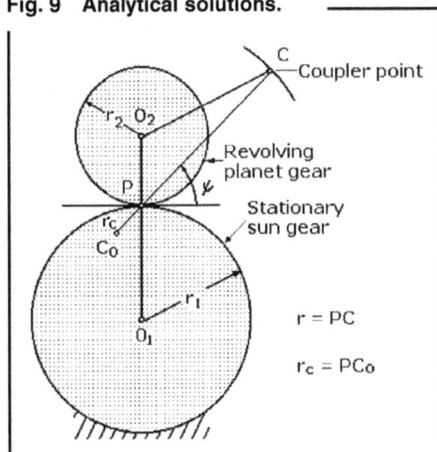

Fig. 8   The center of curvature: a gear rolling iside a gear.

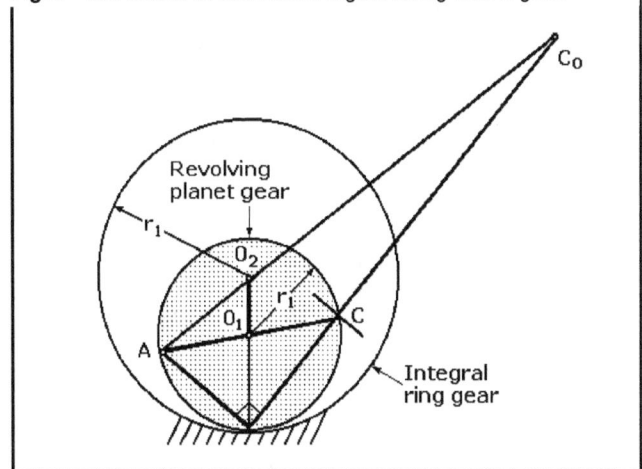

1. Draw extensions of CP and CO_1.
2. Draw a perpendicular of PC at P to locate A.
3. Draw AO_2 to locate C_0.

The center of curvature of a gear rolling on an external gear can be computed directly from the Euler-Savary equation:

$$\left(\frac{1}{r} - \frac{1}{r_c}\right)\sin \psi = \text{constant} \quad (7)$$

where angle ψ and r locate the position of C.

By applying this equation twice, specifically to point O_1 and O_2, which have their own centers of rotation, the following equation is obtained:

$$\left(\frac{1}{r_2} - \frac{1}{r_1}\right)\sin 90° = \left(\frac{1}{r} + \frac{1}{r_c}\right)\sin\psi$$

or

$$\frac{1}{r_2} + \frac{1}{r_1} = \left(\frac{1}{r} + \frac{1}{r_c}\right)\sin\psi$$

This is the final design equation. All factors except r_c are known; hence, solving for r_c leads to the location of C_0.

For a gear rolling inside an internal gear, the Euler-Savary equation is:

$$\left(\frac{1}{r} + \frac{1}{r_c}\right)\sin\psi = \text{constant}$$

which leads to:

$$\frac{1}{r_2} - \frac{1}{r_1} = \left(\frac{1}{r} - \frac{1}{r_c}\right)\sin\psi$$

DESIGN CURVES AND EQUATIONS FOR GEAR-SLIDER MECHANISMS

What is a gear-slider mechanism? It is little more than a crank-and-slider with two gears meshed in line with the crank (Fig. 1). But, because one of the gears (planet gear, 3) is prevented from rotating because it is attached to the connecting rod, the output is taken from the sun gear, not the slider. This produces a variety of cyclic output motions, depending on the proportions of the members.

In his investigation of the capabilities of the mechanism, Professor Preben Jensen of Bridgeport, Connecticut derived the equations defining its motion and acceleration characteristics. He

then devised some variations of his own (Figs. 5 through 8). These, he believes, will outperform the parent type. Jensen illustrated how the output of one of the new mechanisms, Fig. 8, can come to dead stop during each cycle, or progressively oscillate to new positions around the clock. A machine designer, therefore, can obtain a wide variety of intermittent motions from the arrangement and, by combining two of these units, he can tailor the dwell period of the mechanism to fit the automatic feed requirements of a machine.

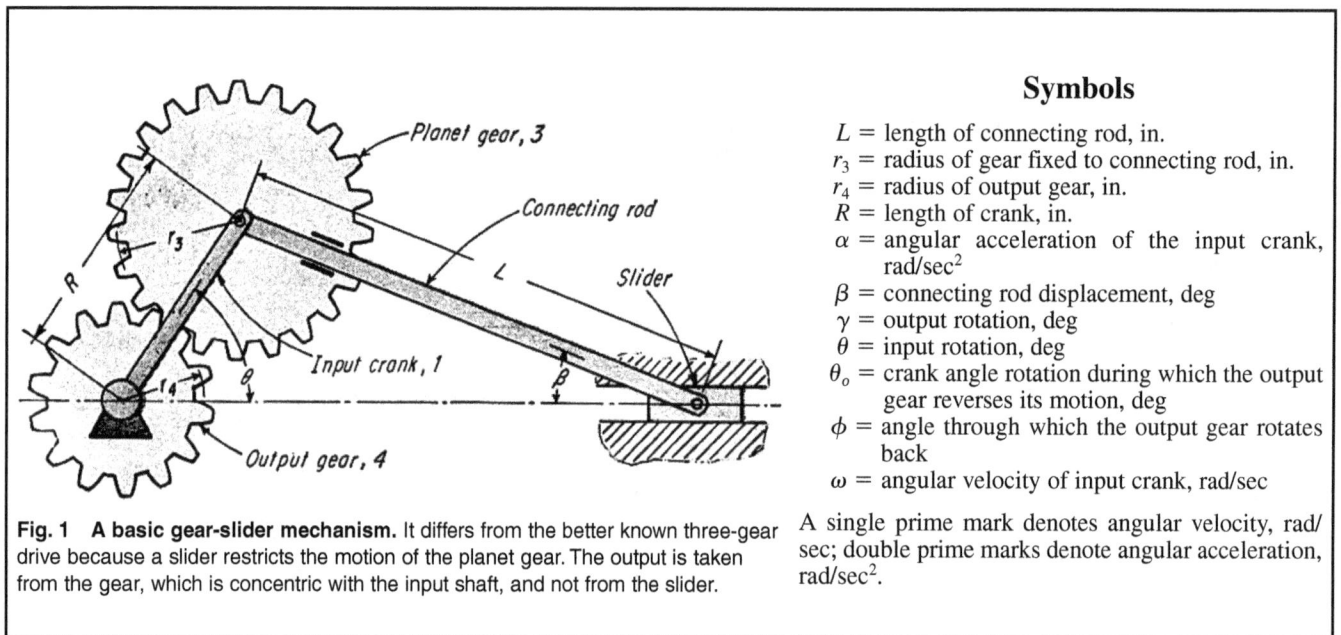

Fig. 1   A basic gear-slider mechanism. It differs from the better known three-gear drive because a slider restricts the motion of the planet gear. The output is taken from the gear, which is concentric with the input shaft, and not from the slider.

Symbols

L = length of connecting rod, in.
r_3 = radius of gear fixed to connecting rod, in.
r_4 = radius of output gear, in.
R = length of crank, in.
α = angular acceleration of the input crank, rad/sec^2
β = connecting rod displacement, deg
γ = output rotation, deg
θ = input rotation, deg
θ_o = crank angle rotation during which the output gear reverses its motion, deg
ϕ = angle through which the output gear rotates back
ω = angular velocity of input crank, rad/sec

A single prime mark denotes angular velocity, rad/sec; double prime marks denote angular acceleration, rad/sec^2.

The Basic Form

The input motion is to crank 1, and the output motion is from gear 4. As the crank rotates, say counterclockwise, it causes planet gear 3 to oscillate while following a satellite path around gear 4. This imparts a varying output motion to gear 4, which rotates twice in the counterclockwise direction (when $r_3 = r_4$) for every revolution of the input.

Jensen's equations for angular displacement, velocity, and acceleration of gear 4, when driven at a speed of ω by crank 1, are as follows:

Angular Displacement

$$\gamma = \theta + \frac{r_3}{r_4}(\theta + \beta) \qquad (1)$$

where β is computed from the following relationship (see the list of symbols in this article):

$$\sin \beta = \frac{R}{L} \sin \theta \qquad (2)$$

Angular Velocity

$$\gamma' = \omega + \frac{r_3}{r_4}(\omega + \beta') \qquad (3)$$

where

$$\frac{\beta'}{\omega} = \frac{R}{L} \frac{\cos \theta}{\left[1 - \left(\frac{R}{L}\right)^2 \sin^2 \theta\right]^{1/2}} \qquad (4)$$

Angular Acceleration

$$\gamma'' = \alpha + \frac{r_3}{r_4}(\alpha + \beta'') \qquad (5)$$

where

$$\frac{\beta''}{\omega^2} = \frac{R}{L} \frac{\sin \theta \left[\left(\frac{R}{L}\right)^2 - 1\right]}{\left[1 - \left(\frac{R}{L}\right)^2 \sin^2 \theta\right]^{3/2}} \qquad (6)$$

For a constant angular velocity, Eq. 5 becomes

$$\gamma'' = \frac{r_3}{r_4}\beta'' \qquad (7)$$

Design Charts

The equations were solved by Professor Jensen for various L/R ratios and positions of the crank angle θ to obtain the design charts in Figs. 2, 3, and 4. Thus, for a mechanism with

$L = 12$ in. $r_3 = 2.5$
$R = 4$ in. $r_4 = 1.5$
$\omega = 1000$ per second
$\quad =$ radians per second

Fig. 2   Angular displacement diagram for the connecting rod.

Fig. 3   Angular velocity curves for various crank angles.

Fig. 4   Angular acceleration curves for various crank angles.

Fig. 5 Modified gear-slider mechanism.

the output velocity at crank angle $\theta = 60°$ can be computed as follows:

$$L/R = 12/4 = 3$$

From Fig. 3 $\beta'/\omega = 0.175$

$$\beta' = 0.175(1000)$$
$$= 175 \text{ radians per second}$$

From Eq. 3

$$\gamma = 2960 \text{ radians per second}$$

Three-Gear Variation

One interesting variation, shown in Fig. 5, is obtained by adding idler gear 5 to the drive. If gears 3 and 4 are then made equal in size, output gear 4 will then oscillate with exactly the same motion as connecting rod 2.

One use for this linkage, Jensen said, is in machinery where a sleeve is to ride concentrically over an input shaft, and yet must oscillate to provide a reciprocat-ing motion. The shaft can drive the sleeve with this mechanism by making the sleeve part of the output gear.

Internal-Gear Variations

By replacing one of the external gears of Fig. 1 with an internal one, two mechanisms are obtained (Figs. 6 and 7) which have wider variable output abilities. But it is the mechanism in Fig. 7 that interested Jensen. This could be proportioned to give either a dwell or a progressive oscillation, that is, one in which the output rotates forward, say 360°, turns back to 30°, moves forward 30°, and then proceeds to repeat the cycle by moving forward again for 360°.

In this mechanism, the crank drives the large ring gear 3 which is fixed to the connecting rod 2. Output is from gear 4. Jensen derived the following equations:

Output Motion

$$\omega_4 = -\left(\frac{L - R - r_4}{Lr_4}\right)R\omega_1 \qquad (8)$$

When $r_4 = L - R$, then $r_4 = 0$ from Eq. 8, and the mechanism is proportioned to give instantaneous dwell. To obtain a progressive oscillation, r_4 must be greater than $L - R$, as shown in Jensen's model (Fig. 8).

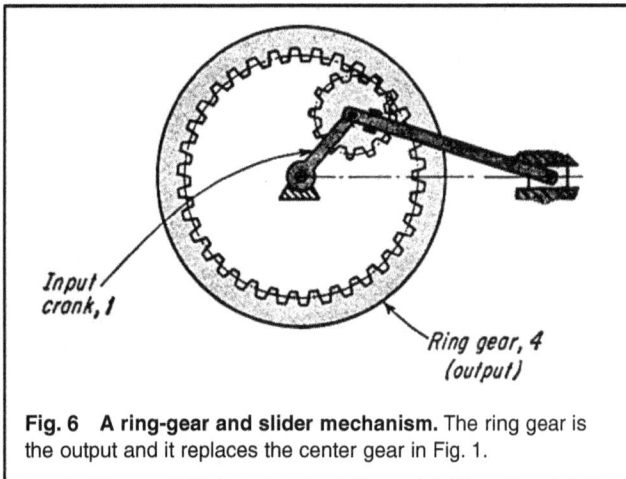

Fig. 6 A ring-gear and slider mechanism. The ring gear is the output and it replaces the center gear in Fig. 1.

Fig. 7 A more practical ring-gear and slider arrangement. The output is now from the smaller gear.

Fig. 8 Jensen's model of the ring-gear and slider mechanism shown in Fig. 7. A progressive oscillation motion is obtained by making r_4 greater than L-R.

176

If gear 4 turns back and then starts moving forward again, there must be two positions where the motion of gear 4 is zero. Those two mechanisms are symmetrical with respect to A_0B. If θ_0 equals the crank-angle rotation (of input), during which the output gear reverses its motion, and ϕ equals the angle through which gear 4 rotates back, then

$$\cos \frac{\theta_0}{2} = \left[\frac{L^2 - R^2}{r_4(2R + r_4)} \right]^{1/2} \quad (9)$$

and

$$\gamma = \theta_0 - \frac{r_3}{r_4}(\theta_0 - \beta_0) \quad (10)$$

where

$$\sin \beta_0 = \frac{R}{L} \sin \frac{\theta_0}{2} \quad (11)$$

Chart for Proportioning

The chart in Fig. 9 helps proportion the mechanism of Fig. 8 to provide a specific kind of progressive oscillation. It is set up for R equals 1 in. For other values of R, convert the chart values for r_4 proportionally, as shown below.

For example, assume that the output gear, during each cycle, is to rotate back 9.2°. Thus $\phi = 9.2°$. Also given is $R = 0.75$ in. and $L = 1.5$ in. Thus $L/R = 2$.

From the right side of the chart, go to the ϕ-curve for $L = 2$, then upward to the θ_0-curve for $L = 2$ in. Read $\theta_0 = 82°$ at the left ordinate.

Now return to the second intersection point and proceed upward to read on the abscissa scale for $L = 2$, a value of $r_4 = 1.5$. Since $R = 0.75$ in., and the chart is for R 1, convert r_4 as follows: $r_4 = 0.75$ (1.5) = 1.13 in.

Thus, if the mechanism is built with an output gear of radius $r_4 = 1.13$ in., then during 82° rotation of the crank, the output gear 4 will go back 9.2°. Of course, during the next 83°, gear 4 will have reversed back to its initial position—and then will keep going forward for the remaining 194° of the crank rotation.

Future Modifications

The mechanism in Fig. 8 is designed to permit changing the output motion easily from progressive oscillation to instantaneous dwell or nonuniform CW or CCW rotation. This is accomplished by shifting the position of the pin which acts as the sliding piece of the centric slider crank. It is also possible to use an eccentric slider crank, a four-bar linkage, or a sliding-block linkage as the basic mechanism.

Two mechanisms in series will give an output with either a prolonged dwell or two separate dwells. The angle between the separated dwells can be adjusted during its operation by interposing a gear differential so that the position of the output shaft of the first mechanism can be changed relative to the position of the input shaft of the second mechanism.

The mechanism can also be improved by introducing an additional link, B-B_0, to guide pin B along a circular arc instead of a linear track. This would result in a slight improvement in the performance of the mechanism.

Fig. 9 **A chart** for proportioning a ring-gear and slider mechanism.

CHAPTER 7
CAM, GENEVA, AND RATCHET DRIVES AND MECHANISMS

CAM-CONTROLLED PLANETARY GEAR SYSTEM

By incorporating a grooved cam, a novel mechanism can produce
a wide variety of output motions.

Construction details of a cam-planetary mechanism used in a film drive.

Do you want more variety in the kinds of output motion given by a planetary gear system? You can have it by controlling the planet with a grooved cam. The method gives the mechanism these additional features:

- Intermittent motion, with long dwells and minimum acceleration and deceleration
- Cyclic variations in velocity
- Two levels, or more, of constant speed during each cycle of the input

The design is not simple because of need to synchronize the output of the planetary system with the cam contour. However, such mechanisms are now at work in film drives and should prove useful in many automatic machines. Here are equations, tables, and a step-by-step sequence that will make the procedure easier.

How the Mechanism Works

The planet gear need not be cut in full—a gear sector will do because the planet is never permitted to make a full revolution. The sun gear is integral with the output gear. The planet arm is fixed to the input shaft, which is coaxial with the output shaft. Attached to the planet is a follower roller which rides in a cam groove. The cam is fixed to the frame.

The planet arm (input) rotates at constant velocity and makes one revolution with each cycle. Sun gear (output) also makes one revolution during each cycle. Its motion is modified, however, by the oscillatory motion of the planet gear relative to the planet arm. It is this motion that is controlled by the cam (a constant-radius cam would not affect the output, and the drive would give only a constant one-to-one ratio).

Comparison with Other Devices

A main feature of this cam-planetary mechanism is its ability to produce a wide range of nonhomogeneous functions. These functions can be defined by no less than two mathematical expressions, each valid for a discrete portion of the range. This feature is not shared by the more widely known intermittent mechanisms: the external and internal genevas, the three-gear drive, and the cardioid drive.

Either three-gear or cardioid can provide a dwell period—but only for a comparatively short period of the cycle. With the cam-planetary, one can obtain over 180° of dwell during a 360° cycle by employing a 4-to-1 gear ratio between planet and sun.

And what about a cam doing the job by itself? This has the disadvantage of producing reciprocating motion. In other words, the output will always reverse during the cycle—a condition unacceptable in many applications.

Design Procedure

The basic equation for an epicyclic gear train is:

$$d\theta_S = d\theta_A - nd\,\theta_{P-A}$$

where: $d\theta_S$ = **rotation of sun gear (output), deg**
$d\theta_A$ = **rotation of planet arm (input), deg**
$d\theta_{P-A}$ = **rotation of planet gear with respect to arm, deg**
n = **ratio of planet to sun gear.**

The required output of the system is usually specified in the form of kinematic curves. Design procedure then is to:

- Select the proper planet-sun gear ratio.
- Develop the equations of the planet motion (which also functions as a cam follower).
- Compute the proper cam contour.

FIVE CAM-STROKE-AMPLIFYING MECHANISMS

Sliding cam

Double-faced cam

Stroke-multiplying mechanism

Cam-and-rack

Cut-out cam

When the pressure angles of stroke-amplifying mechanisms are too high to satisfy the design requirements, and it is undesirable to enlarge the cam size, certain devices can be installed to reduce the pressure angles:

Sliding cam—This mechanism is used on a wire-forming machine. Cam D has a pointed shape because of the special motion required for twisting wires. The machine operates at slow speeds, but the principle employed here is also applicable to high-speed cams.

The original stroke desired was (y_1 + y_2), but this results in a large pressure angle. The stroke therefore is reduced to y_2 on one side of the cam, and a rise of y_1 is added to the other side. Flanges B are attached to cam shaft A. Cam D, a rectangle with the two cam ends (shaded), is shifted upward as it cams off stationary roller R when the cam follower E is

being cammed upward by the other end of cam D.

Stroke-multiplying mechanisms—This mechanism is used in power presses. The opposing slots, the first in a fixed member D, and the second in the movable slide E, multiply the motion of the input slide A driven by the cam. As A moves upward, E moves rapidly to the right.

Double-faced cam—This mechanism doubles the stroke, hence reduces the pressure angles to one-half of their original values. Roller R_1 is stationary. When the cam rotates, its bottom surface lifts itself on R_1, while its top surface adds an additional motion to the movable roller R_2. The output is driven linearly by roller R_2 and thus is approximately the sum of the rise of both of these surfaces.

Cam-and-rack—This mechanism increases the throw of a lever. Cam B

rotates around A. The roller follower travels at distances y_1; during this time, gear segment D rolls on rack E. Thus the output stroke of lever C is the sum of transmission and rotation, giving the magnified stoke y.

Cut-out cam—A rapid rise and fall within 72° was desired. This originally called for the cam contour, D, but produced severe pressure angles. The condition was improved by providing an additional cam C. This cam also rotates around the cam center A, but at five times the speed of cam D because of a 5:1 gearing arrangement (not shown). The original cam was then completely cut away for the 72° (see surfaces E). The desired motion, expanded over 360° (because 72° × 5 = 360°), is now designed into cam C. This results in the same pressure angle as would occur if the original cam rise occurred over 360° instead of 72°.

CAM-CURVE-GENERATING MECHANISMS

It usually doesn't pay to design a complex cam curve if it can't be easily machined—so check these mechanisms before starting your cam design.

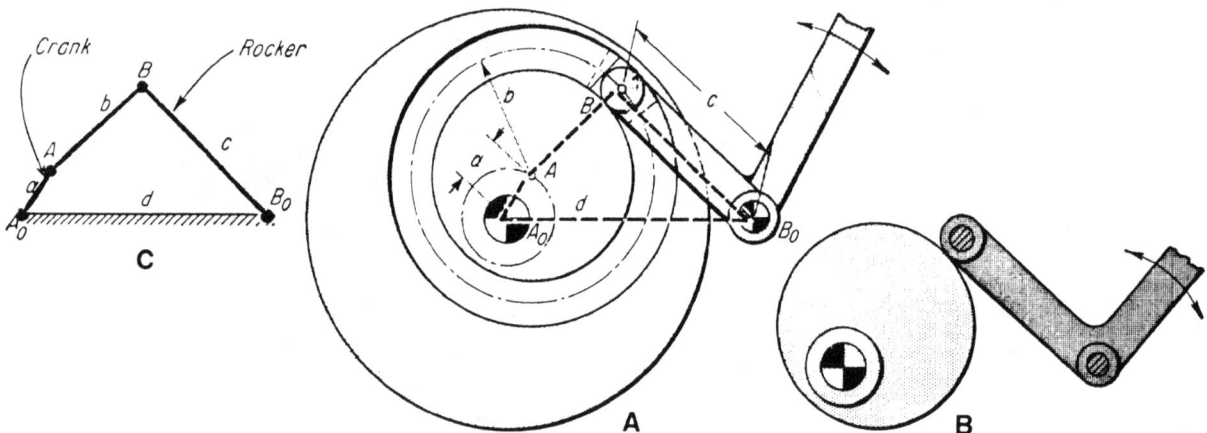

Fig. 1 **A circular cam groove** is easily machined on a turret lathe by mounting the plate eccentrically onto the truck. The plate cam in **(B)** with a spring-load follower produces the same output motion. Many designers are unaware that this type of cam has the same output motion as four-bar linkage **(C)** with the indicated equivalent link lengths. Thus, it's the easiest curve to pick when substituting a cam for an existing linkage.

If you have to machine a cam curve into the metal blank without a master cam, how accurate can you expect it to be? That depends primarily on how precisely the mechanism you use can feed the cutter into the cam blank. The mechanisms described here have been carefully selected for their practicability. They can be employed directly to machine the cams, or to make master cams for producing other cams.

The cam curves are those frequently employed in automatic-feed mechanisms and screw machines They are the circular, constant-velocity, simple-harmonic, cycloidal, modified cycloidal, and circular-arc cam curve, presented in that order.

Circular Cams

This is popular among machinists because of the ease in cutting the groove. The cam (Fig. 1A) has a circular groove whose center, A, is displaced a distance a from the cam-plate center, A_0, and can simply be a plate cam with a spring-loaded follower (Fig. 1B).

Interestingly, with this cam you can easily duplicate the motion of a four-bar linkage (Fig. 1C). Rocker BB_0 in Fig. 1C, therefore, is equivalent to the motion of the swinging follower shown in Fig. 1A.

The cam is machined by mounting the plate eccentrically on a lathe. Consequently, a circular groove can be cut to close tolerances with an excellent surface finish.

If the cam is to operate at low speeds, you can replace the roller with an arc-formed slide. This permits the transmission of high forces. The optimum design of these "power cams" usually requires time-consuming computations.

The disadvantages (or sometimes, the advantage) of the circular-arc cam is that, when traveling from one given point, its follower reaches higher-speed accelerations than with other equivalent cam curves.

Constant-Velocity Cams

A constant-velocity cam profile can be generated by rotating the cam plate and feeding the cutter linearly, both with uniform velocity, along the path the translating roller follower will travel later (Fig. 2A). In the example of a swinging follower, the tracer (cutter) point is placed on an arm whose length is equal to the length of the swinging roller follower, and the arm is rotated with uniform velocity (Fig. 2B).

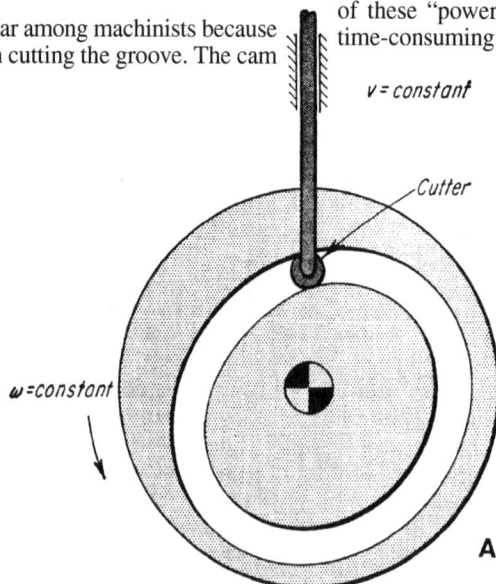

Fig. 2 **A constant-velocity** cam is machined by feeding the cutter and rotating the cam at constant velocity. The cutter is fed linearly **(A)** or circularly **(B)**, depending on the type of follower.

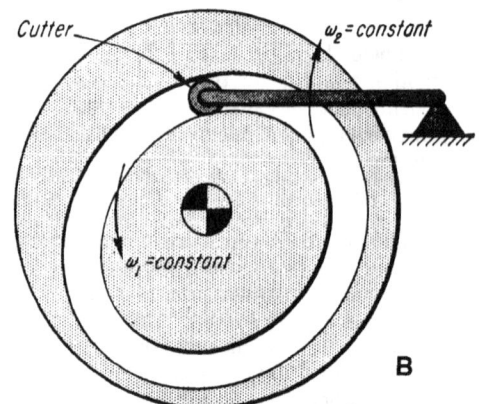

Fig. 3 **For producing simple harmonic curves:**
(**A**) a scotch yoke device feeds the cutter while the gearing arrangement rotates the cam; (**B**) a truncated-cylinder slider for a cylindrical cam; (**C**) a scotch-yoke inversion linkage for avoiding gearing; (**D**) an increase in acceleration when a translating follower is replaced by a swinging follower.

Simple-Harmonic Cams

The cam is generated by rotating it with uniform velocity and moving the cutter with a scotch yoke geared to the rotary motion of the cam. Fig. 3A shows the principle for a radial translating follower; the same principle is applicable for offset translating and the swinging roller follower. The gear ratios and length of the crank working in the scotch yoke control the pressure angles (the angles for the rise or return strokes).

For barrel cams with harmonic motion, the jig in Fig. 3B can easily be set up to do the machining. Here, the barrel cam is shifted axially by the rotating, weight-loaded (or spring-loaded) truncated cylinder.

The scotch-yoke inversion linkage (Fig. 3C) replaces the gearing called for in Fig. 3A. It will cut an approximate simple-harmonic motion curve when the cam has a swinging roller follower, and an exact curve when the cam has a radial or offset translating roller follower. The slotted member is fixed to the machine frame *1*. Crank *2* is driven around the center *0*. This causes link *4* to oscillate back and forward in simple harmonic motion. The sliding piece *5* carries the cam to be cut, and the cam is rotated around the center of *5* with uniform velocity. The length of arm *6* is

made equal to the length of the swinging roller follower of the actual cam mechanism and the device adjusted so that the extreme position of the center of *5* lies on the center line of *4*.

The cutter is placed in a stationary spot somewhere along the centerline of member *4*. If a radial or offset translating roller follower is used, sliding piece *5* is fastened to *4*.

The deviation from simple harmonic motion, when the cam has a swinging follower, causes an increase in acceleration ranging from 0 to 18 percent (Fig. 3D), which depends on the total angle of oscillation of the follower. Note that for a typical total oscillating angle of 45° the increase in acceleration is about 5 percent.

Cycloidal Motion

This curve is perhaps the most desirable from a designer's viewpoint because of its excellent acceleration characteristic. Luckily, this curve is comparatively easy to generate. Before selecting the mechanism, it is worth looking at the underlying theory of cycloids because it is possible to generate not only cycloidal motion but a whole family of similar curves.

The cycloids are based on an offset sinusoidal wave (Fig. 4). Because the

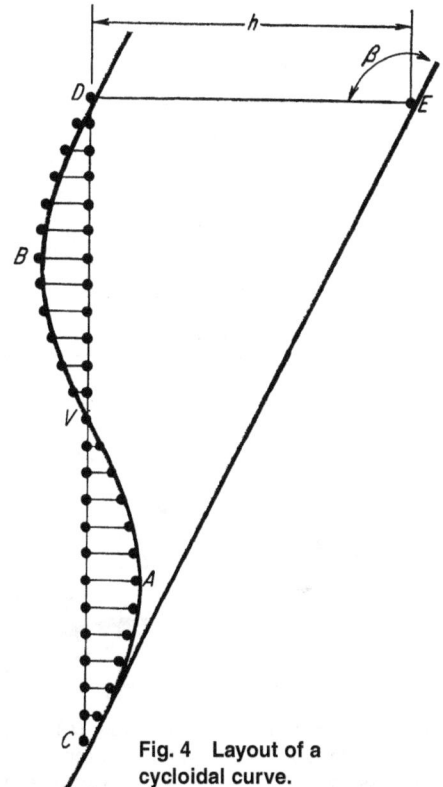

Fig. 4 **Layout of a cycloidal curve.**

183

radii of curvatures in points C, V, and D are infinite (the curve is "flat" at these points), if this curve was a cam groove and moved in the direction of line CVD, a translating roller follower, actuated by this cam, would have zero acceleration at points C, V, and D no matter in what direction the follower is pointed.

Now, if the cam is moved in the direction of CE and the direction of motion of the translating follower is lined up perpendicular to CE, the acceleration of the follower in points, C, V, and D would still be zero. This has now become the basic cycloidal curve, and it can be considered as a sinusoidal curve of a certain amplitude (with the amplitude measured perpendicular to the straight line) superimposed on a straight (constant-velocity) line.

The cycloidal is considered to be the best standard cam contour because of its low dynamic loads and low shock and vibration characteristics. One reason for these outstanding attributes is that sudden changes in acceleration are avoided during the cam cycle. But improved performance is obtainable with certain modified cycloidals.

Modified Cycloids

To modify the cycloid, only the direction and magnitude of the amplitude need to be changed, while keeping the radius of curvature infinite at points C, V, and D.

Comparisons are made in Fig. 5 of some of the modified curves used in industry. The true cycloidal is shown in the cam diagram of Fig. 5A. Note that the sine amplitudes to be added to the constant-velocity line are perpendicular to the base. In the Alt modification shown in Fig. 5B (named after Hermann Alt, a German kinematician who first analyzed it), the sine amplitudes are perpendicular to the constant-velocity line. This results in improved (lower) velocity characteristics (Fig. 5D), but higher acceleration magnitudes (Fig. 5E).

The Wildt modified cycloidal (after Paul Wildt) is constructed by selecting a point w which is 0.57 the distance $T/2$, and then drawing line wp through yp which is midway along OP. The base of the sine curve is then constructed perpendicular to yw. This modification results in a maximum acceleration of $5.88\ h/T^2$. By contrast, the standard cycloidal curve has a maximum acceleration of $6.28\ h/T^2$. This is a 6.8 reduction in acceleration.

(It's a complex task to construct a cycloidal curve to go through a particular point P—where P might be anywhere within the limits of the box in Fig. 5C—and with a specific scope at P. There is a growing demand for this kind of cycloidal modification.

Generating Modified Cycloidals

One of the few methods capable of generating the family of modified cycloidals consists of a double carriage and rack arrangement (Fig. 6A).

The cam blank can pivot around the spindle, which in turn is on the movable carriage I. The cutter center is stationary. If the carriage is now driven at constant speed by the leadscrew in the direction of the arrow, steel bands 1 and 2 will also cause the cam blank to rotate. This rotation-and-translation motion of the cam will cut a spiral groove.

For the modified cycloidals, a second motion must be imposed on the cam to compensate for the deviations from the true cycloidal. This is done by a second

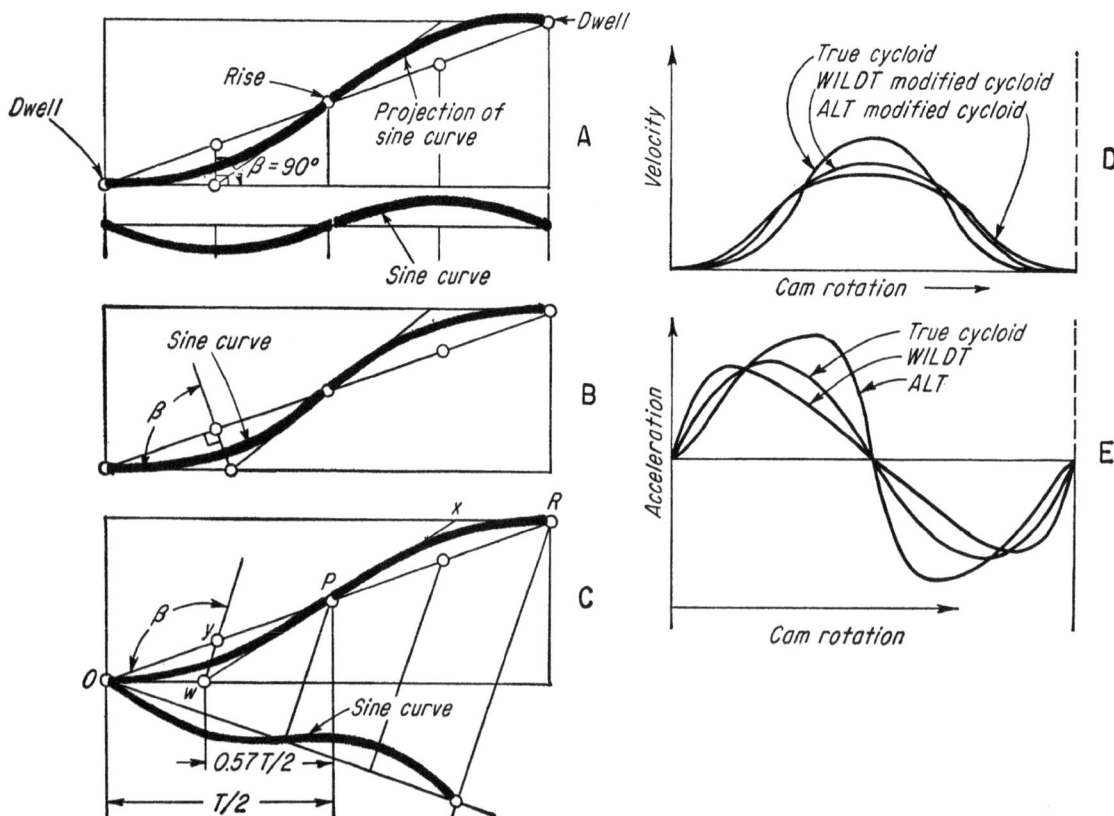

Fig. 5 A family of cycloidal curves: **(A)** a standard cycloidal motion; **(B)** a modification according to H. Alt; **(C)** a modification according to P. Wildt; **(D)** a comparison of velocity characteristics; **(E)** a comparison of acceleration curves.

Fig. 6 Mechanisms for generating (A) modified cycloidal curves, and **(B)** basic cycloidal curves.

Fig. 7 A technique for machining circular-arc cams. Radii r_2 and r_5 are turned on a lathe; hardened templates are added to r_1, r_3, and r_4 for facilitating hand filing.

steel-band arrangement. As carriage I moves, bands 3 and 4 cause the eccentric to rotate. Because of the stationary frame, the slide surrounding the eccentric is actuated horizontally. This slide is part of carriage II. As a result, a sinusoidal motion is imposed on the cam.

Carriage I can be set at various angles β to match angle β in Fig. 5B and C. The mechanism can also be modified to cut cams with swinging followers.

Circular-Arc Cams

In recent years it has become customary to turn to the cycloidal and other similar curves even when speeds are low. However, there are still many applications for circular-arc cams. Those cams are composed of circular arcs, or circular arc and straight lines. For comparatively small cams, the cutting technique illustrated in Fig. 7 produces accurate results.

Assume that the contour is composed of circular arc 1-2 with center at 0_2, arc 3-4 with center at 0_3, arc 4-5 with center at 0_1, arc 5-6 with center at 0_4, arc 7-1 with center at 0_1, and the straight lines 2-3 and 6-7. The method calls for a combination of drilling, lathe turning, and template filing.

First, small holes about 0.1 in. in diameter are drilled at 0_1, 0_3, and 0_4.

Then a hole drilled with the center at 0_2, and radius of r_2. Next the cam is fixed in a turret lathe with the center of rotation at 0_1, and the steel plate is cut until it has a diameter of $2r_5$. This completes the larger convex radius. The straight lines 6-7 and 2-3 are then milled on a milling machine.

Finally, for the smaller convex arcs, hardened pieces are turned with radii r_1, r_3, and r_4. One such piece is shown in Fig. 7. The templates have hubs that fit into the drilled holes at 0_1, 0_3, and 0_4. Next the arcs 7-1, 3-4, and 5-6 are filed with the hardened templates as a guide. The final operation is to drill the enlarged hole at 0_1 to a size that will permit a hub to be fastened to the cam.

This method is usually better than copying from a drawing or filing the scallops from a cam on which a large number of points have been calculated to determine the cam profile.

Compensating for Dwells

One disadvantage with the previous generating machines is that, with the exception of the circular cam, they cannot include a dwell period within the rise-and-fall cam cycle. The mechanisms must be disengaged at the end of the rise, and the cam must be rotated the exact number of degrees to the point where the

A Desired output

B Four-station geneva

C Eight-station geneva

D Double geneva with differential

Fig. 8 **Double genevas with differentials** for obtaining long dwells. The desired output characteristic **(A)** of the cam is obtained by adding the motion **(B)** of a four-station geneva to that of **(C)** an eight-station geneva. The mechanical arrangement of genevas with a differential is shown in **(D)**; the actual device is shown in **(E)**. A wide variety of output dwells **(F)** are obtained by varying the angle between the driving cranks of the genevas.

E Final mechanism

fall cycle begins. This increases the possibility of inaccuracies and slows down production.

There are two mechanisms, however, that permit automatic cam machining through a specific dwell period: the double-geneva drive and the double eccentric mechanism.

Double-Genevas with Differential

Assume that the desired output contains dwells (of specific duration) at both the rise and fall portions, as shown in Fig. 8A. The output of a geneva that is being rotated clockwise will produce an intermittent motion similar to the one shown in Fig. 8B—a rise-dwell-rise-dwell motion. These rise portions are distorted simple-harmonic curves, but are sufficiently close to the pure harmonic to warrant their use in many applications.

If the motion of another geneva, rotating counterclockwise as shown in

(Fig. 8C), is added to that of the clockwise geneva by a differential (Fig. 8D), then the sum will be the desired output shown in (Fig. 8A).

The dwell period of this mechanism is varied by shifting the relative positions between the two input cranks of the genevas.

The mechanical arrangement of the mechanism is shown in Fig. 8D. The two driving shafts are driven by gearing (not shown). Input from the four-star geneva to the differential is through shaft *3*; input from the eight-station geneva is through the spider. The output from the differential, which adds the two inputs, is through shaft *4*.

The actual mechanism is shown in Fig. 8E. The cutter is fixed in space. Output is from the gear segment that rides on a fixed rack. The cam is driven by the motor, which also drives the enclosed genevas. Thus, the entire device reciprocates back and forth on the slide to feed the cam properly into the cutter.

F Various dwell resultants

186

Fig. 9 A four-bar coupler mechanism for replacing the cranks in genevas to obtain smoother acceleration characteristics.

Genevas Driven by Couplers

When a geneva is driven by a constant-speed crank, as shown in Fig. 8D, it has a sudden change in acceleration at the beginning and end of the indexing cycle (as the crank enters or leaves a slot). These abrupt changes can be avoided by employing a four-bar linkage with a coupler in place of the crank. The motion of the coupler point C (Fig. 9) permits its smooth entry into the geneva slot.

Double Eccentric Drive

This is another machine for automatically cutting cams with dwells. The rotation of crank A (Fig. 10) imparts an oscillating motion to the rocker C with a prolonged dwell at both extreme positions. The cam, mounted on the rocker, is rotated by the chain drive and then is fed into the cutter with the proper motion. During the dwells of the rocker, for example, a dwell is cut into the cam.

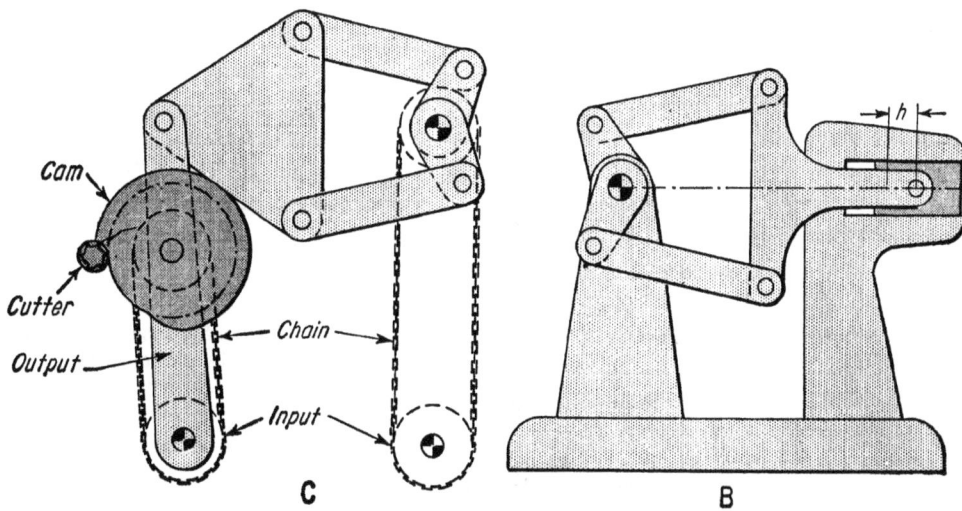

Fig. 10 A double eccentric drive for automatically cutting cams with dwells. The cam is rotated and oscillated, with dwell periods at extreme ends of oscillation corresponding to desired dwell periods in the cam.

FIFTEEN DIFFERENT CAM MECHANISMS

This assortment of devices reflects the variety of ways in which cams can be put to work.

Figs. 1, 2, and 3   A constant-speed rotary motion is converted into a variable, reciprocating motion (Fig. 1); rocking or vibratory motion of a simple forked follower (Fig. 2); or a more robust follower (Fig. 3), which can provide valve-moving mechanisms for steam engines. Vibratory-motion cams must be designed so that their opposite edges are everywhere equidistant when they are measured through their drive-shaft centers.

Fig. 4   An automatic feed for automatic machines. There are two cams, one with circular motion, the other with reciprocating motion. This combination eliminates any trouble caused by the irregularity of feeding and lack of positive control over stock feed.

Fig. 6   This indexing mechanism combines an epicyclic gear and cam. A planetary wheel and cam are fixed relative to one another; the carrier is rotated at uniform speed around the fixed wheel. The index arm has a nonuniform motion with dwell periods.

Fig. 5   A barrel cam with milled grooves is used in sewing machines to guide thread. This kind of cam is also used extensively in textile manufacturing machines such as looms and other intricate fabric-making machines.

Fig. 7   A double eccentric, actuated by a suitable handle, provides powerful clamping action for a machine-tool holding fixture.

Fig. 8   A mixing roller for paint, candy, or food. A mixing drum has a small oscillating motion while rotating.

Fig. 9   A slot cam converts the oscillating motion of a camshaft to a variable but straight-line motion of a rod. According to slot shape, rod motion can be made to suit specific design requirements, such as straight-line and logarithmic motion.

Fig. 10   The continuous rotary motion of a shaft is converted into the reciprocating motion of a slide. This device is used on sewing machines and printing presses.

Fig. 11   Swash-plate cams are feasible for light loads only, such as in a pump. The cam's eccentricity produces forces that cause excessive loads. Multiple followers can ride on a plate, thereby providing smooth pumping action for a multipiston pump.

Fig. 12   This steel-ball cam can convert the high-speed rotary motion of an electric drill into high-frequency vibrations that power the drill core for use as a rotary hammer for cutting masonry and concrete. This attachment can also be designed to fit hand drills.

Fig. 13   This tilting device can be designed so that a lever remains in a tilted position when the cylinder rod is withdrawn, or it can be spring-loaded to return with a cylinder rod.

Fig. 14   This sliding cam in a remote control can shift gears in a position that is otherwise inaccessible on most machines.

Fig. 15   A groove and oval follower form a device that requires two revolutions of a cam for one complete follower cycle.

TEN SPECIAL-FUNCTION CAMS

Fig. 1—A quick drop of the follower is obtained by permitting the cam to be pushed out of the way by the follower itself as it reaches the edge of the cam. Lugs *C* and *C'* are fixed to the camshaft. The cam is free to turn (float) on the camshaft, limited by lug *C* and the adjusting screw. With the cam rotating clockwise, lug C drives the cam through lug *B*. At the position shown, the roller will drop off the edge of the cam, which is then accelerated clockwise until its cam lug *B* strikes the adjusting screw of lug *C'*.

Fig. 2—Instantaneous drop is obtained by the use of two integral cams and followers. The roller follower rides on cam 1. Continued rotation will transfer contact to the flat-faced follower, which drops suddenly off the edge of cam 2. After the desired dwell, the follower is restored to its initial position by cam 1.

Fig. 3—The dwell period of the cam can be varied by changing the distance between the two rollers in the slot.

Fig. 4—A reciprocating pin (not shown) causes the barrel cam to rotate intermittently. The cam is stationary while a pin moves from *1* to *2*. Groove *2-3* is at a lower level; thus, as the pin retracts, it cams the barrel cam; then it climbs the incline from *2* to the new position of *1*.

Fig. 5—A double-groove cam makes two revolutions for one complete movement of the follower. The cam has movable switches, *A* and *B*, which direct the follower alternately in each groove. At the instant shown, *B* is ready to guide the roller follower from slot *1* to slot *2*.

Figs. 6 and 7—Increased stroke is obtained by permitting the cam to shift on the input shaft. Total displacement of the follower is therefore the sum of the cam displacement on the fixed roller plus the follower displacement relative to the cam.

Fig. 1   A quick-acting floating cam.

Fig. 2   A quick-acting dwell cam.

Fig. 3   An adjustable-dwell cam.

Fig. 4   An indexing cam.

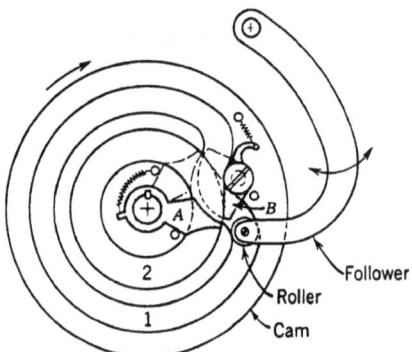

Fig. 5   A double-revolution cam.

Fig. 6   An increased-stroke barrel cam.

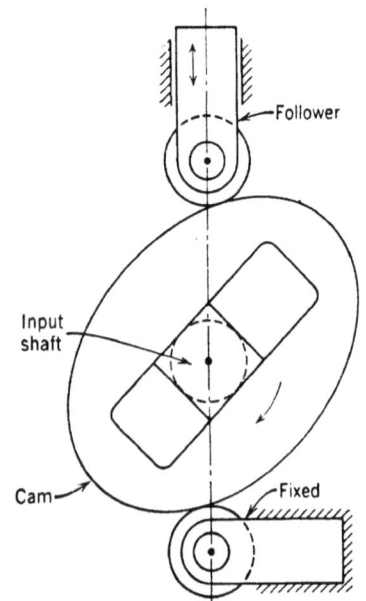

Fig. 7   An increased-stroke plate cam.

Fig. 8   An adjustable roller-position cam.

Fig. 9   An adjustable pivot-point cam.

Fig. 10   An adjustable lug cam.

Fig. 8—The stroke of the follower is adjusted by turning the screw handle which changes distance *AB*.

Fig. 9—The pivot point of the connecting link to the follower is changed from point *D* to point *C* by adjusting the screw.

Fig. 10—Adjustable dwell is obtained by having the main cam, with lug *A*, pinned to the revolving shaft. Lug *A* forces the plunger up into the position shown, and allows the latch to hook over the catch, thus holding the plunger in the up position. The plunger is unlatched by lug *B*. The circular slots in the cam plate permit the shifting of lug *B*, thereby varying the time that the plunger is held in the latched position.

REFERENCE: Rothbart, H. A. *Cams—Design, Dynamics, and Accuracy,* John Wiley and Sons, Inc., New York.

TWENTY GENEVA DRIVES

Locking-arm geneva drive

Output · Cam · Input crank · Driving roller · Locking arm

The driving follower on the rotating input crank of this geneva enters a slot and rapidly indexes the output. In this version, the roller of the locking-arm (shown leaving the slot) enters the slot to prevent the geneva from shifting when it is not indexing.

Planetary gear geneva drive

Output link · Locking surfaces · Planet gear · Ring-gear geneva · Input gear

The output link remains stationary while the input gear drives the planet gear with single tooth on the locking disk. The disk is part of the planet gear, and it meshes with the ring-gear geneva to index the output link one position.

Twin geneva drive

Output · Input · Genevas

The driven member of the first geneva acts as the driver for the second geneva. This produces a wide variety of output motions including very long dwells between rapid indexes.

Groove cam geneva drive

Roller follower · Output geneva · Input link · Locking surface · Driving roller · Groove · Stationary cam

When a geneva is driven by a roller rotating at a constant speed, it tends to have very high acceleration and deceleration characteristics. In this modification, the input link, which contains the driving roller, can move radially while being rotated by the groove cam. Thus, as the driving roller enters the geneva slot, it moves radially inward. This action reduces the geneva acceleration force.

Locking-slide geneva drive

Drive pin · Locking pin · Input crank

One pin locks and unlocks the geneva; the second pin rotates the geneva during the unlocked phase. In the position shown, the drive pin is about to enter the slot to index the geneva. Simultaneously, the locking pin is just clearing the slot.

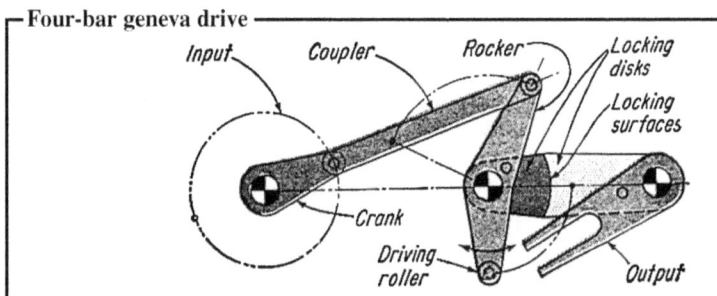

Four-bar geneva drive

Input · Coupler · Rocker · Locking disks · Locking surfaces · Crank · Driving roller · Output

A four-bar geneva produces a long-dwell motion from an oscillating output. The rotation of the input wheel causes a driving roller to reciprocate in and out of the slot of the output link. The two disk surfaces keep the output in the position shown during the dwell period.

Rapid-transfer geneva drive

The coupler point at the extension of the connecting link of the four-bar mechanism describes a curve with two approximately straight lines, 90° apart. This provides a favorable entry situation because there is no motion in the geneva while the driving pin moves deeply into the slot. Then there is an extremely rapid index. A locking cam, which prevents the geneva from shifting when it is not indexing, is connected to the input shaft through gears.

Long-dwell geneva drive

This geneva arrangement has a chain with an extended pin in combination with a standard geneva. This permits a long dwell between each 90° shift in the position of the geneva. The spacing between the sprockets determines the length of dwell. Some of the links have special extensions to lock the geneva in place between stations.

Modified motion geneva drive

The input link of a normal geneva drive rotates at constant velocity, which restricts flexibility in design. That is, for given dimensions and number of stations, the dwell period is determined by the speed of the input shaft. Elliptical gears produce a varying crank rotation that permits either extending or reducing the dwell period.

Dual-track geneva drive

The key consideration in the design of genevas is to have the input roller enter and leave the geneva slots tangentially (as the crank rapidly indexes the output). This is accomplished in the novel mechanism shown with two tracks. The roller enters one track, indexes the geneva 90° (in a four-stage geneva), and then automatically follows the exit slot to leave the geneva.

The associated linkage mechanism locks the geneva when it is not indexing. In the position shown, the locking roller is just about to exit from the geneva.

Internal-groove geneva drive

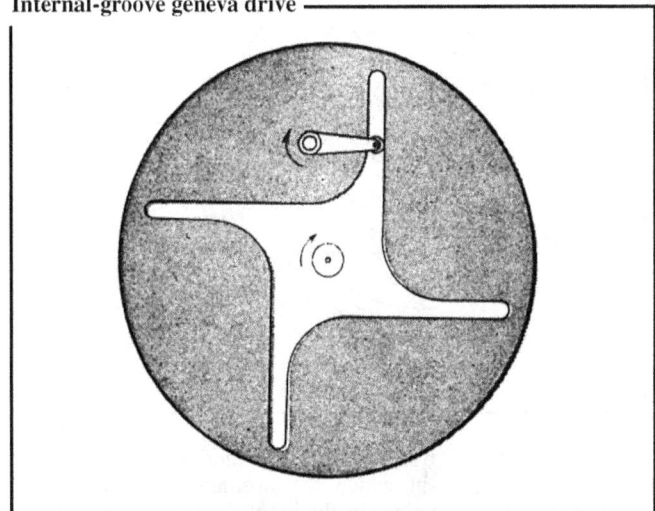

This arrangement permits the roller to exit and enter the driving slots tangentially. In the position shown, the driving roller has just completed indexing the geneva, and it is about to coast for 90° as it goes around the curve. (During this time, a separate locking device might be necessary to prevent an external torque from reversing the geneva.)

The output in this simple mechanism is prevented from turning in either direction—unless it is actuated by the input motion. In operation, the drive lever indexes the output disk by bearing on the pin. The escapement is cammed out of the way during indexing because the slot in the input disk is positioned to permit the escapement tip to enter it. But as the lever leaves the pin, the input disk forces the escapement tip out of its slot and into the notch. That locks the output in both directions.

Progressive oscillating drive

A crank attached to the planet gear can make point *P* describe the double loop curve illustrated. The slotted output crank oscillates briefly at the vertical positions.

Sinusoidal reciprocator drive

This reciprocator transforms rotary motion into a reciprocating motion in which the oscillating output member is in the same plane as the input shaft. The output member has two arms with rollers which contact the surface of the truncated sphere. The rotation of the sphere causes the output to oscillate.

Controlled-output escapement

Parallel-guidance drive

The input crank contains two planet gears. The center sun gear is fixed. By making the three gears equal in diameter and having gear *2* serve as an idler, any member fixed to gear *3* will remain parallel to its previous positions throughout the rotation of the input ring crank.

Rotating-cam reciprocator drive

The high-volume 2500-ton press is designed to shape such parts as connecting rods, tractor track links, and wheel hubs. A simple automatic-feed mechanism makes it possible to produce 2400 forgings per hour.

An external geneva drive. The driver grooves lock the driven wheel pins during dwell. During movement, the driver pin mates with the driven-wheel slot.

An internal geneva drive. The driver and driven wheel rotate in same direction. The duration of dwell is more than 180° of driver rotation.

A spherical geneva drive. The driver and driven wheel are on perpendicular shafts. The duration of dwell is exactly 180° of driver rotation.

An intermittent counter drive. One revolution of the driver advances the driven wheel 120°. The driven-wheel rear teeth are locked on the cam surface during dwell.

SIX MODIFIED GENEVA DRIVES

The mechanisms shown here add a varying velocity component to conventional geneva motion.

Fig. 1 With a conventional external geneva drive, a constant-velocity input produces an output consisting of a varying velocity period plus a dwell. The motion period of the modified geneva shown has a constant-velocity interval which can be varied within limits. When spring-loaded driving roller *a* enters the fixed cam *b*, the output-shaft velocity is zero. As the roller travels along the cam path, the output velocity rises to some constant value, which is less than the maximum output of an unmodified geneva with the same number of slots. The duration of constant-velocity output is arbitrary within limits. When the roller leaves the cam, the output velocity is zero. Then the output shaft dwells until the roller re-enters the cam. The spring produces a variable radial distance of the driving roller from the input shaft, which accounts for the described motions. The locus of the roller's path during the constant-velocity output is based on the velocity-ratio desired.

Fig. 1

Fig. 2

Fig. 2 This design incorporates a planet gear in the drive mechanism. The motion period of the output shaft is decreased, and the maximum angular velocity is increased over that of an unmodified geneva with the same number of slots. Crank wheel *a* drives the unit composed of planet gear *b* and driving roller *c*. The axis of the driving roller coincides with a point on the pitch circle of the planet gear. Because the planet gear rolls around the fixed sun gear *d*, the axis of roller *c* describes a cardioid *e*. To prevent the roller from interfering with the locking disk *f*, the clearance arc *g* must be larger than is required for unmodified genevas.

Fig. 3 A motion curve similar to that of Fig. 2 can be derived by driving a geneva wheel with a two-crank linkage. Input crank *a* drives crank *b* through link *c*. The variable angular velocity of driving roller *d*, mounted on *b*, depends on the center distance *L*, and on the radii *M* and *N* of the crank arms. This velocity is about equivalent to what would be produced if the input shaft were driven by elliptical gears.

Fig. 3

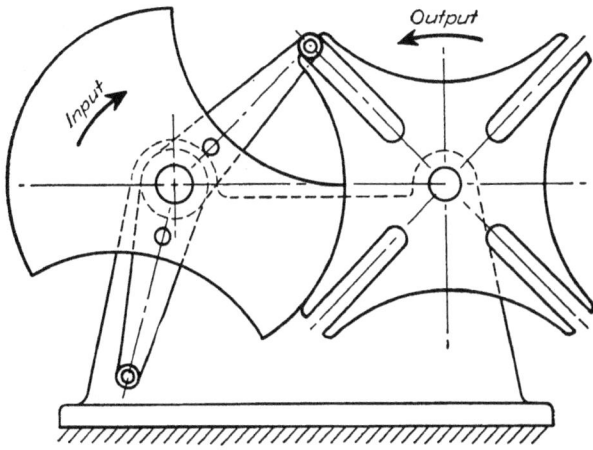

Fig. 4

Fig. 4 The duration of the dwell periods is changed by arranging the driving rollers unsymmetrically around the input shaft. This does not affect the duration of the motion periods. If unequal motion periods and unequal dwell periods are desired, the roller crank-arms must be unequal in length and the star must be suitably modified. This mechanism is called an irregular geneva drive.

Fig. 5 In this intermittent drive, the two rollers drive the output shaft and lock it during dwell periods. For each revolution of the input shaft, the output shaft has two motion periods. The output displacement ϕ is determined by the number of teeth. The driving angle, ψ, can be chosen within limits. Gear a is driven intermittently by two driving rollers mounted on input wheel b, which is bearing-mounted on frame c. During the dwell period, the rollers circle around the top of a tooth. During the motion period, a roller's path d, relative to the driven gear, is a straight line inclined towards the output shaft. The tooth profile is a curve parallel to path d. The top land of a tooth becomes the arc of a circle of radius R, and the arc approximates part of the path of a roller.

Fig. 5

Fig. 6 An intermittent drive with a cylindrical lock. Shortly before and after the engagement of two teeth with driving pin d at the end of the dwell period, the inner cylinder f is unable to cause positive locking of the driven gear. Consequently, a concentric auxiliary cylinder e is added. Only two segments are necessary to obtain positive locking. Their length is determined by the circular pitch of the driven gear.

Fig. 6

KINEMATICS OF EXTERNAL GENEVA WHEELS

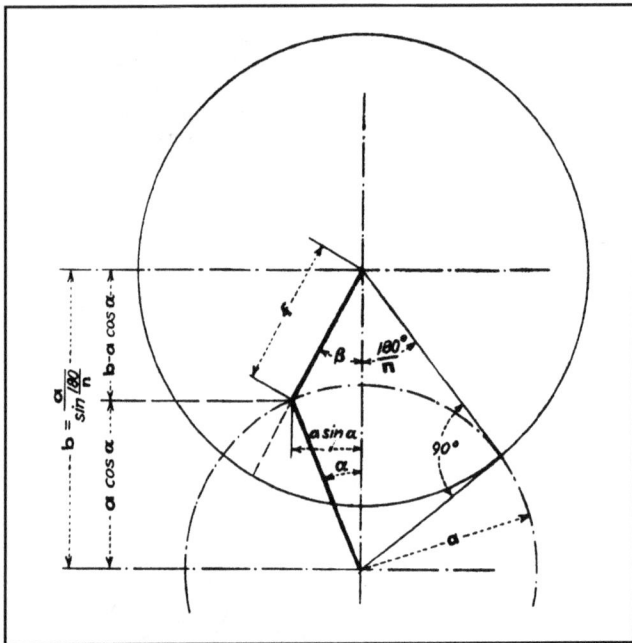

Fig. 1 **A basic outline drawing** for the external geneva wheel. The symbols are identified for application in the basic equations.

Fig. 2 **A schematic drawing** of a six-slot geneva wheel. Roller diameter, d_r, must be considered when determining D.

Table I—Notation and Formulas for the External Geneva Wheel

Assumed or given: a, n, d and p

a = crank radius of driving member

n = number of slots

d_r = roller diameter

p = constant velocity of driving crank in rpm

$$m = \frac{1}{\sin \frac{180}{n}}$$

b = center distance = am

D = diameter of driven member = $2 \sqrt{\frac{d^2_r}{4} + a^2 \cot^2 \frac{180}{n}}$

ω = constant angular velocity of driving crank = $\frac{p\pi}{30}$ radians per sec

α = angular position of driving crank at any time

β = angular displacement of driven member corresponding to crank angle α

$$\cos\beta = \frac{m - \cos\alpha}{\sqrt{1 + m^2 - 2m\cos\alpha}}$$

Angular Velocity of driven member = $\frac{d\beta}{dt}$ = $\omega \left(\frac{m\cos\alpha - 1}{1 + m^2 - 2m\cos\alpha} \right)$

Angular Acceleration of driven member = $\frac{d^2\beta}{dt^2}$ = $\omega^2 \left(\frac{m\sin\alpha\,(1 - m^2)}{(1 + m^2 - 2m\cos\alpha)^2} \right)$

Maximum Angular Acceleration occurs when $\cos\alpha$ =

$$\sqrt{\left(\frac{1 + m^2}{4m}\right)^2 + 2} - \left(\frac{1 + m^2}{4m}\right)$$

Maximum Angular Velocity occurs at α = 0 deg, and equals

$$\frac{\omega}{m - 1}$$ radians per sec

One of the most commonly applied mechanisms for producing intermittent rotary motion from a uniform input speed is the external geneva wheel.

The driven member, or star wheel, contains many slots into which the roller of the driving crank fits. The number of slots determines the ratio between dwell and motion period of the driven shaft. The lowest possible number of slots is three, while the highest number is theoretically unlimited. In practice, the three-slot geneva is seldom used because of the extremely high acceleration values encountered. Genevas with more than 18 slots are also infrequently used because they require wheels with comparatively large diameters.

In external genevas of any number of slots, the dwell period always exceeds the motion period. The opposite is true of the internal geneva. However, for the spherical geneva, both dwell and motion periods are 180°.

For the proper operation of the external geneva, the roller must enter the slot tangentially. In other words, the centerline of the slot and the line connecting the roller center and crank rotation center must form a right angle when the roller enters or leaves the slot.

The calculations given here are based on the conditions stated here.

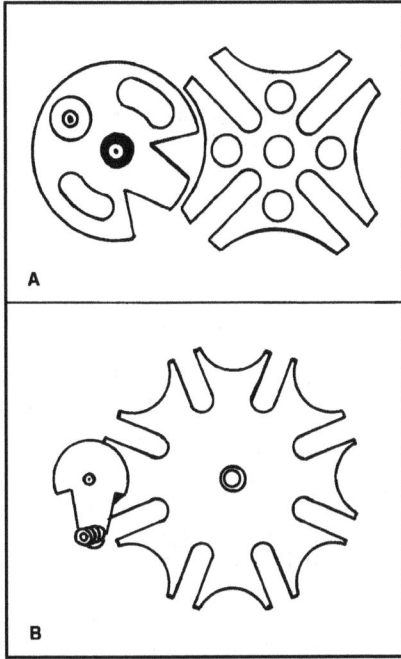

Fig. 3 **A four-slot geneva** (A) and an eight-slot geneva (B). Both have locking devices.

Fig. 4 **Chart** for determining the angular displacement of the driven member.

Consider an external geneva wheel, shown in Fig. 1, in which

n = number of slots
a = crank radius

From

Fig. 1, b = center distance = $\dfrac{a}{\sin \frac{180}{n}}$

Let $\dfrac{1}{\sin \frac{180}{n}} = m$

Then $b = am$

It will simplify the development of the equations of motion to designate the connecting line of the wheel and crank centers as the zero line. This is contrary to the practice of assigning the zero value of α, representing the angular position of the driving crank, to that position of the crank where the roller enters the slot.

Thus, from Fig. 1, the driven crank radius f at any angle is:

$$f = \sqrt{(am - a\cos\alpha)^2 + a^2\sin^2\alpha}$$

$$= a\sqrt{1 + m^2 - 2m\cos\alpha} \qquad (1)$$

and the angular displacement β can be found from:

$$\cos\beta = \frac{m - \cos a}{\sqrt{1 + m^2 - 2m\cos\alpha}} \qquad (2)$$

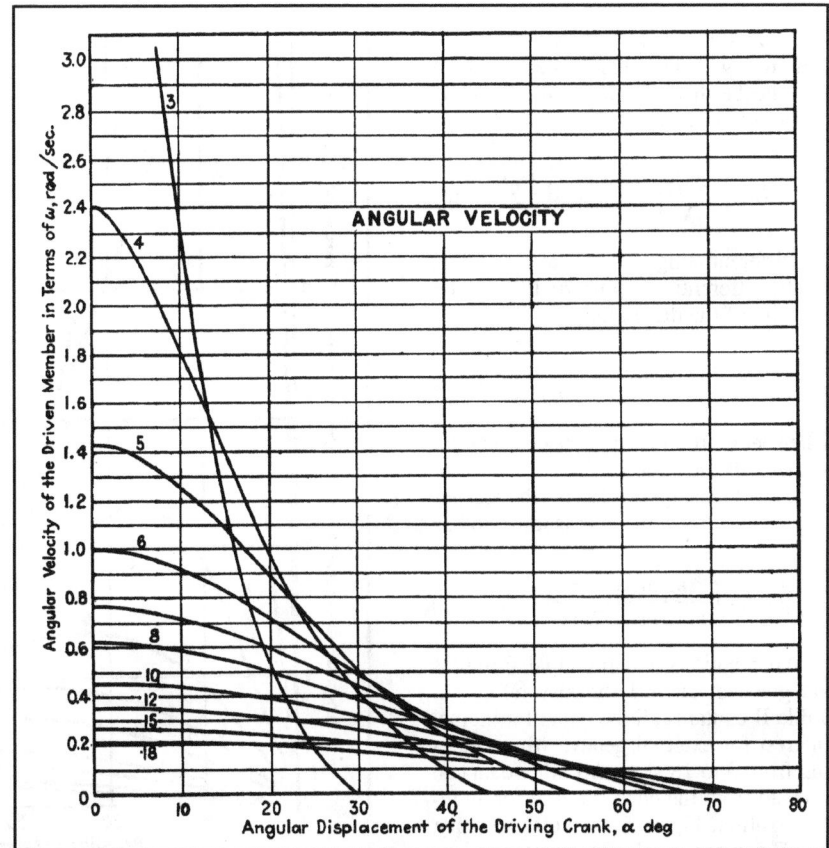

Fig. 5 **Chart** for determining the angular velocity of the driven member.

Table II—Principal Kinematic Data for External Geneva Wheel

No. of Slots	$\dfrac{360°}{n}$	Dwell period	Motion period	m and center-distance for $\alpha = 1$	Maximum angular velocity of driven member, radians per sec. equals ω multiplied by values tabulated. Crank at 0° position	Angular acceleration of driven member when roller enters slot, radians² per sec², equals ω^2 multiplied by values tabulated.			Maximum angular Acceleration of driven member, radians² per sec², equals ω^2 multiplied by values tabulated		
						α	β	Multiplier	α	β	Multiplier
3	120°	300°	60°	1.155	6.458	30°	60°	1.729	4°	27° 58′	29.10
4	90°	270°	90°	1.414	2.407	45°	45°	1.000	11° 28′	25° 11′	5.314
5	72°	252°	108°	1.701	1.425	54°	36°	0.727	17° 31′	21° 53′	2.310
6	60°	240°	120°	2.000	1.000	60°	30°	0.577	22° 55′	19° 51′	1.349
7	51° 25′ 43″	231° 30′	128° 30′	2.305	0.766	64° 17′ 8″	25° 42′ 52″	0.481	27° 41′	18° 11′	0.928
8	45°	225°	135°	2.613	0.620	67° 30′	22° 30′	0.414	31° 38′	16° 32′	0.700
9	40°	220°	140°	2.924	0.520	70°	20°	0.364	35° 16′	15° 15′	0.559
10	36°	216°	144°	3.236	0.447	72°	18°	0.325	38° 30′	14° 16′	0.465
11	32° 43′ 38″	212° 45′	147° 15′	3.549	0.392	73° 38′ 11″	16° 21′ 49″	0.294	41° 22′	13° 16′	0.398
12	30°	210°	150°	3.864	0.349	75°	15°	0.268	44°	12° 26′	0.348
13	27° 41′ 32″	207° 45′	152° 15′	4.179	0.315	76° 9′ 14″	13° 50′ 46″	0.246	46° 23′	11° 44′	0.309
14	25° 42′ 52″	205° 45′	154° 15′	4.494	0.286	77° 8′ 34″	21° 51′ 26″	0.228	48° 32′	11° 3′	0.278
15	24°	204°	156°	4.810	0.263	78°	12°	0.213	50° 30′	10° 27′	0.253
16	22° 30′	202° 30′	157° 30′	5.126	0.242	78° 45′	11° 15′	0.199	52° 24′	9° 57′	0.232
17	21° 10′ 35″	201°	159°	5.442	0.225	79° 24′ 43″	10° 35′ 17″	0.187	53° 58′	9° 26′	0.215
18	20°	200°	160°	5.759	0.210	80°	10°	0.176	55° 30′	8° 59′	0.200

A six-slot geneva is shown schematically in Fig. 2. The outside diameter D of the wheel (when accounting for the effect of the roller diameter d) is found to be:

$$D = 2\sqrt{\frac{d_r^2}{4} + a^2 \cot^2 \frac{180}{n}} \quad (3)$$

Differentiating Eq. (2) and dividing by the differential of time, dt, the angular velocity of the driven member is:

$$\frac{d\beta}{dt} = \omega\left(\frac{m \cos \alpha - 1}{1 + m^2 - 2m \cos \alpha}\right) \quad (4)$$

where ω represents the constant angular velocity of the crank.

By differentiation of Eq. (4) the acceleration of the driven member is found to be:

$$\frac{d^2\beta}{dt^2} = \omega^2\left(\frac{m \sin \alpha(1 - m^2)}{(1 + m^2 - 2m \cos \alpha)^2}\right) \quad (5)$$

All notations and principal formulas are given in Table I for easy reference. Table II contains all the data of principal interest for external geneva wheels having from 3 to 18 slots. All other data can be read from the charts: Fig. 4 for angular position, Fig. 5 for angular velocity, and Fig. 6 for angular acceleration.

Fig. 6 **Chart** for determining the angular acceleration of the driven member.

KINEMATICS OF INTERNAL GENEVA WHEELS

Fig. 1 A four-slot internal geneva wheel incorporating a locking mechanism. The basic sketch is shown in Fig. 3.

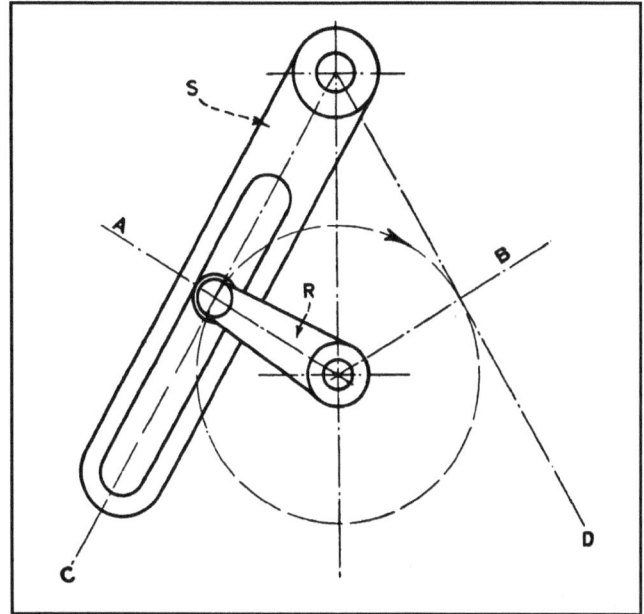

Fig. 2 Slot-crank motion from *A* to *B* represents external geneva action; from *B* to *A* represents internal geneva motion.

Where intermittent drives must provide dwell periods of more than 180°, the external geneva wheel design is satisfactory and is generally the standard device employed. But where the dwell period must be less than 180°, other intermittent drive mechanisms must be used. The internal geneva wheel is one way of obtaining this kind of motion.

The dwell period of all internal genevas is always smaller than 180°. Thus, more time is left for the star wheel to reach maximum velocity, and acceleration is lower. The highest value of angular acceleration occurs when the roller enters or leaves the slot. However, the acceleration occurs when the roller enters or leaves the slot. However, the acceleration curve does not reach a peak within the range of motion of the driven wheel. The geometrical maximum would occur in the continuation of the curve. But this continuation has no significance because the driven member will have entered the dwell phase associated with the high angular displacement of the driving member.

The geometrical maximum lies in the continuation of the curve, falling into the region representing the motion of the external geneva wheel. This can be seen by the following considerations of a crank and slot drive, drawn in Fig. 2.

When the roller crank *R* rotates, slot link *S* will perform an oscillating movement, for which the displacement, angular velocity,

Table I—Notation and Formulas for the Internal Geneva Wheel

Assumed or given: *a, n, d* and *p*

a = crank radius of driving member

n = number of slots

d = roller diameter

p = constant velocity of driving crank in rpm

$$m = \frac{1}{\sin \frac{180°}{n}}$$

b = center distance = $a\,m$

D = inside diameter of driven member = $2\sqrt{\dfrac{d^2}{4} + a^2 \cot^2 \dfrac{180°}{n}}$

ω = constant angular velocity of driving crank in radians per sec = $\dfrac{p\pi}{30}$ radians per sec

α = angular position of driving crank at any time

β = angular displacement of driven member corresponding to crank angle α

$$\cos\beta = \frac{m + \cos\alpha}{\sqrt{1 + m^2 + 2m\cos\alpha}}$$

Angular velocity of driven member = $\dfrac{d\beta}{dt} = \omega\left(\dfrac{1 + m\cos\alpha}{1 + m^2 + 2m\cos\alpha}\right)$

Angular acceleration of driven member = $\dfrac{d^2\beta}{dt^2} = \omega^2\left[\dfrac{m\sin\alpha\,(1 - m^2)}{(1 + m^2 + 2m\cos\alpha)^2}\right]$

Maximum angular velocity occurs at $\alpha = 0°$ and equals = $\dfrac{\omega}{1 + m}$ radians per sec

Maximum angular acceleration occurs when roller enters slot and equals = $\dfrac{\omega^2}{\sqrt{m^2 - 1}}$ radians² per sec²

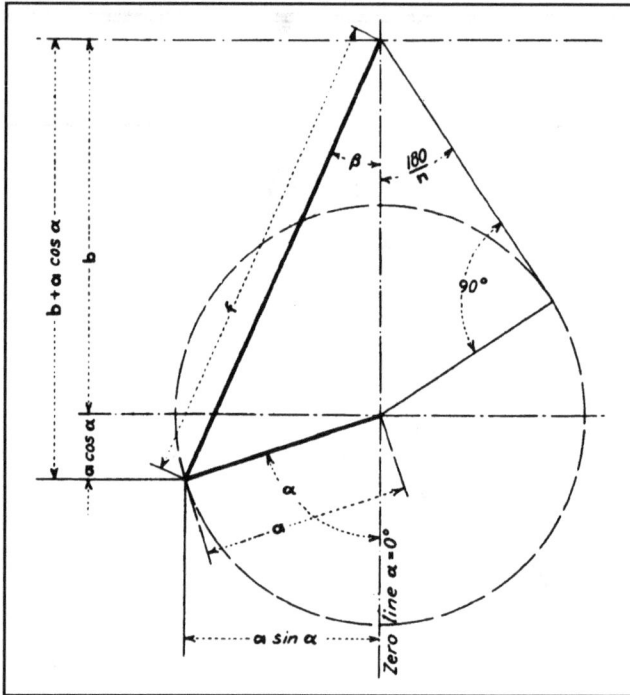

Fig. 3 A basic outline for developing the equations of the internal geneva wheel, based on the notations shown.

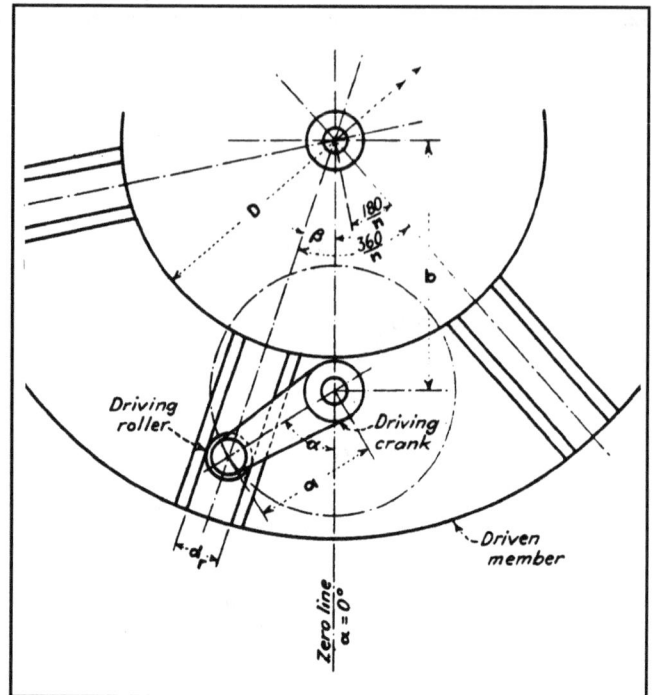

Fig. 4 A drawing of a six-slot internal geneva wheel. The symbols are identified, and the motion equations are given in Table I.

and acceleration can be given in continuous curves.

When the crank *R* rotates from *A* to *B*, then the slot link *S* will move from *C* to *D*, exactly reproducing all moving conditions of an external geneva of equal slot angle. When crank *R* continues its movement from *B* back to *A*, then the slot link *S* will move from *D* back to *C*, this time reproducing exactly (though in a mirror picture with the direction of motion being reversed) the moving conditions of an internal geneva.

Therefore, the characteristic curves of this motion contain both the external and internal geneva wheel conditions; the region of the external geneva lies between *A* and *B*, the region of the internal geneva lies between *B* and *A*.

The geometrical maxima of the acceleration curves lie only in the region between *A* and *B*, representing that portion of the curves which belongs to the external geneva.

The principal advantage of the internal geneva, other than its smooth operation, is its sharply defined dwell period. A disadvantage is the relatively large size of the driven member, which increases the force resisting acceleration. Another feature, which is sometimes a disadvantage, is the cantilever arrangement of the roller crank shaft. This shaft cannot be a through shaft because the crank must be fastened to the overhanging end of the input shaft.

To simplify the equations, the connecting line of the wheel and crank centers is taken as the zero line. The angular

Fig. 5 Angular displacement of the driven member can be determined from this chart.

position of the driving crank α is zero when it is on this line. Then the following relations are developed, based on Fig. 3.

n = number of slots
a = crank radius
b = center distance = $\dfrac{a}{\sin\dfrac{180°}{n}}$

Let

$$\frac{1}{\sin\dfrac{180°}{n}} = m,$$

then, $\qquad b = am$

To find the angular displacement β of the driven member, the driven crank radius f is first calculated from:

$$f = \sqrt{a^2\sin^2\alpha + (am + a\cos\alpha)^2}$$

$$= \alpha\sqrt{1 + m^2 + 2m\cos\alpha} \qquad (1)$$

and because

$$\cos\beta = \frac{m + \cos\alpha}{f}$$

it follows:

$$\cos\beta = \frac{m + \cos\alpha}{\sqrt{1 + m^2 - 2m\cos\alpha}} \qquad (2)$$

From this formula, β, the angular displacement, can be calculated for any angle α, the angle of the mechanism's driving member.

The first derivative of Eq. (2) gives the angular velocity as:

$$\frac{d\beta}{dt} = \omega\left(\frac{1 + m\cos\alpha}{1 + m^2 + 2m\cos\alpha}\right) \qquad (3)$$

where ω designates the uniform speed of the driving crank shaft, namely:

$$\omega = \frac{p\pi}{30}$$

if p equals its number of revolutions per minute.

Differentiating Eq. (3) once more develops the equation for the angular acceleration:

$$\frac{d^2\beta}{dt^2} = \omega^2\left(\frac{m\sin\alpha(1 - m^2)}{(1 + m^2 + 2m\cos\alpha)^2}\right) \qquad (4)$$

The maximum angular velocity occurs, obviously, at $\alpha = 0°$. Its value is found by substituting $0°$ for α in Eq. (3). It is:

$$\frac{d\beta}{dt_{max}} = \frac{\omega}{1 + m} \qquad (5)$$

Fig. 6 Angular velocity of the driven member can be determined from this chart.

Fig. 7 Angular acceleration of the driven member can be determined from this chart.

Table II—Kinematic Data For the Internal Geneva Wheel

Number of slots, n	$\dfrac{360°}{n}$	Dwell period	Motion period	m and center-distance for $a=1$	Maximum angular velocity of driven member equals ω radians per sec. multiplied by values tabulated. Both α and β in 0° position	Angular acceleration of driven member when roller enters slot equals ω^2 radians² per sec² multiplied by values tabulated		
						α	β	Multiplier
3	120°	60°	300°	1.155	0.464	150°	60°	1.729
4	90°	90°	270°	1.414	0.414	135°	45°	1.000
5	72°	108°	252°	1.701	0.370	126°	36°	0.727
6	60°	120°	240°	2.000	0.333	120°	30°	0.577
7	51° 25′ 43″	128° 30′	231° 30′	2.305	0.303	115° 42′ 52″	25° 42′ 52″	0.481
8	45°	135°	225°	2.613	0.277	112° 30′	22° 30′	0.414
9	40°	140°	220°	2.924	0.255	110°	20°	0.364
10	36°	144°	216°	3.236	0.236	108°	18°	0.325
11	32° 43′ 38″	147° 15′	212° 45′	3.549	0.220	106° 21′ 49″	16° 21′ 49″	0.294
12	30°	150°	210°	3.864	0.206	105°	15°	0.268

The highest value of the acceleration is found by substituting $180/n + 980$ for α in Eq. (4):

$$\frac{d^2\beta}{dt^2_{\max}} = \frac{\omega^2}{\sqrt{m^2 - 1}} \qquad (6)$$

A layout drawing for a six-slot internal geneva wheel is shown in Fig. 4. All the symbols in this drawing and throughout the text are compiled in Table I for easy reference.

Table II contains all the data of principal interest on the performance of internal geneva wheels that have from 3 to 18 slots. Other data can be read from the charts: Fig. 5 for angular position, Fig. 6 for angular velocity, and Fig. 7 for angular acceleration.

STAR WHEELS CHALLENGE GENEVA DRIVES FOR INDEXING

Star wheels with circular-arc slots can be analyzed mathematically and manufactured easily.

Star wheels vary in shape, depending on the degree of indexing that must be done during one input revolution.

A family of star wheels with circular instead of the usual epicyclic slots (see drawings) can produce fast start-and-stop indexing with relatively low acceleration forces.

This rapid, jar-free cycling is important in a wide variety of production machines and automatic assembly lines that move parts from one station to another for drilling, cutting, milling, and other processes.

The circular-slot star wheels were invented by Martin Zugel of Cleveland, Ohio.

The motion of older star wheels with epicyclic slots is difficult to analyze and predict, and the wheels are hard to make.

The star wheels with their circular-arc slots are easy to fabricate, and because the slots are true circular arcs, they can be visualized for mathematical analysis as four-bar linkages during the entire period of pin-slot engagement.

Strong points.   With this approach, changes in the radius of the slot can be analyzed and the acceleration curve varied to provide inertia loads below those of the genevas for any practical design requirement.

Another advantage of the star wheels is that they can index a full 360° in a relatively short period (180°). Such one-stop operation is not possible with genevas. In fact, genevas cannot do two-stop operations, and they have difficulty producing three stops per index. Most two-stop indexing devices available are cam-operated, which means they require greater input angles for indexing.

Geared star sector indexes smoothly a full 360° during a 180° rotation of the wheel, then it pauses during the other 180° to allow the wheel to catch up.

The one-stop index motion of the unit can be designed to take longer to complete its indexing, thus reducing its index velocity.

An accelerating pin brings the output wheel up to speed. Gear sectors mesh to keep the output rotating beyond 180°.

Operating sequence. In operation, the input wheel rotates continuously. A sequence starts (see drawing) when the accelerating pin engages the curved slot to start indexing the output wheel clockwise. Simultaneously, the locking surface clears the right side of the output wheel to permit the indexing.

Pin C in the drawings continues to accelerate the output wheel past the midpoint, where a geneva wheel would start deceleration. Not until the pins are symmetrical (see drawing) does the acceleration end and the deceleration begin. Pin D then takes the brunt of the deceleration force.

Adaptable. The angular velocity of the output wheel, at this stage of exit of the acceleration roller from Slot 1, can be varied to suit design requirements. At this point, for example, it is possible either to engage the deceleration roller as described or to start the engagement of a constant-velocity portion of the cycle. Many more degrees of output index can be obtained by interposing gear-element segments between the acceleration and deceleration rollers.

The star wheel at left will stop and start four times in making one revolution, while the input turns four times in the same period. In the starting position, the output link has zero angular velocity, which is a prerequisite condition for any star wheel intended to work at speeds above a near standstill.

In the disengaged position, the angular velocity ratio between the output and input shafts (the "gear" ratio) is entirely dependent upon the design angles α and β and independent of the slot radius, r.

Design comparisons. The slot radius, however, plays an important role in the mode of the acceleration forces. A four-stop geneva provides a good basis for comparison with a four-stage "Cyclo-Index" system.

Assume, for example, that $\alpha = \beta = 22.5°$. Application of trigonometry yields:

$$R = A\left[\frac{\sin \beta}{\sin (\alpha + \beta)}\right]$$

which yields $R = 0.541A$. The only restriction on r is that it be large enough to allow the wheel to pass through its mid-position. This is satisfied if:

$$r > \frac{RA(1 - \cos \alpha)}{A - 2R - A \cos \alpha} \approx 0.1A$$

There is no upper limit on r, so that slot can be straight.

Star-wheel action is improved with curved slots over the radius r, centered on the initial-contact line OP. The units then act as four-bar linkages, OO¹PQ.

The accelerating force of star wheels (curves A, B, C) varies with input rotation. With an optimum slot (curve C), it is lower than for a four-stop geneva.

90° drive, 270° index

This internal star wheel has a radius difference to cushion the indexing shock.

207

RATCHET-TOOTH SPEED-CHANGE DRIVE

An in-line shaft drive, with reduction ratios of 1:1 and 1:16 or 1:28, combined in a single element, was designed by Telefunken of Germany. It consists basically of friction wheels that drip each other elastically.

Crown wheel with a gear ratio of 1:1 provides the coarse adjustment, and friction spur gearing, with a ratio of 1:16 or 1:28, provides the fine or vernier adjustment.

A spring (see diagram) applies pressure to the fine-adjustment pinion, preventing backlash while the coarse adjustment is in use. It uncouples the coarse adjustment when the vernier is brought into play by forward movement of the front shaft. The spring also ensures that the front shaft is always in gear.

MODIFIED RATCHET DRIVE

A ratchet drive was designed to assure movement, one tooth at a time, in only one direction, without overriding. The key element is a small stub that moves along from the bottom of one tooth well, across the top of the tooth, and into an adjacent tooth well, while the pawl remains at the bottom of another tooth well.

The locking link, which carries the stub along with the spring, comprises a system that tends to hold the link and pawl against the outside circumference of the wheel and to push the stub and pawl point toward each other and into differently spaced wells between the teeth. A biasing element, which might be another linkage or solenoid, is provided to move the anchor arm from one side to the other, between the stops, as shown by the double arrow. The pawl will move from one tooth well to the next tooth well only when the stub is at the bottom of a tooth well and is in a position to prevent counter-rotation.

EIGHT TOOTHLESS RATCHETS

Ratchets with springs, rollers, and other devices keep motion
going one way.

Fig. 1

Fig. 2

Fig. 3

Fig. 4

Fig. 5

Fig. 6

Fig. 7

Fig. 8

Fig. 1 Swinging pawls lock on the rim when the lever swings forward, and release on the return stroke. Oversize holes for the supporting stud make sure that both the top and bottom surfaces of the pawls make contact.

Fig. 2 A helical spring grips the shaft because its inner diameter is smaller than the outer diameter of the shaft. During the forward stroke, the spring winds tighter; during the return stroke, it expands.

Fig. 3 A V-belt sheave is pushed around when pawl wedges in the groove. For a snug fit, the bottom of the pawl is tapered like a V-belt.

Fig. 4 Eccentric rollers squeeze a disk on its forward stroke. On the return stroke, rollers rotate backwards and release their grip. Springs keep the rollers in contact with the disk.

Fig. 5 A rack is wedge-shaped so that it jams between the rolling gear and the disk, pushing the shaft forward. When the driving lever makes its return stroke, it carries along the unattached rack by the cross-piece.

Fig. 6 A conical plate moves like a nut back and forth along the threaded center hub of the lever. The light friction of spring-loaded pins keeps the plate from rotating with the hub.

Fig. 7 Flat springs expand against the inside of a drum when a lever moves one way, but they drag loosely when the lever turns the drum in the opposite direction.

Fig. 8 An eccentric cam jams against the disk during the motion half of a cycle. Elongated holes in the levers allow the cam to wedge itself more tightly in place.

ANALYSIS OF RATCHET WHEELS

The ratchet wheel is widely used in machinery, mainly to transmit intermittent motion or to allow shaft rotation in one direction only. Ratchet-wheel teeth can be either on the perimeter of a disk or on the inner edge of a ring.

The pawl, which engages the ratchet teeth, is a beam pivoted at one end; the other end is shaped to fit the ratchet-tooth flank. Usually, a spring or counterweight maintains constant contact between wheel and pawl.

It is desirable, in most designs, to keep the spring force low. It should be just large enough to overcome the separation forces—inertia, weight, and pivot friction. Excess spring force should not be considered for engaging the pawl and holding it against the load.

To ensure that the pawl is automatically pulled in and kept in engagement independently of the spring, a properly drawn tooth flank is necessary.

The requirement for self-engagement is:

$$Pc + M > \mu Pb + P\sqrt{(1 + \mu^2)^{\mu_1 r_1}}$$

Neglecting weight and pivot friction:

$$Pc > \mu Pb$$

but $c/b = r/a = \tan \phi$, and because $\tan \phi$ is approximately equal to $\sin \phi$:

$$c/b = r/R$$

Substituting in required term above

$$rR > \mu$$

For steel on steel, dry, $\mu = 0.15$ Therefore, using

$$r/R = 0.20 \text{ to } 0.25$$

the margin of safety is large; the pawl will slide into engagement easily. For internal teeth with ϕ of 30°, c/b is tan 30° or 0.577, which is larger than μ, and the teeth are therefore self-engaging.

When laying out the ratchet wheel and pawl, locate points O, A and O_1 on the same circle. AO and AO_1 will then be perpendicular to one another; this will ensure that the smallest forces are acting on the system.

Ratchet and pawl dimensions are governed by design sizes and stress. If the tooth, and thus pitch, must be larger than required to be strong enough, a multiple pawl arrangement can be used. The pawls can be arranged so that one of them will engage the ratchet after a rotation of less than the pitch.

A fine feed can be obtained by placing many pawls side by side, with the corresponding ratchet wheels uniformly displaced and interconnected.

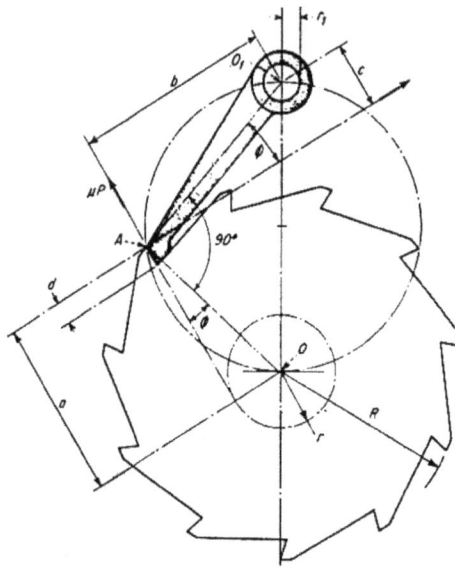

Pawl in compression has tooth pressure P and weight of pawl producing a moment that tends to engage pawl. Friction-force μP and pivot friction tend to oppose pawl engagement.

Inner teeth allow compact assembly.

Pawl in tension has the same forces acting on the unit as other arrangements. The same layout principles apply.

a = moment arm of wheel torque
M = moment about O_1 caused by weight of pawl
$O_1 - O_2$ = ratchet and pawl pivot centers respectively
P = tooth pressure = wheel torque/a
$P\sqrt{(1 + \mu^2)}$ = load on pivot pin
μ, μ_1 = friction coefficients
Other symbols as defined in diagrams.

CHAPTER 8
CLUTCHES AND BRAKES

TWELVE CLUTCHES WITH EXTERNAL OR INTERNAL CONTROL

Both friction and positive clutches are illustrated here. Figures 1 to 7 show externally controlled clutches, and Figures 8 to 12 show internally controlled clutches which are further divided into overload relief, overriding, and centrifugal versions.

Fig. 1

Fig. 2

Fig. 3

Fig. 1 Jaw Clutch: The left sliding half of this clutch is feathered to the driving shaft while the right half rotates freely. The control arm activates the sliding half to engage or disengage the drive. However, this simple, strong clutch is subject to high shock during engagement and the sliding half exhibits high inertia. Moreover, engagement requires long axial motion.

Fig. 2 Sliding Key Clutch: The driven shaft with a keyway carries the freely rotating member with radial slots along its hub. The sliding key is spring-loaded but is restrained from the engaging slots by the control cam. To engage the clutch, the control cam is raised and the key enters one of the slots. To disengage it, the cam is lowered into the path of the key and the rotation of the driven shaft forces the key out of the slot in the driving member. The step on the control cam limits the axial movement of the key.

Fig. 3 Planetary Transmission Clutch: In the disengaged position shown, the driving sun gear causes the freewheeling ring gear to idle counter-clockwise while the driven planet carrier remains motionless. If the control arm blocks ring gear motion, a positive clockwise drive to the driven planet carrier is established.

Fig. 4

Fig. 5

Fig. 6

Fig. 4 Pawl and Ratchet Clutch: (External Control) The driving ratchet of this clutch is keyed to the driving shaft, and the pawl is pinned to the driven gear which can rotate freely on the driving shaft. When the control arm is raised, the spring pulls in the pawl to engage the ratchet and drive the gear. To disengage the clutch the control arm is lowered so that driven gear motion will disengage the pawl and stop the driven assembly against the control member.

Fig. 5 Plate Clutch: The plate clutch transmits power through the friction developed between the mating plate faces. The left sliding plate is fitted with a feather key, and the right plate member is free to rotate on the shaft. Clutch torque capacity depends on the axial force exerted by the control half when it engages the sliding half.

Fig. 6 Cone Clutch: The cone clutch, like the plate clutch, requires axial movement for engagement, but less axial force is required because of the increased friction between mating cones. Friction material is usually applied to only one of the mating conical surfaces. The free member is mounted to resist axial thrust.

Fig.7 Drum — Shoe — Control arm

Fig.8 Driving gear — Driven shaft

Fig.9

Fig. 7 Expanding Shoe Clutch: This clutch is engaged by the motion of the control arm. It operates linkages that force the friction shoes radially outwards so that they contact the inside surface of the drum.

Fig. 8 Spring and Ball Radial Detent Clutch: This clutch will hold the driving gear and driven shaft in a set timing relationship until the torque becomes excessive. At that time the balls will be forced inward against their springs and out of engagement with the holes in the hub. As a result the driving gear will continue rotating while the drive shaft is stationary.

Fig. 9 Cam and Roller Clutch: This over-running clutch is better suited for higher-speed freewheeling than a pawl-and-ratchet clutch. The inner driving member has cam surfaces on its outer rim that hold light springs that force the rollers to wedge between the cam surfaces and the inner cylindrical face of the driven member. While driving, friction rather than springs forces the rollers to wedge tightly between the members to provide positive clockwise drive. The springs ensure fast clutching action. If the driven member should begin to run ahead of the driver, friction will force the rollers out of their tightly wedged positions and the clutch will slip.

Fig.10

Fig.11 Driven member — Driving shaft

Fig.12 Driven member — Driving shaft — Gland — Plates

Fig. 10 Wrapped Spring Clutch: This simple unidirectional clutch consists of two rotating hubs connected by a coil spring that is press-fit over both hubs. In the driving direction the spring tightens around the hubs increasing the friction grip, but if driven in the opposite direction the spring unwinds causing the clutch to slip.

Fig. 11 Expanding Shoe Centrifugal Clutch: This clutch performs in a similar manner to the clutch shown in Fig. 7 except that there is no external control. Two friction shoes, attached to the driving member, are held inward by springs until they reach the "clutch-in" speed.

At that speed centrifugal force drives the shoes outward into contact with the drum. As the drive shaft rotates faster, pressure between the shoes against the drum increases, thus increasing clutch torque.

Fig. 12 Mercury Gland Clutch: This clutch contains two friction plates and a mercury-filled rubber bladder. At rest, mercury fills a ring-shaped cavity around the shaft, but when rotated at a sufficiently high speed, the mercury is forced outward by centrifugal force. The mercury then spreads the rubber bladder axially, forcing the friction plates into contact with the opposing faces of the housing to drive it.

SPRING-WRAPPED CLUTCH SLIPS AT PRESET TORQUE

The simple spring clutch becomes even more useful when designed to slip at a predetermined torque. Unaffected by temperature extremes or variations in friction, these clutches are simple—they can even be "homemade." Information is provided here on two dual-spring, slip-type clutches. Two of the dual-spring clutches are in the tape drive shown.

Fig. 1 **Two dual-spring clutches** are in this tape drive.

Spring clutches are devices for driving a load in one direction and uncoupling it when the output is overdriven or the direction of the input rotation is reversed. A spring clutch was modified to give a predetermined slip in either direction—hence the designation of this type as a "slip clutch." A stepped helical spring was employed to accomplish that modification. Later it was developed further by introducing an intermediate clutch member between two helical springs. This dual-spring innovation was preferred where more output torque accuracy was required.

Most designs employ either a friction-disk clutch or a shoe clutch to obtain a predetermined slip (in which the input drives output without slippage until a certain torque level is reached—then a drag-slippage occurs). But the torque capacity (or slip torque) for friction-disk clutches is the same for both directions of rotation.

By contrast, the stepped-spring slip clutch, pictured on the next page, can be designed to have either the same or different torque capacities for each direction of rotation. Torque levels where slippage occurs are independent of each other, thus providing wide latitude of design.

The element producing slip is the stepped spring. The outside diameter of the large step of the spring is assembled tightly in the bore of the output gear. The inside diameter of the smaller step fits tightly over the shaft. Rotation of the shaft in one direction causes the coils in contact with the shaft to grip tightly, and the coils inside the bore to contract and produce slip. Rotation in the opposite direction reverses the action of the spring parts, and slip is effected on the shaft.

Dual-Spring Slip Clutch

This innovation also permits bidirectional slip and independent torque capacities for the two directions of rotation. It requires two springs, one right-handed and one left-handed, for coupling the input, intermediate, and output members. These members are coaxial, with the intermediate and input free to rotate on the output shaft. The rotation of input in one direction causes the spring, which couples the input and intermediate member, to grip tightly. The second spring, which couples the intermediate and output members, is oppositely wound, tends to expand and slip. The rotation in the opposite direction reverses the action of the two springs so that the spring between the input and intermediate members provides the slip. Because this design permits greater independence in the juggling of dimensions, it is preferred where more accurate slip-torque values are required.

Repeatable Performance

Spring-wrapped slip clutches and brakes have remarkably repeatable slip-torque characteristics which do not change with service temperature. Torque capacity remains constant with or without lubrication, and is unaffected by variations in the coefficient of friction. Thus, break-away torque capacity is equal to the sliding torque capacity. This stability makes it unnecessary to overdesign slip members to obtain reliable operation. These advantages are absent in most slip clutches.

Brake and Clutch Combinations

An interesting example of how slip brakes and clutches worked together to maintain proper tension in a tape drive, in either direction of operation, is pictured above and shown schematically on the next page. A brake here is simply a slip clutch with one side fastened to the frame of the unit. Stepped-spring clutches and brakes are shown for simplicity although, in the actual drive, dual-spring units were installed.

The sprocket wheel drives both the tape and belt. This allows the linear speed of the tape to be constant (one of the requirements). The angular speed of the spools, however, will vary as they wind or unwind. The task here is to maintain proper tension in the tape at all times and in either direction. This is done with a brake-clutch combination. In a counterclockwise direction, for example, the brake might become a "low-torque brake" that resists with a 0.1 in.-lb. torque. The clutch in this direction is a "high-torque clutch"—it will provide a 1-in.-lb torque. Thus, the clutch overrides the brake with a net torque of 0.9 in.-lb.

When the drive is reversed, the same brake might now act as a high-torque brake, resisting with a 1 in.-lb torque, while the clutch acts as a low-torque clutch, resisting with 0.1 in.-lb. Thus, in the first direction the clutch drives the spool, in the other direction,

Stepped-spring clutch

Dual-spring clutch

These two modifications of spring clutches offer independent slip characteristics in either direction of rotation.

| Clockwise: | 0.1 in.-lb | 1 in.-lb |
| Counterclockwise: | 1 in.-lb | 0.1 in.-lb |

| High-torque brake | Low-torque brake |
| Low-torque clutch | High-torque clutch |

| Low-torque brake | High-torque brake |
| High-torque clutch | Low-torque clutch |

This tape drive requires two slip clutches and two brakes to ensure proper tension for bidirectional rotation. The detail of the spool (above) shows a clutch and brake unit.

the brake overcomes the clutch and provides a steady resisting force to provide tension in the tape. Of course, the clutch also permits the pulley that is driven by the belt to overdrive.

Two brake-clutch units are required. The second unit will provide opposing torque values—as listed in the diagram. The drive necessary to advance the tape only in a clockwise direction would be the slip clutch in unit 2 and the brake in unit 1. Advancing the tape in the other direction calls for use of the clutch in unit 1 and the brake in unit 2.

For all practical purposes, the low torque values in the brakes and clutches can be made negligible by specifying minimum interference between the spring and the bore or shaft. The low torque is amplified in the spring clutch at the level necessary to drive the tensioning torques of the brake and slip clutches.

Action thus produced by the simple arrangement of directional slip clutches and brakes cannot otherwise be duplicated without resorting to more complex designs.

Torque capacities of spring-wrapped slip clutches and brakes with round, rectangular, and square wire are, respectively:

$$T = \frac{\pi E d^4 \delta}{32 D^2}; \quad T = \frac{E b t^3 \delta}{6 D^2}; \quad T = \frac{E t^4 \delta}{6 D^2}$$

where E = modules of elasticity, psi; d = wire diameter, inches; D = diameter of shaft or bore, inches; ε = diametral interference

between spring and shaft, or spring and bore, inches; t = wire thickness, inches; b = width of rectangular wire, inches; and T = slip torque capacity, pound-inches.

Minimum interference moment (on the spring gripping lightly) required to drive the slipping spring is:

$$M = \frac{T}{e^{\mu\theta} - 1}$$

where e = natural logarithmic base (e = 2.716); θ = angle of wrap of spring per shaft, radians; μ = coefficient of friction; M = interference moment between spring and shaft, pound-inches.

Design Example

Required: to design a tape drive similar to the one shown above. The torque requirements for the slip clutches and brakes for the two directions of rotation are:

(1) Slip clutch in normal takeup capacity (active function) is 0.5 to 0.8 in.-lb.

(2) Slip clutch in override direction (passive function) is 0.1 in.-lb (maximum).

215

(3) Brake in normal supply capacity (active function) is 0.7 to 1.0 in.-lb.

(4) Brake in override direction (passive function) is 0.1 in.-lb (maximum).

Assume that the dual-spring design shown previously is to include 0.750-in. drum diameters. Also available is an axial length for each spring, equivalent to 12 coils which are divided equally between the bridged shafts. Assuming round wire, calculate the wire diameter of the springs if 0.025 in. is maximum diametral interference desired for the active functions. For the passive functions use round wire that produces a spring index not more than 25.

Slip clutch, active spring:

$$d = 4\sqrt{\frac{32D^2T}{\pi E\delta}} = 4\sqrt{\frac{(32)(0.750)^2(0.8)}{\pi(30 \times 10^6)(0.025)}} = 0.050 \text{ in.}$$

The minimum diametral interference is $(0.025)(0.5)/0.8 = 0.016$ in. Consequently, the ID of the spring will vary from 0.725 to 0.734 in.

Slip clutch, passive spring:

$$\text{Wire dia.} = \frac{\text{drum dia.}}{\text{spring index}} = \frac{0.750}{25} = 0.030 \text{ in.}$$

Diametral interference:

$$\delta = \frac{32D^2T}{\pi E d^4} = \frac{(32)(0.750^2)(0.1)}{\pi(30 \times 10^6)(0.030)^4} = 0.023 \text{ in.}$$

Assuming a minimum coefficient of friction of 0.1, determine the minimum diametral interference for a spring clutch that will drive the maximum slip clutch torque of 0.8 lb-in.

Minimum diametral interference:

$$M = \frac{T}{e^{\mu\theta} - 1} = \frac{0.8}{e^{(0.1\pi)(6)} - 1}$$

ID of the spring is therefore 0.727 to 0.745 in.

$$\text{min.} = 0.023 \times \frac{0.019}{0.1} = 0.0044 \text{ in.}$$

Brake springs

By similar computations the wire diameter of the active brake spring is 0.053 in., with an ID that varies from 0.725 to 0.733 in.; wire diameter of the passive brake spring is 0.030 in., with its ID varying from 0.727 to 0.744 in.

CONTROLLED-SLIP EXPANDS SPRING CLUTCH APPLICATIONS

A remarkably simple change in spring clutches is solving a persistent problem in tape and film drives—how to keep drag tension on the tape constant, as its spool winds or unwinds. Shaft torque has to be varied directly with the tape diameter so many designers resort to adding electrical control systems, but that calls for additional components; an extra motor makes this an expensive solution. The self-adjusting spring brake (Fig. 1) developed by Joseph Kaplan, Farmingdale, NY, gives a constant drag torque ("slip" torque) that is easily and automatically varied by a simple lever

arrangement actuated by the tape spool diameter (Fig. 2). The new brake is also being employed to test the output of motors and solenoids by providing levels of accurate slip torque.

Kaplan used his "controlled-slip" concept in two other products. In the controlled-torque screwdriver (Fig. 3) a stepped spring provides a 11.4-in.-lb slip when turned in either direction. It avoids overtightening machine screws in delicate instrument assemblies. A stepped spring is also the basis for the go/no-go torque gage that permits production inspection of output torques to within 1 percent.

Interfering spring. The three products were the latest in a series of slip clutches, drag brakes, and slip couplings developed by Kaplan for instrument brake drives. All are actually outgrowths of the spring clutch. The spring in this clutch is normally prevented from gripping the shaft by a detent response. Upon release of the detent, the spring will grip the shaft. If the shaft is turning in the proper direction, it is self-energizing. In the other direction, the spring simply overrides. Thus, the spring clutch is a "one-way" clutch.

Fig. 1 Variable-torque drag brake . . .

Fig. 2 . . . holds tension constant on tape

Fig. 3 Constant-torque screwdriver

SPRING BANDS IMPROVE OVERRUNNING CLUTCH

An overrunning clutch that takes up only half the space of most clutches has a series of spiral-wound bands instead of conventional rollers or sprags to transmit high torques. The design (see drawing) also simplifies the assembly, cutting costs as much as 40 percent by eliminating more than half the parts in conventional clutches.

The key to the savings in cost and space is the clutches' freedom from the need for a hardened outer race. Rollers and sprags must have hardened races because they transmit power by a wedging action between the inner and outer races.

Role of spring bands. Overrunning clutches, including the spiral-band type, slip and overrun when reversed (see drawing). This occurs when the outer member is rotated clockwise and the inner ring is the driven member.

The clutch, developed by National Standard Co., Niles, Michigan, contains a set of high-carbon spring-steel bands (six in the design illustrated) that grip the inner member when the clutch is driving.

The outer member simply serves to retain the spring anchors and to play a part in actuating the clutch. Because it isn't subject to wedging action, it can be made of almost any material, and this accounts for much of the cost saving. For example, in the automotive torque converter in the drawing at right, the bands fit into the aluminum die-cast reactor.

Reduced wear. The bands are spring-loaded over the inner member of the clutch, but they are held and rotated by the outer member. The centrifugal force on the bands then releases much of the force on the inner member and considerably decreases the overrunning torque. Wear is consequently greatly reduced.

The inner portion of the bands fits into a V-groove in the inner member. When the outer member is reversed, the bands wrap, creating a wedging action in this V-groove. This action is similar to that of a spring clutch with a helical-coil spring, but the spiral-band type has very little unwind before it overruns, compared with the coil type. Thus, it responds faster.

Edges of the clutch bands carry the entire load, and there is also a compound action of one band upon another. As the torque builds up, each band pushes down on the band beneath it, so each tip is forced more firmly into the V-groove. The bands are rated for torque capacities from 85 to 400 ft.-lb. Applications include their use in auto transmissions, starters, and industrial machinery.

Spiral clutch bands can be purchased separately to fit the user's assembly.

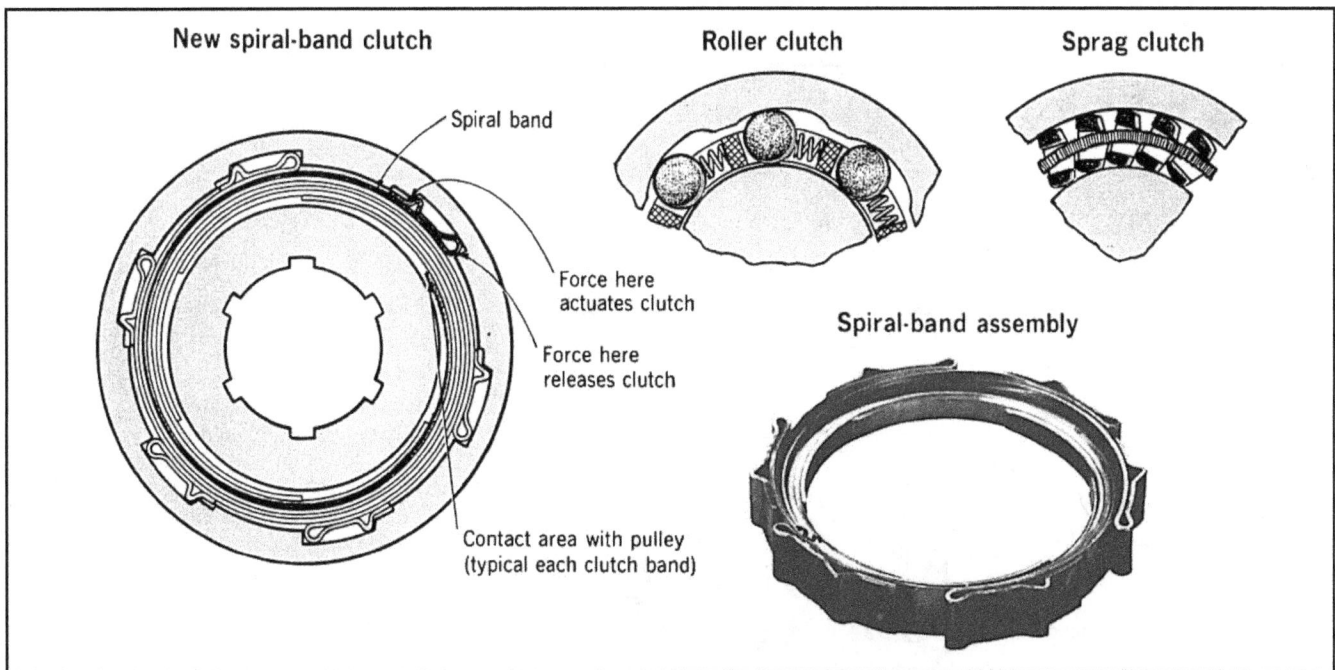

Spiral bands direct the force inward as an outer ring drives counterclockwise. The rollers and sprags direct the force outward.

SLIP AND BIDIRECTIONAL CLUTCHES COMBINE TO CONTROL TORQUE

A torque-limiting knob includes a dual set of miniature clutches—a detent slip clutch in series with a novel bidirectional-locking clutch—to prevent the driven member from backturning the knob. The bidirectional clutch in the knob locks the shaft from backlash torque originating within the panel, and the slip clutch limits the torque transmitted from outside the panel. The clutch was invented by Ted Chanoux, of Medford, N.Y.

The clutch (see drawing) is the result of an attempt to solve a problem that often plagues design engineers. A mechanism behind a panel such as a precision potentiometer or switch must be operated by a shaft that protrudes from the panel. The mechanism, however, must not be able to turn the shaft. Only the operator in front of the knob can turn the shaft, and he must limit the amount of torque he applies.

Solving design problem. This problem showed up in the design of a navigational system for aircraft.

The counter gave a longitudinal or latitudinal readout. When the aircraft was ready to take off, the navigator or pilot set a counter to some nominal figure, depending on the location of his starting point, and he energized the system. The computer then accepts the directional information from the gyro, the air speed from instruments in the wings, plus other data, and feeds a readout at the counter.

The entire mechanism was subjected to vibration, acceleration and deceleration, shock, and other high-torque loads, all of which could feed back through the system and might move the counter. The new knob device positively locks the mechanism shaft against the vibration, shock loads, and accidental turning, and it also limits the input torque to the system to a preset value.

Operation. To turn the shaft, the operator depresses the knob $1/16$ in. and turns it in the desired direction. When it is released, the knob retracts, and the shaft immediately and automatically locks to the panel or frame with zero backlash. Should the shaft torque exceed the preset value because of hitting a mechanical stop after several turns, or should the knob turn in the retracted position, the knob will slip to protect the system mechanism.

Internally, pushing in the knob turns both the detent clutch and the bidirectional-clutch release cage via the keyway. The fingers of the cage extend between the clutch rollers so that the rotation of the cage cams out the rollers, which are usually kept jammed between the clutch cam and the outer race with the roller springs. This action permits rotation of the cam and instrument shaft both clockwise and counterclockwise, but it locks the shaft securely against inside torque up to 30 oz.-in.

Applications. The detent clutch can be adjusted to limit the input torque to the desired values without removing the knob from the shaft. The outside diameter of the shaft is only 0.900 in., and the total length is 0.940 in. The exterior material of the knob is anodized aluminum, black or gray, and all other parts are stainless steel. The device is designed to meet the military requirements of MIL-E-5400, class 3 and MILK-3926 specifications.

Applications were seen in counter and reset switches and controls for machines and machine tools, radar systems, and precision potentiometers.

Eight-Joint Coupler

A novel coupler combines two parallel linkage systems in a three-dimensional arrangement to provide wide angular and lateral off-set movements in pipe joints. By including a bellows between the connecting pipes, the connector can join high-pressure and high-temperature piping such as is found in refineries, steam plants, and stationary power plants.

The key components in the coupler are four pivot levers (drawing) mounted

Miniature knob is easily operated from outside the panel by pushing it in and turning it in the desired direction. When released, the bidirectional clutch automatically locks the shaft against all conditions of shock and vibration.

in two planes. Each pivot lever has provisions for a ball joint at each end. "Twisted" tie rods, with holes in different planes, connect the pivot levers to complete the system. The arrangement permits each pipe face to twist through an appreciable arc and also to shift orthogonally with respect to the other.

Longer tie rods can be formed by joining several bellows together with center tubes.

The connector was developed by Ralph Kuhm Jr. of El Segundo, California.

SLIP CLUTCHES SERVE MANY DESIGN FUNCTIONS

A slip clutch is a mechanical device that permits a driven object to slip with respect to a continuously rotating drive shaft under preset conditions. This property allows for a driven pulley, gear, or output shaft to slip or be stopped, as required for the application. Slip clutches can control tension, torque, or force; prevent equipment overloading; and stop and start driven systems softly to prevent damaging shock transmission. There are four principal kinds of commercial slip clutches: mechanical, pneumatic, one-way, and jaw-type. Of these, the mechanical slip clutches are most popular.

A typical commercial mechanical slip clutch, as shown in the cutaway figure, consists of two subassemblies: cartridge and housing. The cartridge consists of a hub, tension springs, inner and outer plates, friction pads, and an adjusting nut. It is typically fastened to the input shaft by a set screw through the hub to a flat on the input shaft. The housing is most commonly fastened to an output pulley, gear, or output shaft by set screws. Torque is transmitted from the input shaft fastened to the output housing by torque pins mounted in the housing through outer plates keyed to the torque pins.

The amount of torque transmitted can be controlled by the adjusting nut on the cartridge. As it is turned it compresses an array of soft springs. (These springs extend from the back surface of the nut, pass through the cartridge, and terminate on the inner plate.) If the slip clutch is to function with a precise amount of torque, the adjusting nut can be replaced by a permanent collar. (The collar is attached to the hub to keep spring compression constant.) The clutching action is performed by durable friction materials sandwiched between round brass circular plates that remain aligned by the torque pins. Commercial slip clutches are bidirectional. This means that the output gear, pulley, or shaft can become the input component for converting the input shaft to an output shaft.

Commercial or off-the-shelf slip clutches are available with torque values rated from ounce-inches to more than 1000 pound-inches. Clutch capacity is a function of interdependent torque, revolutions per minute, and duty cycle. If any one of these factors is reduced, either or both of the other variables will increase. Excessive heat buildup above the manufacturer's specified values caused by friction can degrade the capability and life of a slip clutch. Heat buildup is measured in watts. (This is expressed as watts = torque [in.-lb] $\times$ rpm $\times$ 0.11.) However, if a friction-plate slip clutch is operated within design limits, it can expect to have an operating life of more than 30 million cycles.

In a shaft-to-pulley or gear installation, the input shaft is inserted into the hub of the cartridge and fastened with set screws. The housing is then inserted over the input shaft so that the torque pins engage the holes in the outer plates. The input shaft keeps the cartridge and housing aligned.

In a shaft-to-shaft installation, the input shaft is also first inserted into the hub of the cartridge and fastened with set screws. Then the output shaft is inserted into the housing and it is also fastened with set screws. The input and output shafts must be properly journaled with centerlines within ± 0.010 TIR. The clutch should not be lubricated in either installation because the friction materials eliminate the need for any lubrication.

There are many applications for commercial mechanical slip clutches because they can provide the following:

- *Tension control*—can keep tension constant on moving wire, paper, film, or thread while it is being wound or unwound. It can also compensate automatically for changes in speed and the diameter of the moving material.

- *Torque control*—can apply the correct tension for attaching screw-type bottle caps, sinking screws in wood or metal, and tightening bolts.

- *Overload protection*—can protect machinery from damage (and operators from injury) by slipping if the machinery is accidentally jammed. Activity can be resumed after the machinery is shut down safely and the fault is corrected.

- *Indexing*—can permit a machine with a continuously running drive motor to be stopped periodically by allowing the clutch to slip. This feature permits stepping motions for the performance of indexing, marking, or labeling.

- *Force control*—can push a product on a production line against a movable gate and hold it there until it is ready to move to the next stage. This feature prevents damage to both the product and the conveyor belt.

- *Soft starting and stopping*—permits a clutch to start up and stop machinery softly by slipping. This eliminates the need for a shock clutch to ensure constant torque or tension.

In addition to the commercial mechanical slip clutch described here, there are also *pneumatic* and *jaw-type* clutches. Pneumatic clutches, actuated by air pressure, have the same friction components as mechanical slip clutches. They can engage or disengage, vary torque during an operation, or adjust the torque remotely at any time. These clutches are recommended for use with servomechanisms because they transmit higher levels of torque than comparably sized mechanical slip clutches. Jaw-type clutches permit simple reliable phase adjustment and engage-release between a shaft and gear, pulley, or roller. The 120 teeth of a standard jaw-type clutch are usually machined from solid steel blanks.

A slip clutch attached to a moving drive shaft can protect against overloads, apply constant tension, control torque, and protect persons and property if the driven equipment malfunctions.

WALKING PRESSURE PLATE DELIVERS CONSTANT TORQUE

This automatic clutch causes the driving plate to move around the surface of the driven plate to prevent the clutch plates from overheating if the load gets too high. The "walking" action enables the clutch to transmit full engine torque for hours without serious damage to the clutch plates or the engine.

The automatic centrifugal clutch, manufactured by K-M Clutch Co., Van Nuys, California, combines the principles of a governor and a wedge to transmit torque from the engine to the drive shaft (see drawing).

How it works. As the engine builds up speed, the weights attached to the levers have a tendency to move towards the rim of the clutch plate, but they are stopped by retaining springs. When the shaft speed reaches 1600 rpm, however, centrifugal force overcomes the resistance of the springs, and the weights move outward. Simultaneously, the tapered end of the lever wedges itself in a slot in pin E, which is attached to the driving clutch plate. The wedging action forces both the pin and the clutch plate to move into contact with the driven plate.

A pulse of energy is transmitted to the clutch each time a cylinder fires. With every pulse, the lever arm moves outward, and there is an increase in pressure between the faces of the clutch. Before the next cylinder fires, both the lever arm and the driving plate return to their original positions. This pressure fluctuation between the two faces is repeated throughout the firing sequence of the engine.

Plate walks. If the load torque exceeds the engine torque, the clutch immediately slips, but full torque transfer is maintained without serious overheating. The pressure plate then momentarily disengages from the driven plate. However, as the plate rotates and builds up torque, it again comes in contact with the driven plate. In effect, the pressure plate "walks" around the contact surface of the driven plate, enabling

the clutch to continuously transmit full engine torque.

Applications. The clutch has undergone hundreds of hours of development testing on 4-stroke engines that ranged from 5 to 9 hp. According to the K-M Clutch Co., the clutch enables designers to use smaller motors than they previously could because of its no-load starting characteristics.

The clutch also acts as a brake to hold engine speeds within safe limits. For example, if the throttle accidentally opens when the driving wheels or driven mechanisms are locked, the clutch will stop.

The clutch can be fitted with sprockets, sheaves, or a stub shaft. It operates in any position and can be driven in both directions. The clutch could be installed in ships so that the applied torque would come from the direction of the driven plate.

The pressure plate was made of cast iron, and the driven-plate casting was made of magnesium. To prevent too much wear, the steel fly weights and fly levers were pre-hardened.

A driving plate moves to plate D, closing the gap, when speed reaches 1600 rpm.

Disengaged

Engaged

When a centrifugal force overcomes the resistance of the spring force, the lever action forces the plates together.

SEVEN OVERRUNNING CLUTCHES

Fig. 1 A lawnmower clutch.

Slides in shaft

Fig. 2 Wedging balls or rollers: internal (A); external (B) clutches.

Housing can be stack of stampings

Fig. 3 Molded sprags (for light duty).

Nylon or similar material

Drive   Idler   Slot   Driven

Fig. 4 A disengaging idler rises in a slot when the drive direction is reversed.

Spring-retaining collars

Grips when driven, slips for overrun

Drive

Fig. 5 A slip-spring coupling.

Fig. 6 An internal ratchet and spring-loaded pawls.

Spline

Spring loaded

Fig. 7 A one-way dog clutch.

ONE-WAY CLUTCH HAS SPRING-LOADED PINS AND SPRAGS

Sprags combined with cylindrical rollers in a bearing assembly can provide a simple, low-cost method for meeting the torque and bearing requirements of most machine applications. Designed and built by Est. Nicot of Paris, this unit gives one-direction-only torque transmission in an overrunning clutch. In addition, it also serves as a roller bearing.

The torque rating of the clutch depends on the number of sprags. A minimum of three, equally spaced around the circumference of the races, is generally necessary to get acceptable distribution of tangential forces on the races.

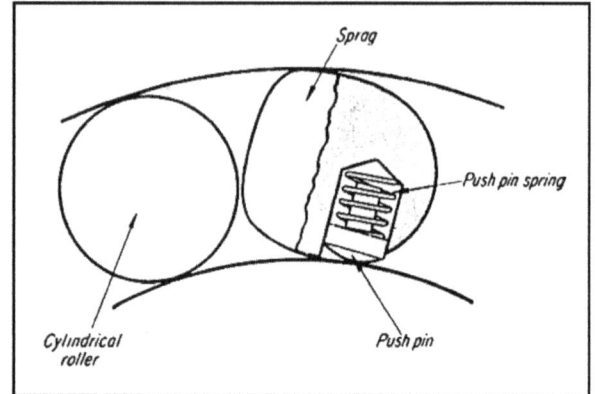

Races are concentric; a locking ramp is provided by the sprag profile, which is composed of two nonconcentric curves of different radius. A spring-loaded pin holds the sprag in the locked position until the torque is applied in the running direction. A stock roller bearing cannot be converted because the hard-steel races of the bearing are too brittle to handle the locking impact of the sprag. The sprags and rollers can be mixed to give any desired torque value.

ROLLER CLUTCH PROVIDES TWO OUTPUT SPEEDS

This clutch can be adapted for either electrical or mechanical actuation, and will control $1/2$ hp at 1500 rpm with only 7 W of power in the solenoid. The rollers are positioned by a cage (integral with the toothed control wheel—see diagram) between the ID of the driving housing and the cammed hub (integral with the output gear).

When the pawl is disengaged, the drag of the housing on the friction spring rotates the cage and wedges the rollers into engagement. This permits the housing to drive the gear through the cam.

When the pawl engages the control wheel while the housing is rotating, the friction spring slips inside the housing and the rollers are kicked back, out of engagement. Power is therefore interrupted.

According to the manufacturer, Tiltman Langley Ltd, Surrey, England, the unit operated over the full temperature range of –40° to 200°F.

A positive drive is provided by this British roller clutch.

Two-speed operation is provided by the new cam clutch

This clutch consists of two rotary members (see diagrams), arranged so that the outer (follower) member acts on its pulley only when the inner member is driving. When the outer member is driving, the inner member idles. One application was in a dry-cleaning machine. The clutch functions as an intermediary between an ordinary and a high-speed motor to provide two output speeds that are used alternately.

SEVEN OVERRIDING CLUTCHES

Fig. 1 Elementary overriding clutches: (A) A ratchet and pawl mechanism converts reciprocating or oscillating movement to intermittent rotary motion. This motion is positive but limited to a multiple of the tooth pitch. (B) A friction-type clutch is quieter, but it requires a spring device to keep the eccentric pawl in constant engagement. (C) Balls or rollers replace the pawls in this device. Motion of the outer race wedges the rollers against the inclined surfaces of the ratchet wheel.

Fig. 2 A commercial overriding clutch has springs that hold the rollers in continuous contact between the cam surfaces and the outer race; thus, there is no backlash or lost motion. This simple design is positive and quiet. For operation in the opposite direction, the roller mechanism can easily be reversed in the housing.

Fig. 3 A centrifugal force can hold the rollers in contact with the cam and outer race. A force is exerted on the lugs of the cage that controls the position of the rollers.

Fig. 4 With cylindrical inner and outer races, sprags can transmit torque. Energizing springs serve as a cage to hold the sprags. (A) Compared to rollers, the shape of a sprag permits a greater number within a limited space; thus higher torque loads are possible. Special cam surfaces are not required, so this version can be installed inside gear or wheel hubs. (B) Rolling action wedges the sprags tightly between the driving and driven members. A relatively large wedging angle ensures positive engagement.

Fig. 5 **A multidisk clutch** is driven by several sintered-bronze friction surfaces. Pressure is exerted by a cam-actuating device that forces a series of balls against a disk plate. A small part of the transmitted torque is carried by the actuating member, so capacity is not limited by the localized deformation of the contacting balls. The slip of the friction surfaces determines the capacity and prevents rapid shock loads. The slight pressure of disk springs ensures uniform engagement.

Fig. 6 **An engaging device** consists of a helical spring that is made up of two sections: a light trigger spring and a heavy coil spring. It is attached to and driven by the inner shaft. The relative motion of the outer member rubbing on the trigger causes this spring to wind up. This action expands the spring diameter, which takes up the small clearance and exerts pressure against the inside surface until the entire spring is tightly engaged. The helix angle of the spring can be changed to reverse the overriding direction.

Fig. 7 **A freewheeling clutch** widely used in power transmission has a series of straight-sided cam surfaces. An engaging angle of about 3° is used; smaller angles tend to become locked and are difficult to disengage while larger ones are not as effective. (A) The inertia of a floating cage wedges the rollers between the cam and outer race. (B) Continual operation causes the wear of surfaces; 0.001 in. wear alters the angle to 8.5° on straight-sided cams. Curved cam surfaces maintain a constant angle.

224

TEN APPLICATIONS FOR OVERRUNNING CLUTCHES

These clutches allow freewheeling, indexing, and backstopping;
they will solve many design problems. Here are examples.

Fig. 1 **Precision sprags** act as wedges and are made of hardened alloy steel. In the formsprag clutch, torque is transmitted from one race to another by the wedging action of sprags between the races in one direction; in the other direction the clutch freewheels.

Fig. 2 **This speed drive** requires input rotation to be reversible. Counterclockwise input (as shown in the diagram) drives gear 1 through clutch 1; the output is counterclockwise; clutch 2 overruns. Clockwise input (schematic) drives gear 2 through clutch 2; the output is still counterclockwise; clutch 1 overruns.

Fig. 3 **This speed drive** for a grinding wheel can be a simple, in-line assembly if the overrunning clutch couples two motors. The outer race of the clutch is driven by a gearmotor; the inner race is keyed to a grinding-wheel shaft. When the gearmotor drives, the clutch is engaged; when the larger motor drives, the inner race overruns.

Fig. 4 **This fan freewheels** when driving power is shut off. Without an overrunning clutch, fan momentum can cause belt breakage. If the driving source is a gearmotor, excessive gear stress can also occur by feedback of kinetic energy from the fan.

Fig. 5 **This indexing table** is keyed to a clutch shaft. The table is rotated by the forward stroke of the rack; power is transmitted through the clutch by its outer-ring gear only during this forward stroke. Indexing is slightly short of the position required. The exact position is then located by a spring-loaded pin that draws the table forward to its final positioning. The pin now holds the table until the next power stroke of the hydraulic cylinder.

225

Fig. 6 This punch press feed is arranged so that the strip is stationary on the downstroke of the punch (clutch freewheels); feed occurs during the upstroke when the clutch transmits torque. The feed mechanism can easily be adjusted to vary the feed amount.

Fig. 7 Indexing and backstopping is done with two clutches arranged so that one drives while the other freewheels. The application shown here is for a capsuling machine; gelatin is fed by the roll and stopped intermittently so the blade can precisely shear the material to form capsules.

Fig. 8 The intermittent motion of a candy machine is adjustable. The clutch ratchets the feed rolls around. This keeps the material in the hopper agitated.

Fig. 9 This double-impulse drive has double eccentrics and drive clutches. Each clutch is indexed 180° out of phase with the other. One revolution of the eccentric produces two drive strokes. Stroke length, and thus the output rotation, can be adjusted from zero to maximum by the control link.

Fig. 10 This anti-backlash device depends on overrunning clutches to ensure that no backlash is left in the unit. Gear *A* drives *B* and shaft *II* with the gear mesh and backlash, as shown in (A). The overrunning clutch in gear *C* permits gear *D* (driven by shaft *II*) to drive gear *C* and results in the mesh and backlash shown in (B). The overrunning clutches never actually overrun. They provide flexible connections (something like split and sprung gears) between shaft *I* and gears *A* and *C* to allow absorption of all backlash.

EIGHT SPRAG CLUTCH APPLICATIONS

Overrunning sprag clutches transmit torque in one direction and reduce speed, rest, hold, or freewheel in the reverse direction. Applications include overrunning, backstopping, and indexing. Their selection—similar to other mechanical devices—requires a review of the torque to be transmitted, overrunning speed, type of lubrication, mounting characteristics, environmental conditions, and shock conditions that might be encountered.

Fig. 1   Overrunning permits torque transmission in one direction and freewheels or overruns in the opposite direction. For example, the gearmotor drives the load by transmitting torque through the overrunning clutch and the high-speed shaft. Energizing the high-speed motor causes the inner member to rotate at the rpm of the high-speed motor. The gear motor continues to drive the inner member, but the clutch is freewheeling.

Fig. 2   Backstopping permits rotation in one direction only. The clutch serves as a counter-rotation holding device. An example is a clutch mounted on the headshaft of a conveyor. The outer race is restrained by torque-arming the stationary frame of the conveyor. If, for any reason, power to the conveyor is interrupted, the back-stopping clutch will prevent the buckets from running backwards and dumping the load.

Fig. 3   Indexing is the transmission of intermittent rotary motion in one direction; an example is the feed rolls of a punch press. On each stroke of the press crankshaft, a feed stroke on the feed roll is accomplished by the rack-and-pinion system. The system feeds the material into the dies of the punch press.

Fig. 4   Unidirectional drives with reverse mechanism incorporate two overrunning clutches into the gears, sheaves, or sprockets. Here, a 1:1 ratio right-angle drive is shown with a reversing input shaft. The output shaft rotates clockwise, regardless of the input shaft direction. By changing gear sizes, combinations of continuous or intermittent unidirectional output relative to the input can be obtained.

Fig. 5 Two-speed unidirectional output is made possible by using spur gears and reversing the direction of the input shaft. The rotation of shaft A transfers the power of gears B, D, and E to the output. Counterclockwise rotation engages the lower clutch, freewheeling the upper clutch because gear C is traveling at a faster rate than the shaft. This is caused by the reduction between gears B and E. Clockwise rotation of A engages the upper clutch, while the lower clutch freewheels because of the speed increase between gears D and E.

Power driven rolls

Fig. 6 A speed-differential or compensation is required where a different speed range for a function is desired, while retaining the same basic speed for all other functions. A series of individually driven power rolls can have different surface speeds because of drive or diameter variations of the rolls. An overrunning clutch permits the rolls with slower peripheral speed to overspeed and adjust to the material speed.

Fig. 7 A speed differential application permits the operation of engine accessories within a narrow speed range while the engine operates over a wide range. Pulley No. 2 contains the overrunning clutch. When the friction or electric clutch is disengaged, the driver pulley drives pulley No. 2 through the overrunning clutch, rotating the driven shaft. The engagement of the friction or electric clutch causes high-speed driven shaft rotation. This causes an overrun condition in the clutch at pulley No. 2.

Engine   Generator

Fig. 8 High inertia dissipation avoids driving back through a power system. In machines with high resistances, it prevents power train damage. If the engine is shut down while the generator is under a no-load condition, it would have a tendency to twist off the generator shaft. The overrunning clutch allows generator deceleration at a slower rate than the engine deceleration.

SIX SMALL CLUTCHES PERFORM PRECISE TASKS

Clutches for small machines must have: (1) quick response—lightweight moving parts; (2) flexibility—permit multiple members to control operation; (3) compactness—for equivalent capacity positive clutches are smaller than friction; (4) dependability; and (5) durability.

Fig. 1   A pawl and ratchet, single-cycle Dennis clutch. The primary parts of this clutch are the driving ratchet B, the driven cam plate C, and the connecting pawl D, which is carried by the cam plate. The pawl is normally held disengaged by the lower tooth of clutch arm A. When activated, arm A rocks counterclockwise until it is out of the path of rim F on cam plate C.

This permits pawl D, under the effect of spring E, to engage with ratchet B. Cam plate C then turns clockwise until, near the end of one cycle, pin G on the plate strikes the upper part of arm A, camming it clockwise back to its normal position. The lower part of A then performs two functions: (1) it cams pawl D out of engagement with the driving ratchet B, and (2) it blocks the further motion of rim F and the cam plate.

Fig. 1

Fig. 2

Fig. 2   A pawl and ratchet, single-cycle, dual-control clutch. The principal parts of this clutch are driving ratchet B, driven crank C, and spring-loaded ratchet pawl D. Driving ratchet B is directly connected to the motor and free to rotate on rod A. Driven crank C is directly connected to the main shaft of the machine and is also free to move on rod A. Spring-loaded ratchet pawl D is carried by crank C, which is normally held disengaged by latch E.

To activate the clutch, arm F is raised, permitting latch E to trip and pawl D to engage with ratchet B. The left arm of clutch latch G, which is in the path of the lug on pawl D, is normally permitted to move out of the way by the rotation of the camming edge of crank C. For certain operations, block H is temporarily lowered. This prevents the motion of latch G, resulting in the disengagement of the clutch after part of the cycle. It remains disengaged until the subsequent raising of block H permits the motion of latch G and the resumption of the cycle.

Fig. 3

(a)

(b)

Fig. 3 Planetary transmission clutch. This is a positive clutch with external control. Two gear trains provide a bidirectional drive to a calculator for cycling the machine and shifting the carriage. Gear A is the driver; gear L, the driven member, is directly connected to the planet carrier F. The planet consists of integral gears B and C. Gear B meshes with freewheeling gear D. Gears D and G carry projecting lugs E and H, respectively. Those lugs can contact formings on arms J and K of the control yoke.

When the machine is at rest, the yoke is centrally positioned so that arms J and K are out of the path of the projecting lugs, permitting both D and G to freewheel. To engage the drive, the yoke rocks clockwise, as shown, until the forming on arm K engages lug H, blocking further motion of ring gear G. A solid gear train is thereby established, driving F and L in the same direction as the drive A. At the same time, the gear train alters the speed of D as it continues counterclockwise. A reversing signal rotates the yoke counterclockwise until arm J encounters lug E, blocking further motion of D. This actuates the other gear train with the same ratio.

Fig. 4

Fig. 4 A multiple-disk friction clutch. Two multiple-disk friction clutches are combined in a single, two-position unit that is shown shifted to the left. A stepped cylindrical housing, C, encloses both clutches. Internal self-lubricated bearings support the housing on coaxial shaft J that is driven by transmission gear H, meshing with housing gear teeth K. At the other end, the housing carries multiple metal disks Q that engage keyways V and can make frictional contact with phenolic laminate disks N. They, in turn, can contact a set of metal disks P that have slotted openings for couplings with flats located on sleeves B and W.

In the position shown, pressure is exerted through rollers L, forcing the housing to the left, making the left clutch compress against adjusting nuts R. Those nuts drive gear A through sleeve B, which is connected to jack shaft J by pin U. When the carriage is to be shifted, rollers L force the housing to the right. However, it first relieves the pressure between the adjoining disks on the left clutch. Then they pass through a neutral position in which both clutches are disengaged, and they finally compress the right clutch against thrust bearing F. That action drives gear G through sleeve W, which rotates freely on the jack shaft.

Fig. 5

Fig. 5 A single-plate friction clutch. The basic parts of this clutch are the phenolic laminate clutch disk A, steel disk B, and drum C. They are normally kept separated by spring washer G. To engage the drive, the left end of a control arm is raised, causing ears F, which are located in slots in plates H, to rock clockwise. This action spreads the plates axially along sleeve P. Sleeves E and P and plate B are keyed to the drive shaft; all other members can rotate freely.

The axial motion loads the assembly to the right through the thrust ball bearings K against plate L and adjusting nut M. It also loads them to the left through friction surfaces on A, B, and C to thrust washer S, sleeve E, and against a shoulder on shaft D. This response then permits phemolic laminate disk A to drive drum C.

Fig. 6

Fig. 6 An overload relief clutch. This is a simple, double-plate, friction coupling with spring loading. Shaft G drives collar E, which drives slotted plates C and D faced with phenolic laminate disks B. Spring H is held in compression by the two adjusting nuts on the threaded end of collar E. These maintain the unit under axial pressure against the shoulder at the left end of the collar.

This enables the phenolic laminate disks B to drive through friction against both faces of the gear, which is free to turn on the collar. This motion of the gear causes output pinion J to rotate. If the machine to which the clutch is attached should jam and pinion J is prevented from turning, the motor can continue to run without overloading. However, slippage can occur between the phenolic laminate clutch plates B and the large gear.

TWELVE DIFFERENT STATION CLUTCHES

Innumerable variations of these station clutches can be designed for starting and stopping machines at selected points in their operation cycles.

Fig. 1(a) The driver and clutch body move in unison with the connector just coming in contact with the extractor. In **Fig. 1(b)** continued movement withdraws the connector.

Fig. 2 A two-station clutch whose stations are 180° apart. Because it has only one extractor arm, this mechanism can function as a one-station clutch.

231

Fig. 3   A modification of the cam extractor shown in Fig. 2.

Fig. 4   A single extractor two-station clutch with the stations that are 180° apart. Only one extractor is required because the connector has two cams.

Fig. 5   This one- or two-station clutch with a dual extractor is compact because there are no parts projecting beyond its body.

Fig. 6   The end and longitudinal section of a station clutch with internal driving recesses.

Fig. 7   This one- or two-station clutch depends on a single or a dual extractor. Its stations are spaced 180° apart.

Fig. 8   This is another one- or two-station clutch. It has a single or dual extractor with stations spaced 180° apart.

232

Fig. 9 A one-station axial connector clutch.

Fig. 10 **A two-station clutch.** The rollers R and R_1 of the extractor can also be arranged on the centerline A-A.

Fig. 11 **A nonselective multistation clutch** for instantaneous stopping in any position.

Fig. 12 **A multistation clutch with remote control.** The extractor pins are actuated by solenoids that either hold the extractor pin in position against spring pressure or release the pin.

233

TWELVE APPLICATIONS FOR ELECTROMAGNETIC CLUTCHES AND BRAKES

Fig. 1 **Coupling or uncoupling power or sensing device.**

Fig. 2 **Calibration protection** (energize to adjust).

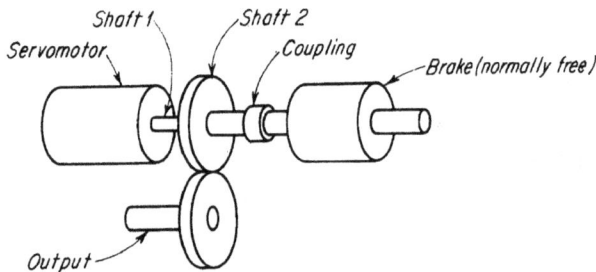

Figs. 3 & 4 **Simple servomotor brakes.**

Note: wire brake in series or parallel with motor

Fig. 5 **Adding or subtracting two inputs.**

Fig. 6 **Controlling output** from a differential.

Magnetic Friction Clutches

The simplest and most adaptable electromagnetic control clutch is the magnetic friction clutch. It works on the same principle as a simple solenoid-operated electric relay with a spring return to normal. Like the relay, it is a straightforward automatic switch for controlling the flow of power (in this cases, torque) through a circuit.

Rotating or Fixed Field?

This is a question primarily of magnetic design. Rotating-field clutches include a rotating coil, energized through brushes and slip rings. Fixed-field units have a stationary coil. Rotating-field units are still more common, but there has been a marked trend toward the fixed-field versions.

Generally speaking, a *rotating-field clutch* is a two-member unit, with the coil carried in the driving (input) member. It can be

mounted directly on a motor or speed-reducer shaft without loading down the driving motor. In the smaller sizes, it offers a better ratio of size to rated output than the fixed-field type, although the rotating coil increases inertia in the larger models.

A *fixed-field clutch,* on the other hand, is a three-member unit with rotating input and output members and a stationary coil housing. It eliminates the need for brushes and slip rings, but it demands additional bearing supports, and it can require close tolerances in mounting.

Purely Magnetic Clutches

Probably less familiar than the friction types are *hysteresis* and *eddy-current clutches.* They operate on straight magnetic principles and do not depend on mechanical contact between their members. The two styles are almost identical in construction, but the magnetic segments of the hysteresis clutch are electrically isolated, and those of the eddy-current clutch are interconnected.

Fig. 7   Simple-speed changing.

Fig. 8   Speed-changing and uncoupling.

Fig. 9   Simple direction-changing.

Fig. 10   Direction-changing and uncoupling.

Fig. 11   Constant tensioning.

Normal: input drives pot, with slip protection
Energized: input disconnected and pot locked

Fig. 12   Potentiometer control.

The magnetic analogy of both styles is similar in that the flux is passed between the two clutch members.

Hysteresis Clutches

The hysteresis clutch is a proportional-torque control device. As its name implies, it exploits the hysteresis effect in a permanent-magnet rotor ring to produce a substantially constant torque that is almost completely independent of speed (except for slight, unavoidable secondary eddy-current torques—which do not seriously reduce performance). It is capable of synchronous driving or continuous slip, with almost no torque variation at any slip differential for a given control current. Its control-power requirement can be met by a transistor drive. Typical applications include wire or tape tensioning, servo-control actuation, and torque control in dynamometers.

Eddy-Current Clutches

Eddy-current clutches on the other hand, are inherently speed-sensitive devices. They exhibit virtually no hysteresis, and develop torque by dissipating eddy currents through the electrical resistance of the rotor ring. This torque is almost a linear function of slip speed. These clutches perform best in speed-control applications, and as oscillation dampers.

Particle and Fluid Magnetic Clutches

There is no real difference between *magnetic-particle* and *magnetic-field clutches*. However, the magnetic medium in the particle clutch is a dry powder; in the fluid clutch it is a powder suspended in oil. In both clutches the ferromagnetic medium is introduced into the airgap between the input and output faces, which do not actually contact one another. When the clutch coil is energized, the particles are excited in the magnetic field between the faces; as they shear against each other, they produce a drag torque between the clutch members.

Theoretically, those clutches can approach the proportional control characteristics of a hysteresis clutch within the small weight and size limits of a comparably rated miniature friction clutch. But in practice, the service life of miniature magnetic-particle clutches has so far been too short for industrial service.

Other Magnetic Clutches

Two sophisticated concepts—neither of them yet developed to the point of practical application—might be of interest to anyone researching this field.

Electrostatic clutches depend on high voltages instead of a magnetic field to create force-producing suspensions.

Magnetostrictive clutches depend on a magnetic force to change the dimensions of a crystal or metal bar poised between two extremely precise facts.

235

CHAPTER 9
LATCHING, FASTENING, AND CLAMPING DEVICES AND MECHANISMS

Diagrams of basic latching and quick-release mechanisms.

Fig. 3 **A latch and cocking lever** is spring-loaded so that latch movement releases the cocking lever. The cocked position can be held indefinitely. Studs in the frame provide stops, pivots, or mounts for the springs.

Fig. 1 **Cam-guided latch** (A) has one cocked, and two relaxed positions, (B) Simple overcenter toggle action. (C) An overcenter toggle with a slotted link. (D) A double toggle action often used in electrical switches.

Fig. 4 **A latch mounted on a cocking lever** allows both levers to be reached at the same time with one hand. After release, the cocking spring initiates clockwise lever movement; then gravity takes over.

Fig. 2 **An identically shaped cocking lever and latch** (A) allow their functions to be interchangeable. The radii of the sliding faces must be dimensioned for a mating fit. The stepped latch (B) offers a choice of several locking positions.

Fig. 5 **A disk-shaped cocking lever** has a tension spring resting against the cylindrical hub. Spring force always acts at a constant radius from the lever pivot point.

Fig. 6 **A sleeve latch** (A) as an L-shaped notch. A pin in the shaft rides in a notch. Cocking requires a simple push and twist action. (B) The latch and plunger depend on axial movement for setting and release. A circular groove is needed if the plunger is to rotate.

Fig. 7 **A geared cocking device** has a ratchet fixed to a pinion. A torsion spring exerts clockwise force on the spur gear; a tension spring holds the gear in mesh. The device is wound by turning the ratchet handle counterclockwise, which in turn winds the torsion spring. Moving the release lever permits the spur gear to unwind to its original position without affecting the ratchet handle.

Fig. 8 **In this overcenter lock** (A) clockwise movement of the latching lever cocks and locks the slide. A counterclockwise movement is required to release the slide. (B) A latching cam cocks and releases the cocking lever with the same counterclockwise movement as (A).

Fig. 9 **A spring-loaded cocking piece** has chamfered corners. Axial movement of the push-rod forces the cocking piece against a spring-loaded ball or pin set in a frame. When cocking builds up enough force to overcome the latch spring, the cocking piece snaps over to the right. The action can be repeated in either direction.

Fig. 10 **A firing-pin mechanism** has a beveled collar on a pin. Pressure on the trigger forces the latch down until it releases the collar when the pin snaps out, under the force of cocking the spring. A reset spring pulls the trigger and pin back. The latch is forced down by a beveled collar on a pin until it snaps back, after overcoming the force of the latch spring. (A latch pin retains the latch if the trigger and firing pin are removed.)

FOURTEEN SNAP-ACTION DEVICES

These diagrams show fourteen ways to produce mechanical snap action.

Mechanical snap action results when a force is applied to a device over a period of time; buildup of this force to a critical level causes a sudden motion to occur. The ideal snap device would have no motion until the force reached a critical level. This, however, is not possible, and the way in which the mechanism approaches this ideal is a measure of its efficiency as a snap device. Some of the designs shown here approach the ideal closely; others do not, but they have other compensating good features.

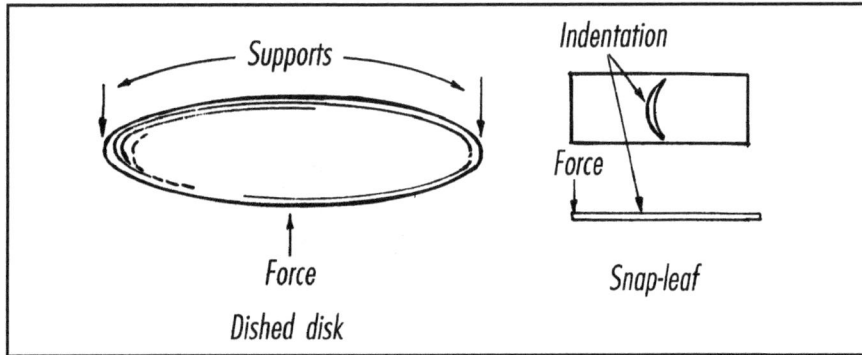

Fig. 1   A dished disk is a simple, common method for producing snap action. A snap leaf made from spring material can have various-shaped impressions stamped at the point where the overcentering action occurs. A "Frog clacker" is, of course, a typical applications. A bimetal element made in this way will reverse itself at a predetermined temperature.

Fig. 2   Friction override can hold against an increasing load until friction is suddenly overcome. This is a useful action for small sensitive devices where large forces and movements are undesirable. This is the way we snap our fingers. That action is probably the original snap mechanism.

Fig. 3   A ratchet-and-pawl combination is probably the most widely used form of snap mechanism. Its many variations are an essential feature in practically every complicated mechanical device. By definition, however, this movement is not true snap-action.

Fig. 4 **Over-centering** mechanisms find many applications in electrical switches. Considerable design ingenuity has been applied to fit this principle into many different mechanisms. It is the basis of most snap-action devices.

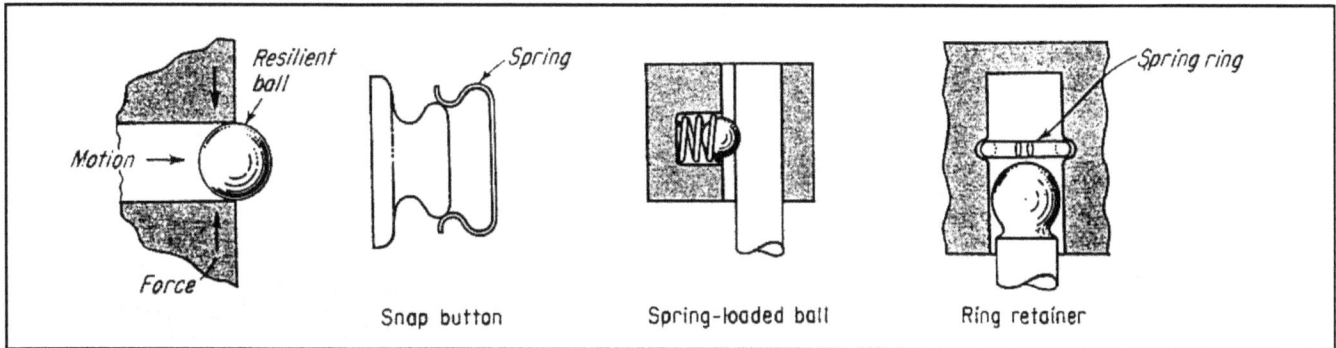

Fig. 5 **The sphere ejection** principle is based on snap buttons, spring-loaded balls and catches, and retaining-rings for fastening that must withstand repeated use. Their action can be designed to provide either easy or difficult removal. Wear can change the force required.

Fig. 6 **A pneumatic dump valve** produces snap action by preventing piston movement until air pressure has built up in the front end of the cylinder to a relatively high pressure. Dump-valve area in the low-pressure end is six times larger than its area on the high-pressure side. Thus the pressure required on the high-pressure side to dislodge the dump valve from its seat is six times that required on the low-pressure side to keep the valve properly seated.

241

Fig. 7 **A torsion ribbon** bent as shown will turn "inside out" at A with a snap action when twisted at B. Design factors are ribbon width, thickness, and bend angle.

Fig. 8 **A collapsing cylinder** has elastic walls that can be deformed gradually until their stress changes from compressive to bending, with the resulting collapse of the cylinder.

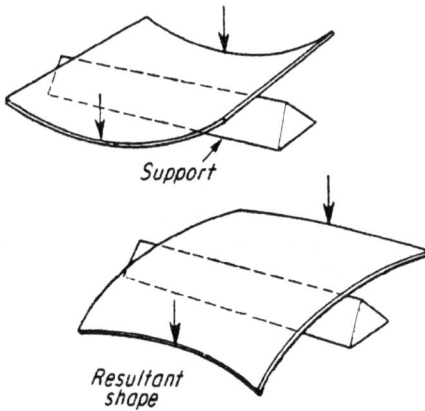

Fig. 9 **A bowed spring** will collapse into a new shape when it is loaded as shown. A "push-pull" steel measuring tape illustrates this action; the curved material stiffens the tape so that it can be held out as a cantilever until excessive weight causes it to collapse suddenly.

Fig. 10 **A flap vane** cuts off air or liquid flow at a limiting velocity. With a regulating valve, the vane will snap shut (because of increased velocity) when pressure is reduced below a design value.

Fig. 11 A sacrificing link is useful where high temperature or corrosive chemicals would be hazardous. If the temperature becomes too high, or atmosphere too corrosive, the link will yield at design conditions. The device usually is required to act only once, although a device like the lower one can be quickly reset. However, it is restricted to temperature control.

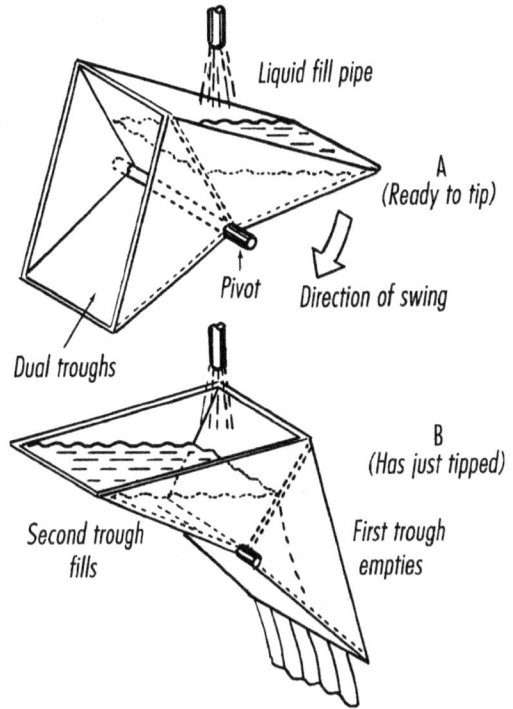

Fig. 12 Gravity-tips, although slower acting than most snap mechanisms, can be called snap mechanisms because they require an accumulation of energy to trigger an automatic release. A tripping trough that spreads sewerage is one example. As shown in A, it is ready to trip. When overbalanced, it trips rapidly, as in B.

Fig. 13 An overcentering tension spring combined with a pivoted contact-strip is one arrangement used in switches. The example shown here is unusual because the actuating force bears on the spring itself.

Fig. 14 An overcentering leaf-spring action is also the basis for many ingenious snap-action switches for electrical control. Sometimes spring action is combined with the thermostatic action of a bimetal strip to make the switch respond to heat or cold, either for control purposes or as a safety feature.

REMOTE CONTROLLED LATCH

This simple mechanism engages and disengages parallel plates carrying couplings and connectors.

Fig. 1 **The latch operation sequence** is shown for locking in steps (a) through (c) and for unlocking in steps (d) through (f).

A new latch mates two parallel plates in one continuous motion (see Fig. 1). On the Space Shuttle, the latch connects (and disconnects) plates carrying 20 fluid couplings and electrical connectors. (The coupling/connector receptacles are one plate, and mating plugs are on the other plate). Designed to lock items in place for handling, storage, or processing under remote control, the mechanism also has a fail-safe feature: It does not allow the plates to separate completely unless both are supported. Thus, plates cannot fall apart and injure people or damage equipment.

The mechanism employs four cam/gear assemblies, one at each corner of the lower plate. The gears on each side of the plate face inward to balance the loading and help align the plates. Worm gears on the cam-gear assemblies are connected to a common drive motor.

Figure 1 illustrates the sequence of movements as a pair of plates is latched and unlatched. Initially, the hook is extended and tilted out. The two plates are brought together, and when they are 4.7 in. (11.9 cm) apart, the drive motor is started (a). The worm gear rotates the hook until it closes on a pin on the opposite plate (b). Further rotation of the worm gear shortens the hook extension and raises the lower plate (c). At that point, the couplings and connectors on the two plates are fully engaged and locked.

To disconnect the plates, the worm gear is turned in the opposite direction. This motion lowers the bottom plate and pulls the couplings apart (d). However, if the bottom plate is unsupported,

the latch safety feature operates. The hook cannot clear the pin if the lower plate hangs freely (e). If the bottom plate is supported, the hook extension lifts the hook clear of the pin (f) so that the plates are completely separated.

This work was done by Clifford J. Barnett, Paul Castiglione, and Leo R. Coda of Rockwell International Corp. for **Johnson Space Center.**

TOGGLE FASTENER INSERTS, LOCKS, AND RELEASES EASILY

A pin-type toggle fastener, invented by C.C. Kubokawa at NASA's Ames Research Center, can be used to fasten plates together, fasten things to walls or decks, or fasten units with surfaces of different curvature, such as a concave shape to a convex surface.

With actuator pin. The cylindrical body of the fastener has a tapered end for easy entry into the hole; the head is threaded to receive a winged locknut and, if desired, a ring for pulling the fastener out again after release. Slots in the body hold two or more toggle wings that respond to an actuator pin. These wings are extended except when the spring-loaded pin is depressed.

For installation, the actuator pin is depressed, retracting the toggle wings. When the fastener is in place, the pin is released, and the unit is then tightened by screwing the locknut down firmly. This exerts a compressive force on the now expanded toggle wings. For removal, the locknut is loosened and the pin is again depressed to retract the toggle wings. Meanwhile, the threaded outer end of the cylindrical body functions as a stud to which a suitable pull ring can be screwed to facilitate removal of the fastener.

This invention has been patented by NASA (U.S. Patent No. 3,534,650).

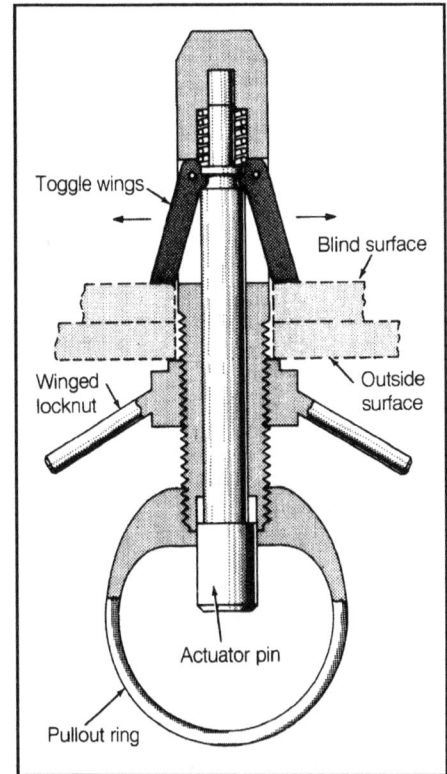

A fastener with controllable toggles can be inserted and locked from only one side.

GRAPPLE FREES LOADS AUTOMATICALLY

A simple grapple mechanism, designed at Argonne National Laboratory in Illinois, engages and releases loads from overhead cranes automatically. This self-releasing mechanism was developed to remove fuel rods from nuclear reactors. It can perform tasks where human intervention is hazardous or inefficient, such as lowering and releasing loads from helicopters.

The mechanism (see drawing) consists of two pieces: a lift knob secured to the load and a grapple member attached to the crane. The sliding latch-release collar under the lift knob is the design's key feature.

Spring magic. The grapple housing, which has a cylindrical inner surface, contains a machined groove fitted with a garter spring and three metal latches. When the grapple is lowered over the lift knob, these latches recede into the

groove as their edges come into contact with the knob. After passing the knob, they spring forward again, locking the grapple to the knob. Now the load can be lifted.

When the load is lowered to the ground again, gravity pull or pressure from above forces the grapple housing down until the latches come into contact with a double cone-shaped release collar. The latches move back into the groove as they pass over the upper cone's surface and move forward again when they slide over the lower cone.

The grapple is then lifted so that the release collar moves up the cylindrical rod until it is housed in a recess in the lift knob. Because the collar can move no farther, the latches are forced by the upward pull to recede again into the groove—allowing the grapple to be lifted free.

A sliding release collar is a key feature of this automatic grapple.

QUICK-RELEASE LOCK PIN HAS A BALL DETENT

A foolproof locking pin releases quickly when the stress on the joint is negligible.

A novel quick-release locking pin has been developed that can be withdrawn to separate the linked members only when stresses on the joint are negligible.

The pin may be the answer to the increasing demand for locking pins and fasteners that will pull out quickly and easily when desired, yet will stay securely in place without chance of unintentional release.

The key to this foolproof pin is a group of detent balls and a matching groove. The ball must be in the groove whenever the pin is either installed or pulled out of the assembly. This is easy to do during installation, but during removal the load must be off the pin to get the balls to drop into the groove.

How it works. The locking pin was developed by T.E. Othman, E.P. Nelson, and L.J. Zmuda under contract to NASA's Marshall Space Flight Center. It consists of a forward-pointing sleeve with a spring-loaded sliding handle as its rear end, housing a sliding plunger that is pushed backward (to its locking position) by a spring within the handle.

To some extend the plunger can slide forward against the plunger spring, and the handle can slide backward against the handle spring. A groove near the front end of the plunger accommodates the detent balls when the plunger is pushed forward by the compression of its spring. When the plunger is released backward, the balls are forced outward into holes in the sleeve, preventing withdrawal of the pin.

To install the pin, the plunger is pressed forward so that the balls fall into their groove and the pin is pushed into the hole. When the plunger is released, the balls lock the sleeve against accidental withdrawal.

To withdraw the pin, the plunger is pressed forward to accommodate the locking balls, and at the same time the handle is pulled backward. If the loading on the pin is negligible, the pin is withdrawn from the joint; if it is considerable, the handle spring is compressed and the plunger is forced backward by the handle so the balls will return to their locking position.

The allowable amount of stress on the joint that will permit its removal can be varied by adjusting the pressure required for compressing the handle spring. If the stresses on the joint are too great for the pin to be withdrawn in the normal manner, hammering on the forward end of the plunger simply ensures that the plunger remains in its rearward position, with the locking balls preventing the withdrawal of the pin. A stop on its forward end prevents the plunger from being driven backward.

AUTOMATIC BRAKE LOCKS HOIST WHEN DRIVING TORQUE CEASES

When torque is removed, the cam is forced into the tapered surface for brake action.

A brake mechanism attached to a chain hoist is helping engineers lift and align equipment accurately by automatically locking it in position when the driving torque is removed from the hoist.

According to the designer, Joseph Pizzo, the brake could also be used on wheeled equipment operating on slopes, to act as an auxiliary brake system.

How it works. When torque is applied to the driveshaft (as shown in the figure), four steel balls try to move up the inclined surfaces of the cam. Although called a cam by the designer, it is really a concentric collar with a cam-like surface on one of its end faces. Because the balls are contained by four cups in the hub, the cam is forced to move forward axially to the left. Because the cam moves away from the tapered surface, the cam and the driveshaft that is keyed to it are now free to rotate.

If the torque is removed, a spring resting against the cam and the driveshaft gear forces the cam back into the tapered surface of the threaded socket for instant braking.

Although this brake mechanism (which can rotate in either direction) was designed for manual operation, the principle can be applied to powered systems.

LIFT-TONG MECHANISM FIRMLY GRIPS OBJECTS

Twin four-bar linkages are the key components in this tong mechanism that can grip with a constant weight-to-grip force ratio any object that fits within its grip range. The tong mechanism relies on a cross-tie between the two sets of linkages to produce equal and opposite linkage movement. The vertical links have extensions with grip pads mounted at their ends, while the horizontal links are so proportioned that their pads move in an inclined straight-line path. The weight of the load being lifted, therefore, wedges the pads against the load with a force that is proportional to the object's weight and independent of its size.

PERPENDICULAR-FORCE LATCH

The installation and removal of equipment modules are simplified.

A latching mechanism simultaneously applies force in two perpendicular directions to install or remove electronic-equipment modules. The mechanism (see Fig. 1) requires only the simple motion of a handle to push or pull an avionic module to insert or withdraw connectors on its rear face into or from spring-loaded mating connectors on a panel and to force the box downward onto or release the box from a mating cold plate that is part of the panel assembly. The concept is also adaptable to hydraulic, pneumatic, and mechanical systems. Mechanisms of this type can simplify the manual installation and removal of modular equipment where a technician's movement is restricted by protective clothing, as in hazardous environments, or where the installation and removal are to be performed by robots or remote manipulators.

Figure 2 shows an installation sequence. In step 1, the handle has been installed on the handle cam and turned downward. In step 2,

Fig. 1 **An avionics box** mates with electrical connectors in the rear and is locked in position on the cold plate when it is installed with the latching mechanism.

Fig. 2 **This installation sequence** shows the positions of the handle and retention cams as the box is moved rearward and downward.

247

the technician or robot pushes the box rearward as slides attached to the rails enter grooves near the bottom of the box. In step 3, as the box continues to move to the rear, the handle cam automatically aligns with the slot in the rail and engages the rail roller.

In step 4, the handle is rotated upward 75°, forcing the box rearward to mate with the electrical connectors. In step 5, the handle is pushed upward an additional 15°, locking the handle cam and the slide. In step 6, the handle is rotated an additional 30°, forcing the box and the mating spring-loaded electrical connectors downward so that the box engages the locking pin and becomes clamped to the cold plate. The sequence for removal is identical except that the motions are reversed.

TWO QUICK-RELEASE MECHANISMS

QUICK-RELEASE MECHANISM

Quick release mechanisms have many applications. Although the design shown here operates as a tripping device for a quick-release hook, the mechanical principles involved have many other applications. Fundamentally, it is a toggle-type mechanism with the characteristic that the greater the load, the more effective the toggle.

The hook is suspended from the shackle, and the load or work is supported by the latch, which is machined to fit the fingers C. The fingers C are pivoted about a pin. Assembled to the fingers are the arms E, pinned at one end and joined at the other by the sliding pin G. Enclosing the entire unit are the side plates H, containing the slot J for guiding the pin G in a vertical movement when the hook is released. The helical spring returns the arms to the bottom position after they have been released.

To trip the hook, the tripping lever is pulled by the cable M until the arms E pass their horizontal centerline. The toggle effect is then broken, releasing the load.

A simple quick-release toggle mechanism was designed for tripping a lifting hook.

POSITIVE LOCKING AND QUICK-RELEASE MECHANISM

The object here was to design a simple device that would hold two objects together securely and quickly release them on demand.

One object, such as a plate, is held to another object, such as a vehicle, by a spring-loaded slotted bolt, which is locked in position by two retainer arms. The retainer arms are constrained from movement by a locking cylinder. To release the plate, a detent is actuated to lift the locking cylinder and rotate the retainer arms free from contact with the slotted bolt head. As a result of this action, the spring-loaded bolt is ejected, and the plate is released from the vehicle.

The actuation of the slidable detent can be initiated by a squib, a fluid-pressure device, or a solenoid. The principle of this mechanism can be applied wherever a positive engagement that can be quickly released on demand is required. Some suggested applications for this mechanism are in coupling devices for load-carrying carts or trucks, hooks or pick-up attachments for cranes, and quick-release mechanisms for remotely controlled manipulators.

This quick-release mechanism is shown locking a vehicle and plate.

SHAPE-MEMORY ALLOY DEVICES RELEASE LATCHES

Two miniature latch-release devices actuated by strips of shape-memory alloy (SMA) were developed by a researcher at Johns Hopkins University's Applied Physics Laboratory for the Goddard Space Flight Center. The objective of this work was to explore potential applications for this class of metal alloys. The shape-memory effect was discovered in the 1930s, but the first serious development of the nickel-titanium (NiTi) alloys with SMA properties was made by the U.S. Naval Ordnance Laboratory in the 1960s. An SMA alloy "remembers" its original cold-formed shape to which it returns after being deformed by heating. This property is being exploited as an alternative to pneumatic or electric actuators for moving some kinds of mechanical components such as robotic limbs.

A typical SMA NiTi alloy is trade named *Flexinol* or *Nitinol*, the latter an acronym for Nickel Titanium Naval Ordnance Laboratories. In preparation for a typical application, a suitably sized and shaped piece of an SMA, typically wire or strips, is deformed by a predetermined amount at the lower of two operating temperatures before being mounted in a test mechanism. When it is desired to have the SMA piece move in one direction, it is heated above its transition temperature to make it return to the "remembered" undeformed state. However, when it is desired to have the SMA piece move in the opposite direction and return to its deformed state, it is allowed to cool below its transition temperature.

The criterion for selecting an SMA alloy for a specific application is to find one that has a transition temperature which has a suitably small increment above the ambient temperature for that application. The devices discussed here, typical of other SMA devices, are heated by resistance to electrical current applied by contacts at the ends of the SMA strips.

At ambient temperature the two bent SMA strips in a container shown in the upper image of Fig. 1 act as two halves of a clamp to retain a bell-shaped knob within the container. The ends of the two bent SMA strips are attached to the base of the container. However, as shown in the lower image, when the two bent SMA strips are heated above their transition temperatures by an electric current, they "remember" their original shape and straighten out to release the knob. This device is redundant in that the knob will be released even if only one of the two

SMA strips straightens out when current is applied.

The upper image in Fig. 2 shows a straight flat SMA strip in a container with a latch pin attached to its upper end. The strip is torsionally deformed to a twist angle of about 90° by a torsion bias coil spring. In this position, the latch pin is in its latching position clamping the cover shut. When the SMA strip in the lower image is heated by an electric current, it rotates to its original untwisted shape, overcoming the bias of the torsion spring, thereby releasing the cover. This concept is suited to a wide variety of

applications requiring a device that can provide low torque actuation through a limited angle.

In addition to NiTi alloys, copper-zinc-aluminum-nickel and copper-aluminum-nickel alloys make up the three main types of shape-memory alloys. But alloys of zinc, copper, gold, and iron also have SMA characteristics. Shape-memory polymers have also been developed, and ferromagnetic shape-memory alloys (FSMA) are also being considered because magnetic response tends to be faster and more efficient than temperature-induced responses.

This work was done by Cliff E. Willey of Johns Hopkins University Applied Physics Laboratory for Goddard Space Flight Center.

Fig. 1 Two deformed SMA strips act as halves of a clamp to retain a bell-shaped knob terminated by a movable shaft. When heated, both strips straighten, releasing the knob, but in this redundant arrangement the knob will be released if only one strip straightens.

Fig. 2 An SMA strip with an attached latch pin is held in a 90° twisted position by a torsion bias coil spring keeping the cover clamped shut. But when the strip is heated, it "remembers" its straight position and overcomes spring torsion to turn the latch pin, releasing the cover.

RING SPRINGS CLAMP PLATFORM ELEVATOR INTO POSITION

A simple yet effective technique keeps a platform elevator locked safely in position without an external clamping force. The platform (see drawing) contains special ring assemblies that grip the four column-shafts with a strong force by the simple physical interaction of two tapered rings.

Thus, unlike conventional platform elevators, no outside power supply is required to hold the platform in position. Conventional jacking power is employed, however, in raising the platform from one position to another.

How the rings work. The ring assemblies are larger versions of the ring springs sometimes installed for shock absorption. In this version, the assembly is made up of an inner nonmetallic ring tapering upward and an outer steel ring tapered downward (see drawing).

The outside ring is linked to the platform, and the inside ring is positioned against the circumference of the column shaft. When the platform is raised to the designed height, the jack force is removed, and the full weight of the platform bears downward on the outside ring with a force that, through a wedging action, is transferred into a horizontal inward force of the inside ring.

Thus, the column shaft is gripped tightly by the inside ring; the heavier the platform, the larger the gripping force produced.

The advantage of the technique is that the shafts do not need notches or threads, and cost is reduced. Moreover, the shafts can be made of reinforced concrete.

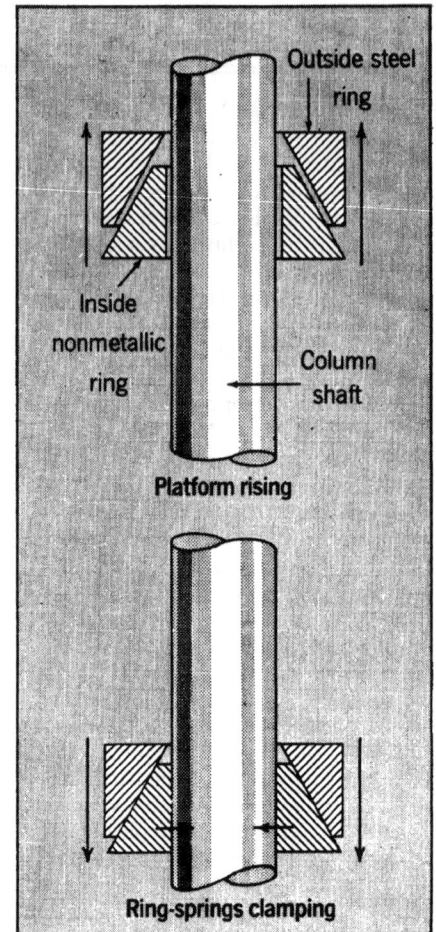

Ring springs unclamp the column as the platform is raised (upper). As soon as the jack power is removed (lower), the column is gripped by the inner ring.

CAMMED JAWS IN HYDRAULIC CYLINDER GRIP SHEET METAL

A single, double-acting hydraulic cylinder in each work holder clamps and unclamps the work and retracts or advances the jaws as required. With the piston rod fully withdrawn into the hydraulic cylinder (A), the jaws of the holder are retracted and open. When the control valve atop the work holder is actuated, the piston rod moves forward a total of 12 in. The first 10 in. of movement (B) brings the sheet-locater bumper into contact with the work. The cammed surface on the rod extension starts to move the trip block upward, and the locking pin starts to drop into position. The next $3/4$ in. of piston-rod travel (C) fully engages the work-holder locking pin and brings the lower jaw of the clamp up to the bottom of the work. The work holder slide is now locked between the forward stop and the locking pin. The last $1\frac{1}{4}$ in. of piston travel (D) clamps the workpiece between the jaws with a pressure of 2500 lbs. No adjustment for work thickness is necessary. A jaws-open limit switch clamps the work holder in position (C) for loading and unloading operations.

QUICK-ACTING CLAMPS FOR MACHINES AND FIXTURES

(A)

(B)

(C)

(D) (a) (b)

(E)

(F)

(G)

(H)

(I)

(J)

(L)

(K)

(M)

(N)

(A) An eccentric clamp. (B) A spindle-clamping bolt. (C) A method for clamping a hollow column to a structure. It permits quick rotary adjustment of the column. (D) (a) A cam catch for clamping a rod or rope. (b) A method for fastening a small cylindrical member to a structure with a thumb nut and clamp jaws. It permits quick longitudinal adjustment of a shaft in the structure. (E) A cam catch can lock a wheel or spindle. (F) A spring handle. Movement of the handle in the vertical or horizontal position provides movement at a. (G) A roller and inclined slot for locking a rod or rope. (H) A method for clamping a light member to a structure. The serrated edge on the structure permits the rapid accommodation of members with different thicknesses. (I) A spring taper holder with a sliding ring. (J) A special clamp for holding member a. (K) The cone, nut, and levers grip member a. The grip can have two or more jaws. With only two jaws, the device serves as a small vise. (L) Two different kinds of cam clamps. (M) A cam cover catch. Movement of the handle downward locks the cover tightly. (N) The sliding member is clamped to the slotted structure with a wedge bolt. This permits the rapid adjustment of a member on the structure.

(A) A method for fastening capacitor plates to a structure with a circular wedge. Rotation of the plates in a clockwise direction locks the plates to the structure. (B) A method for clamping member *a* with a special clamp. Detail *b* pivots on pin *c*. (C) A method for clamping two movable parts so that they can be held in any angular position with a clamping screw. (D) A cam clamp for clamping member *a*. (E) Two methods for clamping a cylindrical member. (F) Two methods for clamping member *a* with a special clamp. (G) A special clamping device that permits the parallel clamping of five parts by the tightening of one bolt. (H) A method for securing a structure with a bolt and a movable detail that provides a quick method for fastening the cover. (I) A method for quickly securing, adjusting, or releasing the center member. (J) A method for securing a bushing in a structure with a clamp screw and thumb nut. (K) A method for securing an attachment to a structure with a bolt and hand lever used as a nut. (L) A method for fastening a member to a structure with a wedge. (M) Two methods for fastening two members to a structure with a spring and one screw. The members can be removed without loosening the screw.

Many different devices for gaining mechanical advantage have been used in the design of friction clamps. These clamps can grip moderately large loads with comparatively small smooth surfaces, and the loads can be tightened or released with simple controls. The clamps illustrated here can be tightened or released with screws, levers, toggles, wedges, and combinations of them.

Clamping skirt

Section A-A

Clamp screw draws shoe and yoke together

Revolving table

Clamping skirt

Stationary table

Clamp shoe

Clamp yoke

Clamp support

A turntable clamp.

The clamp assembly floats on a pin and does not disturb the table setting.

Slide

T-bolt

Clamp screw. Cone tip lifts T-bolt to clamp slide to bed

Bed

A slide clamp.

A

Connecting rod

Clamping yoke

A

Operating position

Section B-B

Elongated hole equalizes clamping action

Clamping lever

Pulley

Spindle

Tapered gib

Feed screw

C

Pivot stud

Spindle housing

Elongated hole

Clamping lever

B

B

Clamp stud

Way gib

Nut

Locknut

C

Section A-A

Clamping lever

Section C-C

A double clamp for a spindle head.

A center-support clamp.

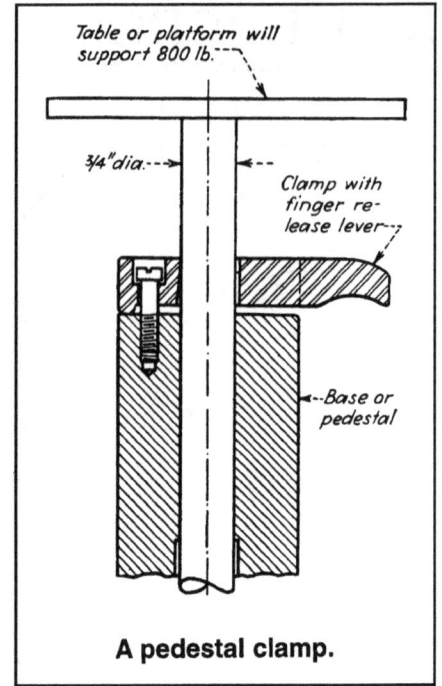

Small handwheel rotates spur gear to adjust center

Large handwheel operates clamp blocks through long sleeve

Sleeve with rack teeth

Support

Center is journaled in bearings within sleeve

Spur gear

Shaft

Long sleeve

Blocks clamp sleeve when pushed together

A pedestal clamp.

Table or platform will support 800 lb.

3/4" dia.

Clamp with finger release lever

Base or pedestal

A right-angle clamp.

Horizontal shaft

Clamp stud tightens two shafts simultaneously

Vertical shaft

Clamp yoke holds shafts in alignment

A slide clamp.

Slide bed

Slide shaft

Clamp spring Lock

Slide

A specimen holder clamp.

A

Clamp handle tightens in less than 30 deg. turn

Specimen clamp screw

Clamp support

Specimen clamp

Clamp stud

Stud lock nut for fine adjustment

Specimen holder has 30-deg. range

Clamping washer with spherical surface

A

Clamping nut with outside threads

Section A-A

Clamp stud key prevents turning

A table clamp.

Table

Clamp nut

Clamp teeth

Clamp

Clamp screw

Frame

254

DETENTS FOR STOPPING MECHANICAL MOVEMENTS

Some of the more robust and practical devices for stopping mechanical movements are illustrated here.

Fixed holding power is constant in both directions.

A domed plunger has long life.

The screw provides adjustable holding.

Wedge action locks the movement in the direction of the arrow.

Friction results in holding force.

Holding power is $R = P \tan \alpha$; for friction coefficient, F. at contact surface $R = P (\tan \alpha\ 1\ F)$

A notch shape dictates the direction of rod motion.

A leaf spring provides limited holding power.

A leaf-spring detent can be removed quickly.

A conical or wedge-ended detent.

A positive detent has a manual release.

.**A leaf spring** for holding flat pieces.

An automatic release occurs in one direction; manual release is needed in the other direction.

Axial positioning (indexing) by means of spaced holes in the index base.

A positive detent has a push-button release for straight rods.

Section on XX

A radially arranged detent holds in slotted index base.

A roller detent positions itself in a notch.

Rise, $S = \dfrac{N\tan a}{2} - R \times \dfrac{1 - \cos}{\cos a}$

Roller Radius, $R = \left(\dfrac{N\tan a}{2} - S\right)\left(\dfrac{\cos a}{1 - \cos a}\right)$

A magnetic detent.

An axial detent for the positioning of the adjustment knob with a manual release.

TWELVE CLAMPING METHODS FOR ALIGNING ADJUSTABLE PARTS

Methods for clamping parts that must be readily movable are as numerous and as varied as the requirements. In many instances, a clamp of any design is satisfactory, provided it has sufficient strength to hold the parts immovable when tightened. However, it is sometimes necessary that the movable part be clamped to maintain accurate alignment with some fixed part. Examples of these clamps are described and illustrated.

Fig. 1 When a nut is tightened, the flange on the edge of the movable part is drawn against the machined edge of the stationary part. This method is effective, but the removal of the clamped part can be difficult if it is heavy or unbalanced.

Fig. 2 The lower edge of the bolt head contacts the angular side of the locating groove, causing the keys to be held tightly against the opposite side of the groove. This design permits easy removal of the clamped part, but it is effective only if the working pressure is directly downward or in a direction against the perpendicular side of the slot.

Fig. 3 The movable part is held against one side of the groove while the T-nut is forced against the other side. The removal of the screw permits easy removal of the clamped part. Heavy pressure toward the side of the key out of contact with the slot can permit slight movement due to the springing of the screw.

Fig. 4 One side of the bolt is machined at an angle to form a side of the dovetail, which tightens in the groove as the nut is drawn tight. The part must be slid the entire length of the slot for removal.

Fig. 5 The angular surface of the nut contacts the angular side of the key and causes it to move outward against the side of the groove. This exerts a downward pull on the clamped part due to the friction of the nut against the side of a groove as the nut is drawn upward by the screw.

257

Fig. 6 and 7 These designs differ only in the depth of the grooves. They cannot withstand heavy pressure in an upward direction but have the advantage of being applicable to narrow grooves.

Fig. 9 The movable member is flanged on one side and carries a conical pointed screw on the other side. A short shaft passes through both members and carries a detent slightly out of alignment with the point of the screw. This shaft is flattened on opposite sides where it passes through the stationary member to prevent it from turning when the movable member is removed. A heavy washer is screwed to the underside of the shaft. When the knurled screw is turned inward, the shaft is drawn upward while the movable member is drawn downward and backward against the flange. The shaft is forced forward against the edge of the slot. The upper member can thus be moved and locked in any position. Withdrawing the point of the screw from the detent in the shaft permits the removal of the upper member.

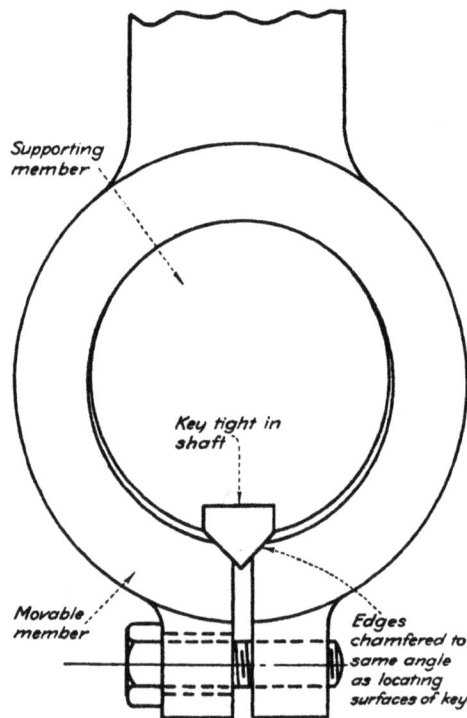

Fig. 8 Screw contact causes the ball to exert an outward pressure against the gib. The gib is loosely pinned to the movable part. This slide can be applied to broad surfaces where it would be impractical to apply adjusting screws through the stationary part.

Fig. 10 One edge of a bar is machined at an angle which fits into mating surfaces on the movable part. When the bolt, which passes through the movable part, is drawn tight, the two parts are clamped firmly together.

Fig. 11 As the screw is tightened, the chamfered edges of the cut tend to ride outward on the angular surfaces of the key. This draws the movable member tightly against the opposite side of the shaft.

Fig. 12 As the screw is turned, it causes the movable side, which forms one side of the dovetail groove, to move until it clamps tightly on the movable member. The movable side should be as narrow as possible, because there is a tendency for this part to ride up on the angular surface of the clamped part.

SPRING-LOADED CHUCKS AND HOLDING FIXTURES

Spring-loaded fixtures for holding work can be preferable to other fixtures. Their advantages are shorter setup time and quick workpiece change. Work distortion is reduced because the spring force can be easily and accurately adjusted.

A spring-loaded nest has radial slots extending into its face. These ensure an even grip on the work, which is pushed over the rim. A slight lead on the rim makes mounting work easier. The principal application of this fixture is for ball-bearing race grinding where only light cutting forces are applied.

A cupped fixture has three leaf-springs equally spaced in a wall. The work, usually to be lacquered, is inserted into the cup during its rotation. Because the work is placed in the fixture by hand, the spindle is usually friction-driven for safety.

This spring clamp has a cam-and-tension spring that applies a clamping force. A tension spring activates the cam through a steel band. When the handle is released, the cam clamps the work against the V-bar. Two stop-pins limit travel when there is no work in the fixture.

A leaf-spring gripper is used mainly to hold work during assembly. One end of a flat coil-spring is anchored in the housing; the other end is held in a bolt. When the bolt is turned, the spring is tightened, and its outside diameter is decreased. After the work is slid over the spring, the bolt handle is released. The spring then presses against the work, holding it tight.

This lathe center is spring loaded and holds the work with spring pressure alone. Eight sharp-edged notches on the conical surface of the driving center bite into the work and drive it. Its spring tension is adjustable.

CHAPTER 10
CHAIN AND BELT DEVICES
AND MECHANISMS

TWELVE VARIABLE-SPEED BELT AND CHAIN DRIVES

Variable-speed drives provide an infinite number of speed ratios within a specific range. They differ from the stepped-pulley drives in that the stepped drives offer only a discrete number of velocity ratios.

Mechanical "all-metal" drives employ friction or preloaded cones, disks, rings, and spheres, which undergo a certain amount of slippage. Belt drives, on the other hand, have little slippage or frictional losses, and chain has none—it maintains a fixed phase relationship between the input and output shafts.

Belt Drives

Belt drives offer high efficiency and are relatively low in price. Most use V-belts, reinforced by steel wires to 3 inches in width.

Speed adjustment in belt drives is obtained through one of the four basic arrangements shown below.

Variable-distance system (Fig. 1). A variable-pitch sheave on the input shaft opposes a solid (fixed-pitch) sheave on the output shaft. To vary the speed, the center distance is varied, usually by an adjustable base, tilting or sliding motor (Fig. 6).

Speed variations up to 4:1 are easily achieved, but torque and horsepower characteristics depend on the location of the variable-diameter sheave.

Fixed-distance system (Fig. 2). Variable-pitch sheaves on both input and output shafts maintain a constant center distance between shafts. The sheaves are controlled by linkage. Either the pitch diameter of one sheave is positively controlled and the disks of the other sheave under spring tension, adjust automatically or the pitch diameters of both sheaves are positively controlled by the linkage system (Fig. 5). Pratt & Whittney has applied the system in Fig. 5 to the spindle drive of numerically controlled machines.

Speed variations up to 11:1 are obtained, which means that with a 1200-rpm motor, the maximum output speed will be $1200\sqrt{11} = 3984$ rpm, and the minimum output speed $= 3984/11 = 362$ rpm.

Double-reduction system (Fig. 3). Solid sheaves are on both the input and output shafts, but both sheaves on the intermediate shaft are of variable-pitch type. The center distance between input and output is constant.

Coaxial shaft system (Fig. 4). The intermediate shaft in this arrangement permits the output shaft to be coaxial with that of the input shaft. To maintain a fixed center distance, all four sheaves must be of the variable-pitch type and controlled by linkage, similar to the system in Fig. 6. Speed variation up to 16:1 is available.

Packaged belt units (Fig. 7). These combine the motor and variable-pitch transmissions as an integral unit. The belts are usually ribbed, and speed ratios can be dialed by a handle.

Fig. 1

Fig. 2

Fig. 3

Fig. 4

Four basic belt arrangements for varying output speed.

Fig. 5 Linkage controlled pulleys.

Fig. 6 Tandem arrangement employs dual belt-system to produce high speed-reduction.

Packaged belt unit.

Fig. 7

Sheave Drives

The axial shifting of variable-pitch sheaves is controlled by one of four methods:

Linkage actuation. The sheave assemblies in Fig. 5 are directly controlled by linkages which, in turn, are manually adjusted.

Spring pressure. The cones of the sheaves in Figs. 2 and 4 are axially loaded by spring force. A typical pulley of this type is illustrated in Fig. 8. These pulleys are used in conjunction with directly controlled sheaves, or with variable center-distance arrangements.

Cam-controlled sheave. The cones of this sheave (Fig. 9) are mounted on a floating sheave, free to rotate on the pulley spindle. Belt force rotates the cones, whose surfaces are cammed by the inclined plane of the spring. The camming action wedges the cones against the belt, thus providing sufficient pressure to prevent slippage at the higher speeds, as shown in the curve.

Centrifugal-force actuator. In this unique sheave arrangement (Fig. 10) the pitch diameter of the driving sheave is controlled by the centrifugal force of steel balls. Another variable-pitch pulley mounted on the driven shaft is responsive to the torque. As the drive speed increases, the centrifugal force of the balls forces the sides of the driving sheave together. With a change in load, the movable flange of the driven sheave rotates in relation to the fixed flange. The differential rotation of the sheave flanges cams them together and forces the V-belt to the outer edge of the driven sheave, which has a lower transmission ratio. The driving sheave is also shifted as the load rises with decreasing speed. With a stall load, it is moved to the idling position. When the torque-responsive sheave is the driving member, any increase in drive speed closes its flanges and opens the flanges of the centrifugal member, thus maintaining a constant output speed. The drive has performed well in transmissions with ratings ranging from 2 to 12 hp.

Fig. 8 Spring-loaded variable-pitch pulley.

Fig. 9 Cam-controlled pulley prevents slippage at high speeds.

Fig. 10 A ball-controlled pulley has its sides pressured by centrifugal force.

Chain Drives

PIV drive (Fig. 11). This chain drive (positive, infinitely variable) eliminates any slippage between the self-forming laminated chain teeth and the chain sheaves. The individual laminations are free to slide laterally to take up the full width of the sheave. The chain runs in radially grooved faces of conical surface sheaves which are located on the input and output shafts. The faces are not straight cones, but have a slight convex curve to maintain proper chain tension at all positions. The pitch diameters of both sheaves are positively controlled by the linkage. Booth action is positive throughout operating range. It is rated to 25 hp with speed variation of 6:1.

Double-roller chain drive (Fig. 12). This specially developed chain is built for capacities to 22 hp. The hardened rollers are wedged between the hardened conical sides of the variable-pitch sheaves. Radial rolling friction results in smooth chain engagement.

Single-roller chain drive (Fig. 13). The double strand of this chain boosts the capacity to 50 hp. The scissor-lever control system maintains the proper proportion of forces at each pair of sheave faces throughout the range.

Fig. 11 A PIV drive chain grips radially grooved faces of a variable-pitch sheave to prevent slippage.

Fig. 13 A single-roller chain drive for high horsepower applications.

Fig. 12 A double-roller chain drive combines strength with ease in changing speed.

BELTS AND CHAINS ARE AVAILABLE IN MANY DIFFERENT FORMS

Imaginative fusions of belts, cables, gears, and chains are expanding the horizons for light-duty synchronous drives.

Belts have long been used for the transfer of mechanical power. Today's familiar flat belts and V-belts are relatively light, quiet, inexpensive, and tolerant of alignment errors. They transmit power solely through frictional contacts. However, they function best at moderate speeds (4000 to 6000 fpm) under static loads. Their efficiencies drop slightly at low speeds, and centrifugal effects limit their capacities at high speeds. Moreover, they are inclined to slip under shock loads or when starting and braking. Even under constant rotation, standard belts tend to creep. Thus, these drives must be kept under tension to function properly, increasing loads on pulley shaft bearings.

Gears and chains, on the other hand, transmit power through bearing forces between positively engaged surfaces. They do not slip or creep, as measured by the relative motions of the driving and driven shafts. But the contacts themselves can slip significantly as the chain rollers and gear teeth move in and out of mesh.

Positive drives are also very sensitive to the geometries of the mating surfaces. A gear's load is borne by one or two teeth, thus magnifying small tooth-to-tooth errors. A chain's load is more widely distributed, but chordal variations in the driving wheel's effective radius produce small oscillations in the chain's velocity.

To withstand these stresses, chains and gears must be carefully made from hard materials and must then be lubricated in operations. Nevertheless, their operating noise betrays sharp impacts and friction between mating surfaces.

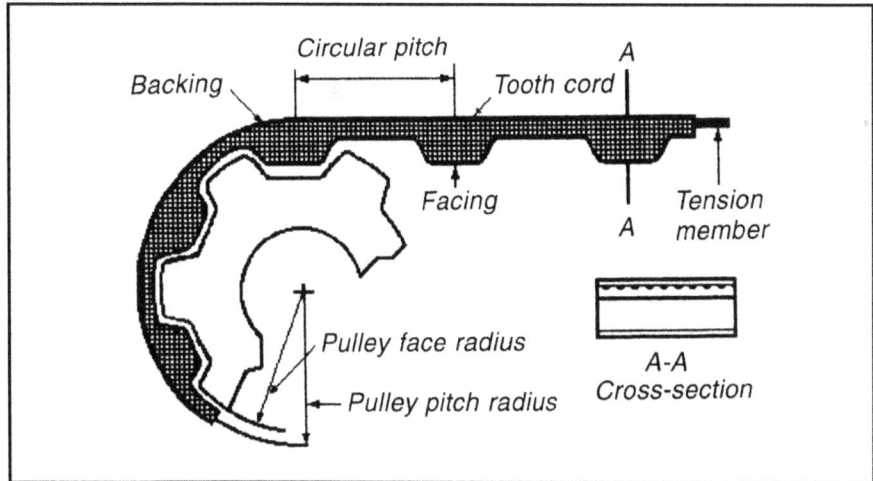

Fig. 1 **Conventional timing belts** have fiberglass or polyester tension members, bodies of neoprene or polyurethane, and trapezoidal tooth profiles.

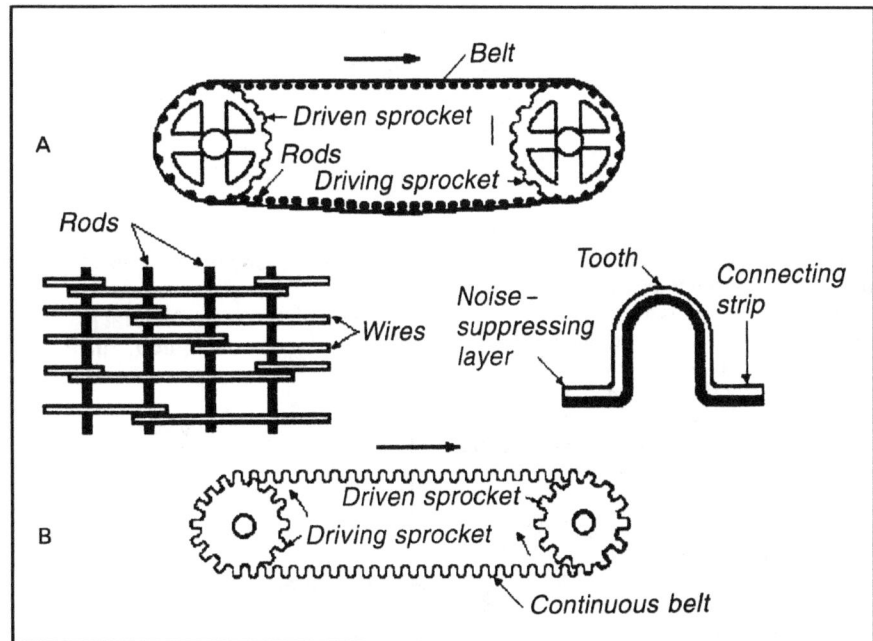

Fig. 2 **NASA metal timing belts** exploit stainless steel's strength and flexibility, and are coated with sound- and friction-reducing plastic.

265

The cogged timing belt, with its trapezoidal teeth (Fig. 1), is the best-known fusion of belt, gear, and chain. Though these well-established timing belts can handle high powers (up to 800 hp), many of the newer ideas in synchronous belting have been incorporated into low and fractional horsepower drives for instruments and business machines.

Steel Belts for Reliability

Researchers at NASA's Goddard Space Flight Center (Greenbelt, MD) turned to steel in the construction of long-lived toothed transmission belts for spacecraft instrument drives.

The NASA engineers looked for a belt design that would retain its strength and hold together for long periods of sustained or intermittent operation in hostile environments, including extremes of heat and cold.

Two steel designs emerged. In the more chain-like version (Fig. 2A), wires running along the length of the belt are wrapped at intervals around heavier rods running across the belt. The rods do double duty, serving as link pins and as teeth that mesh with cylindrical recesses cut into the sprocket. The assembled belt is coated with plastic to reduce noise and wear.

In the second design (Fig. 2B), a strip of steel is bent into a series of U-shaped teeth. The steel is supple enough to flex as it runs around the sprocket with its protruding transverse ridges, but the material resists stretching. This belt, too, is plastic-coated to reduce wear and noise.

The V-belt is best formed from a continuous strip of stainless steel "not much thicker than a razor blade," according to the agency, but a variation can be made by welding several segments together.

NASA has patented both belts, which are now available for commercial licensing. Researchers predict that they will be particularly useful in machines that must be dismantled to uncover the belt pulleys, in permanently encased machines, and in machines installed in remote places. In addition, stainless-steel belts might find a place in high-precision instrument drives because they neither stretch nor slip.

Though plastic-and-cable belts don't have the strength or durability of the NASA steel belts, they do offer versatility and production-line economy. One of the least expensive and most adaptable is the modern version of the bead chain, now common only in key chains and light-switch pull-cords.

The modern bead chain—if chain is the proper word—has no links. It has, instead, a continuous cable of stainless steel or aramid fiber which is covered with polyurethane. At controlled intervals, the plastic coating is molded into a bead (Fig. 3A). The length of the pitches thus formed can be controlled to within 0.001 in.

Fig. 3 Polyurethane-coated steel-cable "chains"—both beaded and 4-pinned—can cope with conditions unsuitable for most conventional belts and chains.

Table 1. Conventional Timing Belts

Type	Circular pitch, in.	Wkg. tension lb/in. width	Centr. loss const., K_c
Standard			
MXL	0.080	32	10×10^{-9}
XL	0.200	41	27×10^{-9}
L	0.375	55	38×10^{-9}
H	0.500	140	53×10^{-9}
40DP	0.0816	13	—
High-torque			
3 mm	0.1181	60	15×10^{-9}
5 mm	0.1968	100	21×10^{-9}
8 mm	0.3150	138	34×10^{-9}

Courtesy Stock Drive Products

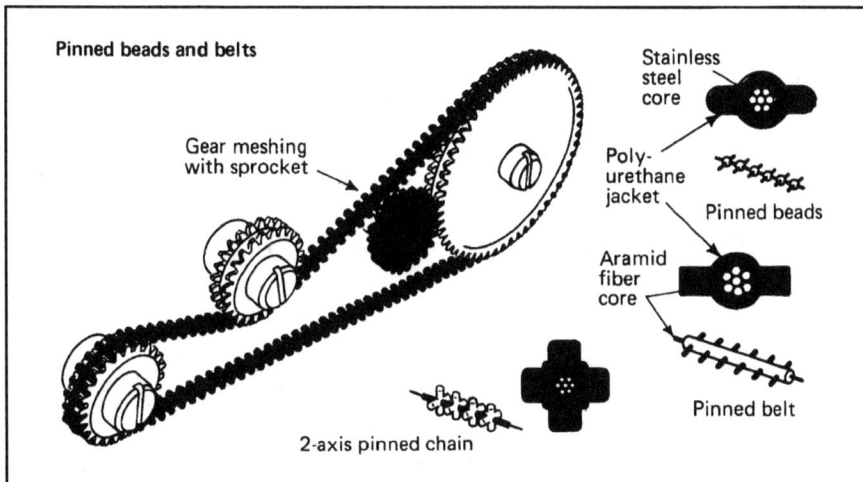

Fig. 4 Plastic pins eliminate the bead chain's tendency to cam out of pulley recesses, and permit greater precision in angular transmission.

In operation, the cable runs in a grooved pulley; the beads seat in conical recesses in the pulley face. The flexibility, axial symmetry, and positive drive of bead chain suit a number of applications, both common and uncommon:

- An inexpensive, high-ratio drive that resists slipping and requires no lubrication (Fig. 3B). As with other chains and belts, the bead chain's capacity is limited by its total tensile strength (typically 40 to 80 lb for a single-strand steel-cable chain), by the speed-change ratio, and by the radii of the sprockets or pulleys.
- Connecting misaligned sprockets. If there is play in the sprockets, or if the sprockets are parallel but lie in different planes, the bead chain can compensate for up to 20° of misalignment (Fig. 3C).

- Skewed shafts, up to 90° out of phase (Fig. 3D).
- Right-angle and remote drives using guides or tubes (Figs. 3E and 3F). These methods are suitable only for low-speed, low-torque applications. Otherwise, frictional losses between the guide and the chain are unacceptable.
- Mechanical timing, using oversize beads at intervals to trip a microswitch (Fig. 3G). The chain can be altered or exchanged to give different timing schemes.
- Accurate rotary-to-linear motion conversion (Fig. 3H).
- Driving two counter-rotating outputs from a single input, using just a single belt (Fig. 3I).
- Rotary-to-oscillatory motion conversion (Fig. 3J).

- Clutched adjustment (Fig. 3K). A regular V-belt pulley without recesses permits the chain to slip when it reaches a pre-set limit. At the same time, bead-pulleys keep the output shafts synchronized. Similarly, a pulley or sprocket with shallow recesses permits the chain to slip one bead at a time when overloaded.
- Inexpensive "gears" or gear segments fashioned by wrapping a bead chain round the perimeter of a disk or solid arc of sheet metal (Fig. 3L). The sprocket then acts as a pinion. (Other designs are better for gear fabrication.)

A More Stable Approach

Unfortunately, bead chains tend to cam out of deep sprocket recesses under high loads. In its first evolutionary step, the simple spherical bead grew limbs—two pins projecting at right angles to the cable axis (Fig. 4). The pulley or sprocket looks like a spur gear grooved to accommodate the belt; in fact, the pulley can mesh with a conventional spur gear of proper pitch.

Versions of the belt are also available with two sets of pins, one projecting vertically and the other horizontally. This arrangement permits the device to drive a series of perpendicular shafts without twisting the cable, like a bead chain but without the bead chain's load limitations. Reducing twist increases the transmission's lifetime and reliability.

These belt-cable-chain hybrids can be sized and connected in the field, using metal crimp-collars. However, nonfactory splices generally reduce the cable's tensile strength by half.

Parallel-Cable Drives

Another species of positive-drive belt uses parallel cables, sacrificing some flexibility for improved stability and greater strength. Here, the cables are connected by rungs molded into the plastic coating, giving the appearance of a ladder (Fig. 6). This "ladder chain" also meshes with toothed pulleys, which need not be grooved.

A cable-and-plastic ladder chain is the basis for the differential drive system in a Hewlett-Packard impact printer (Fig. 5). When the motors rotate in the same direction at the same speed, the carriage moves to the right or left. When they rotate in opposite directions, but at the same speed, the carriage remains stationary and the print-disk rotates. A differential motion of the motors produces a combined translation and rotation of the print-disk.

Fig. 5 A plastic-and-cable ladder chain in an impact-printer drive. In extreme conditions, such hybrids can serve many times longer than steel.

Fig. 6 **A gear chain** can function as a ladder chain, as a wide V-belt, or, as here, a gear surrogate meshing with a standard pinion.

The hybrid ladder chain is also well suited to laboratory of large spur gears from metal plates or pulleys (Fig. 6). Such a "gear" can run quietly in mesh with a pulley or a standard gear pinion of the proper pitch.

Another type of parallel-cable "chain," which mimics the standard chain, weighs just 1.2 oz/ft, requires no lubrication, and runs almost silently.

A Traditional Note

A new high-capacity tooth profile has been tested on conventional cogged belts. It has a standard cord and elastic body construction, but instead of the usual trapezoid, it has curved teeth (Fig. 7). Both 3-mm and 5-mm pitch versions have been introduced.

Fig. 7 **Curved high-torque tooth profiles** (just introduced in 3-mm and 5-mm pitches) increase load capacity of fine-pitch neoprene belts.

CHANGE CENTER DISTANCE WITHOUT ALTERING SPEED RATIO

Increasing the gap between the roller and knife changes chain lengths from F to E. Because the idler moves with the roller sprocket, length G changes to H. The changes in chain length are similar in value but opposite in direction. Chain lengths E minus F closely approximate G minus H. Variations in required chain length occur because the chains do not run parallel. Sprocket offset is required to avoid interference. Slack produced is too minute to affect the drive because it is proportional to changes in the cosine of a small angle ($2°$ to $5°$). For the 72-in. chain, variation is 0.020 in.

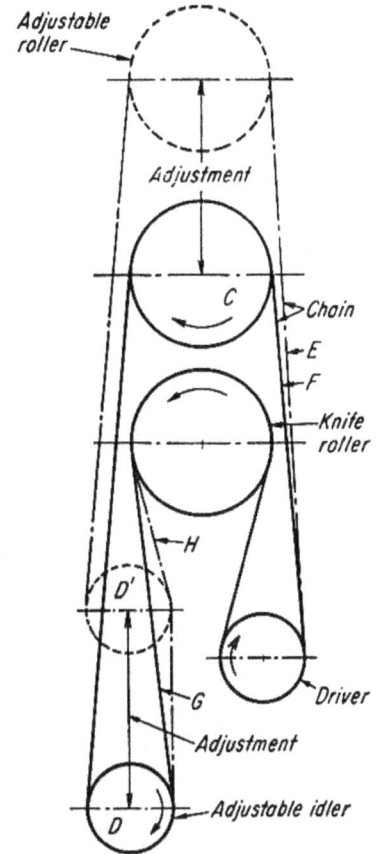

MOTOR MOUNT PIVOTS TO CONTROL BELT TENSION

Belt tensioning proportional to load.

When the agitation cycle is completed, the motor is momentarily idle with the right roller bottomed in the right-hand slot. When spin-dry starts, (A) the starting torque produces a reaction at the stator, pivoting the motor on the bottomed roller. The motor pivots until the opposite roller bottoms in the left-hand slot. The motor now swings out until restrained by the V-belt, which drives the pump and basket.

The motor, momentarily at zero rpm, develops maximum torque and begins to accelerate the load of basket, water, and wash. The motor pivots (B) about the left roller increasing belt tension in proportion to the output torque. When the basket reaches maximum speed, the load is reduced and belt tension relaxes. The agitation cycle produces an identical reaction in the reverse direction.

TEN ROLLER CHAINS AND THEIR ADAPTATIONS

Various roller, side-plate and pin configurations for power transmissions, conveying, and elevating.

STANDARD ROLLER CHAIN—FOR POWER TRANSMISSION AND CONVEYING

SINGLE WIDTH—Sizes ⁵/₈ in. and smaller have a spring-clip connecting link; those ³/₄ in. and larger have a cotter pin.

MULTIPLE WIDTH—Similar to single-width chain. It is made in widths of up to 12 strands.

EXTENDED PITCH CHAIN—FOR CONVEYING

STANDARD ROLLER DIAMETER—made with 1 to 4 in. pitch and cotter-pin-type connecting links.

OVERSIZED ROLLER DIAMETER—Same base chain as standard roller type but not made in multiple widths.

HOLLOW PIN—Made with 1¹/₄ to 15 in. pitch. It is adaptable to a variety of bolted attachments.

OFFSET LINK—Used when length requires an odd number of pitches and to shorten and lengthen a chain by one pitch.

STANDARD PITCH ADAPTATIONS

STRAIGHT LUG—Lugs on one or both sides can be spaced as desired. A standard roller is shown.

BENT LUG—Similar to straight-lug type for adaptations. A standard roller is shown.

EXTENDED PITCH ADAPTATIONS

STRAIGHT LUG—An oversized diameter roller is shown.

BENT LUG—An oversized diameter roller is shown.

HOLLOW PIN

STRAIGHT LUG—Lugs are detachable for field adaptation.

BENT LUG—Similar to straight lug type for adaptations.

EXTENDED PIN CHAINS

STANDARD PITCH—Pins can be extended on either side.

EXTENDED PITCH—Similar to standard for adaptations.

HOLLOW PIN—Pins are designed for field adaptation.

CROSS ROD—The rod can be removed from the hollow pins.

SPECIAL ADAPTATIONS

Used for holding conveyed objects.

Used to keep conveyed object on the center-line of the chain.

Used when flexing is desired in one direction only.

Used for supporting concentrated loads.

TWELVE APPLICATIONS FOR ROLLER CHAIN

This low-cost industrial chain can be applied in a variety of ways to perform tasks other than simply transmitting power.

Fig. 1 This low-cost rack-and-pinion device is easily assembled from standard parts.

Rigid support

Standard angle brackets fastened to support

Solder feed-roll

Feed tube

Part with noncircular periphery clamped to plate (floats free on drive sprockets).

Torch

Drive sprockets

Fig. 2 An extension of the rack-and-pinion principle—This is a soldering fixture for noncircular shells. Positive-action cams can be similarly designed. Standard angle brackets attach the chain to a cam or fixture plate.

Fig. 3 This control-cable direction-changer is extensively used in aircraft.

Drive sprocket

180° Link

Cable

Fig. 4 **The transmission of tipping or rocking motion** can be combined with the previous example (Fig. 3) to transmit this kind of motion to a remote location and around obstructions. The tipping angle should not exceed 40°.

Fig. 5 **This lifting device** is simplified by roller chain.

Fig. 6 **Two examples of indexing and feeding applications** of roller chain are shown here. This setup feeds plywood strips into a machine. The advantages of roller chain as used here are its flexibility and long feed.

Fig. 7 **Simple governor weights** can be attached by means of standard brackets to increase response force when rotation speed is slow.

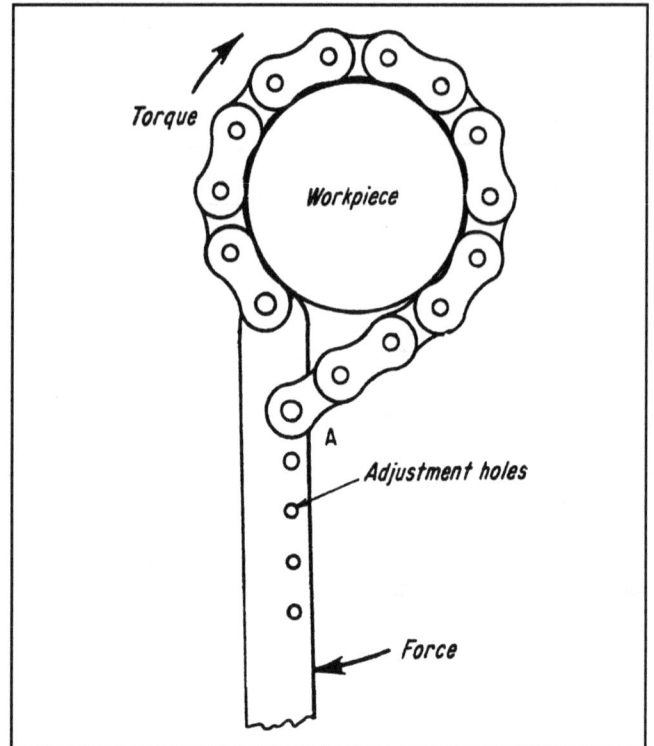

Fig. 8 **Wrench pivot A** can be adjusted to grip a variety of regularly or irregularly shaped objects.

Fig. 9 **Small parts can be conveyed,** fed, or oriented between spaces of roller chain.

Fig. 10 Clamp toggle action is supplied by two chains, thus clearing pin at fulcrum.

Fig. 11 Light-duty trolley conveyors can be made by combining standard roller-chain components with standard curtain-track components. Small gearmotors are used to drive the conveyor.

Fig. 12 Slatted belt, made by attaching wood, plastic, or metal slats, can serve as adjustable safety guard, conveyor belt, fast-acting security-wicket window.

SIX MECHANISMS FOR REDUCING PULSATIONS IN CHAIN DRIVES

Pulsations in chain motion created by the chordal action of chain and sprockets can be minimized or avoided by introducing a compensating cyclic motion in the driving sprocket. Mechanisms for reducing fluctuating dynamic loads in chain drives and the pulsations resulting from them include noncircular gears, eccentric gears, and cam-activated intermediate shafts.

Fig. 1

Fig. 2

Fig. 3

Fig. 4

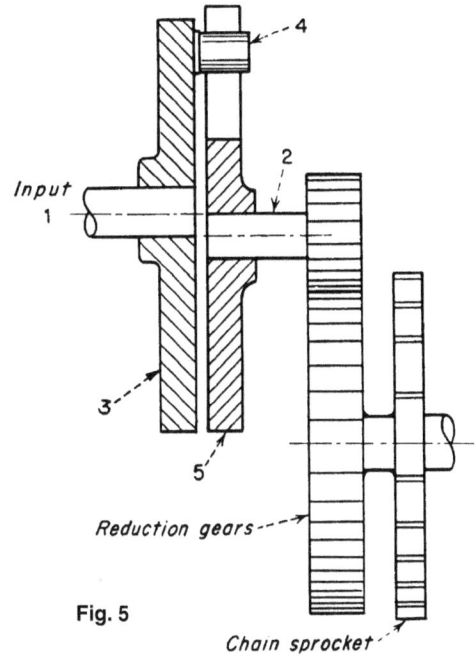

Fig. 5

Reduction gears

Chain sprocket

Fig. 1 The large cast-tooth, noncircular gear, mounted on the chain sprocket shaft, has a wavy outline in which the number of waves equals the number of teeth on a sprocket. The pinion has a corresponding noncircular shape. Although requiring special-shaped gears, the drive completely equalizes the chain pulsations.

Fig. 2 This drive has two eccentrically mounted spur pinions (1 and 2). Input power is through the belt pulley keyed to the same shaft as pinion 1. Pinion 3 (not shown), keyed to the shaft of pinion 2, drives the large gear and sprocket. However, the mechanism does not completely equalize chain velocity unless the pitch lines of pinions 1 and 2 are noncircular instead of eccentric.

Fig. 3 An additional sprocket 2 drives the noncircular sprocket 3 through a fine-pitch chain 1. This imparts pulsating velocity to shaft 6 and to the long-pitch conveyor sprocket 5 through pinion 7 and gear 4. The ratio of the gear pair is made the same as the number of teeth of sprocket 5. Spring-actuated lever and rollers 8 take up the slack. Conveyor motion is equalized, but the mechanism has limited power capacity because the pitch of chain 1 must be kept small. Capacity can be increased by using multiple strands of fine-pitch chain.

Fig. 4 Power is transmitted from shaft 2 to sprocket 6 through chain 4, thus imparting a variable velocity to shaft 3, and through it, to the conveyor sprocket 7. Because chain 4 has a small pitch and sprocket 5 is relatively large, the velocity of 4 is almost constant. This induces an almost constant conveyor velocity. The mechanism requires the rollers to tighten the slack side of the chain, and it has limited power capacity.

Fig. 5 Variable motion to the sprocket is produced by disk 3. It supports pin and roller 4, as well as disk 5, which has a radial slot and is eccentrically mounted on shaft 2. The ratio of rpm of shaft 2 to the sprocket equals the number of teeth in the sprocket. Chain velocity is not completely equalized.

Fig. 6

Fig. 6 The integrated "planetary gear" system (gears 4, 5, 6 and 7) is activated by cam 10, and it transmits a variable velocity to the sprocket synchronized with chain pulsations through shaft 2, thus completely equalizing chain velocity. Cam 10 rides on a circular idler roller 11. Because of the equilibrium of the forces, the cam maintains positive contact with the roller. The unit has standard gears, acts simultaneously as a speed reducer, and can transmit high horsepower.

CHAPTER 11
SPRING AND
SCREW DEVICES
AND MECHANISMS

FLAT SPRINGS IN MECHANISMS

Constant force is approached because of the length of this U-spring. Don't align the studs or the spring will fall.

A flat-wire sprag is straight until the knob is assembled: thus tension helps the sprag to grip for one-way clutching.

Knob turns shaft

Knob slips

Spring is normally straight

Grip springs have preloaded tension

Slide   Anchor bar   Handle

Easy positioning of the slide is possible when the handle pins move a grip spring out of contact with the anchor bar.

Leaf spring

Slide

A spring-loaded slide will always return to its original position unless it is pushed until the spring kicks out.

Nearly constant tension in the spring, as well as the force to activate the slide, is provided by this single coil.

Increasing support area as the load increases on both upper and lower platens is provided by a circular spring.

Upper platen

Lower platen

This volute spring lets the shaft be moved closer to the frame, thus allowing maximum axial movement.

Frame

A return-spring ensures that the operation handle of this two-direction drive will always return to its neutral position.

This cushioning device imparts rapid increase of spring tension because of the small pyramid angle. Its rebound is minimum.

This spring-mounted disk changes its center position as the handle is rotated to move the friction drive. It also acts as a built-in limit stop.

This hold-down clamp has its flat spring assembled with an initial twist to provide a clamping force for thin material.

Indexing is accomplished simply, efficiently, and at low cost by flat-spring arrangement shown here.

TWELVE WAYS TO USE METAL SPRINGS

Variable-rate arrangements, roller positioning, space saving,
and other ingenious ways to get the most from springs.

This setup provides a **variable rate** with a sudden
change from a light load to a heavy load by limiting the
low-rate extension with a spring.

Different-rate springs

This mechanism provides a **three-step rate** change at predetermined positions. The lighter springs will always compress first, regardless of their position.

Lighter spring

Heavier spring

This differential-rate linkage sets the actuator stroke
under light tension at the start, then allows a gradual
transition to heavier tension.

*Support bracket is attached to
a slide, which operates platen*

This compressing mechanism has a dual rate for double-
action compacting. In one direction pressure is high, but in
the reverse direction pressure is low.

Roller — *Spring grips shaft*

Roller positioning by a tightly wound
spring on the shaft is provided by this
assembly. The roller will slide under
excess end thrust.

Wheel-center movement *Slide* *Slide movement*

Spring movement

A short extension of the spring for a long
movement of the slide keeps the tension
change between maximum and minimum low.

Increased tension for the same movement is gained by providing a movable spring mount and gearing it to the other movable lever.

Tube reciprocates in operation

Spring grips pin

Guided wire

This pin grip is a spring that holds a pin by friction against end movement or rotation, but lets the pin be repositioned without tools.

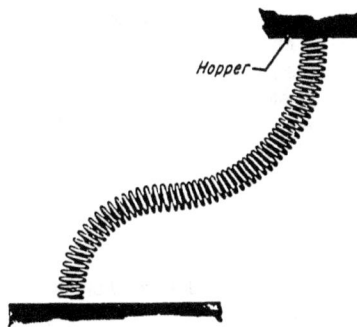

Hopper

A close-wound spring is attached to a hopper, and it will not buckle when it is used a movable feed-duct for nongranular material.

Lever in operating position

Lever in neutral position

Spring pressure holds lever in neutral position

Fixed plate

Toggle action here ensures that the gear-shift lever will not inadvertently be thrown past its neutral position.

Tilting lever

Other end of spring attached to brake lever

Brake-applying lever

Lever tilts

Cam surfaces

Brake-applying shaft

Tension varies at a different rate when the brake-applying lever reaches the position shown. The rate is reduced when the tilting lever tilts.

The spring wheel helps to distribute deflection over more coils than if the spring rested on the corner. The result is less fatigue and longer life.

SEVEN OVERRIDING SPRING MECHANISMS FOR LOW-TORQUE DRIVES

Overriding spring mechanisms are widely used in the design of instruments and controls. All of the arrangements illustrated allow an incoming motion to override the outgoing motion whose limit has been reached. In an instrument, for example, the spring mechanism can be placed between the sensing and indicating elements to provide overrange protection. The dial pointer is driven positively up to its limit before it stops while the input shaft is free to continue its travel. Six of the mechanisms described here are for rotary motion of varying amounts. The last is for small linear movements.

Fig. 1 Unidirectional override. The take-off lever of this mechanism can rotate nearly 360°. Its movement is limited only by one stop pin. In one direction, motion of the driving shaft is also impeded by the stop pin. But in the reverse direction the driving shaft is capable of rotating approximately 270° past the stop pin. In operation, as the driving shaft is turned clockwise, motion is transmitted through the bracket to the take-off lever. The spring holds the bracket against the drive pin. When the take-off lever has traveled the desired limit, it strikes the adjustable stop pin. However, the drive pin can continue its rotation by moving the bracket away from the drive pin and winding up the spring. An overriding mechanism is essential in instruments employing powerful driving elements, such as bimetallic elements, to prevent damage in the overrange regions.

Fig. 2 Two-directional override. This mechanism is similar to that described under Fig. 1, except that two stop pins limit the travel of the take-off lever. Also, the incoming motion can override the outgoing motion in either direction. With this device, only a small part of the total rotation of the driving shaft need be transmitted to the take-off lever, and this small part can be anywhere in the range. The motion of the driving shaft is transmitted through the lower bracket to the lower drive pin, which is held against the bracket by the spring. In turn, the lower drive pin transfers the motion through the upper bracket to the upper drive pin. A second spring holds this pin against the upper drive bracket. Because the upper drive pin is attached to the take-off lever, any rotation of the drive shaft is transmitted to the lever, provided it is not against either stop A or B. When the driving shaft turns in a counterclockwise direction, the take-off lever finally strikes against the adjustable stop A. The upper bracket then moves away from the upper drive pin, and the upper spring starts to wind up. When the driving shaft is rotated in a clock-wise direction, the take-off lever hits adjustable stop B, and the lower bracket moves away from the lower drive pin, winding up the other spring. Although the principal applications for overriding spring arrangements are in instrumentation, it is feasible to apply these devices in the drives of heavy-duty machines by strengthening the springs and other load-bearing members.

Fig. 3 Two-directional, limited-travel override. This mechanism performs the same function as that shown in Fig. 2, except that the maximum override in either direction is limited to about 40°. By contrast, the unit shown in Fig. 2 is capable of 270° movement. This device is suited for applications where most of the incoming motion is to be used, and only a small amount of travel past the stops in either direction is required. As the arbor is rotated, the motion is transmitted through the arbor lever to the bracket. The arbor lever and the bracket are held in contact by spring B. The motion of the bracket is then transmitted to the take-off lever in a similar manner, with spring A holding the take-off lever until the lever engages either stops A or B. When the arbor is rotated in a counterclockwise direction, the take-off lever eventually comes up against the stop B. If the arbor lever continues to drive the bracket, spring A will be put in tension.

Fig. 4 Unidirectional, 90° override. This is a single overriding unit that allows a maximum travel of 90° past its stop. The unit, as shown, is arranged for overtravel in a clockwise direction, but it can also be made for a counterclockwise override. The arbor lever, which is secured to the arbor, transmits the rotation of the arbor to the take-off lever. The spring holds the drive pin against the arbor lever until the take-off lever hits the adjustable stop. Then, if the arbor lever continues to rotate, the spring will be placed in tension. In the counterclockwise direction, the drive pin is in direct contact with the arbor lever so that no overriding is possible.

Fig. 5 Two-directional, 90° override. This double-overriding mechanism allows a maximum overtravel of 90° in either direction. As the arbor turns, the motion is carried from the bracket to the arbor lever, then to the take-off lever. Both the bracket and the take-off lever are held against the arbor lever by spring A and B. When the arbor is rotated counterclockwise, the takeoff lever hits stop A. The arbor lever is held stationary in contact with the take-off lever. The bracket, which is fastened to the arbor, rotates away from the arbor lever, putting spring A in tension. When the arbor is rotated in a clockwise direction, the take-off lever comes against stop B, and the bracket picks up the arbor lever, putting spring B in tension.

Fig. 6 Unidirectional, 90° override. This mechanism operates exactly the same as that shown in Fig. 4. However, it is equipped with a flat spiral spring in place of the helical coil spring used in the previous version. The advantage of the flat spiral spring is that it allows for a greater override and minimizes the space required. The spring holds the take-off lever in contact with the arbor lever. When the take-off lever comes in contact with the stop, the arbor lever can continue to rotate and the arbor winds up the spring.

Fig. 7 Two-directional override, linear motion. The previous mechanisms were overrides for rotary motion. The device in Fig. 7 is primarily a double override for small linear travel, although it could be used on rotary motion. When a force is applied to the input lever, which pivots about point C, the motion is transmitted directly to the take-off lever through the two pivot posts, A and B. The take-off lever is held against these posts by the spring. When the travel causes the take-off lever to hit the adjustable stop A, the take-off lever revolves about pivot post A, pulling away from pivot post B and putting additional tension in the spring. When the force is diminished, the input lever moves in the opposite direction until the take-off lever contacts the stop B. This causes the take-off lever to rotate about pivot post B, and pivot post A is moved away from the take-off lever.

SIX SPRING MOTORS AND ASSOCIATED MECHANISMS

Many applications of spring motors in clocks, motion picture cameras, game machines, and other mechanisms offer practical ideas for adaptation to any mechanism that is intended to operate for an appreciable length of time. While spring motors are usually limited to comparatively small power application where other sources of power are unavailable or impracticable, they might also be useful for intermittent operation requiring comparatively high torque or high speed, using a low-power electric motor or other means for building up energy.

Fig. 1

Gear shift lever

Gear slides in and out of mesh

Drive shaft

Ratchet

Winding shaft

Idler gear — Drive pinion — Internal drive gear

Spring ends

Transmission gearing to turntable shaft

Winding plate

Cam roller allows rotation in one direction only

Governor drive gear

Cam for governor spring adjustment

Winding pinion

Governor adjusted by geared shafting to thumb nut beside turntable

Fig. 2

Pivot

Stop

Weight

Governor

Turntable shaft

Governor adjustment

Turntable shaft

Supporting shafts may be varied in length for different size springs or multiple units

Supporting shaft

Rear frame

Pinion crank arm for winding is stationary during operation. Outer end of spring unwinds 5 turns while rewinding inner end 4 turns

Supporting shaft

Differential double pinion makes 5 to 1 winding and drive ratio

Winding shaft

Drive gear

Governor

Winding ratchet

Pinion crank arm

Section A-A

Fig. 3

The accompanying patented spring motor designs show various methods for the transmission and control of spring-motor power. Flat-coil springs, confined in drums, are most widely used because they are compact, produce torque directly, and permit long angular displacement. Gear trains and feedback mechanisms reduce excess power drain so that power can be applied for a longer time. Governors are commonly used to regulate speed.

Fig. 4

Razor

High-speed shaft

Governor

Series of internal gears and integral pinions operated through idlers

Spring motor

Motion Picture Camera

Gear train

Friction disk

Friction member

Varying tension in power spring adjusts position of friction member on friction disk to vary braking power on governor in accordance with spring motor power

Winding key

Friction plate moved by governor against friction member

Friction member

Fig. 5

Coil spring units

Drive shaft

Pawl

Ratchet

Winding shaft

Drive shaft

Winding shaft

Frame

Winding gear

Governor

Casing rotates around stationary pinion

Drive gear

Fig. 6

287

TWELVE AIR SPRING APPLICATIONS

SIX WAYS TO ACTUATE MECHANISMS WITH AIR SPRINGS

Linear force link: A one- or two-convolution air spring drives the guide rod. The rod is returned by gravity, opposing force, metal spring or, at times, internal stiffness of an air spring.

Rotary force link: A pivoted plate can be driven by a one-convolution or two-convolution spring to 30° of rotation. The limitation on the angle is based on permissible spring misalignment.

Clamp: A jaw is normally held open by a metal spring. Actuation of the air spring then closes the clamp. The amount of opening in the jaws of the clamp can be up to 30° of arc.

Direct-acting press: One-, two-, or three-convolution air springs are assembled singly or in gangs. They are naturally stable when used in groups. Gravity returns the platform to its starting position.

Rotary shaft actuator: The activator shifts the shaft longitudinally while the shaft is rotating. Air springs with one, two, or three convolutions can be used. A standard rotating air fitting is required.

Reciprocating linear force link: It reciprocates with one-, two-, or three-convolution air springs in a back-to-back arrangement. Two- and three-convolution springs might need guides for their force rods.

POPULAR TYPES OF AIR SPRINGS

Air is an ideal load-carrying medium. It is highly elastic, its spring rate can be easily varied, and it is not subject to permanent set.

Air springs are elastic devices that employ compressed air as the spring element. They maintain a soft ride and a constant vehicle height under varying load. In industrial applications they control vibration (isolate or amplify it) and actuate linkages to provide either rotary or linear movement. Three kinds of air springs (bellows, rolling sleeve, and rolling diaphragm) are illustrated.

Pipe threads — *Mounting holes or threaded studs* — *Bead plate with crimped edge* — *Flexible spring*

ONE-CONVOLUTION BELLOW

Bellows Type

A single-convolution spring looks like a tire lying on its side. It has a limited stroke and a relatively high spring rate. Its natural frequency is about 150 cpm without auxiliary volume for most sizes, and as high as 240 cpm for the smallest size. Lateral stiffness is high (about half the vertical rate); therefore the spring is quite stable laterally when used for industrial vibration isolation. It can be filled manually or kept inflated to a constant

Pivot mechanism: It rotates a rod through 145° of rotation. It can accept a 30° mis-alignment because of the circular path of its connecting-link pin. A metal spring or opposing force retracts the link.

Reciprocating rotary motion with one-convolution and two-convolution springs. An arc up to 30° is possible. It can pair a large air spring with a smaller one or a lengthen lever.

(a)

Air line (from compressor)
Valve-control lever
Leveling valve
Air springs

(b)

(c)

Air suspension on vehicle: A view of normal static conditions—air springs at desired height and height-control valve closed (a). When a load is added to the vehicle—the valve opens to admit air to the springs and restore height, but at higher pressure (b). With load removed from the vehicle—valve permits bleeding off excess air pressure to atmosphere and restores its design height (c).

height if is connected to factory air supply through a pressure regulator. This spring will also actuate linkages where short axial length is desirable. It is seldom used in vehicle suspension systems.

Rolling-Sleeve Type

This spring is sometimes called the reversible-sleeve or rolling-lobe type. It has a telescoping action—the lobe at the bottom of the air spring rolls up and down along the piston. The spring is used primarily in vehicle suspensions because lateral stiffness is almost zero.

Rolling-Diaphragm Type

These are laterally stable and can be used as vibration isolators, actuators, or constant-force spring. But because of their

negative effective-area curve, their pressure is not generally maintained by pressure regulators.

Reservoir volume

ROLLING-SLEEVE

ROLLING-DIAPHRAGM

NOVEL APPLICATIONS FOR DIFFERENT SPRINGS

How stops, cams, linkages, and other arrangements can vary the load/deflection ratio during extension or compression.

With tapered-pitch spring the number of effective coils changes with deflection—the coils "bottom" progressively.

A tapered outside diameter and pitch combine to produce a similar effect except that the spring with tapered O.D. will have a shorter solid height.

In dual springs, one spring closes completely before the other.

A cam-and-spring device causes the torque relationship to vary during rotation as the moment arm changes.

Moment arm

Stops can be used with either compression or extension springs.

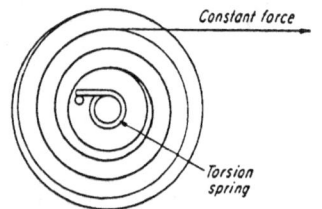

Constant force

Torsion spring

Torsion spring combined with a variable-radius pulley gives a constant force.

Leaf springs can be arranged so that their effective lengths change with deflection.

Adjustment screws

A molded-rubber spring has deflection characteristics that vary with its shape.

A four-bar mechanism in conjunction with a spring has a wide variety of load/deflection characteristics.

These linkage-type arrangements are used in instruments where torque control or anti-vibration suspension is required.

With a tapered mandrel and torsion spring the effective number or coils decreases with torsional deflection.

An arched leaf-spring gives an almost constant force when it is shaped like the one illustrated.

APPLICATIONS FOR BELLEVILLE SPRINGS

Belleville springs are low-profile conical rings with differing height (h) to thickness (t) ratios, as shown in Fig. 1. Four way to stack them are shown in Fig. 2.

Belleville springs lend themselves to a wide variety of applications:

For height to spring ratios of about 0.4—A linear spring rate and high load resistance with small deflections.

For height to spring ratios between 0.8 and 1.0—An almost linear spring rate for fasteners and bearing and in stacks.

For rations of around 1.6—A constant (flat) spring rate starting at about 60 percent of the deflection (relative to the fully compressed flat position) and proceeding to the flat position and, if desired, on to the flipped side to a deflection of about 140 percent. In most applications, the flat position is the limit of travel, and for deflections beyond the flat, the contact elements must be allowed unrestricted travel.

One application of bellevilles with constant spring rate is on live spindles on the tailpiece of a lathe. The work can be loaded on the lathe, and as the piece heats up and begins to expand, the belleville will absorb this change in length without adding any appreciable load.

For high height to spring ratios exceeding about 2.5—The spring is stiff, and as the stability point (high point on the curve) is passed, the spring rate becomes negative causing resistance to drop rapidly. If allowed, the belleville will snap through the flat position. In other words, it will turn itself inside out.

Working in groups. Belleville washers stacked in the parallel arrangement have been used successfully in a variety of applications.

One is a pistol or rifle buffer mechanism (Fig. 3) designed to absorb repeated, high-energy shock loads. A preload nut predeflects the washers to stiffen their resistance. The stacked washers are guided by a central shaft, an outside guide cylinder, guide rings, or a combination of these.

A windup starter mechanism for diesel engines (shown in Fig. 4) replaces a heavy-duty electric starter or auxiliary gas engine. To turn over the engine, energy is manually stored in a stack of bellevilles compressed by a hand crank. When released, the expanding spring pack rotates a pinion meshed with the flywheel ring gear to start the engine.

Figure 5 shows a belleville as a loading spring for a clutch.

Popular arrangements

Fig. 1 Basic belleville

Normal load

Flat load

Fig. 2 Four ways to stack

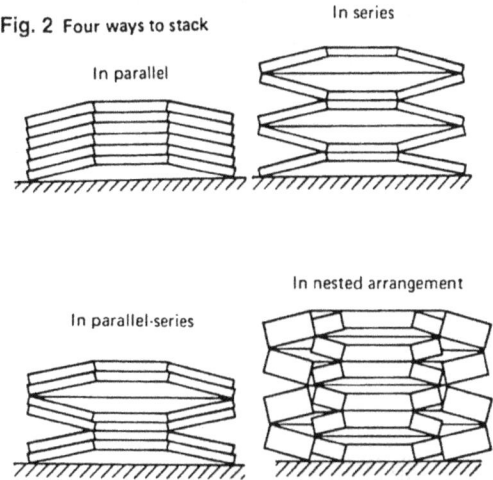

In parallel

In series

In parallel-series

In nested arrangement

Impact surface

Winding shaft

Fig. 4 Windup starter for diesel engines

Trip handle

Pawl

Pinion

Ratchet

Ball nut

Sleeve   Main shaft

Belleville springs

Preload nut

Fig. 3
Buffer mechanism for high-impact energy

Detail of slotted spring

Slotted spring

Driving plate

Release ring

Cover

Friction face

Fig. 5
Loading spring for clutch

VIBRATION CONTROL WITH SPRING LINKAGE

Do you need a buffer between vibrating machinery and the surrounding structure? These isolators, like capable fighters, absorb the light jabs and stand firm against the forces that inflict powerful haymakers

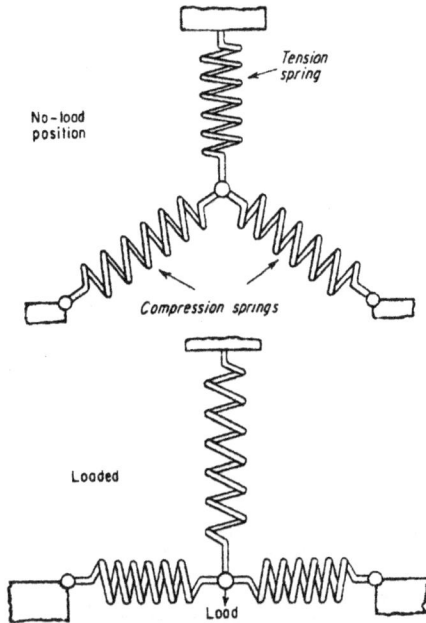

Fig. 1 **This basic spring arrangement** has zero stiffness, and is as "soft as a cloud" when compression springs are in line, as illustrated in the loaded position. But change the weight or compression-spring alignment, and stiffness increases greatly. This support is adequate for vibration isolation because zero stiffness give a greater range or movement than the vibration amplitude generally in the hundredths-of-an-inch range.

Arrangements shown here are highly absorbent when required, yet provide a firm support when large force changes occur. By contrast, isolators that depend upon very "soft" springs, such as the sine spring, are unsatisfactory in many applications; they allow a large movement of the supported load with any slight weight change or large-amplitude displacing force.

Fig. 2 **Alternative arrangements** illustrate adaptability of basic design. Here, instead of the inclined, helical compression springs, wither tension or cantilever springs can serve. Similarly, different types of springs can replace the axial, tension spring. Zero torsional stiffness can also be provided.

Various applications of the principle of vibration isolation show how versatile the design is. Coil spring (Fig. 4) as well as cantilever and torsion-bar suspension of automobiles can all be reduced in stiffness by adding an inclined spring; stiffness of the tractor seat (Fig. 5) and, consequently, transmitted shocks can be similarly reduced. Mechanical tension meter (Fig. 6) provides a sensitive indication of small variations in tension. A weighing scale, for example, could detect small variations in nominally identical objects. A nonlinear torque mete (Fig. 7) provides a sensitive indication of torque variations about a predetermined level.

Fig. 3 **A general-purpose support** is based on basic spring arrangement, except that an axial compression spring is substituted for a tension spring. Inclined compression springs, spaced around a central pillar, carry the component to be isolated. When a load is applied, adjustment might be necessary to bring the inclined springs to zero inclination. Load range that can be supported with zero stiffness on a specific support is determined by the adjustment range and physical limitations of the axial spring.

Fig. 4

Fig. 5

Fig. 6

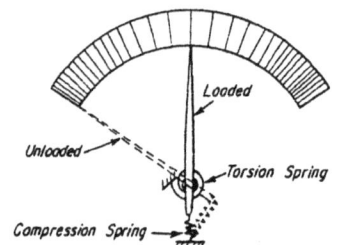

Fig. 7

TWENTY SCREW DEVICES

A threaded shaft and a nut plus some way to make one of these members rotate without translating and the other to translate without rotating are about all you need to do practically all of the adjusting, setting, or locking in a machine design.

Most of these applications have low-precision requirements. That's why the thread might be a coiled wire or a twisted strip; the nut might be a notched ear on a shaft or a slotted disk. Standard screws and nuts from hardware store shelves can often serve at very low cost.

Here are the basic motion transformations possible with screw threads (Fig. 1):

- Transform rotation into linear motion or reverse (A),
- Transform helical motion into linear motion or reverse (B),
- Transform rotation into helical motion or reverse (C).

Of course the screw thread can be combined with other components: in a four-bar linkage (Fig. 2), or with multiple screw elements for force or motion amplification.

Fig. 1 **Motion transformations** of a screw thread include: rotation to translation (A), helical to translation (B), rotation to helical (C). These are reversible if the thread is not self-locking. (The thread is reversible when its efficiency is over 50 percent.)

Fig. 2 **Standard four-bar linkage** has a screw thread substituted for a slider. The output is helical rather than linear.

ROTATION TO TRANSLATION

Fig. 3 **A two-directional lamp adjustment** with screwdriver will move a lamp up and down. A knob adjust (right) rotates the lamp about a pivot.

Fig. 4 **A knife-edge bearing** is raised or lowered by a screw-driven wedge. Two additional screws position the knife edge laterally and lock it.

Fig. 5 **A parallel arrangement** of tandem screw threads raises the projector evenly.

Fig. 6 **Automatic clockwork** is kept wound taut by an electric motor turned on and off by a screw thread and nut. The motor drive must be self-locking or it will permit the clock to unwind as soon as the switch is turned off.

Fig. 7 **A valve stem** has two oppositely moving valve cones. When opening, the upper cone moves up first, until it contacts its stop. Further turning of the valve wheel forces the lower cone out of its seat. The spring is wound up at the same time. When the ratchet is released, the spring pulls both cones into their seats.

Fig. 8 A metal strip or square rod can be twisted to make a long-lead thread. It is ideal for transforming linear into rotary motion. Here a pushbutton mechanism winds a camera. The number of turns or dwell of the output gear is easily altered by changing (or even reversing) the twist of the strip.

Fig. 9 A feeler gage has its motion amplified through a double linkage and then transformed to rotation for moving a dial needle.

Feeler

Fig. 10 The familiar flying propeller-toy is operated by pushing the bushing straight up and off the thread.

Bushing

SELF-LOCKING

Fig. 11 A hairline adjustment for a telescope with two alternative methods for drive and spring return.

Fig. 12 This screw and nut form a self-locking drive for a complex linkage.

(B)

(A)

(C)

Fig. 13 Force translation. The threaded handle in (A) drives a coned bushing that thrusts rods outwardly for balanced pressure. The screw in (B) retains and drives a dowel pin for locking purposes. A right- and left-handed shaft (C) actuates a press.

Fig. 14   Double-threaded screws, when used as differentials, permit very fine adjustment for precision equipment at relatively low cost.

Fig. 15   Differential screws can be made in dozens of forms. Here are two methods: in the upper figure, two opposite-hand threads on a single shaft; in the lower figure, same-hand threads on independent shafts.

Fig. 16   Opposite-hand threads make a high-speed centering clamp out of two moving nuts.

Fig. 17   A measuring table rises very slowly for many turns of the input bevel gear. If the two threads are $1\frac{1}{2}$ to 12 and $\frac{3}{4}$ to 16, in the fine-thread series, the table will rise approximately 0.004 in. per input-gear revolution.

Fig. 18   A lathe turning tool in a drill rod is adjusted by a differential screw. A special double-pin wrench turns the intermediate nut, advancing the nut and retracting the threaded tool simultaneously. The tool is then clamped by a setscrew.

Synchronous motor drive   *Follower-motor drive*

Slide adjusts follower-motor speed

Fig. 19   Any variable-speed motor can be made to follow a small synchronous motor by connecting them to the two shafts of this differential screw. Differences in the number of revolutions between the two motors appear as motion of the traveling nut and slide, thus providing electrical speed compensation.

Fig. 20   A wire fork is the nut in this simple tube-and-screw device.

Two variants of nut

Fig. 21   A mechanical pencil includes a spring as the screw thread and a notched ear or a bent wire as the nut.

TEN APPLICATIONS FOR SCREW MECHANISMS

Three basic components of screw mechanisms are: actuating member (knob, wheel, handle), threaded device (screw-nut set), and sliding device (plunger-guide set).

A nut can rotate but will not move longitudinally. Typical applications: screw jacks, heavy vertically moved doors; floodgates, opera-glass focusing, vernier gages, and Stillson wrenches.

A differential movement is given by threads of different pitch. When the screw is rotated, the nuts move in the same direction but at different speeds.

One screw actuates three gears simultaneously. The axes of gears are at right angles to that of the screw. This mechanism can replace more expensive gear setups where speed reduction and multiple output from a single input is required.

A screw can rotate but only the nut moves longitudinally. Typical applications: lathe tailstock feed, vises, lathe apron.

A screw and plunger are attached to a knob. The nut and guide are stationary. It is used on screw presses, lathe steady-rest jaws for adjustment, and shaper slide regulation.

(A)

(B)

Screw-actuated wedges lock locating pin (A) and hold the work in fixture (B). These are just two of the many tool and die-making applications for these screw actions.

Opposing movement of lateral slides; adjusting members or other screw-actuated parts can be achieved with opposite-hand threads.

Concentric threading also gives differential movement. Such movements are useful wherever rotary mechanical action is required. A typical example is a gas-bottle valve, where slow opening is combined with easy control.

Locking nuts can be placed on opposite sides of a panel to prevent axial screw movement and simultaneously lock against vibrations. Drill-press depth stops and adjustable stops for shearing and cutoff dies are some examples.

(A)   (B)   (C)

Adjustment screws are effectively locked by either a pressure screw (A) or tension screw (B). If the adjusting screw is threaded into a formed sheet-metal component (C), a setscrew can be used to lock the adjustment.

SEVEN SPECIAL SCREW ARRANGEMENTS

Differential, duplex, and other types of screws can provide slow and fast feeds, minute adjustments, and strong clamping action.

Rapid and slow feed. With left- and right-hand threads, slide motion with the nut locked equals L_A plus L_B per turn; with the nut floating, slide motion per turn equals L_B. Extremely fine feed with a rapid return motion is obtained when the threads are differential.

Extremely small movements. Microscopic measurements, for example, are characteristic of this arrangement. Movement A is equal to $N(L_B \times L_t) 12\pi R$, where N equals the number of turns of screw C.

Shock absorbent screw. When the springs coiled as shown are used as worm drives for light loads, they have the advantage of being able to absorb heavy shocks.

Bearing adjustment. This screw arrangement is a handy way for providing bearing adjustment and overload protection.

Backlash elimination. The large screw is locked and all backlash is eliminated when the knurled screw is tightened; finger torque is sufficient.

Differential clamp. This method of using a differential screw to tighten clamp jaws combines rugged threads with high clamping power. Clamping pressure, $P = Te [R(\tan \phi + \tan \alpha)]$, where T = torque at handle, R = mean radius of screw threads, ϕ = angle of friction (approx. 0.1), α = mean pitch angle or screw, and e = efficiency of screw generally about 0.8).

High reduction of rotary motion to fine linear motion is possible here. This arrangement is for low forces. Screws are left- and right-hand. $L_A = L_B$ plus or minus a small increment. When L_B = 1/10 and L_A = 1/10.5, the linear motion of screw A will be 0.05 in. per turn. When screws are the same hand, linear motion equals $L_A + L_B$.

FOURTEEN SPRING AND SCREW ADJUSTING DEVICES

Here is a selection of some basic devices that provide and hold mechanical adjustment.

Fig. 1   A spring-loaded pin supplies a counterforce against which an adjustment force must always act. A leveling foot would work against gravity, but for most other setups a spring is needed to give a counter-force.

Fig. 2   Dual screws provide an inelastic counterforce. Backing-off one screw and tightening the other allows extremely small adjustments to be made. Also, once adjusted, the position remains solid against any forces tending to move the device out of adjustment.

Fig. 3   A differential screw has same-hand threads but with different pitches. The relative distance between the two components can be adjusted with high precision by differential screws.

Figs. 4 and 5   Swivel motion is necessary in (Fig. 4) between the adjusting screw and arm because of a circular locus of female threads in the actuated member. Similar action (Fig. 5) requires either the screw to be pivoted or the arm to be forked.

Fig. 6   This arc-drafting guide is an example of an adjusting device. One of its components, the flat spring, both supplies the counter-force and performs the mechanism's main function—guiding the pencil.

Fig. 7   The worm adjustment shown here is in a device for varying the position of an arm. Measuring instruments, and other tools requiring fine adjustments, include this adjusting device.

Fig. 8

Fig. 9

Figs. 8 and 9   Tierods with opposite-hand threads at their ends (Fig. 8) require only a similarly threaded nut to provide simple, axial adjustment. Flats on the rod ends (Fig. 9) make it unnecessary to restrain both the rods against rotation when the adjusting screw is turned; restraining one rod is enough.

Fig. 10   A split-leg caliper is an example of a simple but highly efficient adjusting device. A tapered screw forces the split leg part, thus enlarging the opening between the two legs.

Fig. 11

Fig. 12

Figs. 11 and 12   Shaft torque is adjusted (Fig. 11) by rotating the spring-holding collar relative to the shaft, and locking the collar at a position of desired torque. Adjusting slots (Fig. 12) accommodate the torsion-spring arm after the spring is wound to the desired torque.

Fig. 13

Fig. 14

Figs. 13 and 14.   Rack and toothed stops (Fig. 13) are frequently used to adjust heavy louvers, boiler doors and similar equipment. The adjustment is not continuous; it depends on the rack pitch. Large counter-adjustment forces might require a weighted rack to prevent tooth disengagement. Indexing holes (Fig. 14) provide a similar adjustment to the rack. The pin locks the members together.

A LONG-STROKE, HIGH-RESOLUTION LINEAR ACTUATOR

This partly schematic cross-section view of the Precision Linear Actuator Direct Drive (PLADD) shows its principal components.

A long-stroke, high-resolution linear actuator developed for coarse positioning in a satellite interferometer is said to have applications in medical imaging and semiconductor device fabrication. Developed by a team of Caltech engineers for NASA's Jet propulsion Laboratory, the *Precision Linear Actuator Direct Drive* or PLADD can perform positioning motions that include submicron increments or slews of about a centimeter. The PLADD designers also say it is capable of both long strokes (120 mm) and high resolution in repeatable increments of 20 nm. They report that it is capable of performing these motions repeatedly for 5 to 10 years. In contrast with other precise linear actuators, the PLADD contains no gears, levers, or hydraulic converters.

The PLADD, as shown in the section view, includes a commercial ball-screw actuator driven directly by a commercial three-phase brushless DC motor. The ball-screw actuator consists of a spring-preloaded ball nut on a ball screw that is restrained against rotation by design. The motor is directly coupled without an intervening gear train to a drive link that, in turn, is coupled to the ball nut. By eliminating the gear train, this direct-drive design eliminates the complexity, backlash, and potential for misalignment associated with a gear train.

A power-off brake that includes flexured levers compressed against the drive link by preload springs prevents inadvertent movement. In addition, there are also piezoelectric stacks that can be activated to oppose the springs and push the levers away from the drive link. This means that power must be applied to the piezoelectric stacks to release the drive link from braking.

To help ensure long operational life, all of the mechanical drive components are immersed in an oil bath within a hermetically sealed bellows. The outer end of the bellows holds the outer end of the ball screw, thereby preventing rotation of the ball screw.

PLADD positioning is controlled electronically by a system that includes both digital and analog subsystems. These subsystems interact with the motor and brake and with two encoders. One is a Hall-effect sensor and rotation encoder and the other is a linear glass-scale encoder. This system carries out a proportional integral derivative control algorithm that results in varying the voltage commands to each of the three pairs of windings of the brushless DC motor.

In one of two alternative control modes, voltages are applied to the windings in a trapezoidal commutation scheme based on timing signals from the Hall-effect sensor. This control mode permits the coarse positioning of 24 steps per motor revolution. The second control mode uses a sinusoidal commutation scheme in which the output of the linear glass-scale encoder is transposed to rotational increments that produce a finer position feedback of more than 400,000 steps per revolution.

This work was done by Brant T. Cook, Donald M. Moore, David F. Braun, John S. Koenig, and Steve M. Hankins of Caltech for NASA's Jet Propulsion Laboratory.

CHAPTER 12
SHAFT COUPLINGS
AND CONNECTIONS

FOUR COUPLINGS FOR PARALLEL SHAFTS

Fig. 1 One method of coupling shafts makes use of gears that can replace chains, pulleys, and friction drives. Its major limitation is the need for adequate center distance. However, an idler can be used for close centers, as shown. This can be a plain pinion or an internal gear. Transmission is at a constant velocity and there is axial freedom.

Fig. 2 This coupling consists of two universal joints and a short shaft. Velocity transmission is constant between the input and output shafts if the shafts remain parallel and if the end yokes are arranged symmetrically. The velocity of the central shaft fluctuates during rotation, but high speed and wide angles can cause vibration. The shaft offset can be varied, but axial freedom requires that one shaft be spline mounted.

Fig. 3 This crossed-axis yoke coupling is a variation of the mechanism shown in Fig. 2. Each shaft has a yoke connected so that it can slide along the arms of a rigid cross member. Transmission is at a constant velocity, but the shafts must remain parallel, although the offset can vary. There is no axial freedom. The central cross member describes a circle and is thus subjected to centrifugal loads.

Fig. 4 This Oldham coupling provides motion at a constant velocity as its central member describes a circle. The shaft offset can vary, but the shafts must remain parallel. A small amount of axial freedom is possible. A tilt in the central member can occur because of the offset of the slots. This can be eliminated by enlarging its diameter and milling the slots in the same transverse plane.

LINKS AND DISKS COUPLE OFFSET SHAFTS

An unorthodox yet remarkably simple arrangement of links and disks forms the basis of a versatile parallel-shaft coupling. This coupling—essentially three disks rotating in unison and interconnected in series by six links (see drawing)—can adapt to wide variations in axial displacement while it is running under load.

Changes in radial displacement do not affect the constant-velocity relationship between the input and output shafts, nor do they affect initial radial reaction forces that might cause imbalance in the system. Those features open up unusual applications for it in automotive, marine, machine-tool, and rolling-mill machinery (see drawings).

How it works. The inventor of the coupling, Richard Schmidt of Madison, Alabama, said that a similar link arrangement had been known to some German engineers for years. But those engineers were discouraged from applying the theory because they erroneously assumed that the center disk had to be retained by its own bearing. Actually, Schmidt found that the center disk is free to assume its own center of rotation. In operation, all three disks rotate with equal velocity.

The bearing-mounted connections of links to disks are equally spaced at 120° on pitch circles of the same diameter. The distance between shafts can be varied steplessly between zero (when the shafts are in line) and a maximum that is twice the length of the links (see drawings.) There is no phase shift between shafts while the coupling is undulating.

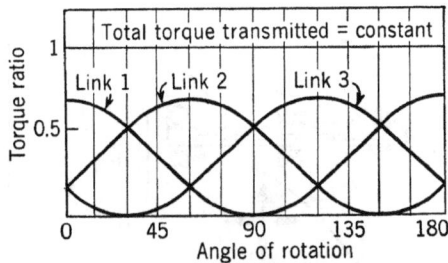

Maximum displacement
(rear view)

Input and output shafts in line
Zero displacement

Parallel-link connections between disks (see upper drawing) exactly duplicate the motion between the input and output shafts—the basis of this principle in coupling. The lower diagrams show three positions of the links as one shaft is shifted with respect to the other shaft in the system.

Midway position

Torque transmitted by three links in the group adds up to a constant value, regardless of the angle of rotation.

Gearshift Clutch Internal gear wheel Schmidt coupling Internal gear housing External housing Input shaft Output shaft Gear wheels Clutch Gearshift Schmidt coupling

The parallelgram-type coupling (above) introduces versatility to a gear-transmission design (left) by permitting both the input and output to clutch in directly to any of the six power gears.

DISK-AND-LINK COUPLINGS SIMPLIFY TORQUE TRANSMISSION

A unique disk-and-link coupling that can handle large axial displacement between shafts, while the shafts are running under load, has opened up new approaches to transmission design. It was developed by Richard Schmidt of Madison, Alabama.

The coupling (drawing, below right) maintains a constant transmission ratio between input and output shafts while the shafts undergo axial shifts in their relative positions. This permits gear-and-belt transmissions to be designed that need fewer gears and pulleys.

Half as many gears. In the internal-gear transmission shown, a Schmidt coupling on the input side permits the input to be plugged in directly to any one of six gears, all of which are in mesh with the internal gear wheel.

On the output side, after the power flows through the gear wheel, a second Schmidt coupling permits a direct power takeoff from any of the same six gears. Thus, any one of 6 × 6 minus 5 or 31 different speed ratios can be selected while the unit is running. A more orthodox design would require almost twice as many gears.

Powerful pump. In the worm-type pump (bottom left), as the input shaft rotates clockwise, the worm rotor is forced to roll around the inside of the gear housing, which has a helical groove running from end to end. Thus, the rotor center-line will rotate counterclockwise to produce a powerful pumping action for moving heavy liquids.

In the belt drive (bottom right), the Schmidt coupling permits the belt to be shifted to a different bottom pulley while remaining on the same top pulley. Normally, because of the constant belt length, the top pulley would have to be shifted too, to provide a choice of only three output speeds. With this arrangement, nine different output speeds can be obtained.

Axial thrust bearing Housing Input shaft (stationary) Gear housing Worm rotor Eccentric displacement of rotor shaft Schmidt coupling

The coupling allows a helically-shaped rotor to oscillate for pumping purposes.

Pulleys Shiftable pulleys and drive shaft Schmidt coupling Stationary output shaft Belt Gear shift Stationary input shaft

This coupling takes up slack when the bottom shifts.

INTERLOCKING SPACE-FRAMES FLEX AS THEY TRANSMIT SHAFT TORQUE

This coupling tolerates unusually high degrees of misalignment, with no variation in the high torque that's being taken from the shaft.

A concept in flexible drive-shaft couplings permits unusually large degrees of misalignment and axial motion during the transmission of high amounts of torque. Moreover, the rotational velocity of the driven member remains constant during transmission at angular misalignments; in other words, cyclic pulsations are not induced as they would be if, say, a universal coupling or a Hooke's joint were employed.

The coupling consists essentially of a series of square space-frames, each bent to provide offsets at the diagonals and each bolted to adjacent members at alternate diagonals. The concept was invented by Robert B. Bossler, Jr. He was granted U.S. Patent No. 3,177,684.

Couplings accommodate the inevitable misalignments between rotating shafts in a driven train. These misalignments are caused by imperfect parts, dimensional variations, temperature changes, and deflections of the supporting structures. The couplings accommodate misalignment either with moving contacts or by flexing.

Most couplings, however, have parts with moving contacts that require lubrication and maintenance. The rubbing parts also absorb power. Moreover, the lubricant and the seals limit the coupling environment and coupling life. Parts wear out, and the coupling can develop a large resistance to movement as the parts deteriorate. Then, too, in many designs, the coupling does not provide true constant velocity.

For flexibility. Bossler studied the various types of couplings on the market and first developed a new one with a moving contact. After exhaustive tests, he became convinced that if there were to be the improvements he wanted, he had to design a coupling that flexed without any sliding or rubbing.

Flexible-coupling behavior, however, is not without design problems. Any flexible coupling can be proportioned with strong, thick, stiff members that easily transmit a design torque and provide the stiffness to operate at design speed. However, misalignment requires flexing of these members. The flexing produces alternating stresses that can limit coupling life. The greater the strength and stiffness of a member, the higher the alternating stress from a given misalignment. Therefore, strength and stiffness provisions that transmit torque at speed will be detrimental to misalignment accommodation capability.

The design problem is to proportion the flexible coupling to accomplish torque transmission and overcome misalignment for the lowest system cost. Bossler looked at a drive shaft, a good example of power transmission—and wondered how he could convert it into one with flexibility.

He began to evolve it by following basic principles. How does a drive shaft transmit torque? By tension and compression. He began paring it down to the important struts that could transmit torque and found that they are curved beams. But a curved beam in tension and compression is not as strong as a straight beam. He ended up with the beams straight in a square space-frame with what might be called a *double helix arrangement*. One helix contained elements in compression; the other helix contained elements in tension.

Flattening the helix. The total number of plates should be an even number to obtain constant velocity characteristics during misalignment. But even with an odd number, the cyclic speed variations are minute, not nearly the magnitude of those in a Hooke's joint.

Although the analysis and resulting equations developed by Bossler are

based on a square-shaped unit, he concluded that the perfect square is not the ideal for the coupling because of the position of the mounting holes. The flatter the helix—in other words the smaller the distance S—the more misalignment the coupling will tolerate.

Hence, Bossler began making the space-frames slightly rectangular instead of square. In this design, the bolt-heads that fasten the plates together are offset from adjoining pairs, providing enough clearance for the design of a "flatter" helix. The difference in stresses between a coupling with square-shaped plates and one with slightly rectangular plates is so insignificant that the square-shape equations can be employed with confidence.

Design equations. By making a few key assumptions and approximations, Bossler boiled the complex analytical relationships down to a series of straightforward design equations and charts. The derivation of the equations and the resulting verification from tests are given in the NASA report *The Bossler Coupling*, CR-1241.

Torque capacity. The ultimate torque capacity of the coupling before buckling that might occur in one of the spaceframe struts under compression is given by Eq. 1. The designer usually knows or establishes the maximum continuous torque that the coupling must transmit. Then he must allow for possible shock loads and overloads. Thus, the clutch should be designed to have an ultimate torque capacity that is at least twice as much, and perhaps three times as much, as the expected continuous torque, according to Bossler.

Induced stress. At first glance, Eq. 1 seems to allow a lot of leeway in selecting the clutch size. The torque capacity is easily boosted, for example, by picking a smaller bolt-circle diameter, d, which

Bent space-frames join to form couplings

Square-plate coupling

Three-plate rectangular coupling

For angular misalignment

For both parallel and angular misalignment

Design equations for the Bossler coupling

Ultimate torque capacity

$$T = 11.62 \frac{Ebt^3}{dn^{0.9}} \qquad (1)$$

Maximum stress per degree of misalignment

$$\sigma_{max} = 0.0276 \; Et/L \qquad (2)$$

Minimum thickness to meet required torque strength

$$t = 0.4415 \left(\frac{dT}{bE}\right)^{1/3} n^{0.3} \qquad (3)$$

Weight of coupling with minimum-thickness plates

$$W = 1.249 \; w \left(\frac{T}{E}\right)^{1/3} d^{4/3} b^{2/3} n^{1.3} \qquad (4)$$

Maximum permissible misalignment

$$\theta_{max} = 54.7 \left[\frac{bd^2}{TE^2}\right]^{1/3} \sigma_c \, n^{0.7} \qquad (5)$$

Maximum permissible misalignment (simplified)

$$\theta/d = 10.9 \frac{n^{0.7}}{T^{1/3}} \qquad (6)$$

Maximum permissible offset-angle

$$\beta = 54.7 \left[\frac{bd^2}{TE^2}\right]^{1/3} \frac{\sigma_e C}{n^{0.3}} \qquad (7)$$

where:

$$\sum_{x=1}^{x=n} \left[1 - (x-1)\frac{S}{S_1}\right]^2$$

Maximum permissible offset-angle (simplified)

$$\beta/d = \frac{10.9C}{T^{1/3}n^{0.3}} \qquad (8)$$

Critical speed frequency

$$f = \frac{60}{2\pi} \left(\frac{k}{M}\right)^{1/2} \qquad (9)$$

where: $k = \dfrac{24(El)_e}{(nS)^3}$ and $(El)_e = 0.886Ebt^3S/L$

List of symbols

b = Width of an element
d = Diameter at the bold circle
E = Modulus of elasticity
f = First critical speed, rpm
l = Flatwise moment of inertia of an element = $bt^3/12$
k = Spring constant for single degree of freedom
L = Effective length of an element. This concept is required because joint details tend to stiffen the ends of the elements. L = 0.667 d is recommended
M = Mass of center shaft plus mass of one coupling with fasteners
n = Number of plates in each coupling
S = Offset distance by which a plate is out of plane
t = Thickness of an element
T = Torque applied to coupling, useful ultimate, usually taken as lowest critical buckling torque
w = Weight per unit volume
W = Total weight of plates in a coupling
$(El)_e$ = Flexural stiffness, the moment that causes one radian of flexural angle change per unit length of coupling
β = Equivalent angle change at each coupling during parallel offset misalignment, deg
υ = Total angular misalignment, deg
σ_c = Characteristic that limits stress for the material: yield stress for static performance, endurance limit stress for fatigue performance

306

makes the clutch smaller, or by making the plates thicker. But either solution would also make the clutch stiffer, hence would restrict the misalignment permitted before the clutch becomes over- stressed. The stress-misalignment relationship is given in Eq. 2, which shows the maximum flat-wise bending stress produced when a plate is misaligned 1° and is then rotated to transmit torque.

Plate thickness. For optimum misalignment capability, the plates should be selected with the least thickness that will provide the required torque strength. To determine the minimum thickness, Bossler found it expedient to rearrange Eq. 1 into the form shown in Eq. 3. The weight of any coupling designed in accordance to the minimum-thickness equation can be determined from Eq. 4.

Maximum misalignment. Angular misalignment occurs when the center-lines of the input and output shafts intersect at some angle—the angle of misalignment. When the characteristic limiting stress is known for the material selected—and for the coupling's dimensions—the maximum allowable angle of misalignment can be computed from Eq. 5.

If this allowance is not satisfactory, the designer might have to juggle the size factors by, say, adding more plates to the unit. To simplify Eq. 5, Bossler made some assumptions in the ratio of endurance limit to modulus and in the ratio of dsb to obtain Eq. 6.

Parallel offset. This condition exists when the input and output shafts remain parallel but are displaced laterally. As with Eq. 6, Eq. 7 is a performance equation and can be reduced to design curves. Bossler obtained Eq. 8 by making the same assumptions as in the previous case.

Critical speed. Because of the noncircular configurations of the coupling, it is important that the operating speed of the unit be higher than its critical speed. It should not only be higher but also should avoid an integer relationship.

Bossler worked out a handy relationship for critical speed (Eq. 9) that employs a somewhat idealized value for the spring constant k.

Bossler also made other recommendations where weight reduction is vital:

- **Size of plates.** Use the largest d consistent with envelope and centrifugal force loading. Usually, centrifugal force loading will not be a problem below 300 ft/s tip speed.

- **Number of plates.** Pick the least n consistent with the required performance.

- **Thickness of plates.** Select the smallest t consistent with the required ultimate torque.

- **Joint details.** Be conservative; use high-strength tension fasteners with high preload. Provide fretting protection. Make element centerlines and bolt centerlines intersect at a point.

- **Offset distance.** Use the smallest S consistent with clearance.

COUPLING WITH OFF-CENTER PINS CONNECTS MISALIGNED SHAFTS

Two Hungarian engineers developed an all-metal coupling (see drawing) for connecting shafts where alignment is not exact—that is, where the degree of misalignment does not exceed the magnitude of the shaft radius.

The coupling is applied to shafts that are being connected for either high-torque or high-speed operation and that must operate at maximum efficiency. Knuckle joints are too expensive, and they have too much play; elastic joints are too vulnerable to the influences of high loads and vibrations.

How it's made. In essence, the coupling consists of two disks, each keyed to a splined shaft. One disk bears four fixed-mounted steel studs at equal spacing; the other disk has large-diameter holes drilled at points facing the studs.

Each large hole is fitted with a bearing that rotates freely inside it on rollers or needles. The bore of the bearings, however, is off-center. The amount of eccentricity of the bearing bore is identical to the deviation of the two shaft center lines.

In operation, input and output shafts can be misaligned, yet they still rotate with the same angular relationship they would have if perfectly aligned.

Eccentrically bored bearings rotate to make up for misalignment between shafts.

UNIVERSAL JOINT TRANSMITS TORQUE 45° AT CONSTANT SPEED

A novel arrangement of pivots and ball-socket joints transmits uniform motion.

A universal joint that transmits power at constant speeds through angles up to 45° was designed by Malton Miller of Minnesota.

Models of the true-speed drive that can transmit up to 20 hp have been developed.

It had not been possible to transmit power at constant speeds with only one universal joint. Engineers had to specify an intermediate shaft between two Hooke's joints or use a Rzappa-type joint to get the desired effect.

Ball-and-socket. Basically, the True-Speed joint is a system of ball-and-socket connections with large contact areas (low unit pressure) to transmit torsional forces across the joint. This arrangement minimizes problems when high bearing pressures build up against running surfaces. The low-friction bearings also increase efficiency. The joint is balanced to keep vibration at high speeds to a minimum.

The joint consists of driving and driven halves. Each half has a coupling sleeve at its end of the driveshaft, a pair of driving

arms opposite each other and pivoted on a cross pin that extends through the coupling sleeve, and a ball-and-socket coupling at the end of each driving arm.

As the joint rotates, angular flexure in one plane of the joint is accommodated by the swiveling of the all-and-socket couplings and, in the 90° plane, by the oscillation of the driving arms about the transverse pin. As rotation occurs, torsion is transmitted from one half of the joint to the other half through the swiveling ball-and-socket couplings and the oscillating driving arms.

Balancing. Each half of the joint, in effect, rotates about its own center shaft, so each half is considered separate for balancing. The center ball-and-socket coupling serves only to align and secure the intersection point of the two shafts. It does not transmit any forces to the entire drive unit.

Balancing for rotation is achieved by equalizing the weight of the two driving arms of each half of the joint. Balancing the acceleration forces due

to the oscillation of the ball-and-socket couplings, which are offset from their swiveling axes, is achieved by the use of counterweights extending from the opposite side of each driving arm.

The outer ball-and-socket couplings work in two planes of motion, swiveling widely in the plane perpendicular to the main shaft and swiveling slightly about the transverse pin in the plane parallel to the main shaft. In this coupling configuration, the angular displacement of the driving shaft is exactly duplicated in the driven shaft, providing constant rotational velocity and constant torque at all shaft intersection angles.

Bearings. The only bearing parts are the ball-and-socket couplings and the driving arms on the transverse pins. Needle bearings support the driving arms on the transverse pin, which is hardened and ground. A high-pressure grease lubricant coats the bearing surfaces of the ball-and-socket couplings. Under maximum rated loadings of 600 psi on the ball-and-socket surfaces, there is no appreciable heating or power loss due to friction.

Capabilities. Units have been laboratory-tested at all rated angles of drive under dynamometer loadings. Although the first available units were for smaller capacities, a unit designed for 20 hp at 550 rpm, suitable for tractor power take-off drive, has been tested.

Similar couplings have been designed as pump couplings. But the True-Speed drive differs in that the speed and transfer elements are positive. With the pump coupling, on the other hand, the speed might fluctuate because of spring bounce.

An earlier version for angled shafts required spring-loaded sliding rods.

TEN UNIVERSAL SHAFT COUPLINGS

Hooke's Joints

The commonest form of a universal coupling is a *Hooke's joint*. It can transmit torque efficiently up to a maximum shaft alignment angle of about 36°. At slow speeds, on hand-operated mechanisms, the permissible angle can reach 45°. The simplest arrangement for a Hooke's joint is two forked shaft-ends coupled by a cross-shaped piece. There are many variations and a few of them are included here.

Fig. 1   The Hooke's joint can transmit heavy loads. Anti-friction bearings are a refinement often used.

Fig. 2 A   pinned sphere shaft coupling replaces a cross-piece. The result is a more compact joint.

Fig. 3   A grooved-sphere joint is a modification of a pinned sphere. Torques on fastening sleeves are bent over the sphere on the assembly. Greater sliding contact of the torques in grooves makes simple lubrication essential at high torques and alignment angles.

Fig. 4   A pinned-sleeve shaft-coupling is fastened to one shaft that engages the forked, spherical end on the other shaft to provide a joint which also allows for axial shaft movement. In this example, however, the angle between shafts must be small. Also, the joint is only suitable for low torques.

Constant-Velocity Couplings

The disadvantages of a single Hooke's joint is that the velocity of the driven shaft varies. Its maximum velocity can be found by multiplying driving-shaft speed by the secant of the shaft angle; for minimum speed, multiply by the cosine. An example of speed variation: a driving shaft rotates at 100 rpm; the angle between the shafts is 20°. The minimum output is 100×0.9397, which equals 93.9 rpm; the maximum output is 1.0642×100, or 106.4 rpm. Thus, the difference is 12.43 rpm. When output speed is high, output torque is low, and vice versa. This is an objectionable feature in some mechanisms. However, two universal joints connected by an intermediate shaft solve this speed-torque objection.

Fig. 5   A constant-velocity joint is made by coupling two Hooke's joints. They must have equal input and output angles to work correctly. Also, the forks must be assembled so that they will always be in the same plane. The shaft-alignment angle can be double that for a single joint.

This single constant-velocity coupling is based on the principle (Fig. 6) that the contact point of the two members must always lie on the homokinetic plane. Their rotation speed will then always be equal because the radius to the contact point of each member will always be equal. Such simple couplings are ideal for toys, instruments, and other light-duty mechanisms. For heavy duty, such as the front-wheel drives of military vehicles, a more complex coupling is shown diagrammatically in Fig. 7A. It has two joints close-coupled with a sliding member between them. The exploded view (Fig. 7B) shows these members. There are other designs for heavy-duty universal couplings; one, known as the *Rzeppa*, consists of a cage that keeps six balls in the homokinetic plane at all times. Another constant-velocity joint, the *Bendix-Weiss*, also incorporates balls.

Fig. 6

Fig. 7

Triple-strand spring

Fig. 8 This flexible shaft permits any shaft angle. These shafts, if long, should be supported to prevent backlash and coiling.

Fig. 9 This pump-type coupling has the reciprocating action of sliding rods that can drive pistons in cylinders.

Fig. 10 This light-duty coupling is ideal for many simple, low-cost mechanisms. The sliding swivel-rod must be kept well lubricated at all times.

NINETEEN METHODS FOR COUPLING ROTATING SHAFTS

Methods for coupling rotating shafts vary from simple bolted flange assembles to complex spring and synthetic rubber assembles. Those including chain belts, splines, bands, and rollers are shown here.

Gasket between housing flanges retains lubricant

Steel grid transmits power and absorbs shock and vibration

Flanges of housing bolted together

Fig. 1

Hubs keyed for shafts

Neoprene seal rings retain lubricant

Filler hole for lubricant

Tapered grooves for grid provide bearing surface. Grid bears in grooves in proportion to load

Floating sleeve, carrying generated internal splines at each end. The splines of the sleeve permanently engage the splines of each hub

Each set of splines in mesh around entire circumference. Assembly revolves as one unit. Lateral and angular play allowed between spline faces

Hubs are pressed on and keyed to each shaft. Each hub carries generated splines, cut at maximum distance from shaft end

Misalignment of connected shafts is compensated by sleeve assuming neutral position between shaft hubs

Oil filler plug

Sleeve jack holes to facilitate separation of center flanges

Clearance between casing and hub to allow lateral float for each shaft

Load is carried by oil film

Hub jack holes to facilitate installation or removal of hubs

Sleeve bearings carried on bearing rings. Rings located on transverse center line of hub spline faces

Oil levels. Coupling is filled when standing. In operation centrifugal force distributes oil to immerse splines

Bolt heads and nuts in flanges counterbored as safety precaution

Oil film between splines eliminates metal-to-metal contact

Fig. 2

Oil hole with safety screw plug

Floating housing shell cut with internal gears at each end

Generated spherical gears on hubs

Boltholes counterbored as safety precaution

Hubs splined for shafts

Gasket between flanges to ensure oil tight seal

Oil seal of flexible composition material

Clearance space between hubs to allow for end play

Fig. 3

Tapered bores do not run completely through hubs

Double-tapered jaws held by keyseats in end of hub

Shaft

Flanged hubs

Bolts draw flanged hubs together

Jaws machined on inner surface to radius less than shaft. Shaft gripped by jaws when flanges are drawn together by bolts

Fig. 4

Steel rims, one of which is of smaller diameter than the other

Rims mounted on flanges

Rims

Endless leather belt laced through alternate rim slots

Axial slots on each rim

Flanges keyed to shaft

Fig. 5

Casing provided with lugs that fit loosely in the twin-arms of the spider and bear against spring plugs

Removable access plate to springs

Casing and spider keyed to shaft

Spider

Shaft

Helical springs in constant compression

Case-hardened plugs fitted into pockets between twin-arms of spider

Clearance between ends of spring plugs less than maximum deflection of spring Sudden overload can not break springs

Helical springs

Smooth exterior for safety

Fig. 6

Cylindrical sleeve with eccentric chambers on inside

Turning coupling forces rollers up inclined sides of eccentric chamber to lock coupling to shaft

Side clearance provided between chain and teeth for accomodation of angular displacement between shafts

Teeth cut on flanges of hubs

Roller chain over teeth on hub flanges. All rollers in contact with teeth for equal distribution of transmitted load

Smooth exterior for safety

Fig. 7

With rollers located in largest part of eccentric chamber, coupling can be slipped over end of shaft

Two steel rollers held parallel by light wire frame

Fig. 8

Chain provided with master link for removal

Hubs keyed to shafts

Metal housing over neoprene biscuits

T flange

Neoprene center designed for uniform stress, linear deflection and absorption of vibration

Shaft keyed to flange

Shaft

Metallic screen core

Keyed

T flange

Outer fabric ring impregnated with neoprene, provides support for center section

Fig. 9

Compensating member provides connection between hub and outer sleeve

Flanged hub keyed to shaft

Tapped holes facilitate assembly and disassembly

Generated external and internal gear teeth

Flanged outer sleeve bolted directly to flanged hub

Wide face of internal gear teeth permits full end float without disengagement

Geared hub keyed to shaft

Flexible, oil-resistant packing retains oil inside the coupling and excludes dirt, grit and moisture

Fig. 12

Metal facing plates riveted to flexible disk prevent excessive wear

Flange hubs keyed to shafts

Bosses on metal flanges pass through holes in flexible disk and enter elongated holes in mating flange

Flexible disk of vulcanized rubbered fabric

Elongated holes receive boss from mating flange

Fig. 10

Setscrews secure hubs to shaft

Large number of teeth produce very large bearing surface

Fig. 11

Long gear teeth in sleeve prevent hub from disengaging

Clearance between sleeve and hub permits free end float

Solid metal under gear teeth gives added strength and durability

Flexible, oil resistant packing retains oil inside the coupling and excludes dirt, grit and moisture

Generated external and internal gear teeth

Safety flange with countersunk holes for fitted bolts and self-locking nuts

Two tapped holes in each half of sleeve facilitate assembly and removal

Two tapped holes in each hub facilitate assembly and removal

Gasket prevents oil leakage

Load cushioned by oil film between the gear teeth

Spherical contour of hub teeth permits free sliding and rocking motion

Oil chamber

Machined bands on each hub facilitate accurate alignment

Oil-supply replenished through either of two plugged holes

Fig. 13

Rubber vulcanized to steel plates

Plates bolted to flanges

Shaft

Flange keyed to shaft

Fig. 14

Outer fabric ring impregnated with neoprene, provides support for center section

Metallic screen core

Trunnion pins fitted into outer diameter of hub and welded

Neoprene center designed for uniform stress

Keyway

Boltheads recessed as safety precaution

Neoprene biscuit pressed over trunnion blocks

Two-piece housing clamps around neoprene biscuits. Face of housing attaches to standard flange

Fig. 15

Hubs secured to shafts by keys and setscrews

Non-metallic insert between jaws of coupling absorbs vibration and shocks

Fig. 16

Ball bearing insert permits unrestricted end play

Hubs secured to shafts by keys and setscrews

Fig. 17

Flexible laminated pin units compensate for misalignment of connected shafts. One end held by spring retaining ring, other end moves laterally in bushing

Spring retaining ring holds flexible pin units in flange

Flange hubs secured to shaft with setscrews and keys

Steel laminations swivel on cross pins in slot of keeper

Cross pin

Wax-impregnated, self lubricating bronze bushings

Fig. 18

No lubrication is required on this coupling

Projection on retaining cover holds balls in proper position

Balls contained in pockets in flange

Hubs keyed for shafts

Faceplate retains balls

Power transmitted by six rubber balls

Small holes in flange to aid in removing balls

Fig. 19

FIVE DIFFERENT PIN-AND-LINK COUPLINGS

Fig. 1

Fig. 1 If constant velocity is not required, a pin and slot coupling can be used. Velocity transmission is irregular because the effective radius of operation is continually changing. The shafts must remain parallel unless a ball joint is placed between the slot and pin. Axial freedom is possible, but any change in the shaft offset will further affect the fluctuation of velocity transmission.

Fig. 2

Fig. 2 This parallel-crank coupling drives the overhead camshaft on engines. Each shaft has at least two cranks connected by links. Each must have full symmetry for constant velocity action and to avoid dead points. By attaching ball joints at the ends of the links, displacement between the crank assembles is possible.

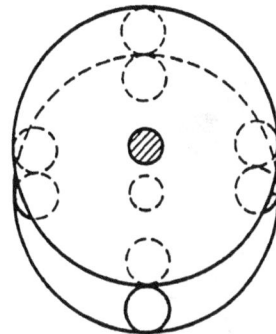

Fig. 3

Fig. 3 This coupling is kinematically equivalent to Fig. 2. It can be made by substituting two circular and contacting pins for each link. Each shaft has a disk carrying three or more projecting pins. The sum of the radii of the pins is equal to the eccentricity of offset of the shafts. The center lines between each pair of pins remain parallel as the coupling rotates. The pins need not have equal diameters. Transmission is at a constant velocity, and axial freedom is possible.

Fig. 4

Fig. 4 This coupling is similar to the coupling shown in Fig. 3. However, holes replace one set of pins. The difference in radii is equal to the eccentricity or offset. Velocity transmission is constant; axial freedom is possible, but as in Fig. 3, the shaft axes must remain fixed. This type of coupling can be installed in epicyclic reduction gear boxes.

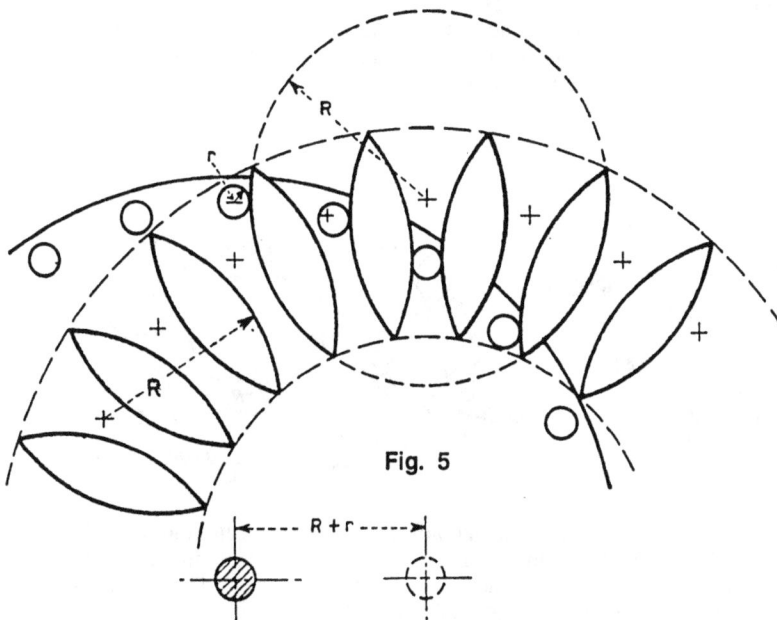

Fig. 5

Fig. 5 An unusual development in pin coupling is shown. A large number of pins engages the lenticular or shield-shaped sections formed from segments of theo-retical large pins. The axes forming the lenticular sections are struck from the pith points of the coupling, and the distance $R + r$ is equal to the eccentricity between the shaft centers. Velocity transmission is constant; axial freedom is possible, but the shafts must remain parallel.

TEN DIFFERENT SPLINED CONNECTIONS

1. SQUARE SPLINES make simple connections. They are used mainly for transmitting light loads, where accurate positioning is not critical. This spline is commonly used on machine tools; a cap screw is required to hold the enveloping member.

Taper 0.75" per foot

Tooth Proportions

Addendum (ext.) $A = \dfrac{0.50}{DP}$

Addendum (int.) $a = \dfrac{0.30}{DP}$

Total depth $h = \dfrac{1.00}{DP}$

2. SERRATIONS of small size are used mostly for transmitting light loads. This shaft forced into a hole of softer material makes an inexpensive connection. Originally straight-sided and limited to small pitches, 45° serrations have been standardized (SAE) with large pitches up to 10 in. dia. For tight fits, the serrations are tapered.

SAE STANDARD SPLINE PROPORTIONS

No. of splines		4	6	10 or 16
Spline width	W	0.241D	0.250D	0.156D ✷
Tight fit	h	0.075D	0.050D	0.045D
Sliding fit	Unloaded h	0.125D	0.075D	0.070D
	loaded h	——	0.100D	0.095D
✷ Width of 16-tooth spline is 0.098D				
Root dia, d = D − 2h				

3. STRAIGHT-SIDED splines have been widely used in the automotive field. Such splines are often used for sliding members. The sharp corner at the root limits the torque capacity to pressures of approximately 1,000 psi on the spline projected area. For different applications, tooth height is altered, as shown in the table above.

Tooth Proportions
$A = 0.25B$
$C = 0.3A$

External

Internal member

Ground lands

4. MACHINE-TOOL splines have wide gaps between splines to permit accurate cylindrical grinding of the lands—for precise positioning. Internal parts can be ground readily so that they will fit closely with the lands of the external member.

Major diameter fit

(A)

Minor diameter fit

(B)
Tooth width fit

30°

Addendum $A = \dfrac{0.500}{\text{Diam pitch}}$

5. INVOLUTE-FORM splines are used where high loads are to be transmitted. Tooth proportions are based on a 30° stub tooth form. (A) Splined members can be positioned either by close fitting major or minor diameters. (B) Use of the tooth width or side positioning has the advantage of a full fillet radius at the roots. Splines can be parallel or helical. Contact stresses of 4,000 psi are used for accurate, hardened splines. The diametral pitch shown is the ratio of teeth to the pitch diameter.

6. SPECIAL INVOLUTE splines are made by using gear tooth proportions. With full depth teeth, greater contact area is possible. A compound pinion is shown made by cropping the smaller pinion teeth and internally splining the larger pinion.

7. TAPER-ROOT splines are for drivers that require positive positioning. This method holds mating parts securely. With a 30° involute stub tooth, this type is stronger than parallel root splines and can be hobbed with a range of tapers.

FACE SPLINES

8. MILLED SLOTS in hubs or shafts make inexpensive connections. This spline is limited to moderate loads and requires a locking device to maintain positive engagement. A pin and sleeve method is used for light torques and where accurate positioning is not required.

9. RADIAL SERRATIONS made by milling or shaping the teeth form simple connections. (A) Tooth proportions decrease radially. (B) Teeth can be straight-sided (castellated) or inclined; a 90° angle is common.

10. CURVIC COUPLING teeth are machined by a face-mill cutter. When hardened parts are used that require accurate positioning, the teeth can be ground. (A) This process produces teeth with uniform depth. They can be cut at any pressure angle, although 30° is most common. (B) Due to the cutting action, the shape of the teeth will be concave (hour-glass) on one member and convex on the other—the member with which it will be assembled.

317

FOURTEEN WAYS TO FASTEN HUBS TO SHAFTS

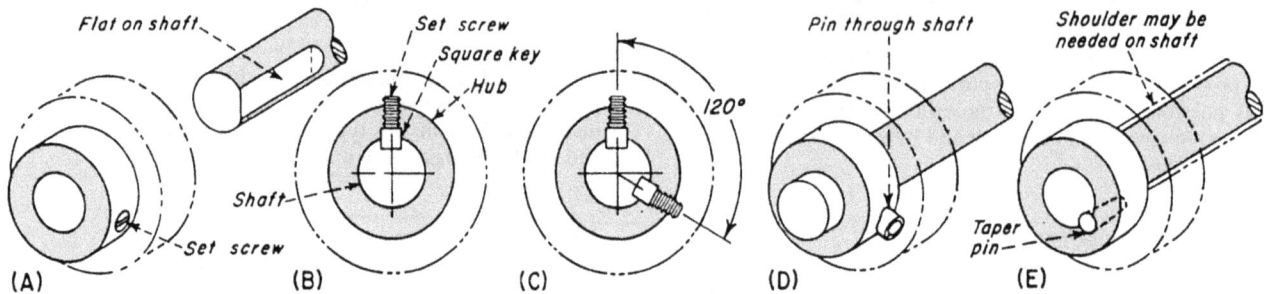

Fig. 1 A cup-point set screw in hub (A) bears against a flat on a shaft. This fastening is suitable for fractional horsepower drives with low shock loads but is unsuitable when frequent removal and assembly are necessary. The key with set screw (B) prevents shaft marring from frequent removal and assembly.

It can withstand high shock loads. Two keys 120° apart (C) transmit extra heavy loads. Straight or tapered pin (D) prevents end play. For experimental setups, an expanding pin is suitable yet easy to remove. Taper pin (E) parallel to shaft might require a shoulder on the shaft. It can be used when a gear or pulley has no hub.

Fig. 2 A tapered shaft with a key and threaded end is a rigid concentric assembly. It is suitable for heavy-duty applications, yet it can be easily disassembled.

Fig. 3 A feather key (A) allows axial gear movement. A keyway must be milled to the end of the shaft. For a blind keyway (B) the hub and key must be drilled and tapped, but the design allows the gear to be mounted anywhere on the shaft with only a short keyway.

Fig. 4 Splined shafts are frequently used when a gear must slide. Square splines can be ground to close minor diameter gaps, but involute splines are self-centering and stronger. Non-sliding gears can be pinned to the shaft if it is provided with a hub.

Fig. 5   A retaining ring allows quick gear removal in light-load applications. A shoulder on the shaft is necessary. A shear pin can secure the gear to the shaft if protection against an excessive load is required.

Fig. 6   A stamped gear and formed wire shaft can be used in light-duty application. Lugs stamped on both legs of the wire prevent dis-assembly. The bend radii of the shaft should be small enough to allow the gear to seat.

Fig. 7   Interlocking tapered rings hold the hub tightly to the shaft when the nut is tight-ened. Coarse machining of the hub and shaft does not affect concentricity as in pinned and keyed assemblies. A shoulder is required (A) for end-of-shaft mounting. End plates and four bolts (B) allow the hub to be mounted anywhere on the shaft.

Fig. 8   This split bushing has a tapered outer diameter. Split holes in the bushing align with split holes in the hub. For tightening, the hub half of the hole is tapped, and the bushing half is un-tapped. A screw pulls the bushing into the hub as it is tightened, and it is removed by reversing the procedure.

Fig. 9   The split hub of a stock precision gear is clamped onto a shaft with a separate hub clamp. Manufacturers list correctly dimen-sioned hubs and clamps so that they can be efficiently fastened to a precision-ground shaft.

319

POLYGON SHAPES PROVIDE SUPERIOR CONNECTIONS

Drive shafts and hubs with precisely ground polygon profiles can be joined to make far stronger connections than those with splines, keys, and flats, according to engineers in companies that grind these forms. The most popular out-of-round polygon shapes have either three or four lobes or peaks. A three-lobed profile, basically a triangle with rounded points, is typically used in fixed (non-sliding) and tapered connections. By contrast, a four-lobed profile, basically a square with rounded corners, is usually used in either sliding or fixed connections. Figure 1 shows how a three-lobed shaft locks with a mating hub as torque increases.

In ancient times, the ends of wooden axles were carved into tapered polygonal-shapes and inserted into matching chiseled holes in wooden wheels. One wheel was fastened at each end of the axle with wooden wedges to form a rigid assembly. These assemblies were attached to carts that would be pulled by humans or work animals. The assemblies rotated in wooden bearings when the carts were moved. It was found that these out-of-round connections could withstand the stress and strain without failure when loaded carts were pulled. Artisans hand-crafted these assemblies with crude tools before iron rings were available to permit wheels rotation on fixed axles. Unfortunately, after that, the advantages of out-of-round profiles to form rigid connections made with any materials were lost to history.

Later when metal shafts and hubs were to be connected permanently, it was easier and cheaper to machine splines, keys, and flats on mating parts than grinding non-circular mating surfaces. However, out-of-round connections were revived, developed, and standardized in Austria during World War II, some 70 years ago. After the war they were adopted in France and in Germany, where polygon shapes were included in German DIN Standards. These standards included data for grinding three- and four-lobe metric connections (DIN 32711 and 32712). Subsequently, the DIN dimensions were converted into English units and the concept was introduced to American industry.

Polygon connections became more acceptable and less expensive with the invention of the Fortuna grinding machine in Germany. This machine can grind matched polygonal shaft and bore diameters because it included a wheelhead with a separate mechanical drive system. This allowed the wheelhead to be set so it could move the grinding wheel in an elliptical path while keeping the wheel normal to the centerline of the rotating workpiece. By setting a ratio between the cycles of wheelhead movement and part rotation, polygons could be generated with an unlimited number of lobes starting with two to achieve an oval shape. Also, three- and four-lobe forms can be ground for press or slide fitting.

With this precision grinding technology available, new applications were found where ground polygon drive shaft connections had advantages over milled spline, key, flat, and serrated shaft connections. Three-lobed polygons can be ground with straight shafts (Fig. 2) or tapered shafts (Fig. 3), and hubs can have matching polygon bores for interlocking engagement to provide zero backlash and perfect angular orientation. This feature makes repetitive assembly and disassembly easier. The shaft ends can be threaded and locked in place with nuts and washers. (Four-lobed polygon shapes are preferred for sliding connections.)

Commercial companies are now organized to perform this grinding service for customers. Engineers in those firms say polygon-shaped shafts and bores or hubs offer advantages over conventional methods for making fixed or sliding connections because of their greater load-carrying capacities for connecting all lengths and diameters of shafts and bores. On sliding drive connections, for

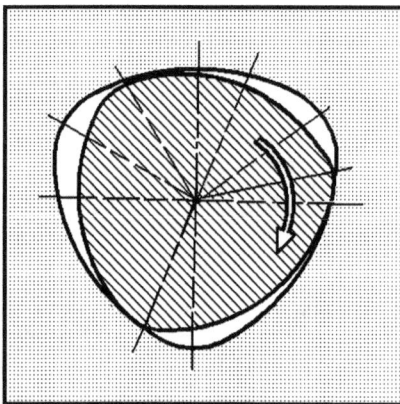

Fig. 1 Exaggerated view illustrates how a polygon profile locks with a hub as torque increases.

Fig. 2 A straight three-lobe polygon profile shaft matches a mating hub with a similar profile.

Fig. 3 A tapered three-lobe polygon profile shaft matches a mating hub with a similar profile.

example, smoother movement is provided with less allowance for sliding clearance, thereby reducing backlash. Moreover, the connections are capable of withstanding greater shock loads under extreme torque reversal conditions, such as occur with brakes and transmissions.

Engineers also claim that polygon-shaped shafts and bores provide the highest torque carrying capacity, within the shaft diameter, of any drive connection. This has been demonstrated by torsion tests to the failure point. While they insist that polygons provide optimum results, they admit that they are not usually the first choice of designers if lower performing and less expensive connections will meet their requirements.

The reasons why polygon connections should be specified are because they:

- Overcome problems with spline connections such as failure caused by torsional effects.
- Reduce backlash to near-zero.
- Offer optimum control of angular positioning (grinding after hardening permits mating parts to be oriented within 2 sec of arc over 24-in. shaft lengths).
- Permit finish machining after hardening (parts can withstand higher torque after hardening because heat treating after machining can introduce distortions in splined shafts).
- Lower manufacturing costs (this applies especially on large-volume production runs when inspection and straightening are critical).

What are the functional advantages of polygons?

- Offer highest shear strength in a given section—twice that of involute splines
- Produce shaft shoulders without runout
- Offer highest torque carrying capacity within the shaft diameter (it carries the load over the entire section rather than on a few teeth of splined shafts)
- Provide concentricity unequalled by other drive connection (polygons and bearing diameters are ground in the same operation)
- Permit fitting of parts subject to extreme interference to those that are loose-sliding
- Permit shorter connections (half the length of splines) and yet can carry equal shear loads
- Solve space and weight limitation problems
- Provide self-centering (even in close-fit connections), simplifying assembly and disassembly
- Offer optimum angular positioning accuracy
- Permit precision grinding ($\pm$ 0.0002 in. on a 1-in. dia, $\pm$ 0.0005 in. on a 5-in. dia, and profile accuracy of $\pm$ 0.0003 in.)

CHAPTER 13
MOTION-SPECIFIC DEVICES, MECHANISMS, AND MACHINES

TIMING BELTS, FOUR-BAR LINKAGE TEAM UP FOR SMOOTH INDEXING

A class of intermittent mechanisms based on timing belts, pulleys, and linkages (see drawing) instead of the usual genevas or cams is capable of cyclic start-and-stop motions with smooth acceleration and deceleration.

Developed by Eric S. Buhayar and Eugene E. Brown of the Engineering Research Division, Scott Paper Co. (Philadelphia), the mechanisms are employed in automatic assembly lines.

These mechanisms, moreover, can function as phase adjusters in which the rotational position of the input shaft can be shifted as desired in relation to the output shaft. Such phase adjusters have been used in the textile and printing industries to change the "register" of one roll with that of another, when both rolls are driven by the same input.

Outgrowth from chains. Intermittent-motion mechanisms typically have ingenious shapes and configurations. They have been used in watches and in production machines for many years. There has been interest in the chain type of intermittent mechanism (see drawing), which ingeniously routes a chain around four sprockets to produce a dwell-and-index output.

The input shaft of such a device has a sprocket eccentrically fixed to it. The input also drives another shaft through one-to-one gearing. This second shaft mounts a similar eccentric sprocket that is, however, free to rotate. The chain passes first around an idler pulley and then around a second pulley, which is the output.

As the input gear rotates, it also pulls the chain around with it, producing a modulated output rotation. Two spring-loaded shoes,

Output displacement, pulley angle, degrees

Output velocity, angular, deg./deg. of input

One crank angle

Output velocity

Unmodulated output velocity

Output displacement

Dwell period

Input crank angle, degrees

however, must be employed because the perimeter of the pulleys is not a constant figure, so the drive has varying slack built into it.

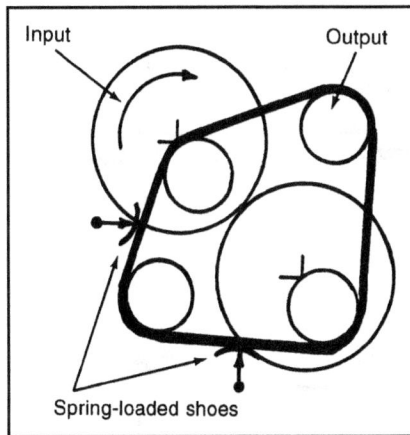

Commercial type. A chain also links the elements of a commercial phase-adjuster drive. A handle is moved to change the phase between the input and output shafts. The theoretical chain length is constant.

In trying to improve this chain device, Scott engineers decided to keep the input and output pulleys at fixed positions and maintain the two idlers on a swing frame. The variation in wraparound length turned out to be surprisingly little, enabling them to install a timing belt without spring-loaded tensioners instead of a chain.

If the swing frame is held in one position, the intermittent mechanism produces a constant-speed output. Shifting the swing frame to a new position automatically shifts the phase relationship between the input and output.

Computer consulted. To obtain intermittent motion, a four-bar linkage is superimposed on the mechanism by adding a crank to the input shaft and a connecting rod to the swing frame. The developers chose an iterative program on a computer to optimize certain variables of the four-bar version.

In the design of one two-stop drive, a dwell period of approximately 50° is obtained. The output displacement moves slowly at first, coming to a "pseudo dwell," in which it is virtually stationary. The output then picks up speed smoothly until almost two-thirds of the input rotation has elapsed (240°). After the input crank completes a full circle of rotation, it continues at a slower rate and begins to repeat its slowdown—dwell—speed-up cycle.

TEN INDEXING AND INTERMITTENT MECHANISMS

This mechanism transmits intermittent motion between two skewed shafts. The shafts need not be at right angles to one another. Angular displacement of the output shaft per revolution of input shaft equals the circular pitch of the output gear wheel divided by its pitch radius. The duration of the motion period depends on the length of the angular joint a of the locking disks b.

A "mutilated tooth" intermittent drive mechanism. Driver b is a circular disk of width w with a cutout d on its circumference. It carries a pin c close to the cutout. The driven gear, a, of width 2w has an even number of standard spur gear teeth. They alternately have full and half-width (mutilated) teeth. During the dwell period, two full-width teeth are in contact with the circumference of the driving disk, thus locking it. The mutilated tooth between them is behind the driver. At the end of the dwell period, pin c contacts the mutilated tooth and turns the driven gear one circular pitch. Then, the full-width tooth engages the cutout d, and the driven gear moves one more pitch. Then the dwell period starts again and the cycle is repeated.

An operating cycle of 180° motion and 180° dwell is produced by this mechanism. The input shaft drives the rack, which is engaged with the output shaft gear during half the cycle. When the rack engages, the lock teeth at the lower end of the coulisse are disengaged and, conversely, when the rack is disengaged, the coulisse teeth are engaged. This action locks the output shaft positively. The changeover points occur at the dead-center positions, so that the motion of the gear is continuously and positively governed. By varying the radius R and the diameter of the gear, the number of revolutions made by the output shaft during the operating half of the cycle can be varied to suit many differing requirements.

A cam-driven ratchet.

A six-sided Maltese cross and double driver give a 3:1 ratio.

A cam operated escapement on a taximeter (a). A solenoid-operated escapement (b).

An escapement on an electric meter.

A solenoid-operated ratchet with a solenoid-resetting mechanism A sliding washer engages the teeth.

A plate oscillating across the plane of a ratchet-gear escapement carries stationary and spring-held pawls.

A worm drive, compensated by a cam on a work shaft, produces intermittent motion of the gear.

TWENTY-SEVEN ROTARY-TO-RECIPROCATING MOTION AND DWELL MECHANISMS

Four-bar slider mechanism

Space-velocity graph

Input

Output

With proper dimensions, the rotation of the input link can impart an almost-constant velocity motion to the slider within the slot.

Oscillating-chain mechanism

Input

Output

Approximate straight line

The rotary motion of the input arm is translated into linear motion of the linkage end. The linkage is fixed to the smaller sprocket, and the larger sprocket is fixed to the frame.

Three-gear stroke multiplier mechanism

Gears

Output

Input

The rotation of the input gear causes the connecting link, attached to the machine frame, to oscillate. This action produces a large-stroke reciprocating motion in the output slider.

Rack and gear sector mechanism

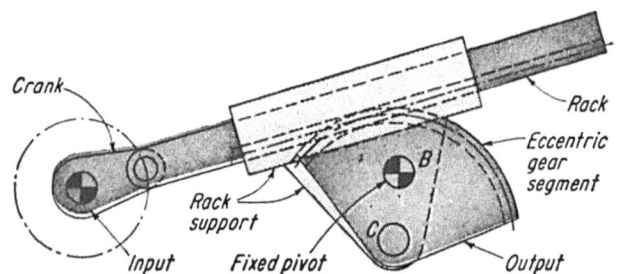

Crank

Rack

Eccentric gear segment

Rack support

B

C

Input   Fixed pivot   Output

The rotary motion of the input shaft is translated into an oscillating motion of the output gear segment. The rack support and gear sector are pinned at *C*, but the gear itself oscillates around *B*.

Linear reciprocator mechanism

This linear reciprocator converts a rotary motion into a reciprocating motion that is *in line* with the input shaft. Rotation of the shaft drives the worm gear which is attached to the machine frame with a rod. Thus input rotation causes the worm gear to draw itself (and the worm) to the right—thus providing a reciprocating motion.

Disk and roller drive mechanism

A hardened disk in this drive, riding at an angle to the axis of an input roller, transforms the rotary motion into linear motion parallel to the axis of the input. The roller is pressed against the input shaft by flat spring *F*. The feed rate is easily varied by changing the angle of the disk. This arrangement can produce an extremely slow feed with a built-in safety factor in case of possible jamming.

Bearing and roller drive mechanism

This drive arrangement avoids large Hertzian stresses between the disk and roller by including three ball bearings in place of the single disk. The inner races of the bearings make contact on one side or the other. Hence a gearing arrangement is required to alternate the angle of the bearings. This arrangement also reduces the bending moment on the shaft.

Reciprocating space crank mechanism

The rotary input of this crank causes the bottom surface of link *A* to wobble with respect to the center link. Link *B* is free of link *A*, but it is restrained from rotating by the slot. This causes the output member to reciprocate linearly.

Oscillating crank and planetary drive mechanism

The planet gear is driven with a stop-and-go motion. The driving roller is shown entering the circular-arc slot on the planet link. The link and the planet remain stationary while the roller travels along this section of the slot. As a result, the output sun gear has a rotating output motion with a progressive oscillation.

Chain-slider drive mechanism

The chain link drives a lever that oscillates. A slowdown-dwell occurs when the chain pin passes around the left sprocket.

Chain-oscillating drive mechanism

The output shaft reciprocates with a constant velocity, but it reaches a long dwell at both ends as the chain lever, whose length is equal to the radius of the sprockets, goes around both sprockets.

Chain and slider dirve mechanism

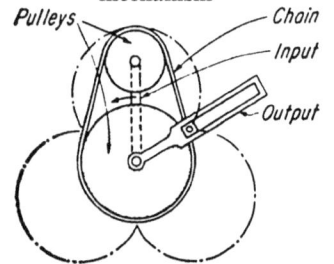

The input crank causes the small pulley to orbit around the stationary larger pulley. A pivot point attached to the chain slides inside the slot of the output link. In the position shown, the output is about to start a long dwell period of about 120°.

Epicyclic dwell mechanism

The output crank pulsates back and forth with a long dwell at its extreme right position. The input shaft rotates the planet gear with a crank. The pin on the planet gear traces the epicyclic three-lobe curve shown. The right side of the curve is a near circular arc of radius R. If the connecting rod length equals R, the output crank reaches a virtual standstill during a third of the total rotation of the input crank. The crank then reverses, stops at its left position, reverses, and repeats its dwell.

Cam-worm dwell mechanism

Without the barrel cam, the input shaft would drive the output gear by the worm gear at constant speed. The worm and the barrel cam, however, can slide linearly on the input shaft. The rotation of the input shaft now causes the worm gear to be cammed back and forth, thus adding or subtracting motion to the output. If barrel cam angle α is equal to the worm angle β, the output stops during the limits of rotation shown. It then speeds up to make up for lost time.

Cam-helical dwell mechanism

When one helical gear is shifted linearly (but prevented from rotating) it will impart rotary motion to the mating gear because of the helix angle. This principle is applied in the mechanism illustrated. The rotation of the input shaft causes the intermediate shaft to shift to the left, which in turn adds or subtracts from the rotation of the output shaft.

Cam-roller dwell mechanism

A steel strip is fed at constant linear velocity in this mechanism. But at the die station (illustrated), it is desired to stop the strip so that the punching operation can be performed. The strip passes over movable rollers which, when shifted to the right, cause the strip to move to the right. Since the strip is normally fed to the left, proper design of the cam can nullify the linear feed rates so that the strip stops, and then speeds to catch up to the normal rate.

Double-crank dwell mechanism

Six-bar dwell mechanism

The rotation of the input crank causes the output bar to oscillate with a long dwell at its extreme right position. This occurs because point C describes a curve that is approximately a circular arc (from C to C' with its center at P). The output is almost stationary during that part of the curve.

Three-gear drive mechanism

This is actually a four-bar linkage combined with three gears. As the input crank rotates, it turns the input gear which drives the output gear through the idler. Various output motions are possible. Depending on the relative diameters of the gears, the output gear can pulsate, reach a short dwell, or even reverse itself briefly.

Both cranks are connected to a common shaft which also acts as the input shaft. Thus the cranks always remain a constant distance apart from each other. There are only two frame points—the center of the input shaft and the guide for the output slider. As the output slider reaches the end of its stroke (to the right), it remains at a virtual standstill while one crank rotates through angle PP'.

331

Dwell Mechanisms (*continued*)

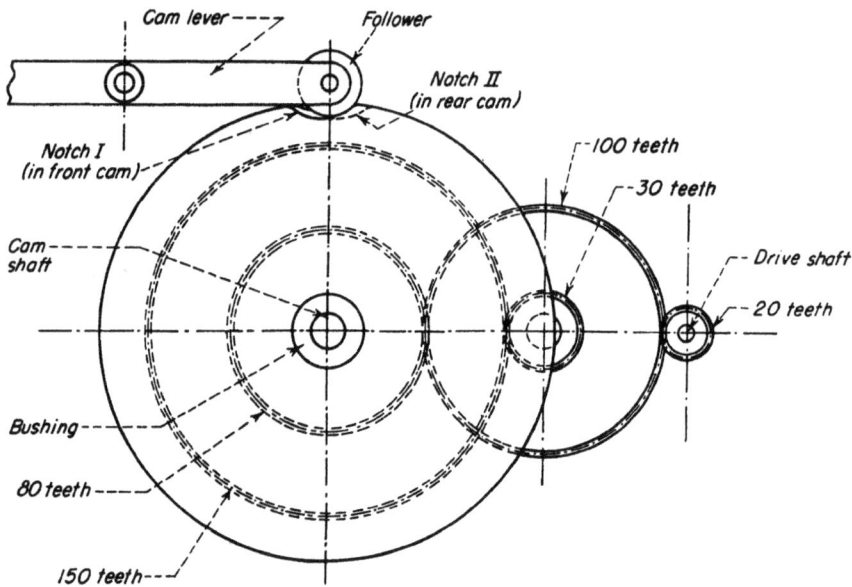

Fast Cam-Follower Motion

Fast cam action every *n* cycles (where *n* is a relatively large number) can be obtained with this manifold cam and gear mechanism. A single notched cam geared $1/n$ to a shaft turning once per cycle moves relatively slowly under the follower. The double notched-cam arrangement shown is designed to operate the lever once in 100 cycles, imparting a rapid movement to it. One of the two identical cams and the 150-tooth gear are keyed to the bushing which turns freely around the cam shaft. The cam shaft carries the second cam and the 80-tooth gear. The 30- and 100-tooth gears are integral, while the 20-tooth gear is attached to the one-cycle drive shaft. One of the cams turns in the ratio of 20/80 or 1/4; the other turns in the ratio 20/100 times 30/150 or 1/25. The notches therefore coincide once every 100 cycles (4 × 25). Lever movement is the equivalent of a cam turning in a ratio of 1 to 4 in relation to the drive shaft. To obtain fast cam action, *n* must be reduced to prime factors. For example, if 100 were factored into 5 and 20, the notches would coincide after every 20 cycles.

Intermittent Motion

This mechanism can be adapted to produce a stop, a variable speed without stop, or a variable speed with momentary reverse motion. A uniformly rotating input shaft drives the chain around the sprocket and idler. The arm serves as a link between the chain and the end of the output shaft crank. The sprocket drive must be in the ratio N/n with the cycle of the machine, where *n* is the number of teeth on the sprocket and *N* the number of links in the chain. When point *P* travels around the sprocket from point *A* to position *B*, the crank rotates uniformly. Between *B* and *C*, *P* decelerates; between *C* and *A* it accelerates; and at *C* there is a momentary dwell. By changing the size and position of the idler, or the lengths of the arm and crank, a variety of motions can be obtained. If the length of the crank is shortened, a brief reverse period will occur in the vicinity of *C;* if the crank is lengthened, the output velocity will vary between a maximum and minimum without reaching zero.

SHORT-DWELL MECHANISMS

Gear-slider crank mechanism

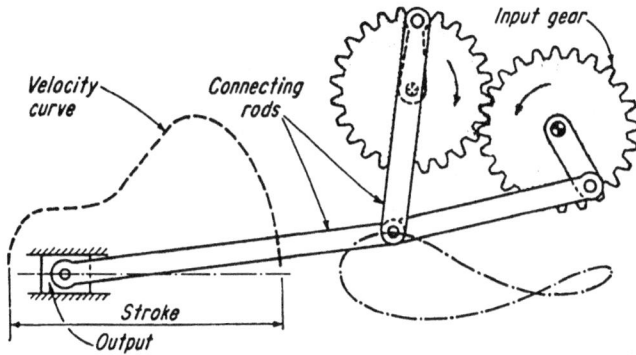

The input shaft drives both gears which, in turn, drive the connecting rods to produce the velocity curve shown. The piston moves with a low constant velocity.

Curve-slider drive mechanism

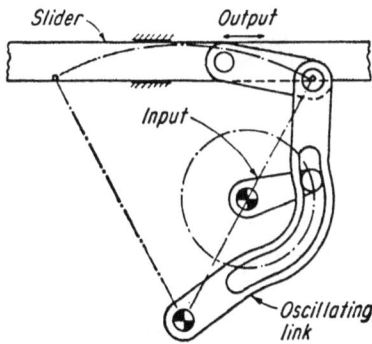

The circular arc on the oscillating link permits the link to reach a dwell during the right position of the output slider.

Whitworth quick-return drive mechanism

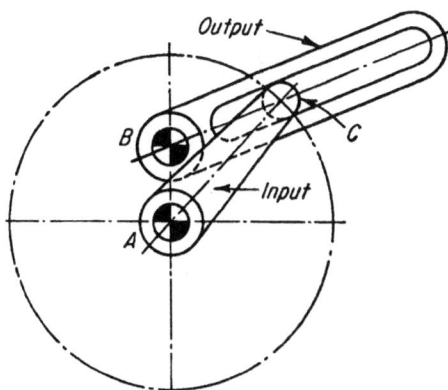

Varying motion can be imparted simply to output shaft B. However, the axes, A and B, are not colinear.

Gear oscillating crank mechanism

In this arrangement, the curve described by the pin connection has two parts, C_1 and C_2, which are very close to circular arc with its centers at A_1 and A_2. Consequently the driven link will have a dwell at both of its extreme positions.

Triple-harmonic drive mechanism

The input shaft drives three gears with connecting rods. A wide variety of reciprocating output motions can be obtained by selecting different lengths for the linkages. In addition, one to several dwells can be obtained per cycle.

Wheel and slider drive mechanism

For each revolution of the input disk, the slider moves in to engage the wheel and index it one tooth width. A flat spring keeps the wheel locked while it is stationary.

FIVE FRICTION MECHANISMS FOR INTERMITTENT ROTARY MOTION

Friction mechanisms do not have the disadvantages inherent in conventional pawl and ratchet drives such as: (1) noisy operation; (2) backlash needed for pawl engagement; (3) load concentrated on one tooth of the ratchet; and (4) pawl engagement dependent on an external spring. Each of the five mechanisms presented here converts the reciprocating motion of a connecting rod into an intermittent rotary motion. The connecting rod stroke to the left drives a shaft counterclockwise and that shaft is uncoupled. It remains stationary during the return stroke of the connecting rod to the right.

Fig. 1 The wedge and disk mechanism consists of shaft A supported in bearing block J; ring C is keyed to A and it contains an annular groove G; body B, which can pivot around the shoulders of C; lever D, which can pivot about E; and connecting rod R, which is driven by an eccentric (not shown). Lever D is rotated counter-clockwise about E by the connecting rod moving to the left until surface F wedges into groove G. Continued rotation of D causes A, B, and D to rotate counterclockwise as a unit about A. The reversal of input motion instantly swivels F out of G, thus unlocking the shaft, which remains stationary during its return stroke because of friction induced by its load. As D continues to rotate clockwise about E, node H, which is hardened and polished to reduce friction, bears against the bottom of G to restrain further swiveling. Lever D now rotates with B around A until the end of the stroke.

Fig. 2 The pin and disk mechanism: Lever D, which pivots around E, contains pin F in an elongated hole K. The hole permits slight vertical movement of the pin, but set screw J prevents horizontal movement. Body B can rotate freely about shaft A. Cut-outs L and H in body B allow clearances for pin F and lever D, respectively. Ring C, which is keyed to shaft A, has an annular groove G to permit clearance for the tip of lever D. Counterclockwise motion of lever D, actuated by the connecting rod, jams a pin between C and the top of cut-out L. This occurs about 7° from the vertical axis. A, B, and D are now locked together and rotate about A. The return stroke of R pivots D clockwise around E and unwedges the pin until it strikes the side of L. Continued motion of R to the right rotates B and D clockwise around A, while the uncoupled shaft remains stationary because of its load.

Fig. 3

Fig. 3 The sliding pin and disk mechanism: The counterclockwise movement of body B about shaft A draws pin D to the right with respect to body B, aided by spring pressure, until the flat bottom F of the pin is wedged against the annular groove E of ring C. The bottom of the pin is inclined about 5° for optimum wedging action. Ring C is keyed to A, and parts, A, C, D, and B now rotate counterclockwise as a unit until the end of the connecting rod's stroke. The reversal of B draws the pin out of engagement so that A remains stationary while the body completes its clockwise rotation.

Fig. 4 The toggle link and disk mechanism: The input stroke of connecting rod R (to the left) wedges block F in groove G by straightening toggle links D and E. Body B, toggle links, and ring C, which is keyed to shaft A, rotate counterclockwise together about A until the end of the stroke. The reversal of connecting rod motion lifts the block, thus uncoupling the shaft, while body B continues clockwise rotation until the end of stroke.

Fig. 4

Fig. 5 The rocker arm and disk mechanism: Lever D, activated by the reciprocating bar R moving to the left, rotates counterclockwise on pivot E, thus wedging block F into groove G of disk C. Shaft A is keyed to C and rotates counterclockwise as a unit with body B and lever D. The return stroke of R to the right pivots D clockwise about E and withdraws the block from the groove so that shaft is uncoupled while D, striking adjusting screw H, travels with B about A until the completion of stroke. Adjusting screw J prevents wedging block F from jamming in the groove.

Fig. 5

335

NINE DIFFERENT BALL SLIDES FOR LINEAR MOTION

Fig. 1 V-grooves and flat surface make a simple horizontal ball slide for reciprocating motion where no side forces are present and a heavy slide is required to keep the balls in continuous contact. The ball cage ensures the proper spacing of the balls and its contacting surfaces are hardened and lapped.

Fig. 2 Double V-grooves are necessary where the slide is in a vertical position or when transverse loads are present. Screw adjustment or spring force is required to minimize any looseness in the slide. Metal-to-metal contact between the balls and grooves ensure accurate motion.

Fig. 3 The ball cartridge has the advantage of unlimited travel because the balls are free to recirculate. Cartridges are best suited for vertical loads. (A) Where lateral restraint is also required, this type is used with a side preload. (B) For flat surfaces the cartridge is easily adjusted.

Fig. 4 Commercial ball bearings can be used to make a reciprocating slide. Adjustments are necessary to prevent looseness of the slide. (A) Slide with beveled ends, (B) Rectangular-shaped slide.

Fig. 5 This sleeve bearing, consisting of a hardened sleeve, balls, and retainer, can be used for reciprocating as well as oscillating motion. Travel is limited in a way similar to that of Fig. 6. This bearing can withstand transverse loads in any direction.

Fig. 6 This ball reciprocating bearing is designed for rotating, reciprocating, or oscillating motion. A formed-wire retainer holds the balls in a helical path. The stroke is about equal to twice the difference between the outer sleeve and the retainer length.

Fig. 7 This ball bushing has several recirculating systems of balls that permit unlimited linear travel. Very compact, this bushing requires only a bored hole for installation. For maximum load capacity, a hardened shaft should be used.

Fig. 8 Cylindrical shafts can be held by commercial ball bearings that are assembled to make a guide. These bearings must be held tightly against the shaft to prevent any looseness.

Fig. 9 Curvilinear motion in a plane is possible with this device when the radius of curvature is large. However, uniform spacing between its grooves is important. Circular-sectioned grooves decrease contact stresses.

337

BALL-BEARING SCREWS CONVERT ROTARY TO LINEAR MOTION

This cartridge-operated rotary actuator quickly retracts the webbing to separate a pilot forcibly from his seat as the seat is ejected in emergencies. It eliminates the tendency of both pilot and seat to tumble together after ejection, preventing the opening of the chute. Gas pressure from the ejection device fires the cartridge in the actuator to force the ball-bearing screw to move axially. The linear motion of the screw is translated into the rotary motion of a ball nut. This motion rapidly rolls up the webbing (stretching it as shown) so that the pilot is snapped out of his seat.

Before retraction

After retraction

This time-delay switching device integrates a time function with a missile's linear travel. Its purpose is to arm the warhead safely. A strict "minimum G-time" system might arm a slow missile too soon for the adequate protection of friendly forces because a fast missile might arrive before the warhead is fused. The weight of the nut plus the inertia under acceleration will rotate the ball-bearing screw which has a flywheel on its end. The screw pitch is selected so that the revolutions of the flywheel represent the distance the missile has traveled.

Reset cam

Guide rod

Reset plunger

Sealing screw for reset mechanism

Ball nut

Roller

Flywheel assembly gear

Electrical connector

Pin

Set-back mass

b/b screw assembly

Acceleration force

Reset rod

Flywheel assembly

Distance-switch deck

Contact cam

Switch actuator

Hand-wheel

Rotating screw

Stationary ball-nut

Swivel joint

Gate

Fast, easy, and accurate control of fluid flow through a valve is obtained by the rotary motion of a screw in the stationary ball nut. The screw produces linear movement of the gate. The swivel joint eliminates rotary motion between the screw and the gate.

NINETEEN ARRANGEMENTS FOR CHANGING LINEAR MOTION

These arrangements of linkages, slides, friction drives, gears, cams, pistons, and solenoids permit linear motion to be changed.

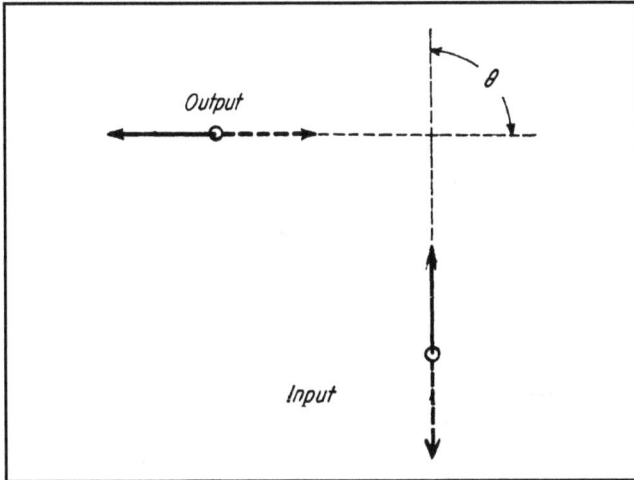

Fig. 1 **Basic problem** (θ is generally close to 90°).

Fig. 2 **Slotted lever.**

Fig. 3 **Spherical bearings.**

Fig. 4 **Spring-loaded lever.**

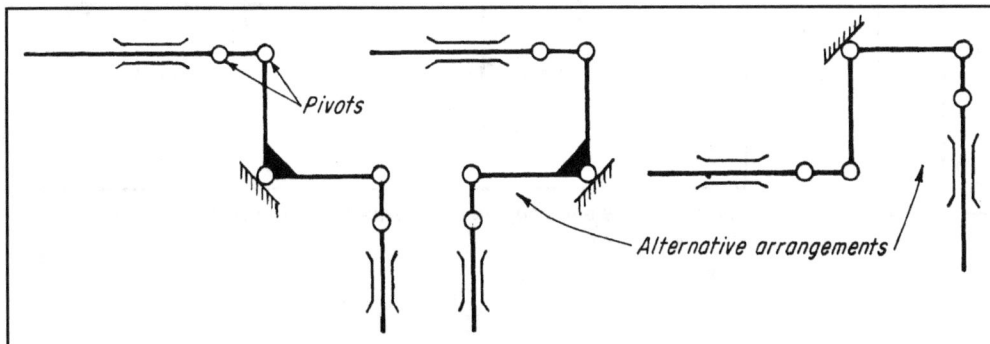

Fig. 5 **Pivoted levers** with alternative arrangements.

Fig. 6 Single connecting rod (left) is relocated (right) to eliminate the need for extra guides.

Fig. 7 Inclined bearing-guide.

Fig. 8 A belt, steel band, or rope around the drum is fastened to the driving and driven members; sprocket-wheels and chain can replace the drum and belt.

Fig. 9 Matching gear-segments.

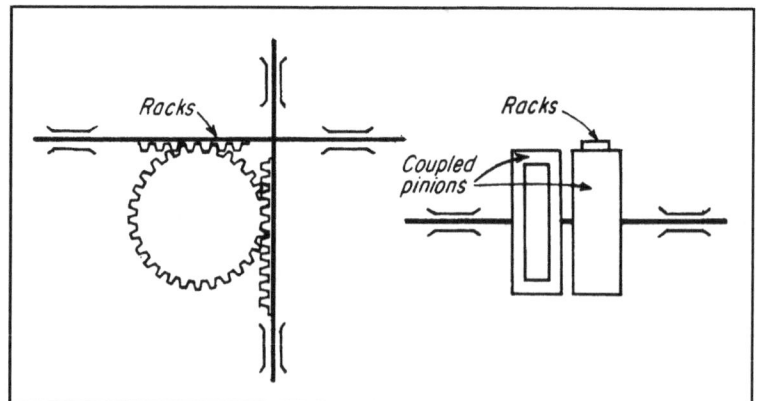

Fig. 10 Racks and coupled pinions (can be substituted as friction surfaces for a low-cost setup).

Fig. 11 An axial screw with a rack-actuated gear (A) and an articulated driving rod (B) are both irreversible movements, i.e., the driver must always drive.

Fig. 12 A rack-actuated gear with associated bevel gears is reversible.

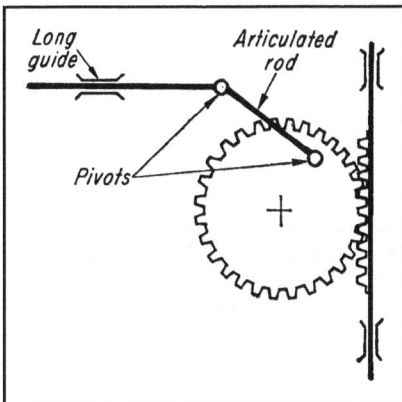

Fig. 13 An articulated rod on a crank-type gear with a rack driver. Its action is restricted to comparatively short movements.

Fig. 14 A cam and spring-loaded follower allows an input/output ratio to be varied according to cam rise. The movement is usually irreversible.

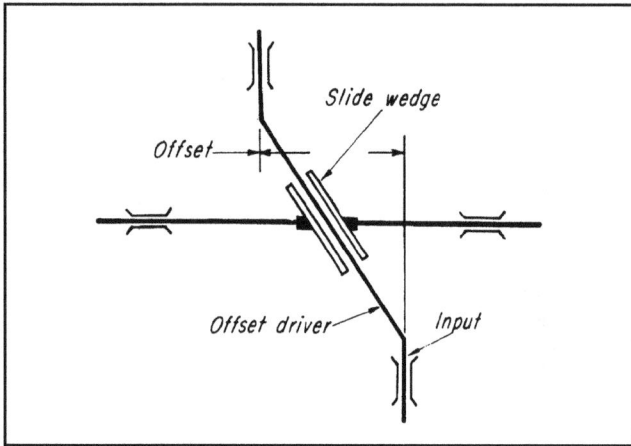

Fig. 15 An offset driver actuates a driven member by wedge action. Lubrication and materials with a low coefficient of friction permit the offset to be maximized.

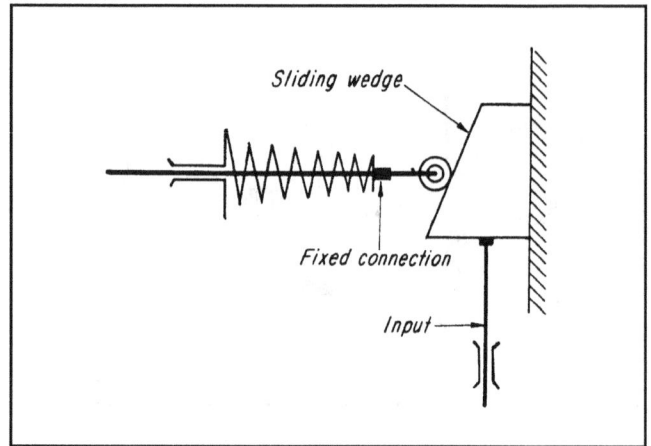

Fig. 16 A sliding wedge is similar to an offset driver but it requires a spring-loaded follower; also, low friction is less critical with a roller follower.

Fig. 17 A fluid coupling allows motion to be transmitted through any angle. Leak problems and accurate piston-fitting can make this method more expensive than it appears to be. Also, although the action is reversible, it must always be compressive for the best results.

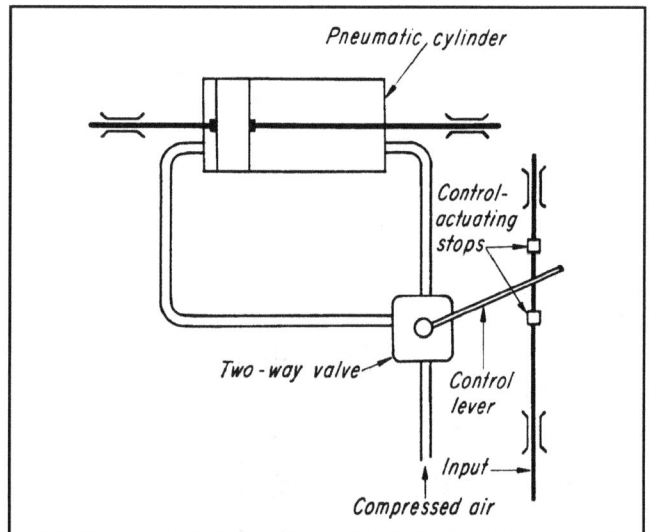

Fig. 18 A pneumatic system with a two-way valve is ideal when only two extreme positions are required. The action is irreversible. The speed of a driven member can be adjusted by controlling the input of air to the cylinder.

Fig. 19 Solenoids and a two-way switch are organized as an analogy of a pneumatic system. Contact with the energized solenoid is broken at the end of each stroke. The action is irreversible.

EIGHT ADJUSTABLE-OUTPUT MECHANISMS

Linkage-motion adjuster

Here the motion and timing of the output link can be varied during its operation by shifting the pivot point of the intermediate link of the six-bar linkage illustrated. Rotation of the input crank causes point C to oscillate around the pivot point P. This, in turn, imparts an oscillating motion to the output crank. A screw device shifts point P.

Cam-motion adjuster

The output motion of the cam follower is varied by linearly shifting the input shaft to the right or left during its operation. The cam has a square hole which fits over the square cross section of the crank shaft. Rotation of the input shaft causes eccentric motion in the cam. Shifting the input shaft to the right, for instance, causes the cam to move radially outward, thus increasing the stroke of the follower.

Double-cam adjuster

This is a simple but effective mechanism for changing the timing of a cam. The follower can be adjusted in the horizontal plane, but it is restricted in the vertical plane. The plate cam contains two or more cam tracks.

Valve-stroke adjuster

This mechanism adjusts the stroke of valves of combustion engines. One link has a curved surface and pivots around an adjustable pivot point. Rotating the adjusting link changes the proportion of strokes or points A and B and hence of the valve. The center of curvature of the curve link is at point Q.

Three-dimensional adjuster

Output motions of four followers can be varied during the rotation by shifting the quadruple 3D cam to the right or left. A linear shift can be made with the adjustment lever, which can be released in any of the six positions.

Piston-stroke adjuster

The input crank oscillates the slotted link to drive the piston up and down. The position of the pivot point can be adjusted with the screw mechanism even when the piston is under full load.

Shaft-synchronizer adjuster

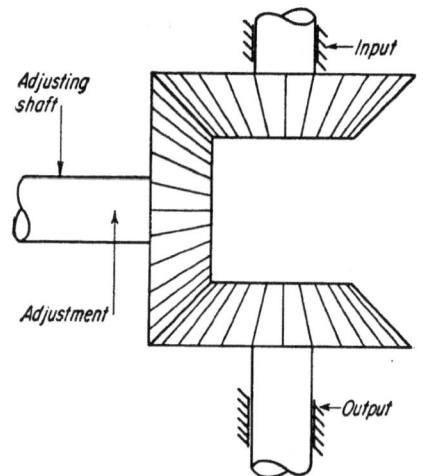

The actual position of the adjusting shaft is normally kept constant. The input then drives the output with the bevel gears. Rotating the adjusting shaft in a plane at right angles to the input-output line changes the relative radial position of the input and output shafts. They introduce a torque into the system while running, synchronizing the input and output shafts, or changing the timing of a cam on the output shaft.

Eccentric pivot-point adjuster

Rotation of the input crank causes the piston to reciprocate. The stroke length depends on the position of the pivot point which is easily adjusted, even during rotation, by rotating the eccentric shaft.

FOUR DIFFERENT REVERSING MECHANISMS

Double-link reverser

This mechanism automatically reverses the output drive for every 180° rotation of the input. The input disk has a press-fit pin which strikes link A to drive it clockwise. Link A in turn drives link B counterclockwise with its gear segments (or gears pinned to the links). The output shaft and the output link (which can be the working member) are connected to link B.

After approximately 180° of rotation, the pin slides past link A to strike link B coming to meet it—and thus reverses the direction of link B (and of the output). Then after another 180° rotation the pin slips past link B to strike link A and starts the cycle over again.

Toggle-link reverser

This mechanism also employs a striking pin—but here the pin is on the output member. The input bevel gear drives two follower bevels which are free to rotate on their common shaft. The ratchet clutch, however, is spline-connected to the shaft—although free to slide linearly. As shown, it is the right follower gear that is locked to the drive shaft. Hence the output gear rotates clockwise until the pin strikes the reversing level to shift the toggle to the left. Once past its center, the toggle spring snaps the ratchet to the left to engage the left follower gear. This instantly reverses the output, which now rotates counterclockwise until the pin again strikes the reversing level. Thus the mechanism reverses itself for every 360° rotation of the input.

Modified-Watt's reverser

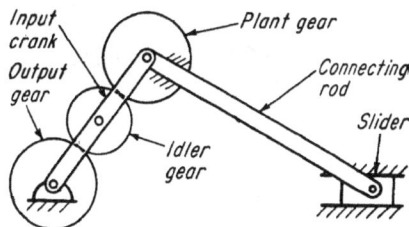

This is a modification of the well-known Watt crank mechanism. The input crank causes the planet gear to revolve around the output gear. But because the planet gear is fixed to the connecting rod, it causes the output gear to continually reverse itself. If the radii of the two gears are equal, each full rotation of the input link will cause the output gear to oscillate through the same angle as the rod.

Automatically switching from one pivot point to another in midstroke.

Two pivots and the intermediary flange govern the cutting sequence. The flange is connected to the press frame at the upper pivot, and the cutting ram is connected to the flange at the lower pivot. In the first part of the cycle, the ram turns around the lower pivot and shears the plate with the square-cut blade; the motion of the intermediary flange is restrained by the flange-holding piston.

After the shearing cut, the ram stop bottoms on the flange. This overcomes the restraining force of the flange-holding piston, and the ram turns around the upper pivot. This brings the beveling blade into contact with the plate for the bevel cut.

TEN MECHANICAL COMPUTING MECHANISMS

Analog computing mechanisms are capable of almost instantaneous response to minute variations in input. Basic units, similar to the examples shown, are combined to form the final mechanism. These mechanisms add, subtract, resolve vectors, or solve special or trigonometric functions.

(A) Bevel-gear differential

(B) Sliding-link differential

(C) Rotating-link differential

Fig. 1 Addition and subtraction is usually based on the differential principle; variations depend on whether inputs: (A) rotate shafts, (B) translate links, or (C) angularly displace links. Mechanisms can solve the equation: $z = c (x \pm y)$, where c is the scale factor, x and y are inputs, and z is the output. The motion of x and y in the same direction performs addition; in the opposite direction it performs subtraction.

(A)

(B)

(C)

Fig. 2 Functional generators mechanize specific equations. (A) A reciprocal cam converts a number into its reciprocal. This simplifies division by permitting simple multiplication between a numerator and its denominator. The cam is rotated to a position corresponding to the denominator. The distance between the center of the cam to the center of the follower pin corresponds to a reciprocal. (B) A function-slot cam is ideal for performing complex functions involving one variable. (C) A function is plotted on a large sheet attached to a table. The x lead screw is turned at a constant speed by an analyzer. An operator or photoelectric follower turns the y output to keep the aligning sight on the curve.

(A) 3-d cam

(B)

Fig. 3 (A) **A three-dimensional cam** generates functions with two variables: $z = f(x, y)$. A cam is rotated by the y-input; the x-input shifts a follower along a pivot rod. The contour of the cam causes a follower to rotate, giving angular displacement to the z-output gear. (B) **A conical cam** for squaring positive or negative inputs: $y = c (\pm x)^2$. The radius of a cone at any point is proportional to the length of string to the right of the point; therefore, cylinder rotation is proportional to the square of cone rotation. The output is fed through a gear differential to convert it to a positive number.

(A)   **(B)**   **(C)**

Mechanical Computing Mechanisms (*continued*)

Fig. 4 Trigonometric functions. (A) A Scotch-yoke mechanism for sine and cosine functions. A crank rotates about fixed point P, generating angle a and giving motion to the arms: $y = c \sin a$; $x = c \cos a$. (B) A tangent-cotangent mechanism generates $x = c \tan a$ or $x = c$ cot β. (C) The eccentric and follower is easily manufactured, but sine and cosine functions are approximate. The maximum error is zero at 90° and 270°; l is the length of the link, and c is the length of the crank.

(A)   **(B)**

(C)   **(D)**   **(E)**

Fig. 5 Component resolvers determine x and y components of vectors that are continuously changing in both angle and magnitude. Equations are $x = z \cos a$, $y = z \sin a$, where z is magnitude of vector, and a is vector angle. Mechanisms can also combine components to obtain a resultant. Inputs in (A) are through bevel gears and lead screws for z-input, and through spur gears for a-input. Compensating gear differential (B) prevents the a-input from affecting the z-input. This problem is solved in (C) with constant-lead cams (D) and (E).

Fig. 6(A)

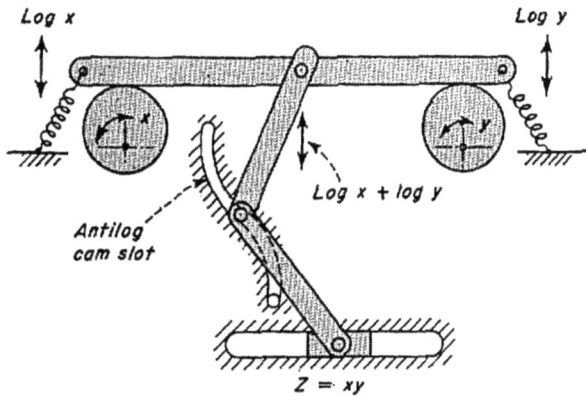

Fig. 6(B)

Fig. 6 **The multiplication of two tables,** x and y, can usually be solved by either: (A) The similar triangle method, or (B) the logarithmic method. In (A), lengths x' and y' are proportional to the rotation of input gears x and y. Distance c is constant. By similar triangles: $z/x = y/c$ or $z = xy/c$, where z is vertical displacement of output rack. The mechanism can be modified to accept negative variables. In (B), the input variables are fed through logarithmic cams to give linear displacements of log x and log y. The functions are then added by a differential link giving $z = \log x + \log xy$ (neglecting scale factors). The result is fed through the antilog cam so that the motion of the follower represents $z = xy$.

Fig. 7(A)

Fig. 7(B)

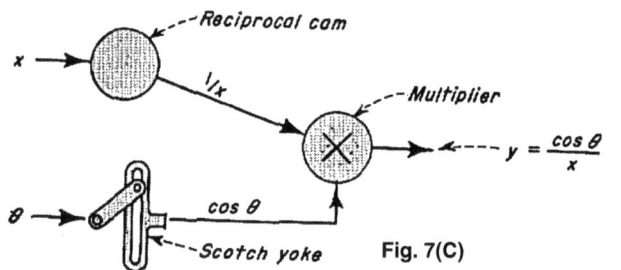

Fig. 7(C)

Fig. 7 **Multiplication of complex functions** can be accomplished by substituting cams in place of input slides and racks of the mechanism in Fig. 6. The principle of similar triangles still applies. The mechanism in (A) solves the equation: $z = f(y)x^2$. The schematic is shown in (B). Division of two variables can be done by feeding one of the variables through a reciprocal cam and then multiplying it by the other. The schematic in (C) shows the solution of $y = \cos\theta/x$.

(A)

(B)

Fig. 8 Integrators are essentially variable-speed drives. The x-input shaft in Fig. 8 (A) rotates the disk which, in turn, rotates the friction wheel on the y-input shaft which is perpendicular to the x-input shaft. As the friction wheel turns, it rotates a spline on the movable y-input shaft. The gear on the end of the parallel z-output shaft drives that shaft.

Moving the y-input shaft along the radius dimension of the disk changes the rotational speed of the friction wheel from zero at the center of the disk to a maximum at the periphery. The z-axis output is thus a function of the rotational speed of the x-input, the diameter of the friction wheel, and y, the radius distance of the wheel on the disk.

In the integrator shown in Fig. 8 (B), two balls replace the friction wheel and spline of the y-input axis, and a roller replaces the gear on the z-output shaft to provide a variable-speed output as the y-input shaft is moved across the entire diameter of the disk.

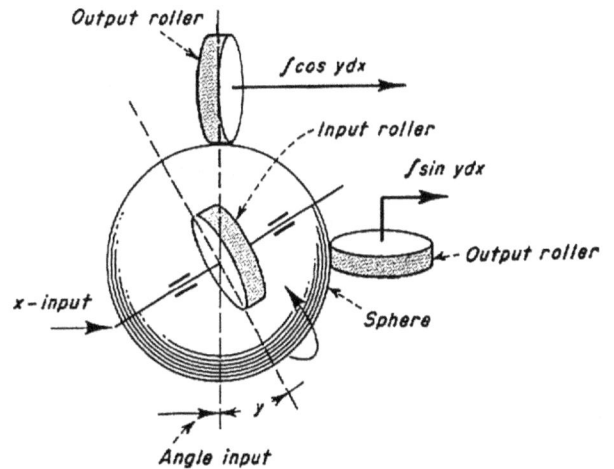

Fig. 9 A component integrator has three disks to obtain the x and y components of a differential equation. The input roller on the x-input shaft spins the sphere, and the y-input lever arm changes the angle of the roller with respect to the sphere. The sine and cosine output rollers provide integrals of components that parallel the x and y axes.

Fig. 10 This differentiator is based on the principle that a viscous drag force in a thin layer of fluid is proportional to the velocity of a rotating x-input shaft. The drag force is counteracted by resisting springs in tension. Spring length is regulated by a servomotor controlled by electrical contacts at the base of the housing. A change in shaft velocity causes a change in viscous torque. A shift in the housing closes one set of electrical contacts, causing the motor shaft to turn. This repositions a rack which adjusts the spring tension and balances the system. The total rotation of the servomotor gear is proportional to dx/dt.

349

NINE DIFFERENT MECHANICAL POWER AMPLIFIERS

Precise positioning and movement of heavy loads are two
basic jobs for this all-mechanical torque booster.

This mechanical power amplifier has a fast response. Power
from its continuously rotating drums is instantaneously available.
When used for position-control applications, pneumatic,
hydraulic, and electrical systems—even those with continuously
running power sources—require transducers to change signals
from one energy form to another. The mechanical power amplifier, on the other hand, permits direct sensing of the controlled
motion.

Four major advantages of this all-mechanical device are:

1. Kinetic energy of the power source is continuously available
   for rapid response.
2. Motion can be duplicated and power amplified without converting energy forms.
3. Position and rate feedback are inherent design characteristics.
4. Zero slip between input and output eliminates the possibility
   of cumulative error.

One other important advantage is the ease with which this
device can be adapted to perform special functions—jobs for
which other types of systems would require the addition of more
costly and perhaps less reliable components. The six applications
which follow illustrate how those advantages have been put to
work in solving widely divergent problems.

The capstan principle is the basis for the mechanical power amplifier described here that combines two counterrotating drums. The
drums are continuously rotating but only transmit torque when the
input shaft is rotated to tighten the band on drum A. Overrun of output
is stopped by drum B, when overrun tightens the band on this drum.

A capstan is a simple mechanical amplifier—rope wound on a motor-driven drum
slips until slack is taken up on the free end. The force needed on the free end to lift
the load depends on the coefficient of friction and the number of turns of rope. By
connecting bands A and B to an input shaft and arm, the power amplifier provides an
output in both directions, plus accurate angular positioning. When the input shaft is
turned clockwise, the input arm takes up the slack on band A, locking it to its drum.
Because the load end of locked band A is connected to the output arm, it transmits
the CW motion of the driven drum on which it is wound to the output shaft. B and B
therefore slacks off and slips on its drum. When the CW motion of the input shaft
stops, tension on band A is released and it slips on its own drum. If the output shaft
tries to overrun, the output arm will apply tension to band B, causing it to tighten on
the CCW rotating drum and stop the shaft.

Nonlinear Broaching

Problem: In broaching large-bore rifles, the twist given to the lands and grooves represents a nonlinear function of barrel length. Development work on such rifles usually requires some experimentation with this function. At present, rotation of the broaching head is performed by a purely mechanical arrangement consisting of a long, heavy wedge-type cam and appropriate gearing. For steep twist angles, however, the forces acting on this mechanism become extremely high.

Solution: A suitable mechanical power amplifier, with its inherent position feedback, was added to the existing mechanical arrangement, as shown in Fig. 1. The cam and follower, instead of having to drive the broaching head, simply furnish enough torque to position the input shaft of the amplifier.

Hydraulic Winch Control

Problem: Hydraulic pump-motor systems are excellent for controlling position and motion at high power levels. In the 10- to 150-hp range, for example, the usual approach is to vary the output of a positive displacement pump in a closed-loop hydraulic circuit. In many of the systems that might be able to control this displacement, however, a force feedback proportional to system pressure can lead to serious errors or even oscillations.

Solution: Figure 2 shows an external view of the complete package. The output shaft of the mechanical power amplifier controls pump displacement, while its input is controlled by hand. In a more recent development requiring remote manual control, a servomotor replaces this local handwheel. Approximately 10 lb-in. torque drives a 600 lb-in. load. If this system had to transmit 600 lb-in., the equipment would be more expensive and more dangerous to operate.

Load Positioning

Problem: It was necessary for a 750-lb load to be accelerated from standstill in 0.5 s and brought into speed and position synchronization with a reference linear motion. It was also necessary that the source of control motion be permitted to accelerate more rapidly than the load itself. Torque applied to the load could not be limited by any kind of slipping device.

Solution: A system with a single mechanical power amplifier provided the solution (Fig. 3). A mechanical memory device, preloaded for either rotation, drives the input shaft of the amplifier. This permits the input source to accelerate as rapidly as desired. The total control input travel minus the input travel of the amplifier shaft is temporarily stored. After 0.5 seconds, the load reaches proper speed, and the memory device transmits position information in exact synchronization with the input.

Tensile Testing Machine

Problem: On a hydraulic tensile testing machine, the stroke of the power cylinder had to be controlled as a function of two variables: tension in, and extension of, the test specimen. A programming device, designed to provide a control signal proportional to these variables, had an output power level of about 0.001 hp—too low to drive the pressure regulator controlling the flow to the cylinder.

Solution: An analysis of the problem revealed three requirements: the output of the programmer had to be amplified about 60 times, position accuracy had to be within 2°, and acceleration had to be held at a very low value. A mechanical power amplifier satisfied all three requirements. Figure 4 illustrates the completed system. Its design is based principally on steady-state characteristics.

Remote Metering and Counting

Problem: For a remote, liquid-metering job, synchro systems had been used to transmit remote meter readings to a central station and repeat this information on local indicating counters. The operation involved a large number of meters and indicators. As new equipment (e.g. ticket printers) was added, the torque requirement also grew.

Solution: Mechanical power amplifiers in the central station indicators not only supplied extra output torque but also made it possible to specify synchros that were even smaller than those originally selected to drive the indicators alone (see Fig. 5).

The synchro transmitters selected operate at a maximum speed of 600 rpm and produce only about 3 oz-in. of torque. The mechanical power amplifiers furnish up to 100 lb-in. of torque, and are designed to fit in the bottom of the registers shown in Fig. 5. Total accuracy is within 0.25 gallon, and error is noncumulative.

Irregular Routing

Problem: To control remotely the table position of a routing machine from information stored on a film strip. The servoloop developed to interpret this information produced only about 1 oz.-in. of torque. About 20 lb.-ft was required at the table feedscrew.

Solution: Figure 6 shows how a mechanical power amplifier supplied the necessary torque at the remote table location. A position transmitter converts the rotary motion output of the servoloop to a proportional electrical signal and sends it to a differential amplifier at the machine location. A position receiver, geared to the output shaft, provides a signal proportional to table position. The differential amplifier compares these, amplifies the difference, and sends a signal to either counterrotating electromagnetic clutch, which drives the input shaft of the mechanical power amplifier.

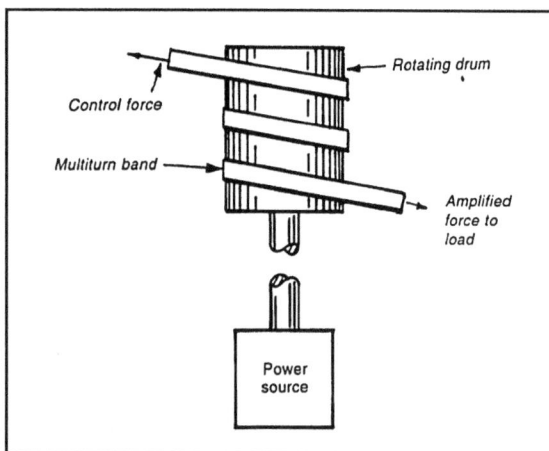

A mechanical power amplifier that drives a crossfeed slide is based on the principle of the windlass. By varying the control force, all or any part of power to the drum can be used.

Two drums mounted back to back supply the bi-directional power needed in servo systems. Replacing the operator with a two-phase induction servomotor permits electronic or magnetic signal amplification. A rotating input avoids a linear input and output of the simple windlass. Control and output ends of the multiturn bands are both connected to gears mounted concentrically with the drum axis.

When the servomotor rotates the control gear, it locks the band-drum combination, forcing output gear to rotate with it. Clockwise rotation of the servomotor produces CW power output while the second drum idles. Varying the servo speed, by changing servo voltage, varies output speed.

FORTY-THREE VARIABLE-SPEED DRIVES AND TRANSMISSIONS

CONE DRIVES

The simpler cone drives in this group have a cone or tapered roller in combination with a wheel or belt (Fig. 1). They have evolved from the stepped-pulley system. Even the more sophisticated designs are capable of only a limited (although infinite) speed range, and generally must be spring-loaded to reduce slippage.

Adjustable-cone drive (Fig. 1A). This is perhaps the oldest variable-speed friction system, and is usually custom built. Power from the motor-driven cone is transferred to the output shaft by the friction wheel, which is adjustable along the cone side to change the output speed. The speed depends upon the ratio of diameters at point of contact.

Two-cone drive (Fig. 1B). The adjustable wheel is the power transfer element, but this drive is difficult to preload because both input and output shafts would have to be spring loaded. The second cone, however, doubles the speed reduction range.

Cone-belt drives (Fig. 1C and D). In Fig. 1C the belt envelopes both cones; in Fig. 1D a long-loop endless belt runs between the cones. Stepless speed adjustment is obtained by shifting the belt along the cones. The cross section of the belt must be large enough to transmit the rated force, but the width must be kept to a minimum to avoid a large speed differential over the belt width.

Electrically coupled cones (Fig. 2). This drive is composed of thin laminates of paramagnetic material. The laminates are separated with semidielectric materials which also localize the effect of the inductive field. There is a field generating device within the driving cone. Adjacent to the cone is a positioning motor for the field generating device. The field created in a particular section of the driving cone induces a magnetic effect in the surrounding lamination. This causes the laminate and its opposing lamination to couple and rotate with the drive shaft. The ratio of diameters of the cones, at the point selected by positioning the field-generating component, determines the speed ratio.

Fig. 1

Fig. 2

Section AA

Fig. 3

Graham drive (Fig. 3). This commercial unit combines a planetary-gear set and three tapered rollers (only one of which is shown). The ring is positioned axially by a cam and gear arrangement. The drive shaft rotates the carrier with the tapered rollers, which are inclined at an angle equal to their taper so that their outer edges are parallel to the centerline of the assembly. Traction pressure between the rollers and ring is created by centrifugal force, or spring loading of the rollers. At the end of each roller a pinion meshes with a ring gear. The ring gear is part of the planetary gear system and is coupled to the output shaft.

The speed ratio depends on the ratio of the diameter of the fixed ring to the effective diameter of the roller at the point of contact, and is set by the axial position of the ring. The output speed, even at its maximum, is always reduced to about one-third of input speed because of the differential feature. When the angular speed of the driving motor equals the angular speed of the centers of the tapered rollers around their mutual centerline (which is set by the axial position of the nonrotating friction ring), the output speed is zero. This drive is manufactured in ratings up to 3 hp; efficiency reaches 85 percent.

Cone-and-ring drive (Fig. 4). Here, two cones are encircled by a preloaded ring. Shifting the ring axially varies the output speed. This principle is similar to that of the cone-and-belt drive (Fig. 1C). In this case, however, the contact pressure between ring and cones increases with load to limit slippage.

Planetary-cone drive (Fig. 5). This is basically a planetary gear system but with cones in place of gears. The planet cones are rotated by the sun cone which, in turn, is driven by the motor. The planet cones are pressed between an outer non-rotating ring and the planet hold. Axial adjustment of the ring varies the rotational speed of the cones around their mutual axis. This varies the speed of the planet holder and the output shaft. Thus, the mechanism resembles that of the Graham drive (Fig. 3).

The speed adjustment range of the unit illustrated is from 4:1 to 24:1. The system is built in Japan in ratings up to 2 hp.

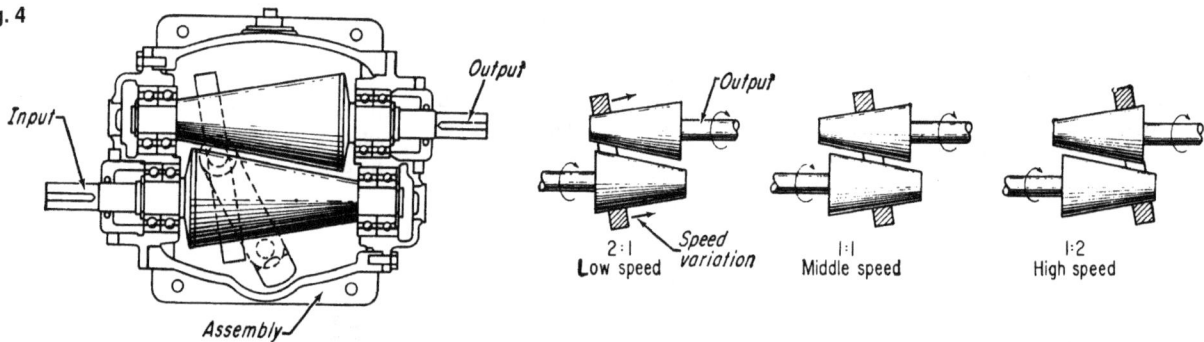

Fig. 4

Input / Output / Assembly

2:1 Low speed — Speed variation

1:1 Middle speed

1:2 High speed

Fig. 5

Outer ring / Planet cones / Output / Pressure applier / Planet holder

Low reduction

Medium reduction

High reduction

Speed control wheel / Motor drive / Planet cone / Sun cone / B / Pressure applier / Output / Effective diameter / Planet holder / Assembly

Adjustable disk drives (Figs. 6A and 6B). The output shaft in Fig. 6A is perpendicular to the input shaft. If the driving power, the friction force, and the efficiency stay constant, the output torque decreases in proportion to increasing output speed. The wheel is made of a high-friction material, and the disk is made of steel. Because of relatively high slippage, only small torques can be transmitted. The wheel can move over the center of the disk because this system has infinite speed adjustment.

To increase the speed, a second disk can be added. This arrangement (Fig. 6B) also makes the input and output shafts parallel.

Spring-loaded disk drive (Fig. 7). To reduce slippage, the contact force between the rolls and disks in this commercial drive is increased with the spring assembly in the output shaft. Speed adjustments are made by rotating the leadscrew to shift the cone roller in the vertical direction. The drive illustrated has a 4-hp capacity. Drives rated up to 20 hp

can have a double assembly of rollers. Efficiency can be as high as 92 percent. Standard speed range is 6:1, but units of 10:1 have been built. The power transferring-ring components, which are made of hardened steel, operate in an oil mist, thus minimizing wear.

Planetary disk drive (Fig. 8). Four planet disks replace planet gears in this friction drive. Planets are mounted on levers which control radial position and therefore control the orbit. Ring and sun disks are spring-loaded.

Fig. 6

(A) Single-disk drive

(B) Double-disk drive

Fig. 7

Fig. 8

Ring-and-pulley drive (Fig. 9). A thick steel ring in this drive encircles two variable-pitch (actually variable-width) pulleys. A novel gear-and-linkage system simultaneously changes the width of both pulleys (see Fig. 9B). For example, when the top pulley opens, the sides of the bottom pulley close up. This reduces the effective pitch diameter of the top pulley and increases that of the bottom pulley, thus varying the output speed.

Normally, the ring engages the pulleys at points A and B. However, under load,

the driven pulley resists rotation and the contact point moves from B to D because of the very small elastic deformation of the ring. The original circular shape of the ring is changed to a slightly oval form, and the distance between points of contact decreases. This wedges the ring between the pulley cones and increases the contact pressure between ring and pulleys in proportion to the load applied, so that constant horsepower at all speeds is obtained. The drive can have up to 3-hp capacity; speed variations can be 16:1, with a practical range of about 8:1.

Some manufacturers install rings with unusual cross sections (Fig. 10) formed by inverting one of the sets of sheaves.

Double-ring drive (Fig. 11). Power transmission is through two steel traction rings that engage two sets of disks mounted on separate shafts. This drive requires that the outer disks be under a compression load by a spring system (not illustrated). The rings are hardened and convex-ground to reduce wear. Speed is changed by tilting the ring support cage, forcing the rings to move to the desired position.

Fig. 9

a

Operating principle

b

Fig. 10

With inverted pulleys

Fig. 11

Sphere-and-disk drives (Figs. 12 and 13). The speed variations in the drive shown in Fig. 12 are obtained by changing the angle that the rollers make in contacting spherical disks. As illustrated, the left spherical disk is keyed to the driving shaft and the right disk contains the output gear. The sheaves are loaded together by a helical spring.

One commercial unit, shown in Fig. 13, is a coaxial input and output shaft-version of the Fig. 12 arrangement. The rollers are free to rotate on bearings and can be adjusted to any speed between the limits of 6:1 and 10:1. An automatic device regulates the contact pressure of the rollers, maintaining the pressure exactly in proportion to the imposed torque load.

Double-sphere drive (Fig. 14). Higher speed reductions are obtained by grouping a second set of spherical disks and rollers. This also reduces operating stresses and wear. The input shaft runs through the unit and carries two opposing spherical disks. The disks drive the double-sided output disk through two sets of three rollers. To change the ratio, the angle of the rollers is varied. The disks are axially loaded by hydraulic pressure.

Tilting-ball drive (Fig. 15). Power is transmitted between disks by steel balls whose rotational axes can be tilted to change the relative lengths of the two contact paths around the balls, and hence the output speed. The ball axes can be tilted uniformly in either direction; the effective rolling radii of balls and disks produce speed variations up to 3:1 increase, or 1:3 decrease, with the total up to 9:1 variation in output speed.

Tilt is controlled by a cam plate through which all ball axes project. To prevent slippage under starting or shock load, torque responsive mechanisms are located on the input and output sides of the drive. The axial pressure created is proportional to the applied torque. A worm drive positions the plate. The drives have been manufactured with capacities to 15-hp. The drive's efficiency is plotted in the chart.

Sphere and roller drive (Fig. 16). The roller, with spherical end surfaces, is

Fig. 12

Mechanism for pivoting disks — Spherical disks — Roller — Output gear — Input shaft

Fig. 13

Angle of:
Lowest speed ratio
One to one ratio
Highest speed ratio

Fig. 14

Double-sided output disks — Input — Output bell — Output housing — Rollers

Fig. 15

Efficiency of tilting-ball drive

1:1 ratio   3:1 decrease   1:3 increase

Fig. 16

Fig. 17

eccentrically mounted between the coaxial input and output spherical disks. Changes in speed ratio are made by changing the angular position of the roller.

The output disk rotates at the same speed as the input disk when the roller centerline is parallel to the disk center-line, as in Fig. 16A. When the contact point is nearer the centerline on the output disk and further from the centerline on the input disk, as in Fig. 16B, the output speed exceeds that of the input. Conversely, when the roller contacts the output disk at a large radius, as in Fig. 16C, the output speed is reduced.

A loading cam maintains the necessary contact force between the disks and power roller. The speed range reaches 9 to 1; efficiency is close to 90 percent.

Ball-and-cone drive (Fig. 17). In this simple drive the input and output shafts are offset. Two opposing cones with 90° internal vertex angles are fixed to each shaft. The shafts are preloaded against each other. Speed variation is obtained by positioning the ball that contacts the cones. The ball can shift laterally in relation to the ball plate. The conical cavities, as well as the ball, have hardened surfaces, and the drive operates in an oil bath.

Ball-and-disk drive (Fig. 18). Friction disks are mounted on splined shafts to allow axial movement. The steel balls carried by swing arms rotate on guide rollers, and are in contact with driving and driven disks. Belleville springs provide the loading force between the balls and the disks. The position of the balls controls the ratio of contact radii, and thus the speed.

Only one pair of disks is required to provide the desired speed ratio; the multiple disks increase the torque capacity. If the load changes, a centrifugal loading device increases or decreases the axial pressure in proportion to the speed. The helical gears permit the output shaft to be coaxial with respect to the input shaft. Output to input speed ratios are from 1 to 1 to 1 to 5, and the drive's efficiency can reach 92 percent. Small ball and disk drives are rated to 9 hp, and large ball and disk drives are rated to 38 hp.

Oil-coated disks (Fig. 19). Power is transmitted without metal-to-metal contact at 85 percent efficiency. The interleaved disk sets are coated with oil when operating. At their points of contact, axial pressure applied by the rim disks compresses the oil film, increasing its viscosity. The cone disks transmit motion to the rim disks by shearing the molecules of the high-viscosity oil film.

Three stacks of cone disks (only one stack is shown) surround the central rim stack. Speed is changed by moving the cones radially toward the rim disks (output speed increases) or away from the rim disks (output speed decreases). A spring and cam on the output shaft maintain the pressure of the disks at all times.

Drives with ratings in excess of 60 hp have been built. The small drives are cooled, but water cooling is required for the larger units.

Under normal conditions, the drive can transmit its rated power with a 1 percent slip at high speeds and 3 percent slip at low speeds.

Fig. 18

Fig. 19

Variable-stroke drive (Fig. 20). This drive is a combination of a four-bar linkage with a one-way clutch or ratchet. The driving member rotates the eccentric that, through the linkage, causes the output link to rotate a fixed amount. On the return stroke, the output link overrides the output shaft. Thus a pulsating motion is transmitted to the output shaft which in many applications, such as feeders and mixers, is a distinct advantage. Shifting the adjustable pivot varies the speed ratio. By adding eccentrics, cranks, and clutches in the system, the frequency of pulsations per revolution can be increased to produce a smoother drive.

Morse drive (Fig. 21). The oscillating motion of the eccentric on the output shaft imparts motion to the input link, which in turn rotates the output gears. The travel of the input link is regulated by the control link that oscillates around its pivot and carries the roller, which rides in the eccentric cam track. Usually, three linkage systems and gear assemblies overlap the motions: two linkages on return, while the third is driving. Turning the handle repositions the control link and changes the oscillation angles of the input link, intermediate gear, and input gear. This is a constant-torque drive with limited range. The maximum torque output is 175 ft-lb at the maximum input speed of 180 rpm. Speed can be varied between 4.5 to 1 and 120 to 1.

Zero-Max drive (Fig. 22). This drive is also based on the variable-stroke principle. With an 1800-rpm input, it will deliver 7200 or more impulses per minute to the output shaft at all speed ratings above zero. The pulsations of this drive are damped by several parallel sets of mechanisms between the input and output shafts. (Figure 22 shows only one of these sets.)

At zero input speed, the eccentric on the input shaft moves the connecting rod up and down through an arc. The main link has no reciprocating motion. To set the output speed, the pivot is moved (upward in the figure), thus changing the direction of the connecting rod motion and imparting an oscillatory motion to the main link. The one-way clutch mounted on the output shaft provides the ratchet action. Reversing the input shaft rotation does not reverse the output. However, the drive can be reversed in two ways: (1) with a special reversible clutch, or (2) with a bellcrank mechanism in gear-head models.

This drive is classified as an infinite-speed range drive because its output speed passes through zero. Its maximum speed is 2000 rpm, and its speed range is from zero to one-quarter of its input speed. It has a maximum rated capacity of $^3/_4$ hp.

Fig. 20

Fig. 21

Fig. 22

At zero speed

Fig. 23

Fig. 23 **The Sellers' disks** consist of a mechanism for transmitting power between fixed parallel shafts. Convex disks are mounted freely on a rocker arm, and they are pressed firmly against the flanges of the shaft wheels by a coiled spring to form the intermediate sheave. The speed ratio is changed by moving the rocker lever. No reverse is possible, but the driven shaft can rotate above or below the driver speed. The convex disk must be mounted on self-aligning bearings to ensure good contact in all positions.

Fig. 24 **A curved disk device** is formed by attaching a motor that is swung on its pivot so that it changes the effective diameters of the contact circles. This forms a compact drive for a small drill press.

Fig. 25 **This is another motorized modification** of the older mechanism shown in Fig. 2. It works on the principle that is similar that of Fig. 2, but it has only two shafts. Its ratio is changed by sliding the motor in vee guides.

Fig. 26 **Two cones mounted close together** and making contact through a squeezed belt permit the speed ratio to be changed by shifting the belt longitudinally. The taper on the cones must be moderate to avoid excessive wear on the sides of the belt.

Fig. 27 **These cones** are mounted at any convenient distance apart. They are connected by a belt whose outside edges consist of an envelope of tough, flexible rubberized fabric that is wear-resistant. It will withstand the wear caused by the belt edge traveling at a slightly different velocity than that part of the cone it actually contacts. The mechanism's speed ratio is changed by sliding the belt longitudinally.

Fig. 24

Fig. 26

Fig. 25

Fig. 27

361

Fig. 28 **This drive** avoids belt "creep" and wear in speed-cone transmissions. The inner bands are tapered on the inside, and they present a flat or crowned contact surface for the belt in all positions. The speed ratio is changed by moving the inner bands rather than the main belts.

Fig. 29 **This drive** avoids belt wear when the drive has speed cones. However, the creeping action of the belt is not eliminated, and the universal joints present ongoing maintenance problems.

Fig. 30 **This drive** is a modification of the drive shown in Fig. 7. A roller is substituted for the belt, reducing the overall size of the drive.

Fig. 31 **The main component** of this drive is a hollow internal cone driven by a conical pulley on the motor shaft. Its speed ratio can be changed by sliding the motor and pulley up or down in the vee slide. When the conical pulley on the motor shaft is moved to the center of the driving cone, the motor and cone run at the same speed. This feature makes the system attractive in applications where heavy torque requirements are met at the motor's rated speed and it is useful to have lower speeds for light preliminary operations.

Fig. 32 **In this transmission,** the driving pulley cone and driven cone are mounted on the same shaft with their small diameters directed toward each other. The driving pulley (at right) is keyed to the common shaft, and the driven cone (at left) is mounted on a sleeve. Power is transmitted by a series of rocking shafts with rollers mounted on their ends. The shafts are free to slide while they are pivoted within sleeves within a disk that is perpendicular to the driven-cone mounting sleeve. The speed ratio can be changed by pivoting the rocking shafts and allowing them to slide across the conical surfaces of the driving pulley and driven cone.

Fig. 33 **This transmission** has curved surfaces on its planetary rollers and races. The cone-shaped inner races revolve with the drive shaft, but are free to slide longitudinally on sliding keys. Strong compression springs keeps the races in firm contact with the three planetary rollers.

Fig. 34 **This Graham transmission** has only five major parts. Three tapered rollers are carried by a spider fastened to the drive shaft. Each roller has a pinion that meshes with a ring gear connected to the output shaft. The speed of the rollers as well as the speed of the output shaft is varied by moving the contact ring longitudinally. This movement changes the ratio of the contacting diameters.

Fig. 30

Fig. 31

Fig. 32

Fig. 33

Fig. 28

Fig. 29

Fig. 34

Fig. 35

Fig. 36

Fig. 37

Fig. 38

These ratchet and inertial drives provide variable-speed driving of heavy and light loads.

Fig. 35 This **variable-speed drive** is suitable only for very light duty in a laboratory or for experimental work. The drive rod receives motion from the drive shaft and it rocks the lever. A friction clutch is formed in a lathe by winding wire around a drill rod whose diameter is slightly larger than the diameter of the driven shaft. The speed ratio can be changed when the drive is stationary by varying the length of the rods or the throw of the eccentric.

Fig. 36 This **Torrington lubricator drive** illustrates the general principles of ratchet transmission drives. Reciprocating motion from a convenient sliding part, or from an eccentric, rocks the ratchet lever. That motion gives the variable-speed shaft an intermittent unidirectional motion. The speed ratio can be changed only when the unit is stationary. The throw of the ratchet lever can be varied by placing a fork of the driving rod in a different hole.

Fig. 37 This **drive** is an extension of the principle illustrated in Fig. 2. The Lenney transmission replaces the ratchet with an over-running clutch. The speed of the driven shaft can be varied while the unit is in motion by changing the position of the connecting-lever fulcrum.

Fig. 38 This **transmission** is based on the principle shown in Fig. 3. The crank disk imparts motion to the connecting rod. The crosshead moves toggle levers which, in turn, give unidirectional motion to the clutch wheel when the friction pawls engage in a groove. The speed ratio is changed by varying the throw of the crank with the aid of a rack and pinion.

Fig. 39 This is a **variable speed transmission** for gasoline-powered railroad section cars. The connecting rod from the crank, mounted on a constant-speed shaft, rocks the oscillating lever and actuates the over-running clutch. This gives intermittent but unidirectional motion to the variable-speed shaft. The toggle link keeps the oscillating lever within the prescribed path. The speed ratio is changed by swinging the bell crank toward the position shown in the dotted lines, around the pivot attached to the frame. This varies the movement of the over-running clutch. Several units must be out-of-phase with each other for continuous shaft motion.

Fig. 39

Fig. 40   This Thomas transmission is an integral part of an auto-mobile engine whose piston motion is transferred by a conventional connecting rod to the long arm of the bellcrank lever oscillating about a fixed fulcrum. A horizontal connecting rod, which rotates the crankshaft, is attached to the short arm of the bellcrank. Crankshaft motion is steadily and continuously maintained by a flywheel. However, no power other than that required to drive auxiliaries is taken from this shaft. The main power output is transferred from the bellcrank lever to the over-running clutch by a third connecting rod. The speed ratio is changed by sliding the top end of the third con-necting rod within the bellcrank lever with a crosshead and guide mechanism. The highest ratio is obtained when the crosshead is far-thest from the fulcrum, and movement of the crosshead toward the fulcrum reduces the ratio until a "neutral" position is reached. That occurs when the center line of the connecting rod coincides with the fulcrum.

Fig. 41   This Constantino torque converter is another automotive transmission system designed and built as part of the engine. It fea-tures an inherently automatic change of speed ratio that tracks the speed and load on the engine. The constant-speed shaft rotates a crank which, in turn, drives two oscillating levers with inertia weights at their ends. The other ends are attached by links to the rocking levers. These rocking levers include over-running clutches. At low engine speeds, the inertia weights oscillate through a wide angle. As a result, the reaction of the inertia force on the other end of the lever is very slight, and the link imparts no motion to the rocker lever. Engine speed increases cause the inertia weight reaction to increase. This rocks the small end of the oscillating lever as the crank rotates. The resulting motion rocks the rocking lever through the link, and the variable shaft is driven in one direction.

Fig. 42   This transmission has a differential gear with an adjustable escapement. This arrangement bypasses a variable por-tion of the drive-shaft revolutions. A constant-speed shaft rotates a freely mounted worm wheel that carries two pinion shafts. The firmly fixed pinions on these shafts, in turn, rotate the sun gear that meshes with other planetary gears. This mechanism rotates the small worm gear attached to the variable-speed output shaft.

Fig. 43   This Morse transmission has an eccentric cam integral with its constant-speed input shaft. It rocks three ratchet clutches through a series of linkage systems containing three rollers that run in a circular groove cut in the cam face. Unidirectional motion is transmitted to the output shaft from the clutches by planetary gear-ing. The speed ratio is changed by rotating an anchor ring containing a fulcrum of links, thus varying the stroke of the levers.

Fig. 40

Fig. 41

Fig. 42

Fig. 43

TEN VARIABLE-SPEED FRICTION DRIVES

These drives can be used to transmit both high torque, as on industrial machines, and low torque, as in laboratory instruments. All perform best if they are used to reduce and not to increase speed. All friction drives have a certain amount of slip due to imperfect rolling of the friction members, but with effective design this slip can be held constant, resulting in constant speed of the driven member. Compensation for variations in load can be achieved by placing inertia masses on the driven end. Springs or similar elastic members can be used to keep the friction parts in constant contact and exert the force necessary to create the friction. In some cases, gravity will take the place of such members. Custom-made friction materials are generally recommended, but neoprene or rubber can be satisfactory. Normally only one of the friction members is made or lined with this material, while the other is metal.

Fig. 1 A disk and roller drive. The roller is moved radially on the disk. Its speed ratio depends upon the operating diameter of the disk. The direction of relative rotation of the shafts is reversed when the roller is moved past the center of the disk, as indicated by dotted lines.

Fig. 2 Two disks have a free-spinning, movable roller between them. This drive can change speed rapidly because the operating diameters of the disks change in an inverse ratio.

Fig. 3 Two disks are mounted on the same shaft and a roller is mounted on a threaded spindle. Roller contact can be changed from one disk to the other to change the direction of rotation. Rotation can be accelerated or decelerated by moving the screw.

Fig. 4 A disk contacts two differential rollers. The rollers and their bevel gears are free to rotate on shaft S_2. The other two bevel gears are free to rotate on pins connected by S_2. This drive is suitable for the accurate adjustment of speed. S_2 will have the differential speed of the two rollers. The differential assembly is movable across the face of the disk.

Fig. 5 This drive is a drum and roller. A change of speed is performed by skewing the roller relative to the drum.

Fig. 6 This drive consists of two spherical cones on intersecting shafts and a free roller.

Fig. 7 This drive consists of a spherical cone and groove with a roller. It can be used for small adjustments in speed.

Fig. 8 This drive consists of two disks with torus contours and a free rotating roller.

Fig. 9 This drive consists of two disks with a spherical free rotating roller.

Fig. 10 This drive has split pulleys for V belts. The effective diameter of the belt grip can be adjusted by controlling the distance between the two parts of he pulley.

FOUR DRIVES CONVERT OSCILLATING MOTION TO ONE-WAY ROTATION

These four drives change oscillating motion into
one-way rotation to perform feeding tasks and counting.

Input — Conventional spring clutch — Reverse-lock spring — Output

Stationary housing

Fig. 1 A double-spring clutch drive.

Stationary housing — Bevel gears — Input — Reverse-lock spring — Output

Spring clutch

Fig. 3 A full-wave rectification drive.

Input — A B — Output

Fig. 2 A basic spring clutch.

Helical gear — Core — Stationary magnet — Hysteresis brake (no wearing parts) — Clamped shaft

Drive pin — Input — A B

Spur gears — Shuttle pinion Contains both spur and helical pinions

Side view of gears 2,4 — Output

Fig. 4 A shuttle-pinion drive.

Fig. 5   A reciprocating-ball drive.

Double-Spring Clutch Drive

The one way drive, shown in Fig. 1, was designed for a printer. It was necessary to convert the oscillating motion of the input crank (20° in this case) into one-way motion to advance an inking ribbon. The drive has a spring clutch, a helical spring joining two co-linear butting shafts (Fig. 2). The spring can be made of square or rectangular cross section wire.

This clutch transmits torque in one direction only because it overrides when it is reversed. The helical spring, which bridges both shafts, need not be fastened at either end; a slight interference fit is acceptable. Rotating the input shaft in the direction tending to wind the spring (direction A in Fig. 2) causes the spring to grip both shafts and then transmit motion from the input to the output shaft. Reversing the input unwinds the spring, and it overrides the output shaft with a drag—but this drag, slight as it was, caused a problem in operation.

Basic Spring Clutch

The spring clutch (Fig. 2) did not provide enough friction in the tape drive to allow the spring clutch to slip on the shafts on the return stroke. Thus the output moved in sympathy with the input, and the desired one-way drive was not achieved.

At first, an attempt was made to add friction artificially to the output, but this resulted in an awkward design. Finally the problem was elegantly solved (Fig. 1) by installing a second helical spring, slightly larger than the first that served

exactly the same purpose: transmission of motion in one direction only. This spring, however joined the output shaft and a stationary cylinder. In this way, with the two springs of the same hand, the undesirable return motion of the ribbon drive was immediately arrested, and a positive one-way drive was obtained quite simply.

This compact drive can be considered to be a mechanical *half-wave rectifier* in that it transmits motion in one direction only while it suppresses motion in the reverse direction.

Full-Wave Rectification Drive

The principles described will also produce a mechanical *full-wave rectifier* by introducing some reversing gears, Fig. 3. In this application the input drive in one direction is directly transmitted to the output, as before, but on the reverse stroke the input is passed through reversing gears so that the output appears in the opposite sense. In other words, the original sense of the output is maintained. Thus, the output moves forward twice for each back-and-forth movement of the input.

Shuttle-Pinion Drive

Earlier, a one-way drive was developed that harnessed the axial thrust of a pair of helical gears to shift a pinion, Fig. 4. Although at first glance, it might look somewhat complicated, the drive is inexpensive to make and has been operating successfully with little wear.

When the input rotates in direction *A*, it drives the output through spur gears *1*

and *2*. The shuttle pinion is also driving the helical gear whose rotation is resisted by the magnetic flux built up between the stationary permanent magnet and the rotating core. This magnet-core arrangement is actually a hysteresis brake, and its constant resisting torque produces an axial thrust in mesh of the helical pinion acting to the left. Reversing the input reverses the direction of thrust, which shifts the shuttle pinion to the right. The drive then operates through gears *1, 3,* and *4,* which nullifies the reversion to produce output in the same direction.

Reciprocating-Ball Drive

When the input rotates in direction A, Fig. 5, the drive ball trails to the right, and its upper half engages one of the radial projections in the right ring gear to drive it in the same direction as the input. The slot for the ball is milled at 45° to the shaft axes and extends to the flanges on each side.

When the input is reversed, the ball extends to the flanges on each side, trails to the left and deflects to permit the ball to ride over to the left ring gear, and engage its radial projection to drive the gear in the direction of the input.

Each gear, however, is constantly in mesh with a pinion, which in turn is in mesh with the other gear. Thus, regardless of the direction the input is turned, the ball positions itself under one or another ring gear, and the gears will maintain their respective sense of rotation (the rotation shown in Fig. 5). Hence, an output gear in mesh with one of the ring gears will rotate in one direction only.

EIGHTEEN DIFFERENT LIQUID AND VACUUM PUMPS

1. ROTARY REVERSIBLE PUMP: The driving and idler rotors of this pump turn in the same direction and rotate against neoprene cushions. These cushions can withstand degradation from liquids including oil, kerosene, and gasoline while also allowing sand or grit that enters the pump to flow freely through it without causing damage. This pump, capable of developing constant pressure, can deliver a uniform flow of liquids at any given speed, even when positioned at different angles. Pumps of this type can be driven by gasoline or diesel engines as well as electric motors. They have found applications in vehicular water circulating and oil cooling systems as well as for applying cooling oil to machine tools.

2. IDLER/TOOTH ROTARY PUMP: This pump operates when the idler mounted by a pin to the casing meshes with the rotor's teeth. Capable of operating in either direction, this pump is designed to prevent liquids from the splashing, foaming, or churning that causes pounding noise. Filtered liquids with a wide range of viscosities can be transferred with this iron and bronze pump.

3. SWINGING VANE ROTARY PUMP: The hinged vanes or buckets on the eccentric rotor of this pump swing out by centrifugal force to contact the inner housing wall as the shaft turns. They fit loosely in the rotor recesses and their tips make only light contact with the housing wall, minimizing wear. Liquid pressure increases as the vanes pass between the input (suction) port and output (discharge) port.

4. SWASHPLATE PUMP: These pumps contain a rotating barrel with multiple parallel pistons in cylinders positioned radially around the barrel axis. The pistons are attached to a hold-down plate by universal joints to form a hold-down assembly. This assembly rotates within the barrel as it rotates within the housing driven by the shaft. Springs push the pistons and hold-down plate against the sloping face of a stationary cam plate or swashplate at one end of the housing. As the cylinder barrel is rotated, the pistons are forced to follow the hold-down plate in close contact with the swashplate. This forces the pistons to reciprocate axially as their cylinders suck liquid in during the first half of a revolution and expel it during the second half. The greater the angle on the swashplate, the farther the piston travels, and the larger the amount of liquid transferred.

369

5. HYDRAULIC GEAR PUMP: The gear shafts of this pump are mounted on tapered roller bearings that position the gears accurately, decrease end play, and minimize wear. Heavy-duty pumps of this type can operate at pressures up to 1000 psi. These pumps are made with either single- or double-end shafts, and they can be foot- or flange-mounted. The drive shaft entrance packing is made from oil-resistant material, and the gear shafts are made from hardened molybdenum steel.

6. TWIN PISTON HYDRAULIC PUMP: This pump can build pressures from 100 to 4000 psi at speeds from 600 to 1200 rpm. Pumps of this type can be operated continuously at 900 rpm and 2500 psi with 1.37 gpm delivered. They can be mounted at any angle, and they are typically used with small diameter oil lines and rams as well as compact valves. Pumps of this type contain a pressure-adjusting valve that can be factory set to bypass at predetermined pressures.

7. CENTRIFUGAL PUMP: This pedestal-mounted pump (shown in cross-section) has a five-vaned open bronze impeller capable of transferring high volumes of liquids containing solids. The impeller shaft is aligned and supported by two sets of ball bearings mounted in the pedestal casting. A seal on the shaft prevents any liquid leakage from reaching the bearings. Its suction eye is shown as the threaded opening at the front of the pump casing, and its discharge port is the threaded pipe shown behind the top of the casing. Pumps built with this design are capable of volume ratings up to 500 gpm.

8. EXTERNAL ROTOR PUMP: Rotor (or gear) pumps are positive displacement units used to transfer, meter, or proportion liquids. One of its two rotors is driven, and the lobes of both rotors rotate to make contact with the liquid and move it in the gaps between the lobes and the inner surface of the housing. This raises the liquid pressure as it is transferred from the inlet port to the outlet port. Automatic controls compensate for pump wear to maintain pump volumetric efficiency. These pumps, made from stainless steel, can transfer 5 to 300 gpm without churning or foaming, and they need no lubrication.

These pumps are used to transfer liquids and semisolids, pump vacuums, and boost oil pressure.

9. ROTATING-CYLINDER BLOCK PUMP: The drive shaft and rotating cylinder block of this pump are mounted eccentrically in the pump housing. Each of the four pistons within the block is connected by a rod whose other end is connected to an eccentric ring. As the cylinder block is rotated, the pistons within the block alternately advance or retract as the rod ends slide around the ring. These reciprocating pistons perform pumping as they retract to form suction chambers and extend to form compression chambers. The housing is divided so liquids are drawn into the inlet or suction side and expelled from the outlet or compression side.

10. PERISTALTIC PUMP: These bidirectional pumps are based on continuously squeezing a flexible neoprene or other synthetic rubber tube connected between the input or suction port and the discharge or output port. Liquids drawn into the tube on the suction side by a partial vacuum are forced to move along its length as the tube is sequentially compressed by blocks driven by an eccentric cam. The tube regains its original diameter as the cams pass by. These pumps are widely used to transfer corrosive liquids because the tube isolates the liquid from internal pump parts that could be damaged by it. The bronze pump shown can transfer liquids at a rate of 15 gpm. It develops a vacuum of 35 in. Hg and will work against pressures of 50 lb/in.[2].

11. HIGH-VACUUM PUMP: There are three moving parts in this pump used for "roughing" the first stage of pumping a high vacuum. They are used in the manufacture of various vacuum tubes for lighting, microwave power generation, and other functions in scientific instruments. The exhaust tube from the vessel being evacuated is then transferred to a diffusion pump for finish pumping. A single high-vacuum pump can draw a vacuum from 2 to 5 μm, but when coupled in series, the vacuum can reach 0.5 μm. These pumps are liquid cooled to dissipate heat generated during the pumping action.

12. HELICAL-ROTOR PUMP: This positive displacement pump contains a helical rotor that rotates within a helical stator cavity to transfer liquids or pastes. The rotor is made from low-alloy or tool steel and the stator, a double internal helix, is made from natural rubber or an elastomeric material. The rotor contour matches that of the stator. When the rotor rotates, it can force liquids or paste in either direction. Self-priming with flow that is continuous and smooth, this pump can transfer free-flowing liquids or viscous pastes. Either substance can be clean or contaminated with abrasives, chemically inert or active, and homogeneous or containing solid particles up to 7/8 in. (2.2 cm) in diameter. Simple in design, this pump has no valves or timing gears. The largest standard pump of this design can transfer fluid at flow rates of 150 gpm at up to 200 psi.

13. THREE-SCREW PUMP: These pumps transfer liquids such as fuel oil, lubrication oil, and sea water. Pumps with three screws (as shown) have a central screw that causes the two adjacent screws to rotate in opposite directions. All multi-screw pumps transfer liquid by coordinated action of all screws meshed together in a casing provided with channels to accommodate the number of screws. By contrast, twin-screw pumps have timing gears to control the relative motion of the screws.

14. TWO-SCREW PUMP: These pumps transfer liquid by the rotation of two screws that mesh together in a housing with provision for the adequate flow of liquids. Self-centering herringbone gears control the relative motions of the two-screw pump and prevent side wear of the screws while they are pumping. Some models take in liquid at both ends of the pump body and move it to the center for discharge; others have both input and output ports at the center. These pumps are recommended as fuel pumps for firing marine and stationary boilers because they can deliver a smooth flow of oil. Their characteristically low vibration and lack of pulsing reduces the strain on coupled pipes, hoses, and fittings. These self-priming pumps are quiet running, highly reliable, and most efficient when driven at less than 1200 rpm by an electric motor or less than 1300 rpm by a steam turbine.

15. RAMELLI PUMP (A) has spring-loaded vanes to ensure contact with the wall; vane ends are rounded for line contact. Pump (B) has two vanes pivot in the housing and are driven by an eccentrically mounted disk; vanes slide in glands and are always radial to the housing, thus providing surface contact. Pump (C) has a housing with a cardioid curve allows the use of a single vane because opposing points on the housing in line with the disk center are equidistant.

16. ROTARY COMPRESSOR PUMP has a disk mounted eccentrically on the drive shaft displaces liquid in continuous flow. A spring-loaded gland separates the inlet from the outlet except when the disk is at the top of stroke.

17. ROTARY COMPRESSOR PUMP has a link separating its suction and compression sides. The link is hinged to a ring which oscillates while it is driven by the disk. Oscillating action pumps the liquid in a continuous flow.

18. ROOTS COMPRESSOR PUMP has two identical impellers with specially shaped teeth. The shafts are connected by external gearing to ensure constant contact between the impellers.

TEN DIFFERENT PUMP DESIGNS EXPLAINED

Fig. 1   Flexible-vane rotary pumps have rotors with radial flexible vanes that contact the inner walls of the housing to seal the input sides of the pumps from their output sides so suction can draw in liquid. The offset plate in the version shown allows the rotor drive shaft to be centered in the housing, keeping the rotor balanced during rotation. The synthetic rubber rotor can easily be replaced after a recommended service period. Able to function if driven in either direction, the pump can also be mounted at any angle. Flexible-vane pumps are limited to pumping non-corrosive liquids. They typically run at rotor speeds from 100 to 2000 rpm, and they can deliver up to 55 gal/min (208 L/min) against 25 lb^2/ in. (1.7 bar).

Fig. 2   Sliding-piston rotary pumps have rotors that include radial spring-loaded sliding pistons positioned in cavities. They advance and retract to remain in constant contact with the inner walls of the housing. The pump shown has an eccentric rotor mounted on an offset drive shaft. The pistons in this version are fully extended at the top of the housing and fully retracted at the bottom. This permits the pump to suction liquids from its input port and discharge them from its output port. Both its input and output ports are located at the center of a hollow drive shaft.

Fig. 3   Sliding-vane rotary pumps have rotors with radial spring-loaded sliding vanes positioned in cavities. They advance and retract to remain in constant contact with the inner walls of the housing. This pump has its rotor mounted on an offset drive shaft so that the vanes are compressed at the top of the housing and extended at the bottom as it turns. Air trapped between any two blades passing the inlet port is pressurized as its volume decreases. Air pressure will exceed housing pressure before it escapes through the outlet port. When operated as a vacuum pump, its performance is similar except that air pressure decreases from outlet to inlet. These pumps can produce free air flow up to 10 ft^3/min (283 L/min) with maximum pressure to 15 $lb/in.^2$ (1 bar), and vacuums to 26 in. Hg (−880 mbar).

Fig. 4   Peristaltic pumps with rollers squeeze flexible tubing against the inner walls of housings as they rotate, forcing liquid to move from the inlet to the outlet. Each of the three rollers in the version shown draws more liquid in as it moves past a contact point on the inlet side of the tubing. At least one roller compresses the tubing at all times, eliminating the need for a backflow prevention valve. The rollers can be driven directly by a motor that controls liquid flow by increasing or decreasing its speed, but a constant speed motor with a gearbox can increase or decrease liquid flow at a constant speed. These pumps are recommended for pumping both corrosive and viscous liquids; the tubing isolates the liquid being pumped from contact with any internal pump to prevent corrosion damage. Tubing compatible with a wide range of chemicals is available, and it can easily be replaced if it loses resilience. Both AC and DC motors are used to drive these pumps, but DC-driven pumps are reversible. These pumps can produce free liquid flow rates of 3000 mL/min), suction heights of 8 m H_2O, and pressure heights to 10 m H_2O.

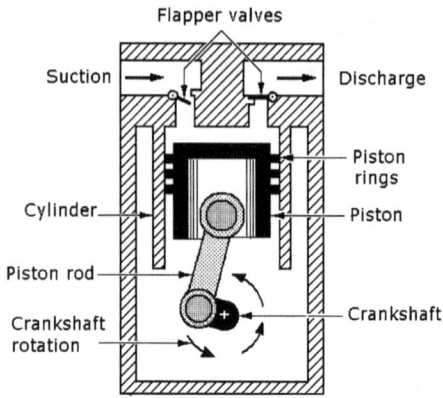

Fig. 5 Articulated piston pump/compressors, used primarily as compressors, operate on the same principle as the internal combustion engine. Rotating camshaft motion is transformed into the reciprocal motion of a connecting rod that drives the piston to compress air in the cylinder. As the piston descends in the cylinder, air is drawn in through a one-way valve; on the ascending stroke, higher pressure air is discharged through another one-way valve. Spring-loaded piston rings seal gas in the cylinder. As compressors these pumps can perform tasks requiring relatively high flow rates at high pressure. They can serve as roughing vacuum pumps by drawing air in from a closed container through the intake valve and exhausting it through the discharge valve. These pump/compressors, driven by AC or DC motors, achieve air volume flow as high as 10 ft³/min (283 L/min) with pressure as high as 175 lb/in.² (12.1 bar). As vacuum pumps, they can produce up to 27.5 in. Hg (−930 mbar).

Fig. 6 Diaphragm pump/compressors can function as air or liquid pumps and compressors. A flexible diaphragm at one end is flexed back and forth by a connecting rod driven by an eccentric on a motor shaft. On down strokes, air is drawn into the pump chamber through an input flapper valve, and it is expelled on the upstroke through the output flapper valve. These efficient gas-tight machines can sample air in portable gas detectors and gas analyzers, and are used in medical autoclaves and other laboratory equipment. Driven by AC or DC motors, diaphragm pumps can produce free air flow to 3.2 ft³/min (86 L/min), pressure up to 41 lb/in.² (2.8 bar), and vacuums up to 7.5 torr (10 mbar abs). Liquid diaphragm pumps can produce free liquid flows up to 3 ft³/min (85 L/min), suction heights of 6 m H₂O, and pressure heights of 80 m H₂O.

Fig. 7 Linear pumps can function as vacuum and liquid pumps and air compressors. An electromagnetic coil, on the version shown, applies a force to the combined magnetized armature and piston section. Both are pulled down by the force against the coil spring within the cylinder during each AC cycle. As a result, air is drawn into the cylinder through the inlet port flapper valves. In the next part in the AC cycle the spring recoils, pushing the piston back to the cylinder top. Air that has been compressed is discharged through the output port. Piston reciprocation speed within the cylinder is determined by the interaction between the electromagnetic fields and the applied AC frequency. The pump's ability to transfer air is set by the strength of its magnetic fields; this generally limits it to low differential pressure applications. These pumps can produce free air volumes to 13 ft³/min (369 L/min), pressures to 10 lb/in.² (0.7 bar), and vacuums to 14 in. Hg (−474 mbar). Liquid linear diaphragm pumps can produce free liquid flows of 260 mL/min, suction heights of 4.7 m of H₂O, and pressure heights of 8 m of H₂O.

Fig. 8 Bent-axis pumps have rotating cylinder blocks containing multiple pistons in parallel, arranged radially around their axes. The pistons are connected by universal couplings to the thrust plate on the end of the drive shaft, offset from the axis of the block by 30°. The drive shaft revolves within the end of the housing. Connecting rods from the pistons are hinged by ball joints or universal couplings to the thrust plates so they also revolve. As the shafts rotate the cylinder blocks, the pistons are forced to reciprocate, and this changes their stroke lengths as the angles of the thrust plates change with respect to each piston. As the cylinder block revolves, half of the pistons draw in liquid; as they pass the input port, the rest of the pistons discharge their liquid through the output port. The end view of the pump housing shows its semicircular input and output ports.

Fig. 9 Wobble-plate pumps have stationary cylinder blocks containing five or more (odd number) pistons arranged radially around their axes. The ends of the pistons have strong spring-loaded shoes that clamp them against the wobble plate. But they are allowed to slide in place as the plate revolves. The face of the wobble plate is angled with respect to its base so its thickness varies around its axis. When the wobble plate rotates, the pistons reciprocate with fixed stroke lengths within each cylinder. Liquid is drawn into the cavity between the wobble plate and block from a separate input channel. Each piston has an internal spring-loaded one-way ball valve that controls liquid flow, and the block contains many of these ball valves. Liquid under pressure passes through the piston valves before it is discharged from a common outlet port. These pumps can produce pressures of up to 1000 lb/in.^{2} (70 bar).

Fig. 10 Centrifugal/impeller pumps convert the energy of their drive motors into the velocity or kinetic energy of the liquid entering the pumps. Their eccentrically mounted rotors revolving at high speed spin the liquid between their vanes outward and impart centrifugal acceleration to it. (See the direction of impeller rotation.) The impeller slings the liquid out of the volute rather than scooping it upward. As the liquid passes the vane tips it encounters resistance at the volute casing, which slows the liquid and converts its kinetic energy into pressure. Liquid is forced from the discharge port by this pressure as a powerful jet. Pressure (and jet height) is set by impeller diameter and its rotating shaft speed. As liquid leaves the suction eye of the impeller, a low-pressure zone creates suction that draws more liquid into the pump.

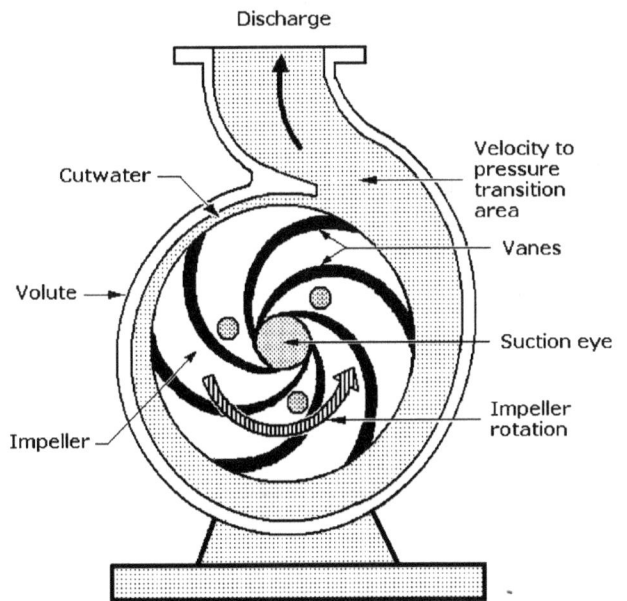

GLOSSARY OF PUMP TERMS

absolute pressure: In a vacuum system, the pressure above a perfect vacuum.

air flow capacity: A measure of displacement of a vacuum pump measured in cubic feet per minute (cfm or ft^3/min).

flow rate: A measure specified in the following units: (1) gallons per hour—gal/h, (2) gallons per minute—gal/min, (3) cubic feet per hour—ft^3/h, (4) cubic feet per minute—ft^3/min or cfm, (5) liters per hour—L/h, (6) liters per minute—L/min, and (7) cubic centimeters per minute—cm^3/min.

gage pressure: In a vacuum system, the difference between evacuated system and atmospheric pressures.

positive displacement pump: A pump that draws liquid from a reservoir and discharges it under pressure.

pressure: A measure specified in the following common units: (1) pounds per square inch—lb/in^2 or psi, (2) atmospheres—atm or bar, (3) kilopascal per square centimeter—kPa/cm^2, (4) inches of mercury—in. Hg, (5) micrometers of mercury—μm Hg, (5) torr—mm Hg.

torr: A unit of pressure equal to 1/760 atm or 1 mm Hg.

vacuum: A volume containing air at a pressure below atmosphere (0 ft^2/min gage).

vacuum levels: (1) coarse—0 to 28 in. Hg, (2) fine—28 to 29 in. Hg, (3) high—29 to 29.2 in. Hg.

vacuum pumps: Three basic vacuum pump types are: diaphragm, piston, and rotary vane.

diaphragm: Includes an elastic diaphragm that is flexed back and forth by a reciprocating shaft, creating a piston-like motion inside the housing.

piston: Contains one or more cylindrical pistons that reciprocate inside one or more cylinders.

rotary vane: Includes a rotor with vanes set in slots so they slide out as the rotor revolves. The vanes contact and revolve inside the housing walls because of spring pressure or centrifugal force.

volume: A measure specified in the following common units: (1) cubic meters—m^3, (2) cubic centimeters—cm^3, (3) liters—L, (4) milliliters—ml, (5) cubic feet—ft^3, (6) cubic inches—$in.^3$, (7) gallons—gal, (8) quarts—qt.

BEARINGLESS MOTOR-GENERATORS HAVE HIGHER SPEED AND LONGER LIFE

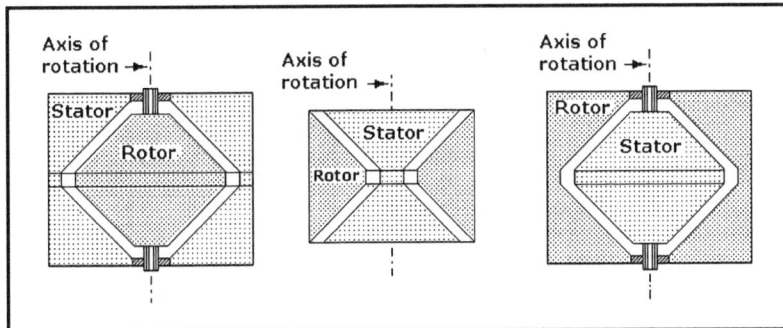

These simplified cross-section views show three design concepts for conical bearingless motor-generators (M-Gs). The rotors are levitated and rotated by their stators in all three concepts shown. However, while stators enclose rotors in most motor generators, the conical bearingless principle allows the rotor to enclose the stator, as in the third diagram.

Bearingless motor-generators can have either radial or axial magnetic bearings, but a more recent design features conical magnetic bearings. Machines with these conical magnetic bearings are seen as an improvement over those with either radial or axial bearings. Because the conical magnetic rotors induce both radial and axial magnetic forces and their design is simpler and more compact, these machines can rotate at higher speeds and have longer operational lives than earlier designs.

Both earlier and present conical magnetic rotor machines are in the same class of "bearingless" or "self-bearing" machines because neither contains mechanical bearings. Both classes provide the torque and force allowing them to function as either motors or generators. Because they do not have mechanical bearings that are subject to wear and failure, the operational lives of both classes are expected to be longer than those with bearings. Moreover, because they do not need lubrication, they can operate over wider speed and temperature ranges. These include extreme temperature conditions in which the lubricants would be dissipated or degraded, causing the mechanical bearings to fail.

Three typical configurations for the improved conical bearingless motor-generators are diagrammed in the figure. The principal components of all three versions are concentric conically tapered rotors and stators which face each other across gaps. Because a conical motor-generator induces both radial and axial magnetic forces, it acts like a combination of an axial and a radial magnetic bearing. Consequently, only two conical motor generators, one at each end of a rotor, are needed for complete magnetic levitation of the rotor. By contrast, previous bearingless motor generator designs required a combination of an axial and a radial magnetic bearing at each end of the rotor for complete magnetic levitation as well as a separate motor to provide torque.

The stator part of each motor-generator includes a magnetic core and iron shell with many teeth and slots at the gap with electromagnetic coils wound in the slots. The rotor is essentially a magnetic core but could include permanent magnets, depending on the design. Except for their conical configurations, the rotor and stator resemble those of other radial magnetic bearings and bearingless motors.

In previous magnetic bearing and bearingless motors, electric currents flow through the coils producing magnetic fluxes in the core and rotor that result in torques and forces on the rotor. A feedback control system adjusts the current waveforms in response to the positions of the ends of the rotor, as measured by sensors. These waveforms produce the radial and axial forces needed to keep the rotor levitating at the desired radial and axial position. For either motor or generator operation, the timing for the electronic switching of the input or output current coil connections is determined by the rotation angle, as measured by another sensor.

Unlike in earlier bearingless motors, this design eliminates the need for one set of coils for levitation interspersed with another set of coils for motor-generator operation. Instead, an improved control scheme permits the use of a single set of coils for both levitation and motor-generator operation. The scheme requires at least six coils (for three or more pole pairs) in each bearing. In its motor mode, the scheme permits simultaneous operation of the equivalent of three overlapping three-phase motors. Moreover, this motor configuration provides some fault tolerance; if any of the pole-pair coils or the drive circuits for a pole pair should fail, the remaining 2 three-phase subsystems can continue to provide magnetic levitation and motor-generator operation.

This work was done by P. Kascak and R. Jansen of the University of Toledo and T. Dever of QSS Group Inc. for John H. Glenn Research Center.

ENERGY EXCHANGE IN SEAWATER DESALINATION BOOSTS EFFICIENCY

Piping for desalinating seawater in a seawater reverse osmosis (SWRO) system requires a continuous flow of high-pressure seawater be directed against polymer membranes to force fresh water out the other side. However, the energy remaining in the brine exiting at high pressurize is lost in these systems, lowering their efficiency and keeping operating cost high. Energy Recovery Inc., San Leandro, California, has developed its PX energy recovery device for recovering this energy and using it to boost the pressure of incoming seawater. As a result, the energy exchange process permits seawater at lower pressure to be pumped into the membranes for desalination because it is supplemented by seawater pressurized by the energy recovered from the brine. This feature increases system efficiency while lowering operating cost.

The PX recovery exchanger device is shown schematically in Fig. 1. Essentially a water-powered pump, it performs the pressure/volume exchange when the high-pressure brine waste enters (upper right) and impacts low-pressure seawater entering (lower left) into the multiple ducts of the rapidly spinning rotor. Energy is imparted to the seawater to boost its pressure before it leaves the outlet (upper left). Low-pressure brine waste leaves the outlet drain (lower right). The rotor, made from Corundum, a strong aluminum-oxide ceramic, has 12 peripheral ducts in parallel with its axis (making it look like an oversized revolver magazine); it is the only moving part of the exchanger. The rotor is enclosed in a Corundum sleeve with Corundum end covers, each with an input and output pipe. This complete assembly is housed in a fiberglass pressure vessel.

The free-spinning rotor is levitated by the entering high-pressure waste brine stream at a rate proportional to that flow. At nominal speed, the rotor rotates at 1200 rpm, completing 20 revolutions every second. When driven at this speed, its energy is delivered to the entering low-pressure seawater as hammer-like blows within the ducts. Because water is incompressible, and these impacts are so fast, energy transfer occurs with almost no mixing of sea and fresh water.

At any given time, one half of the rotor ducts are exposed to the high-pressure waste stream while the other half admits low-pressure seawater. As the rotor turns, the ducts pass a sealing area (shown as parallel lines in the middle of Fig. 1) that separates high and low pressures. Thus, the ducts containing high-pressure waste brine are separated from the adjacent ducts containing low-pressure seawater by a seal formed by the rotor's ribs and its ceramic end covers. This pressure exchange process is repeated for each duct with every rotation of the rotor, so the ducts are continuously filling and discharging.

Clearances are so precise that when the gap between the rotor and housing is filled with high-pressure water, an almost frictionless hydrodynamic bearing is formed. According to Energy Recovery, the PX pressure exchanger has a near 98 percent efficiency. The company selected Corundum in its search for a suitable material to make the rotor and its container because of its strength and ability to withstand the corrosive effects of saltwater.

Fig. 1 This exploded view of the pressure exchange unit (PSU) shows water flow paths. High-pressure brine impacts low-pressure seawater in the multiple ducts of the rapidly spinning rotor, imparting its energy to the seawater, boosting its pressure. This high-pressure seawater is then sent to the input side of the membranes, and the low-pressure brine is expelled.

Fig. 2 Low-pressure seawater enters the pressure exchanger (lower left) as high-pressure waste brine from the membranes enters the pressure exchanger and potable water exits (right). After the energy exchange between the brine and seawater, pressurized seawater goes to the circulation pump and the brine, now low in pressure, exits. High-pressure seawater from the circulation pump mixes with low-pressure seawater from the high-pressure pump (left), providing enough pressure for desalination. The energy exchange reduces the power requirement for the high-pressure pump, lowering system operating costs and increasing its efficiency.

Figure 2 illustrates the flow path of a PX energy recovery device installed in an SWRO system. In conventional SWRO systems, 100 percent of the seawater input to the desalination membranes must be provided by the high-pressure pump (left). About 40 percent of the incoming pressure is lost by forcing seawater through the membranes to obtain the output of potable water (right). The residual brine at a pressure of 60 percent of input is returned to the sea and its energy is lost. The installation of an energy recovery device converts the SWRO system from open-loop to a closed-loop energy-recovery circuit.

Low-pressure potable water leaves the RO membrane as high-pressure brine leaves to enter the exchanger (right). Meanwhile, low-pressure seawater is diverted from the high-pressure pump (far left) to the exchanger (left). After energy conversion, high-pressure seawater from the exchanger (left) passes the circulation pump with 60 percent of the pressure required for SWRO operation and merges with seawater at 40 percent of the required SWRO pressure to merge for entry into the RO membrane stream. This merger boosts SWRO system efficiency to about 57 percent.

TWO-CYCLE ENGINE IMPROVES EFFICIENCY AND PERFORMANCE

An improved two-cycle engine. This design for a two-cycle internal combustion engine is said to offer improved performance while reducing its mechanical complexity.

A two-cycle internal combustion engine designed for NASA's John H. Glenn Research Center offers improvements in volumetric efficiency and performance. This engine may be competitive with four-cycle engines for many applications. Simpler and with fewer parts than existing two-cycle engines, it is supercharged with variable compression. These features suggest that it could be manufactured at lower cost and be more reliable than comparable two-cycle engines. Some of the major improvements in the design shown in the cross-section diagram are as follows.

Reciprocating Shuttle Inlet Valve

The reciprocating shuttle inlet valve and its operating gear are made as a single unit. This shuttle valve is actuated in a sequence in which, at the end of its stroke, extensions of the valve stop against abutments at the ends of grooves in a piston skirt. This shuttle-valve design eliminates the more complex conventional valve mechanism that is actuated by an engine crankshaft or camshaft, yet it could be effective on every type of two-cycle engine from small high-speed, single-cylinder engines to large low-speed, multiple-cylinder engines.

Variable Compression Ratio

The piston is made in a stepped configuration with a narrow power section in its upper part and a wider compressor-supercharger flange or skirt at its lower end. This variable compression ratio mechanism includes a high-pressure oil lubrication circuit that acts in unison with the pulsating flow and pressure of the air caused by the reciprocating piston skirt. Simply stated, this mechanism operates because of the interaction between air, oil, and combustion gas pressures and flows that change the axial position of a floating combustion bowl in the power section of the piston. As a result, this interaction changes the compression ratio. The mechanism is designed so that when the throttle opening is suddenly changed, the compression ratio rapidly adjusts to a value for the most efficient engine operation.

Supercharging Helps to Scavenge and Cool the Cylinder

The stepped-piston configuration eliminates the complication and high cost of "add-on" supercharging mechanisms such as those on conventional two-cycle engines. During the compression stroke, the motion of the compressor-supercharger section of the piston causes high-pressure air to flow from the compressor cylinder through one-way transfer valves and a plenum before entering the power cylinder. This flow contributes to the scavenging and cooling of the power cylinder. Highly compressed air continues to enter the plenum and power cylinder after the exhaust ports are closed and cylinder supercharging has been completed.

The compressed air that continues to enter the plenum after the inlet ports are covered by the rising power piston is retained in the plenum under pressure until the end of the expansion stroke when the receding piston opens the exhaust ports. After the exhaust ports open, the abutments in the piston skirt make contact with the projections on the reciprocating shuttle inlet valve, forcing the valve to open. Then the compressed air rushes from the plenum into the power cylinder, causing the initial scavenging. The stepped-piston configuration has an additional benefit: It isolates the blow-back gasses and particulate matter that escape past the power piston rings from the crankcase and returns them to the power cylinder on the following stroke.

This work was done by Bernard Wiesen of Wiesen Engine Technologies for NASA's John H. Glenn Research Center.

CHAPTER 14

PACKAGING, CONVEYING, HANDLING, AND SAFETY MECHANISMS AND MACHINES

FIFTEEN DEVICES THAT SORT, FEED, OR WEIGH

ORIENTING DEVICES

Orienting short, tubular parts

Here's a common problem: Parts arrive in either open-end or closed-end first; you need a device that will orient all the parts so they feed out, facing the same way. In Fig. A, when a part comes in open-end first, it is pivoted by the swinging lever so that the open end is up. When it comes in closed-end first, the part brushes away the lever to flip over headfirst. Figs. B and C show a simpler arrangement with pin in place of lever.

Orienting dish-like parts

A part with its open-end facing to the right (part 1) falls on a matching projection as the indexing wheel begins to rotate clockwise. The projection retains the part for 230° to point A where it falls away from the projection to slide down the outlet chute, open-end up. An incoming part facing the other way (2) is not retained by the projection, hence it slides *through* the indexing wheel so that it too, passes through the outlet with its open-end up.

Orienting pointed-end parts

The important point here is that the built-in magnet cannot hold on to a part as it passes by if the part has its pointed end facing the magnet. Such a correctly oriented part (part 1) will fall through the chute as the wheel indexes to a stop. An incorrectly oriented part (part 2) is briefly held by the magnet until the indexing wheel continues on past the magnet position. The wheel and the core with the slot must be made from some nonmagnetic material.

Orienting U-shaped parts

The key to this device is two pins that reciprocate one after another in the horizontal direction. The parts come down the chute with the bottom of the "U" facing either to the right or left. All pieces first strike and rest on pin 2. Pin 1 now moves into the passage way, and if the bottom of the "U" is facing to the right, the pin would kick over the part as shown by the dotted lines. If, on the other hand, the bottom of the "U" had been to the left, the motion of pin 1 would have no effect, and as pin 2 withdrew to the right, the part would be allowed to pass down through the main chute.

Orienting cone-shaped parts

Regardless of which end of the cone faces forward as the cones slide down the cylindrical rods, the fact that both rods rotate in opposite directions causes the cones to assume the position shown in section *A-A* (above). When the cones reach the thinned-down section of the rods, they fall down into the chute, as illustrated.

In the second method of orienting cone-shaped parts (left), if the part comes down small end first, it will fit into the recess. The reciprocating rod, moving to the right, will then kick the cone over into the exit chute. But if the cone comes down with its large end first, it sits on top of the plate (instead of inside the recess), and the rod simply pushes it into the chute without turning it over.

Orienting stepped-disk parts

Parts rolling down the top rail to the left drop to the next rail which has a circular segment. The part, therefore, continue to roll on in the original direction, but their faces have now been rotated 180°. The idea of dropping one level might seem oversimplified, but it avoids the cam-based mechanisms more commonly used for accomplishing this job.

SIMPLE FEEDING DEVICES

Feeding a fixed number of parts

The oscillating sector picks up the desired number of parts, left diagram, and feeds them by pivoting the required number of degrees. The device for oscillating the sector must be able to produce dwells at both ends of the stroke to allow sufficient time for the parts to fall in and out of the sector.

The circular parts feed down the chute by gravity, and they are separated by the reciprocating rod. The parts first roll to station 3 during the downward stroke of the reciprocator, then to station 1 during the upward stroke; hence the time span between parts is almost equivalent to the time it takes for the reciprocator to make one complete oscillation.

The device in Fig. B is similar to the one in Fig. A, except that the reciprocator is replaced by an oscillating member.

One-by-one separating device

(A) (B)

Mixing different parts together

Two counter rotating wheels form a simple device for alternating the feed of two different workpieces.

Pausing until actuated

Each gear in this device is held up by a pivotable cam sector until the gear ahead of it moves forward. Thus, gear 3, rolling down the chute, kicks down its sector cam but is held up by the previous cam. When gear 1 is picked off (either manually or mechanically), its sector cam pivots clockwise because of its own weight. This permits gear 2 to move into place of gear 1—and frees cam 2 to pivot clockwise. Thus, all gears in the row move forward one station.

SORTING DEVICES

In the simple device (A) the balls run down two inclined and slightly divergent rails. The smallest balls, therefore, will fall into the left chamber, the medium-size ones into the middle-size chamber, and the largest ones into the right chamber.

In the more complicated arrangement (B), the balls come down the hopper and must pass a gate which also acts as a latch for the trap-door. The proper-size balls pass through without touching (actuating) the gate. Larger balls, however, brush against the gate which releases the catch on the bottom of the trap-door, and fall through into the special trough for the rejects.

Sorting balls according to size

(A) (B)

Sorting according to height

Workpieces of varying heights are placed on this slowly rotating cross-platform. Bars 1, 2, and 3 have been set at decreasing heights beginning with the highest bar (bar 1), down to the lowest bar (bar 3). The workpiece is therefore knocked off the platform at either station 1, 2, or 3, depending on its height.

WEIGHT-REGULATING ARRANGEMENTS

By varying the vibration amplitudes

The material in the hopper is fed to a conveyor by the vibration of the reciprocating slider. The pulsating force of the slider is transmitted through the rubber wedge and on to the actuating rod. The amplitude of this force can be varied by moving the wedge up or down. This is done automatically by making the conveyor pivot around a central point. As the conveyor becomes overloaded, it pivots clockwise to raise the wedge, which reduces the amplitude of the force and slows the feed rate of the material.

Further adjustments in feed rate can be made by shifting the adjustable weight or by changing the speed of the conveyor belt.

By linkage arrangement

The loose material falls down the hopper and is fed to the right by the conveyor system which can pivot about the center point. The frame of the conveyor system also actuates the hopper gate so that if the amount of material on the belt exceeds the required amount, the conveyor pivots clockwise and closes the gate. The position of the counterweight on a frame determines the feed rate of the system.

By electric-eye and balancer

The indexing table automatically stops at the feed station. As the material drops into the container, its weight pivots the screen upward to cut off the light beam to the photocell relay. This in turn shuts the feed gate. The reactuation of the indexing table can be automatic after a time delay or by the cutoff response of the electric eye.

SEVEN CUTTING MECHANISMS

By pressing down on the foot pedal of this mechanism, the top knife and the clamp will be moved downward. However, when the clamp presses on the material, both it and link *EDO* will be unable to move further. Link *AC* will now begin to pivot around point *B*, drawing the lower knife up to begin the cutting action.

Four-bar mechanism

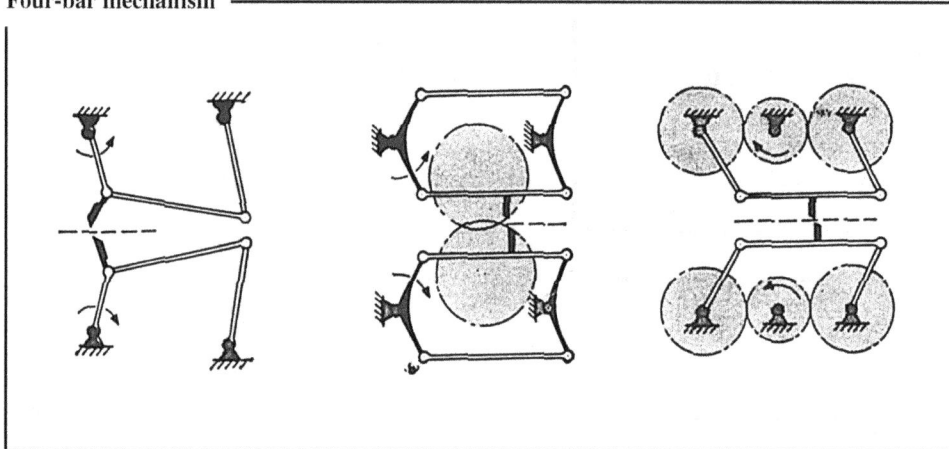

These 3 four-bar cutters provide a stable, strong, cutting action by coupling two sets of links to chain four-bar arrangements.

Parallel cutter mechanisms

The cutting edges of the knives in the four mechanisms move parallel to each other, and they also remain vertical at all times to cut the material while it is in motion. The two cranks are rotated with constant velocity by a 1 to 1 gear system (not shown), which also feeds the material through the mechanism.

Curved-motion cutter mechanism

(a)

The material is cut while in motion by the reciprocating action of the horizontal bar. As the bar with the bottom knife moves to the right, the top knife will arc downward to perform the cutting operation.

Vertical cutter motion mechanism

(b)

The top knife in this arrangement remains parallel to the bottom knife at all times during cutting to provide a true scissor-like action, but friction in the sliding member can limit the cutting force.

Slicing mechanism

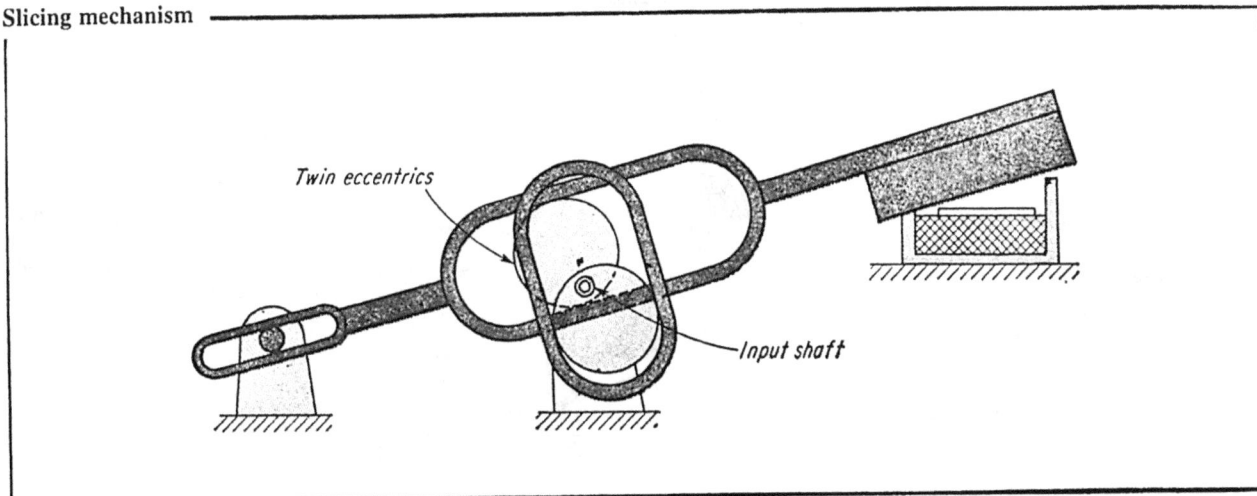

Slicing motion is obtained from the synchronized effort of two eccentric disks. The two looped rings actuated by the disks are welded together. In the position shown, the bottom eccentric disk provides the horizontal cutting movement, and the top disk provides the up-and-down force necessary for the cutting action.

Web-cutting mechanism

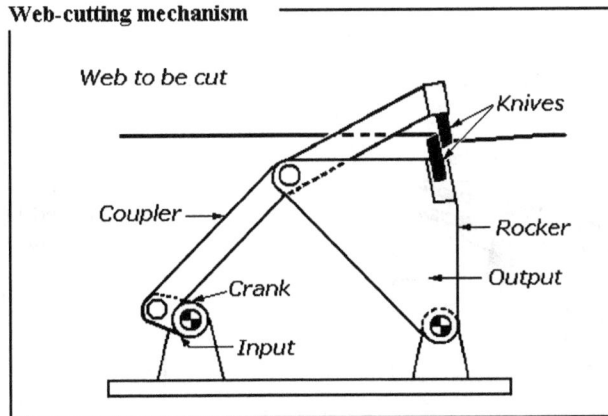

This four-bar linkage with an extended coupler can cut a web on the run at high speeds. The four-bar linkage shown is dimensioned to give the knife a velocity during the cutting operation that is equal to the linear velocity of the web.

TWO FLIPPING MECHANISMS

This mechanism can turn over a flat piece by driving two four-bar linkages from one double crank. The two flippers are actually extensions of the fourth members of the four-bar linkages. Link proportions are selected so that both flippers rise up at the same time to meet a line slightly off the vertical to transfer the piece from one flipper to the other by the momentum of the piece.

Turnover mechanism

Input crank

Upside-down flipper mechanism

This is a four-bar linkage (links *a*, *b*, *c*, *d*) in which the part to be turned over is coupler *c* of the linkage. For the proportions shown, the 180° rotation of link *c* is accomplished during the 90° rotation of the input link.

ONE VIBRATING MECHANISM

Vibrating mechanism

Output

Input crank

As the input crank rotates, the slotted link, which is fastened to the frame with an intermediate link, oscillates to vibrate the output table up and down.

SEVEN BASIC PARTS SELECTORS

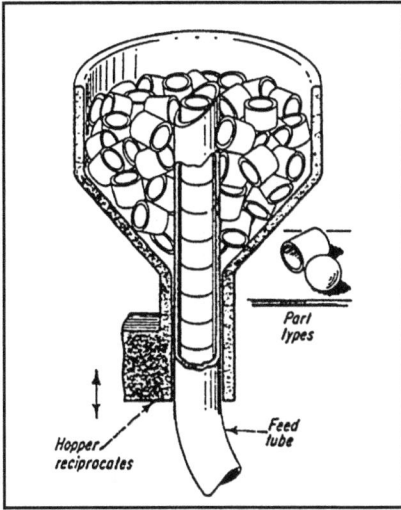

A reciprocating feed for spheres or short cyclinders is one of the simplest feed mehanisms. Either the hopper or the tube reciprocates. The hopper must be kept topped-up with parts unless the tube can be adjusted to the parts level.

A centerboard selector is similar to reciprocating feed. The centerboard top can be milled to various section shapes to pick up moderately complex parts. It works best, however, with cylinders that are too long to be led with the reciprocating hopper. The feed can be continuous or as required.

A rotary screw-feed handles screws, headed pings, shouldered shafts, and similar parts in most hopper feeds; random selection of chance-oriented parts calls for additional machinery if the parts must be fed in only one specific position. Here, however, all screws are fed in the same orientation (except for slot position) without separate machinery.

Rotary centerblades catch small U-shaped parts effectively if their legs are not too long. The parts must also be resilient enough to resist permanent set from displacement forces as the blades cut through a pile of parts. The feed is usually continuous.

A paddle wheel is effective for feeding disk-shaped parts if they are stable enough. Thin, weak parts would bend and jam. Avoid these designs, if possible—especially if automatic assembly methods will be employed.

A long-cylinder feeder is a variation of the first two hoppers. If the cylinders have similar ends, the parts can be fed without proposition, thus assisting automatic assembly. A cylinder with differently shaped ends requires extra machinery to orientate the part before it can be assembled.

A barrel hopper is useful if parts tend to become entangled. The parts drop free of the rotating-barrel sides. By chance selection, some of them fall onto the vibrating rack and are fed out of the barrel. The parts should be stiff enough to resist excessive bending because the tumbling action can subject them to relatively severe loads. The tumbling can help to remove sharp burrs.

ELEVEN PARTS-HANDLING MECHANISMS

Gravity feed for rods. Single rods of a given length are transferred from the hopper to the lower guide cylinder by means of an intermittently rotating disk with a notched circumference. The guide cylinder, moved by a lever, delivers the rod when the outlet moves free of the regulating plate.

Feeding electronic components. Capacitors, for example, can be delivered by a pair of intermittently rotating gearlike disks with notched circumferences. Then a pick-up arm lifts the capacitor and it is carried to the required position by the action of a cam and follower.

Feeding headed rivets. Headed rivets, correctly oriented, are supplied from a parts-feeder in a given direction. They are dropped, one by one, by the relative movement of a pair of slide shutters. Then the rivet falls through a guide cylinder to a clamp. Clamp pairs drop two rivets into corresponding holes.

Label feed. Labels are taken out of the hopper by a carrying arm with a vacuum unit to hold the label. The label is then placed into the required position, and the vacuum hold is released.

Horizontal feed for fixed-length rods. Single rods of a given length are brought from the hopper to the slot of a fixed plate by a moving plate. After being gauged in the notched portion of the fixed plate, each rod is moved to the chute by means of a lever, and is removed from the chute by a vibrating table.

Pin inserter. Pins, supplied from the parts-feeder, are raised to a vertical position by a magnet arm. The pin drops through a guide cylinder when the electromagnet is turned off.

Cutoff and transfer devices for glass tubes. The upper part of a rotating glass tube is held by a chuck (not shown). When the cutter cuts the tube to a given length, the mandrel comes down and a spring member (not shown) drops the tube on the chute.

Vertical feed for wires. Wires of fixed length are stacked vertically, as illustrated. They are removed, one by one, as blocks A and B are slid by a cam and lever (not shown) while the wires are pressed into the hopper by a spring.

Feeding special-shaped parts. Parts of such special shapes as shown are removed, one by one, in a given direction, and are then moved individually into the corresponding indents on transfer platforms.

Lateral feed for plain strips. Strips supplied from the parts-feeder are put into the required position, one by one, by an arm that is part of a D-drive linkage.

Vertical feed for rods. Rods supplied from the parts-feeder are fed vertically by a direction drum and a pushing bar. The rod is then drawn away by a chucking lever.

SEVEN AUTOMATIC-FEED MECHANISMS

The design of feed mechanisms for automatic or semiautomatic machines depends largely upon such factors as size, shape, and character of the materials or parts that are to be fed into a machine, and upon the kinds of operations to be performed. Feed mechanisms can be simple conveyors that give positive guidance, or they might include secure holding devices if the parts are subjected to processing operations while being fed through a machine. One of the functions of a feed mechanism is to extract single pieces from a stack or unassorted supply of stock. If the stock is a continuous strip of metal, roll of paper, long bar, or tube, the mechanism must maintain intermittent motion between processing operations. These conditions are illustrated in the accompanying drawings of feed mechanisms.

Fig. 1

Rocker cam in disengaged position

Latch

Stop bridge

Strip stock

Guide bars

Sliding carriage

Hand lever operates latch to shift rocker arm and rotate threaded head, thus lowering jawplate to operating position

Connecting rod to power crankshaft

Stop bridge

Guide bars

Strip stock

Oscillator block

Lift spring

Gripper plate

Sliding-jaw plate fulcrums around pin to engage and disengage strip on feed and retract strokes

Pat. No. 2,278,921

Section A-A

Centerline of turret

Spring-biased wedge gives scissors grip on fork

Jaws

One of 8 radial arms on turret

Capscrew head slides over cam, raising open jaws to receive fork

Roller and cam surface for releasing spring-held jaws

Pin slides over cam surface to raise fork for processing operation

Pat. No. 2,292,966.

Centerline of turret

Stack of forks

Finger height adjustment

Stop allows removal of only one fork

Jaws after closing on fork handle

One of two supporting fingers

Outer end of intermittently moving radial arm turret

Fig. 2

Spring blades, rotated by spiral bevel gears, align tubes by pushing them against a stop

One or more saws adjustably mounted on drive shaft

Unassorted supply non-uniform length tubes

Four clamp jaws adjustably mounted on each slotted crossbar spanning machine between two drive chains

Fig. 3

Pat. No. 2,298,955

Four spring-operated fingers pick up flat or flexed sheet metal or fiber blanks from magazine

Rack

Rollers deliver blanks to reciprocating slide bars carrying dogs

Leaf springs lift fingers

Drive chain

Segment gear
Operating link

Eccentric

Fig. 4

Pat. No. 2,270,871

Fig. 5

Chain-tightening adjustment

Stock of varying thickness is grooved by cutterhead

Drive chain

Cutterhead shaft

Fingers feed stock over cutterhead, while straightening and flattening warped stock

Solenoid circuit energized by cam-operated switch unless opened by electric eye

Fixed

Paper length adjustment

Latch

Paper

Silent ratchet

Paper-feed rolls

Correct adjustment

Spring is extended by crank on drive shaft when mechanism is latched

Drive shaft

Light source

Cell

When paper shuts off light, cell opens solenoid circuit, thus latching feed mechanism

Fig. 6

Intermittent Feed Gearing

19 teeth

$10\frac{9}{16}$"

21 teeth

Drive gear

$6\frac{1}{2}$"

Fig. 7

FIFTEEN CONVEYOR SYSTEMS FOR PRODUCTION MACHINES

Conveyor systems can be divided into two classes: those that are part of a machine for processing a product, and those that move products in various stages of fabrication. The movement might be from one worker to another or from one part of a plant to another. Most of the conveyors shown here are components in processing machines. Both continuous and intermittently moving equipment are illustrated.

Intermittently moving grooved bar links convey pasteboard tubes through a drying chamber.

A rotary conveyor transfers articles from one belt conveyor to another without disturbing their relative positions.

Co-acting cams in the paths of follower rollers open and close tongs over bottlenecks by a wedging action.

Hooks on a chain-driven conveyor move articles through a plating bath.

A rotating disk carries food cans in a spiral path between stationary guides for presealing heat treatment.

Hooks on a cable-driven conveyor and an automatic cradle for removing coils.

A double belt sandwiches shoe soles during their cycle around a spiral system and then separates to discharge the soles.

A matchbook carrier links with holding clips that are moved intermittently by sprockets.

Bottle clips with release bars for automatic operation.

An intermittent rotary conveyor inverts electrical capacitors that are to be sealed at both ends by engaging radial pins which have holding clips attached.

This pasteurizer carrier links lock bottles in place on straight ways.

Conveyor belts on sides and bottom

Tapering guides shape paper sacks

Paper sacks

Pat. No. 1,998,287

The wedging action of the side belts shapes paper sacks for wrapping and packing.

Paddle arm slides in sleeve

Cam rollers

Cams on drive shaft raise and lower two arms causing paddle to follow dotted path

Drive shaft

Eccentric on drive shaft swings both arms on lower shaft

Fork guide

Pat. No. 1,726,418

A reciprocating pusher plate is activated by an eccentric disk and two cams on a drive shaft.

Counterweight

Drive chain section

Pusher on radially adjustable arm may be used as guide roller

Pat. No. 2,287,196

A pusher-type conveyor can have a drive on either side.

Co-acting side bars on synchronous chains

Pat. No. 2,228,932

Synchronous chains with side arms grasp and move packages.

SEVEN TRAVERSING MECHANISMS FOR WINDING MACHINES

The seven mechanisms shown are parts of different yarn- and coil-winding machines. Their fundamentals, however, might be applicable to other machines that require similar changes of motion. Except for the leadscrews found on lathes, these seven represent the operating principles of all well-known, mechanical traversing devices.

Fig. 1 A package is mounted on a belt-driven shaft on this precision winding mechanism. A camshaft imparts reciprocating motion to a traverse bar with a cam roll that runs in a cam groove. Gears determine the speed ratio between the cam and package. A thread guide is attached to the traverse bar, and a counter-weight keeps the thread guide against the package.

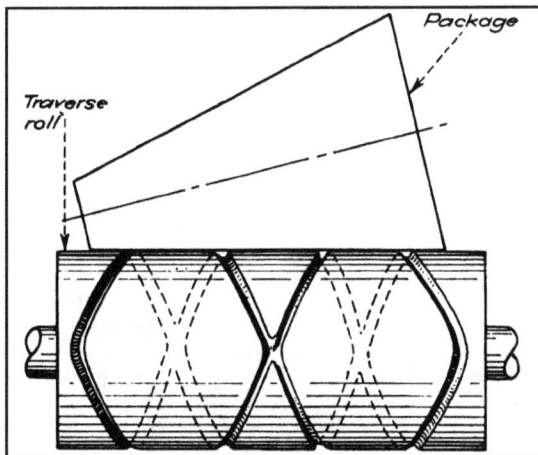

Fig. 2 A package is friction-driven from a traverse roll. Yarn is drawn from the supply source by traverse roll and is transferred to a package from the continuous groove in the roll. Different winds are obtained by varying the grooved path.

Fig. 3 Reversing bevel gears that are driven by a common bevel gear drive the shaft carrying the traverse screw. A traverse nut mates with this screw and is connected to the yarn guide. The guide slides along the reversing rod. When the nut reaches the end of its travel, the thread guide compresses the spring that actuates the pawl and the reversing lever. This action engages the clutch that rotates the traverse screw in the opposite direction. As indicated by the large pitch on the screw, this mechanism is limited to low speeds, but it permits longer lengths of traverse than most of the others shown.

Fig. 4 A drum drives the package by friction. A pointed cam shoe, which pivots in the bottom side of the thread guide assembly, rides in cam grooves and produces a reciprocating motion of the thread guide assembly on the traverse bar. Plastic cams have proved to be satisfactory even with fast traverse speeds. Interchangeable cams permit a wide variety of winding methods.

Fig. 5 A roll that rides in a heart-shaped cam groove engages a slot in a traverse bar driver which is attached to the traverse bar. Maximum traverse is obtained when the adjusting guide is perpendicular to the driver. As the angle between the guide and driver is decreased, traverse decreases proportionately. Inertia effects limit this mechanism to slow speeds.

Fig. 6 The two cam rolls that engage this heart-shaped cam are attached to the slide. The slide has a driver roll that engages a slot in the traverse bar driver. Maximum traverse (to the capacity of the cam) occurs when the adjusting disk is set so the slide is parallel to the traverse bar. As the angle between the traverse bar and slide increases, traverse decreases. At 90° traverse is zero.

Fig. 7 A traverse cam imparts reciprocating motion to a cam follower that drives thread guides on traverse guide rods. The package is friction driven from the drum. Yarn is drawn from the supply source through a thread guide and transferred to the drum-driven package. The speed of this mechanism is determined by the weight of its reciprocating parts.

400

VACUUM PICKUP FOR POSITIONING PILLS

This pickup carries tablet cores to moving dies, places cores accurately in coating granulation, and prevents the formation of tablets without cores.

Cores are hopper fed to a rotating feeder disk through a tablet duster. This disk is vibrated clockwise under a slotted pick-up ring which rotates counter-clockwise. Each slot in the pickup ring holds two cores and lets broken tablets fall through to an area under the feeder table. Cores are picked from ring slots, carried to tablet press dies, and deposited in dies by vacuum nozzles fastened to a chain driven by the press die table. This chain also drives the pickup ring to synchronize the motion of ring slots and pickup nozzles. Coating granulation is fed into the dies ahead of and after the station where a vacuum pickup deposits a core in each die. Compressing rolls are at the left side of the machine. The principal design objective here was to develop a machine to apply dry coatings at speeds that lowered costs below those of liquid coating techniques.

MACHINE APPLIES LABELS FROM STACKS OR ROLLERS

The flow of containers through this labeler is shown by the top-view drawing of the machine. Bottle spacers ensure that containers remain $7^1/_2$ in. apart on the conveyor. Dual label-transfer turrets allow for the simultaneous application of front and back labels.

This labeling machine can perform either conventional glue-label application or it can heat-seal labels in cut or roll form. The machine labels the front and back of round or odd-shaped containers at speeds of 60 to 160 containers per minute. The containers handled range from 1 in. diameter or thickness to $4^1/_4$ in. diameter by $5^1/_2$ in. wide. Container height can vary from 2 to 14 inches. The unit handles labels ranging from $7/_8$ to $5^1/_2$ in.

wide and $7/_8$ to $6^1/_2$ in. high. The label hopper is designed for labels that are generally rectangular in shape, although it can be modified to handle irregular shapes. Provision has been made in design of the unit, according to the manufacturer, to allow labels to be placed at varying heights on the containers. The unit's cut-and-stacked label capacity is 4,500. An electric eye is provided for cutting labels in web-roll form.

TWENTY HIGH-SPEED MACHINES FOR APPLYING ADHESIVES

Fig. 1 A gravity spreader has an open bottom and a grooved lip.

Viscous liquid adhesives are used to glue fabrics and paper, apply paper labels, make cardboard and wooden boxes and shoes, and bind books. Specially designed machines are required if the application of adhesives with different characteristics is to be satisfactorily controlled. The methods and machines shown here have been adapted to the application of adhesives in mass production. They might also work well for the application of liquid finishes such as primers, paint, and lacquer.

Fig. 2 Spiral conveyors feed the applicator roller by the force of gravity.

Fig. 3 An applicator wheel is fed by a spiral conveyor.

Pat. 2,195,323

Fig. 4 An adhesive pattern produced by raised faces on the applicator roll.

Pat. 2,276,997

Fig. 5 A gravity spreader with flow from its bottom holes.

Pat. 1,652,697

Fig. 6 Vibrating brushes spread the coating after the application by a cylindrical brush.

Pat. 1,747,104

Fig. 7 This applicator wheel is fed by a transfer disk.

Labels: Rotary applicator · Transfer disk · Adjusting screw · Closed reservoir · Glue · Compressible material such as rubber, felt, or brush · Pat. 2,096,364

Fig. 8 This applicator surface, consisting of a series of plate edges, is rotated by a Geneva mechanism in the glue pot.

Labels: Applicator rotated by Geneva mechanism · Picker tube · Rubber suction cup transfers labels from stack to applicator · Stack of labels · Rubber suction cup transfers label from applicator to bottle (not shown) · Applicator · Glue pot · Pat. 2,289,237

Fig. 9 A rotating applicator disk fed by a trough collector on a transfer drum.

Labels: Rotating applicator · Hinge pin · Trough collector · Adjustable scraper controls quantity · Pat. 2,268,420

Fig. 10 Cam-controlled transfer rollers supply applicator wheel pads with two kinds of adhesive.

Labels in figure:
- Applicator wheel
- Cam controlled transfer rollers
- Glue pot
- Glue pot
- Wheel takes adhesive when no blank envelopes are in position
- Pat. 2,111,317
- Water

Fig. 11 The bottom label is spread with glue by two abutting glue-coated picker plates, which separate during contact with label stack, then carry the label to the bottle.

Labels in figure:
- Label stack
- Twin picker plates on two arms
- Transfer roller
- Glue roller
- Pat. 2,257,785
- Stripper finger moves between picker plates

405

Fig. 12 Measured quantities of cement are forced through perforations in specially designed upper and lower die plates, which are closed hydraulically over zippers. Only the lower die is shown.

Fig. 13 A brush applicator is fed through passages between bristle tufts by a spring-operated plunger.

Fig. 14 A shoulder on a valve stem in a glue chamber retains glue until pressure on the tip opens the bottom valve.

Fig. 15 Glue is applied to label stock by a spray nozzle.

Fig. 16 A rocker shaft on a rack, which is moved vertically by a sector gear, carries glue on a contact bar from the roll to the label stack.

Fig. 17 Glue is extruded through nozzle on work.

Fig. 18 Pin applicators reciprocate vertically, first immersing themselves in glue, then contacting the undersides of carton flaps in a desired pattern.

406

Agitates and feeds paste toward pulley

Shifts paste bead from top to bottom of belt

Scraper leaves paste bead on belt

Wound helix wire belt applicator

Label stack

Guide mechanism keeps paste from edge of label stack

Pat. 2,206,964

Fig. 19 A paste belt applicator passes around the pulley in a pastepot and slides over the label stack.

Bar on cranks depresses dauber

Holder for inverted glue bottle

Rubber plug dauber

Glue pot

Pin on gear produces horizontal movement of dauber assembly

Pat. 2,237,359

Fig. 20 A dauber assembly is moved horizontally between a glue pot and work by an eccentric pin on a gear. Vertical movements are produced by a crank-operated bar over the dauber shaft.

TWENTY-FOUR AUTOMATIC MECHANISMS FOR STOPPING UNSAFE MACHINES

Automatic stopping mechanisms that prevent machines from damaging themselves or destroying work in process are based on the principles of mechanics, electricity, hydraulics, and pneumatics.

Fig. 1 A repetition of the machine cycle is prevented if a pedal remains depressed. The latch carried by the left slide pushes the right slide downward with a curved shoulder until the latch is disengaged by trip member.

Fig. 2 The gumming of the suction picker and label carrier when the label is not picked up by the suction is prevented by insufficient suction on a latch-operating cylinder, caused by open suction holes on the picker. When a latch-operating cylinder does not operate, the gum box holding latch returns to its holding position after cyclic removal by the cam and roller, thus preventing the gum box and rolls from rocking to make contact with the picker face.

Fig. 3 Damage to the milling cutter, work, or fixtures is prevented by the shroud around the cutter. Upon contact with the work, the shroud closes the electric circuit through the relay, thus closing contact A. This causes contact B to close, thus energizing relay C to operate a stop valve. It also closes a circuit through relay D, thus reversing the selector valve by means of a shifter rod so that bed travel will reverse on starting. Simultaneously, relay F opens the circuit of relay E and closes a holding circuit that was broken by the shifter lever at K. Relay G also closes a holding circuit and opens a circuit through relay D. The starting lever, released by push button H, releases contact A and returns the circuit to normal. If contact is made with the shroud when the bed travel is reversed, interchange the positions of D and E, with F and G in the sequence of operations.

408

Fig. 4 A high-speed press is stopped when a metal strip advances improperly. A hole punched in the strip fails to match with the opening in the die block to permit a light beam to pass. When the light beam to the photoelectric cell is blocked, the solenoid which withdraws the clutch pin is activated.

Fig. 5 A broken thread allows the contact bar to drop, thereby closing the electronic relay circuit; this stops the beamer reeling equipment.

Fig. 6 A nozzle on the packaging machine does not open when the container is improperly positioned.

Fig. 7 An obstruction under the explorer foot of a wire-stitching machine raises a vertical plunger which releases a latch lever. A rotary cam then raises a lever that retains the clutch-operating plunger.

409

Fig. 8

Circuit opened here by movement of plunger

Stop handle

Solenoid operates stop handle by lever system

Detector closes contact when thread breaks

Thread

Fig. 9

Continuous steam cooker and canning machine

Tripping arm is raised by can standing on end, thus releasing latch

Cans

Can conveyor track

Clutch shifting lever

Clutch mechanism

Spring releases clutch when shift lever latch is released

Drive shaft

Fig. 10

Rack moves in fixed slide

Ratchet

Paper feed

Drive shaft

Electromagnetic locking piston

Flexible wiring

Adjustable pivot for length of feed

Paper

Lamp

Photoelectric switch prevents energization of electromagnetic locking piston and feeding of paper if paper is not properly located and if mercury switch has not been tripped by package

Trip

Package

Conveyor platform fixed

Pivot

Pushers

Mercury switch

Pivot

D-C generator belt driven by main A-C motor

Fig. 11

Pivot

Mercury switch

Switch supporting pin carried by hub and keyed to square shaft

Stationary segment

Manual switch support

Switch box

Detector shoe

Paper web

Square shaft rotates when paper breaks, and opens mercury switch thus stopping printing press

Detector shoe

Adjustable detector arm

Web supporting shaft

Paper web

Lever operated contacts closed when cross-head is raised, open when cross-head is down

Plunger operated switch closed when pad is depressed

Solenoid opens line switch if pad is held depressed by sticking part that is not ejected

Pivot

Pad

Fig. 12

Yarn roll

Electric contacts close when roll is predetermined size thus energizing solenoid and stopping machine

Adjustable terminal

Solenoid

Armature

Stop wheel

Stop lever

Fig. 14

Cylindrical contact plate

One of many drop wires that close solenoid circuit when thread breaks

Yarn roll size control contact

Solenoid

Stop lever

Contact ring

Wire ring

Drop wire

Turret head

Tube support for "Cheesing tree"

Wiring Diagram

Fig. 13

Light source

Yarn

Targets suspended by threads drop and intercept beam, which causes machine to stop if any thread breaks

Slotted tube or coil spring

Photocell

Adjusting screw

Pointer shaped shutter moves to intercept light beams through slots in double wall separating lamp and light sensitive cell, which opens power switch

Flat spring pivots detector lever

Stationary support bracket

Work

Adjusting screws

Grinder

Shutter lever is released by detector arm

Fig. 15

Cam opens valve just before indexing mold carrier. There will be no air pressure in line unless pressing plunger has not raised and closed safety valve

Clutch

Drive shaft

Pneumatic safety valve

Clutch lever

Pneumatic cylinder releases drive shaft clutch and applies mold carrier brake only if cam-operated valve and safety valve are open simultaneously

Brake shoe

Pressing plunger

Mold carrier

Mold

Fig. 16

411

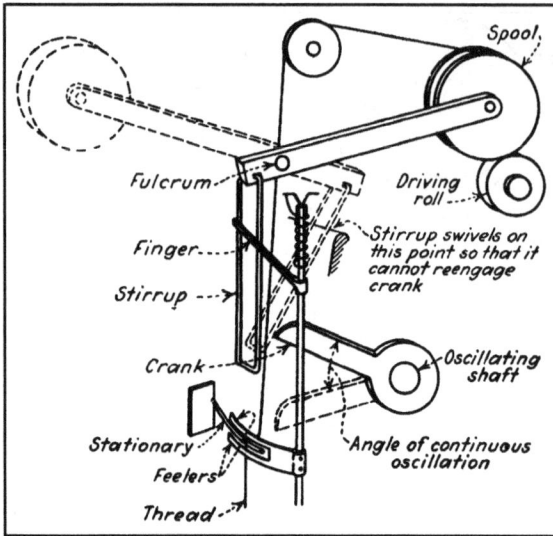

Fig. 17 A mechanism on a spooler. When a thread breaks, the feelers are released and the spiral spring causes the spindle with finger to rotate. The finger throws the stirrup into the path of the oscillating crank, which on its downward stroke throws the spool into the position shown dotted. The stirrup is then thrown out of the path of the oscillating crank.

Fig. 18 A mechanism in use on tubular braiding machines. When the machine is braiding, tension on the wire or thread lifts the idler carrier which then releases the pawl from the ratchet on the spool flange and allows the spool to turn and unwind. When the machine stops, the tension on the wire decreases, allowing the idler carrier to fall so that the pawl can engage the ratchet. If a wire breaks while the machine is running, the unsupported idler carrier falls to the base of the standard, and when the standard arrives at the station in the raceway adjacent to the cam *C*, the lug *L* on the idler carrier strikes the cam *C*, rotating it far enough to disengage a clutch on the driving shaft, thereby stopping the machine.

Fig. 19 When thread breaks, the stop drops and intercepts the reciprocating bar. On the next counter-clockwise oscillation of the eccentric arm, the bar B is raised. A feature of this design is that it permits the arm B to move up or down independently for a limited distance.

Fig. 20 A diagram of a mechanism that causes a bobbin changer to operate. If the contact arm does not slip on the bobbin, lever *A* will rotate to the position shown. But if contact with the bobbin center slips, when the bobbin is empty, lever *A* will not rotate to the position indicated by the dashed line. This will cause the bobbin changer to operate.

Fig. 21 A simple stop mechanism for limiting the stroke of a reciprocating machine member. Arrows indicate the direction of movements.

Fig. 22 When the predetermined weight of material has been poured on the pan, the movement of the scale beam pushes the latch out of engagement. This allows the paddle wheel to rotate and dump the load. The scale beam drops, thereby returning the latch to the holding position and stopping the wheel when the next vane hits the latch.

Fig. 23 In this textile machine, any movement that will rotate the stop lever counter-clockwise will move it into the path of the continuously reciprocating shaft. This will cause the catch lever to be pushed counterclockwise, freeing the hardened steel stop on the clutch control shaft. A spiral spring then impels the clutch control shaft to rotate clockwise. That movement throws out the clutch and applies the brake. The initial movement of the stop lever can be caused by a breaking thread or a moving dog.

Fig. 24 Some package-loading machines have provisions to stop the machine if a package passes the loading station without receiving an insert. Pawl finger *F* has a rocking motion imparted by the crankshaft, timed so that it enters the unsealed packages and is stopped against the contents. If the box is not filled, the finger enters a long distance. The pawl end at the bottom engages and holds a ratchet wheel on the driving clutch which disengages the machine-driving shaft.

SIX AUTOMATIC ELECTRICAL CIRCUITS FOR STOPPING TEXTILE MACHINES

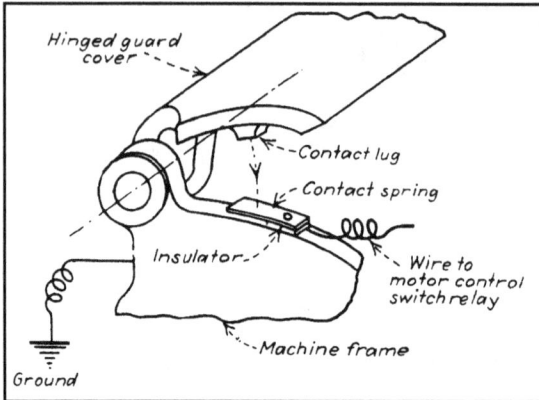

Fig. 1 A safety mechanism on some machines stops the motor when a guard cover is lifted. The circuit is complete only when the cover is down. In that position a contact lug establishes a metal-to-metal connection with the contact spring, completing the relay circuit.

Fig. 2 An electrical three-point wedging "warp stop" is shown after a thread has broken and a drop wire has fallen and tilted to close the circuit. Dotted lines indicate the normal position of the drop wire when it is riding on a thread. When a thread breaks, the drop wire falls and strikes the top of the terminal blade B, the inclined top of the slot. This causes a wedging action that tilts the drop wire against the terminal strip at C and D, reinforcing the circuit-closing action.

Fig. 3 Bobbin changer. When a bobbin is empty, the feeders contact the metal bobbin center, completing the circuit through a solenoid which pulls a latch. That causes the bobbin-changing mechanism to operator and put a new bobbin in the shuttle. As long as the solenoid remains deenergized, the pawl on the lay bar is raised clear of the hook on the bobbin-changing mechanism.

Fig. 4 Control for the automatic shear. When a seam of two thicknesses of cloth passes between the rolls, the swing roller is moved outward and closes a sensitive switch which energizes a solenoid. The solenoid pulls in an armature whose outer end is attached to the hinged ring where a cam plate is also fastened. The cam plate depresses the pins in a rotating plate. As the plate rotates, the depressed pins lift a hinged cam arm on which two mercury switches are mounted. When tilted, the switches complete circuits in two motor controls. A resetting cam for pushing the depressed pins back to their original position is fastened on the machine frame. The two motors are stopped and reversed until the seam has passed through rollers before they are stopped and reversed again.

Fig. 5 Electric stop for a loom. When a thread breaks or slackens, the drop wire falls and contact A rides on contact C. The drop wire, supported off-center, swings so that contact B is pulled against the inner terminal strip D, completing the solenoid circuit.

Fig. 6 This automatic stop for a folder or yarder machine always stops the machine in the same position when a seam in the cloth passes between the rolls. A seam passing between the rolls causes the swivel-mounted roll to lift slightly. This motion closes contacts in a sensitive switch that opens a relay in the control box. The next time the cam closes the limit switch, the power of the motor with the integral magnetic brake is shut off. The brake always stops the machine in the same place.

SIX AUTOMATIC MECHANISMS FOR ASSURING SAFE MACHINE OPERATION

The best automatic safety mechanisms are those that have been designed specifically for the machine on which they will be installed. When properly designed, automatic safety devices (1) do not reduce the operator's visibility, (2) do not interfere with the operator's performance, (3) do not make physical contact with the operator to prevent injury (e.g., by knocking his hand out of the way), (4) are fail-safe, (5) are sensitive enough to operate instantly, and (6) render the machine inoperative if anyone attempts to tamper with them or remove them.

Safety mechanisms can range from those that keep both of the operator's hands occupied on controls away from the work area to shields that completely enclose the work in progress on the machine and prevent machine operation unless the shield is securely in place. Many modern safety systems are triggered if any person or object breaks a light beam between a photoemitter and photoreceiver.

Fig. 1
Cam roller normally operates guard mechanism
Machine frame
Clutch operating member
Interlocking member slides into slot unless guard is improperly positioned
Reciprocating rod causes disengagement of clutch if held in path of operating member by lifted guard or control rod
Operator's control moved up to start machine
Guard interference causes disengagement of clutch
Work line
Pat. No. 2,257,273

Fig. 2
Guideways for rollers on transparent guard
Mating molds. Lower mold rises in operation of mechanism
Handle for lowering shield operates clamping lever on work and clutch operating mechanism
Counterweighted arm lifts guard when released automatically by latch
Stop lever on bell crank. When lowered by raising safety shield, prevents operation of clutch
Movable transparent shield extends across work area
Clutch operating lever
Guard latch is released by contact with rising mold at time of mold contact
Bell crank
Brackets
Pat. No. 2,238,290

Perspective of Slide
Pat. No. 2,301,817
Clutch mechanism trip lever arm
Fig. 3
Latch disk
To left-hand lever
To right-hand lever
Unless both slide blocks are moved and engage two latch disks simultaneously, disks are displaced into races of slide block and trip lever arm is not moved
Latch disks
Section **A-A**
Box assembly slides in stationary housing when slide blocks move together

Spring operates ring guard and releases latch

Cams

Hinge pin

Clutch

A

Cam

Spring operates ring guard and releases latch

Link

Latch release arm

Section A-A

Trip

Latch

Latch release arm

Guard arm positioned by latch release arm

Cam-released link permits spring to snap guard ring to lower position and release latch

Cam roller on lever arm moves arm when unlatched. When obstruction in work area prevents unlatching, cam moves to left, releasing clutch as shown, thus preventing stitching operation.

Stitching element

Ring guard over work area descends before stitching elements. Obstructions cause release of operating clutch

Pat. No. 2,271,359

Fig. 4

Cutter carrier

Pins on guard arms distort guard wire, which throws microswitch and disengages clutch if operator's hands are in path of cutter mechanism or contact wire directly

Turret

Hinges

Guard arm

Guard arm

Cutter

Rollers

Knives

Pat. No. 2,333,675

Fig. 5

Microswitch, operated by lever arm to which wire is attached, controls stop mechanism

Section A-A

Cutter carrier frame moves over turret

Cutter rollers

Turret

Roller

Slot in plate

Section A-A

To clutch

Clutch rod is lowered to engage clutch

Slots in two plates permit disengagement of clutch when rod is in any position

When solenoid pulls roller to this position, downward thrust on foot pedal carries roller through slot in plate

Roller

Fig. 6

Slot in plate

Clutch rod

Solenoid activated by "electric eye" beam across work zone

Pat. No. 2,311,441

CHAPTER 15
TORQUE, SPEED, TENSION, AND LIMIT CONTROL SYSTEMS

APPLICATIONS OF THE DIFFERENTIAL WINCH TO CONTROL SYSTEMS

Known for its mechanical advantage, the differential winch is a control mechanism that can supplement the gear and rack and four-bar linkage systems in changing rotary motion into linear. It can magnify displacement to meet the needs of delicate instruments or be varied almost at will to fulfill uncommon equations of motion.

Fig. 1 A standard differential winch consists of two drums, D_1 and D_2, and a cable or chain which is anchored on both ends and wound clockwise around one drum and counterclockwise around the other. The cable supports a load-carrying sheave, and if the shaft is rotated clockwise, the cable, which unwinds from D_1 on to D_2, will raise the sheave a distance

$$\text{Sheave rise/rev} = \frac{2\pi R - 2\pi r}{2} = \pi(R - r)$$

The winch, which is not in equilibrium exerts a counterclockwise torque.

$$\text{Unbalanced torque} = \frac{P}{2}(R - r)$$

Fig. 2(A) Hulse Differential Winch*. Two drums, which are in the form of worm threads contoured to guide the cables, concentrically occupy the same logitudinal space. This keeps the cables approximately at right angles to the shaft and eliminates cable shifting and rubbing, especially when used with variable cross sections as in Fig. 2(B). Any equation of motion can be satisfied by choosing suitable cross sections for the drums. Methods for resisting or supporting the axial thrust should be considered in some installations. Fig. 2(C) shows typical reductions in displacement. *Pat. No. 2,590,623

$$\frac{x}{y} = \frac{R-r}{2R}$$

R	r	x/y
3	$2\frac{15}{16}$	$1/96$
1	$\frac{63}{64}$	$1/128$

Fig. 3(B)

Fig. 3(A)

Fig. 3(A) A Hulse Winch with opposing sheaves. This arrangement, which uses two separate cables and four anchor points, can be considered as two winches back-to-back with one common set of drums. Variations in motion can be obtained by: (1) restraining in the sheaves so that when the system is rotated the drums will travel toward one of the sheaves; (2) restraining the drums and allowing the sheaves to travel. The distance between the sheaves will remain constant and is usually connected by a bar; (3) permitting the drums to move axially while restraining them transversely. When the system is rotated, drums will travel axially one pitch per revolution, and sheaves remain in the same plane perpendicular to the drum axis. This variation can be reversed by allowing sheaves to move axially; and (4) sheaves need not be opposite but can be arranged as in Fig. 3(B) to rotate a wheel.

Fig. 4(A) **Pressures and temperature indicators.** A pressure change causes the diaphragm and sheave to move vertically and the pointer radially. Equilibrium occurs when the spring force balances the actuating torque. Replacing the diaphragm with a thermal element changes the instrument into a temperature indicator. Two sheaves and a reciprocating carriage, Fig. 4(B), are based on the principle shown in Fig. 3(A). A carriage is activated by pressure or temperature and is balanced by a spring force in the opposite end. Further magnification can be obtained, Fig. 4(C), by wrapping a cable around the roller to which the pointer is attached.

Fig. 5 **A hydraulic control system,** actuated by a differential winch, performs remote precision positioning of a control rod with a minimum of applied torque. The sending piston, retained in a cylinder block, reciprocates back and forth from a torque applied to the winch shaft. Fluid is forced out from one end of the cylinder through the pipe lines to displace the receiving piston, which in turn activates a control rod. The receiver simultaneously displaces a similar amount of fluid from the opposite end back to the sender. By suitable valving, the sender can become a double-acting pump.

421

SIX WAYS TO PREVENT REVERSE ROTATION

Fig. 1   An eccentric latch allows the shaft to rotate in one direction; any attempted reversal immediately causes the latch to wedge against the disk wall.

Fig. 2   A lug on a shaft pushes the notched disk free during normal rotation. The disk periphery stops the lug to prevent reverse rotation.

Fig. 3   A latch on the rim of the pulley is free only when the rotation is in the direction shown. This arrangement is ideal for conveyor-belt pulleys.

Fig. 4   Spring-loaded friction pads contact the right gear. The idler meshes and locks the gear set when the rotation is reversed.

Fig. 5   A fixed wedge and sliding wedge tend to disengage when the gear is turning clockwise. The wedges jam in the reverse direction.

Fig. 6   A sliding key has a tooth which engages the worm threads. In reverse rotation, the key is pulled in until its shoulders contact the block.

CALIPER BRAKES KEEP PAPER TENSION IN WEB PRESSES

A simple cam-and-linkage arrangement (drawing) works in a team with two caliper disk brakes to provide automatic tension control for paper feeds on a web press.

In the feed system controlled tension must be maintained on the paper that's being drawn off at 1200 fpm from a roll up to 42 in. wide and 36 in. in diameter. Such rolls, when full, weigh 2000 lb. The press must also be able to make nearly instantaneous stops.

Friction-disk brakes are subject to lining wear, but they can make millions of stops before they need relining.

In the system, two pneumatic disk brakes made by Tol-O-Matic, Inc., Minneapolis, were mounted on each roll, gripping two separate 12-in. disks that provide maximum heat dissipation. To provide the desired constant-drag tension on the rolls, the brakes are always under air pressure. A dancer roll riding on the paper web can, however, override the brakes at any time. It operates a cam that adjusts a pressure regulator for controlling brake effort.

If the web should break or the paper run out on the roll, the dancer roll will allow maximum braking. The press can be stopped in less than one revolution.

This linkage system works in combination with a regulator and caliper disk brakes to stop a press rapidly from a high speed if the web should break.

CONTROL SYSTEM FOR PAPER CUTTING

Two clutch/brake systems, teamed with magnetic pickup sensors, cut paper sheets into exact lengths. One magnetic pickup senses the teeth on a rotating sprocket. The resulting pulses, which are related to the paper length, are counted, and a cutter wheel is actuated by the second clutch/brake system. The flywheel on the second system enhances the cutting force.

This control system makes cutting sheets to desired lengths and counting how many cuts are made simpler.

WARNING SYSTEM PREVENTS OVERLOADING OF BOOM

Cranes can now be protected against unsafe loading by a device whose movable electrical contacts are shifted by a combination of fluidic power and cam-and-gear arrangement (see drawing).

The device takes into consideration the two key factors in the safe loading of a crane boom: the boom angle (low angles create a greater overturning torque than high angles) and the compression load on the boom, which is greatest at high boom angles. Both factors are translated into inputs that are integrated to actuate the electrical warning system, which alerts the crane operator that a load is unsafe to lift.

How it works. In a prototype built for Thew-Lorain Inc. by U.S. Gauge, Sellersville, Pennsylvania, a tension-to-pressure transducer (see drawing) senses the load on the cable and converts it into a hydraulic pressure that is proportional to the tension. This pressure is applied to a Bourdon-tube pressure gage with a rotating pointer that carries a small permanent magnet (see details in drawing). Two miniature magnetic reed switches are carried by another arm that moves on the same center as the pointer.

This arm is positioned by a gear and rack controlled by a cam, with a sinusoidal profile, that is attached to the cab. As the boom is raised or lowered, the cam shifts the position of the reed switches so they will come into close proximity with the magnet on the pointer and, sooner or later, make contact. The timing of this contact depends partly on the movement of the pointer that carries the magnet. On an independent path, the hydraulic pressure representing cable tension is shifting the pointer to the right or left on the dial.

When the magnet contacts the reed switches, the alarm circuit is closed, and it remains closed during a continuing pressure increase without retarding the movement of the point. In the unit built for Thew-Lorain, the switches were arranged in two stages: the first to trigger an amber warning light and second to light a red bulb and also sound an alarm bell.

Over-the-side or over-the-rear loading requires a different setting of the Bourdon pressure-gage unit than does over-the-front loading. A cam built into the cab pivot post actuates a selector switch.

LEVER SYSTEM MONITORS CABLE TENSION

A simple lever system solved the problem of how to keep track of varying tension loads on a cable as it is wound on its drum.

Thomas Grubbs of NASA's Manned Spacecraft Center in Houston devised the system, built around two pulleys mounted on a pivoted lever. The cable is passed between the pulleys (drawing) so an increase in cable tension causes the lever to pivot. This, in turn, pulls linearly on a flat metal tongue to which a strain gage has been cemented. Load on the lower pulley is proportional to tension on the cable. The stretching of the strain gage changes the electrical current that gives a continuous, direct reading of the cable tension.

The two pulleys on the pivoting lever are free to translate on the axes of rotation to allow proper positioning of the cable as it traverses the take-up drum.

A third pulley might be added to the two-pulley assembly to give some degree of adjustment to strain-gage sensitivity. Located in the plane of the other two pulleys, it would be positioned to reduce the strain on the tongue (for heavy loads) or increase the strain (for light loads).

A cam on the cab positions an arm with reed switches according to boom angle; the pressure pointer reacts to cable tension.

A load on the lower pulley varies with tension on the cable, and the pivoting of the lever gives a direct reading with a strain gage.

424

EIGHT TORQUE-LIMITERS PROTECT LIGHT-DUTY DRIVES

Light-duty drives break down when they are overloaded. These eight devices disconnect them from dangerous torque surges.

Fig. 1 Permanent magnets transmit torque in accordance with their numbers and size around the circumference of the clutch plate. Control of the drive in place is limited to removing magnets to reduce the drive's torque capacity.

Fig. 2 Arms hold rollers in the slots that are cut across the disks mounted on the ends of butting shafts. Springs keep the roller in the slots, but excessive torque forces them out.

Fig. 3 A cone clutch is formed by mating a taper on the shaft to a beveled central hole in the gear. Increasing compression on the spring by tightening the nut increases the drive's torque capacity.

Fig. 4 A flexible belt wrapped around four pins transmits only the lightest loads. The outer pins are smaller than the inner pins to ensure contact.

Fig. 5 Springs inside the block grip the shaft because they are distorted when the gear is mounted to the box on the shaft.

Fig. 6 The ring resists the natural tendency of the rollers to jump out of the grooves in the reduced end of one shaft. The slotted end of the hollow shaft acts as a cage.

Fig. 7 Sliding wedges clamp down on the flattened end of the shaft. They spread apart when torque becomes excessive. The strength of the springs in tension that hold the wedges together sets the torque limit.

Fig. 8 Friction disks are compressed by an adjustable spring. Square disks lock into the square hole in the left shaft, and round disks lock onto the square rod on the right shaft.

THIRTEEN LIMITERS PREVENT OVERLOADING

These 13 "safety valves" give way if machinery jams, thus preventing serious damage.

Fig. 1 A shear pin is a simple and reliable torque limiter. However, after an overload, removing the sheared pin stubs and replacing them with a new pin can be time-consuming. Be sure that spare shear pins are available in a convenient location.

Fig. 2 Friction clutch torque limiter. Adjustable spring tension holds the two friction surfaces together to set the overload limit. As soon as an overload is removed, the clutch reengages. A drawback to this design is that a slipping clutch can destroy itself if it goes undetected.

Fig. 3 Mechanical keys. A spring holds a ball in a dimple in the opposite face of this torque limiter until an overload forces it out. Once a slip begins, clutch face wear can be rapid. Thus, this limiter is not recommended for machines where overload is common.

Fig. 4 A cylinder cut at an angle forms a torque limiter. A spring clamps the opposing-angled cylinder faces together, and they separate from angular alignment under overload conditions. The spring tension sets the load limit.

Fig. 5 A retracting key limits the torque in this clutch. The ramped sides of the keyway force the key outward against an adjustable spring. As the key moves outward, a rubber pad or another spring forces the key into a slot in the sheave. This holds the key out of engagement and prevents wear. To reset the mechanism, the key is pushed out of the slot with a tool in the reset hole of the sheave.

Fig. 6 Disengaging gears. The axial forces of a spring and driving arm are in balance in this torque limiter. An overload condition overcomes the force of the spring to slide the gears out of engagement. After the overload condition is removed, the gears must be held apart to prevent them from being stripped. With the driver off, the gears can safely be reset.

Fig. 7 A cammed sleeve connects the input and output shafts of this torque limiter. A driven pin pushes the sleeve to the right against the spring. When an overload occurs, the driving pin drops into the slot to keep the shaft disengaged. The limiter is reset by turning the output shaft backwards.

427

Fig. 8 A magnetic fluid is the coupler in this torque limiter. The case is filled with a mixture of iron or nickel powder in oil. The magnetic flux passed through the mixture can be controlled to vary the viscosity of the slurry. The ability to change viscosity permits the load limit to be varied over a wide range. Slip rings carry electric current to the vanes to create the magnetic field.

Fig. 9 A fluid is the coupling in this torque limiter. Internal vanes circulate the fluid in the case. The viscosity and level of the fluid can be varied for close control of the maximum load. The advantages of this coupling include smooth torque transmission and low heat rise during slip.

Fig. 10 The shearing of a pin releases tension in this coupling. A toggle-operated blade shears a soft pin so that the jaws open and release an excessive load. In an alternative design, a spring that keeps the jaws from spreading replaces the shear pin.

Fig. 11 A spring plunger provides reciprocating motion in this coupling. Overload can occur only when the rod is moving to the left. The spring is compressed under an overload condition.

Fig. 12 Steel shot transmits more torque in this coupling as input shaft speed is increased. Centrifugal force compresses the steel shot against the outer surfaces of the case, increasing the coupling's resistance to slip. The addition of more steel shot also increases the coupling's resistance to slip.

Fig. 13 A piezoelectric crystal produces an electric signal that varies with pressure in this metal-forming press. When the amplified output of the piezoelectric crystal reaches a present value corresponding to the pressure limit, the electric clutch disengages. A yielding ring controls the compression of the piezoelectric crystal.

428

SEVEN WAYS TO LIMIT SHAFT ROTATION

Traveling nuts, clutch plates, gear fingers, and pinned members
form the basis of these ingenious mechanisms.

Mechanical stops are often required in automatic machinery and servomechanisms to limit shaft rotation to a given number of turns. Protection must be provided against excessive forces caused by abrupt stops and large torque requirements when machine rotation is reversed after being stopped.

Fig. 1 **A traveling nut** moves along the threaded shaft until the frame prevents further rotation. This is a simple device, but the traveling nut can jam so tightly that a large torque is required to move the shaft from its stopped position. This fault is overcome at the expense of increased device length by providing a stop pin in the traveling nut.

Fig. 2 The engagement between the pin and the rotating finger must be shorter than the thread pitch so the pin can clear the finger on the first reverse-turn. The rubber ring and grommet lessen the impact and provide a sliding surface. The grommet can be oil-impregnated metal.

Fig. 3 **Clutch plates** tighten and stop their rotation as the rotating shaft moves the nut against the washer. When rotation is reversed, the clutch plates can turn with the shaft from A to B. During this movement, comparatively low torque is required to free the nut from the clutch plates. Thereafter, subsequent movement is free of clutch friction until the action is repeated at the other end of the shaft. The device is recommended for large torques because the clutch plates absorb energy well.

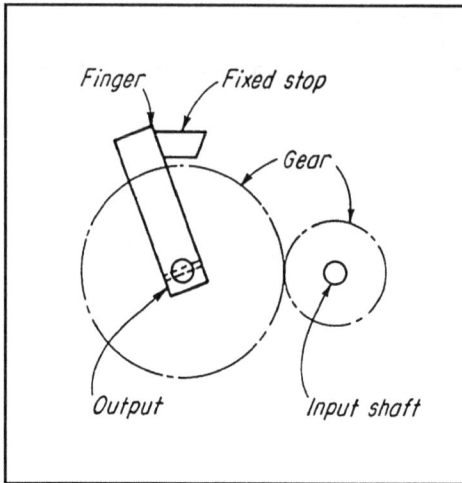

Fig. 4 **A shaft finger** on the output shaft hits the resilient stop after making less than one revolution. The force on the stop depends upon the gear ratio. The device is, therefore, limited to low ratios and few turns, unless a worm-gear setup is used.

Fig. 5 **Two fingers** butt together at the initial and final positions to prevent rotation beyond these limits. A rubber shock-mount absorbs the impact load. A gear ratio of almost 1:1 ensures that the fingers will be out-of-phase with one another until they meet on the final turn. Example: Gears with 30 to 32 teeth limit shaft rotation to 25 turns. Space is saved here, but these gears are expensive.

Fig. 6 **A large gear ratio** limits the idler gear to less than one turn. Stop fingers can be added to the existing gears in a train, making this design the simplest of all. The input gear, however, is limited to a maximum of about five turns.

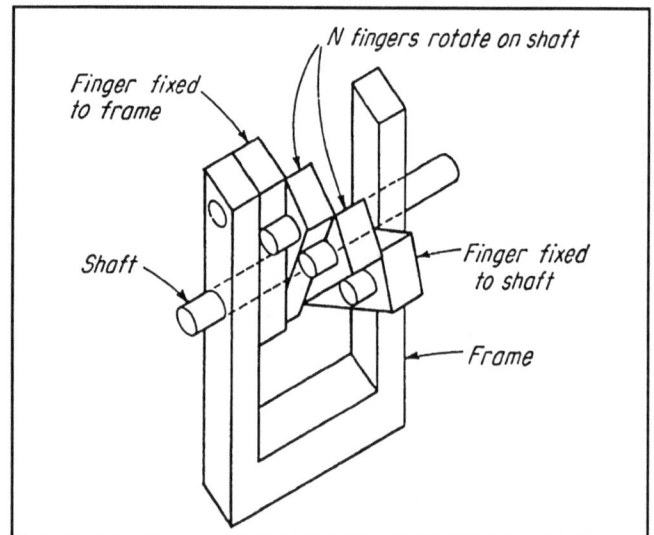

Fig. 7 **Pinned fingers** limit shaft turns to approximately N + 1 revolutions in either direction. Resilient pin-bushings would help reduce the impact force.

MECHANICAL SYSTEMS FOR CONTROLLING TENSION AND SPEED

The key to the successful operation of any continuous-processing system that is linked together by the material being processed is positive speed synchronization of the individual driving mechanisms. Typical examples of such a system are steel mill strip lines, textile processing equipment, paper machines, rubber and plastic processers, and printing presses. In each of these examples, the material will become wrinkled, marred, stretched or otherwise damaged if precise control is not maintained.

FIG. 1—PRIMARY INDICATORS

Floating Roll

Web

To controller

Use: constant tension winding; registry control; section synchronizing.

Contact Rolls

Strip →

To controller

Use: velocity control; cutter feed control.

Compensating Loop

Wire

To controller

Use: line-tension control; section synchronizing.

Flying Caliper

To controller

Use: thickness control; diameter control.

FIG. 2—SECONDARY INDICATORS

Follower Roll

To controller

Windup stand

Use: winder control; periphery cutting:

Mechanical Power

Tension idler

Motor

Load

Control belt

Gear differential

To controller

Use: winding tension; feeder control.

Electric Power

Motor

Load

Electric instrument

Supply

To controller

Use: winding tension; feeder control.

FIG. 3—CONTROLLERS AND ACTUATORS

On-and-Off Controller

Indicator

Limit switches

Slow

Fast

Controlled output shaft

Reversing motor

Timer

Input shaft

Actuator

Use: delivery rate; storage capacity; rough section synchronizing.

Proportioning Controller

Indicator

Adjustable speed actuator

Controller

Input drive

Controlled output drive

Use: winding controls; section synchronizing; cutter feeds.

The automatic control for such a system contains three basic elements: The *signal device* or *indicator*, which senses the error to be corrected; the *controller*, which interprets the indicator signal and amplifies it, if necessary, to initiate control action; and the *transmission*, which operates from the controller to change the speed of the driving mechanism to correct the error.

Signal indicators for continuous systems are divided in two general classifications: *Primary indicators* that measure the change in speed or tension of the material by direct contact with the material; and *secondary indicators* that measure a change in the material from some reaction in the system that is proportional to the change.

The primary type is inherently more accurate because of its direct contact with the material. These indicators take the form of contact rolls, floating or compensating rolls, resistance bridges, and flying calipers, as illustrated in Fig. 1. In each case, any change in the tension, velocity, or pressure of the material is indicated directly and immediately by a displacement or change in position of the indicator element. The primary indicator, therefore, shows deviation from an established norm, regardless of the factors that have caused the change.

Secondary indicators, shown in Fig. 2, are used in systems where the material cannot be in direct contact with the indicator or when the space limitations of a particular application make their use undesirable. This type of indicator introduces a basic inaccuracy into the control system which is the result of measuring an error in the material from a reaction that is not exactly proportional to the error. The control follows the summation of the errors in the material and the indicator itself.

The controlling devices, which are operated by the indicators, determine the degree of speed change required to correct the error, the rate at which the correction must be made, and the stopping point of the control action after the error has been corrected. The manner in which the corrective action of the controller is stopped determines both the accuracy of the control system and the kind of control equipment required.

Three general types of control action are illustrated in Fig. 3. Their selection for any individual application is based on the degree of control action required, the amount of power available for initiating the control, that is, the torque amplification required, and the space limitations of the equipment.

The on-and-off control with timing action is the simplest of the three types. It functions in this way: when the indicator is displaced, the timer contact energizes the control in the proper direction for correcting the error. The control action continues until the timer stops the action. After a short interval, the timer again energizes the control system and, if the error still exists, control action is continued in the same direction. Thus, the control process is a step-by-step response to make the correction and to stop the operation of the controller.

The proportioning controller corrects an error in the system, as shown by the indicator, by continuously adjusting the actuator to a speed that is in exact proportion to the displacement of the indicator. The diagram in Fig. 3 shows the proportioning controller in its simplest form as a direct link connection between the indicator and the actuating drive. However, the force amplification between the indicator and the drive is relatively low; thus it limits

Fluid Pressure

Use: fluid level control; constant pressure control; filtering rate control.

Liquid Level

Use: pumping rate control; system pressure control.

Temperature

Use: annealing; drum dryers; kilns.

Pressure

Use: fluid density; feeding rate; flow rate.

Proportioning-Throttling Controller

Use: constant tension winding; registry control; exact section synchronizing.

432

Fig. 4 **Floating rolls** are direct indicators of speed and tension in the paper web. Controller-actuators adjust feed and windup rolls to maintain registry during printing.

Fig. 5 **Dimension control** of extruded materials calls for primary indicators like the contact rolls shown. Their movements actuate conveyor control mechanisms.

this controller to applications where the indicator has sufficient operating force to adjust the speed of the variable-speed transmission directly.

The most accurate controller is the proportioning type with throttling action. Here, operation is in response to the rate or error indication. This controller, as shown in Fig. 3, is connected to a throttling valve, which operates a hydraulic servomechanism for adjusting the variable-speed transmission.

The throttling action of the valve provides a slow control action for small error correction or for continuous correction at a slow rate. For following large error, as shown by the indicator, the valve opens to the full position and makes the correction as rapidly as the variable-speed transmission will allow.

Many continuous processing systems can be automatically controlled with a packaged unit consisting of a simple, mechanical, variable-speed transmission and an accurate hydraulic controller.

This controller-transmission package can change the speed relationship at the driving points in the continuous system from any indicator that signals for correction by a displacement. It has anti-hunting

characteristics because of the throttling action on the control valve, and is self-neutralizing because the control valve is part of the transmission adjustment system.

The rotary printing press is an example of a continuous processing system that requires automatic control. When making billing forms on a press, the printing plates are rubber, and the forms are printed on a continuous web or paper. The paper varies in texture, moisture content, flatness, elasticity, and finish. In addition, the length of the paper changes as the ink is applied.

In a typical application of this kind, the accuracy required for proper registry of the printing and hole punching must be held to a differential of $1/32$ in. in 15 ft of web. For this degree of accuracy, a floating or compensating roll, as shown in Fig. 4, serves as the indicator because it is the most accurate way to indicate changes in the length of the web by displacement. In this case, two floating rolls are combined with two separate controllers and actuators. The first controls the in-feed speed and tension of the paper stock, and the second controls the wind-up.

The in-feed is controlled by maintaining the turning speed of a set of feeding rolls that pull the paper off the stock roll. The second floating roll controls the speed of the wind-up mandrel. The web of paper is held to an exact value of tension between the feed rolls and the punching cylinder of the press by the in-feed control. It is also held between the punching cylinder and the wind-up roll. Hence, it is possible to control the tension in the web of different grades of paper and also adjust the relative length at these two points, thereby maintaining proper registry.

The secondary function of maintaining exact control of the tension in the paper as it is rewound after printing is to condition the paper and obtain a uniformly wound roll. This makes the web ready for subsequent operations.

The control of dimension or weight by tension and velocity regulation can be illustrated by applying the same general type of controller actuator to the take-off conveyors in a extruder line such as those used in rubber and plastics processing. Two problems must be solved: First, to set the speed of the take-away conveyor at the extruder to match the variation in extrusion rate; and, second, to set the speeds of the subsequent conveyor sections to match the movement of the stock as it cools and tends to change dimension.

One way to solve these problems is to use the pivoted idlers or contact rolls as indicators, as shown in Fig. 5. The rolls contact the extruded material between each of the conveyor sections and control the speed of the driving mechanism of the following section. The material forms a slight catenary between the stations, and the change in the catenary length indicates errors in driving speeds.

The plasticity of the material prevents the use of a complete control loop. Thus, the contract roll must operate with very little resistance or force through a small operating angle.

The difficulties in winding or coiling a strip of thin steel that has been plated or pre-coated for painting on a continuous basis is typical of processing systems whose primary indicators cannot be used. While it is important that no contact be made with the prepared surface of the steel, it also desirable to rewind the strip after preparation in a coil that is sound and slip-free. An automatic, constant-tension winding control and a secondary indicator initiate the control action.

The control system shown in Fig. 6 is used in winding coils from 16 in. core diameter to 48 in. maximum diameter. The power to wind the coil is the controlling medium because, by maintaining constant winding power as the coil builds

433

up, a constant value of strip tension can be held within the limits required. Actually, this method is so inaccurate that the losses in the driving equipment (which are a factor in the power being measured) are not constant; hence the strip tension changes slightly. This same factor enters into any control system that uses winding power as an index of control.

A torque-measuring belt that operates a differential controller measures the power of the winder. Then, in turn, the controller adjusts the variable-speed transmission. The change in speed between the source of power and the transmission is measured by the three-shaft gear differential, which is driven in tandem with the control belt. Any change in load across the control belt produces a change in speed between the driving and driven ends of the belt. The differential acts as the controller, because any change in speed between the two outside shafts of the differential results in a rotation or displacement of the center or control shaft. By connecting the control shaft of the differential directly to a screw-controlled variable-speed transmission, it is possible to adjust the transmission to correct any change in speed and power delivered by the belt.

This system is made completely automatic by establishing a neutralizing speed between the two input shafts of the differential (within the creep value of the belt). When there is no tension in the strip (e.g., when it is cut), the input speed to the actuator side of the differential is higher on the driven side than it is on the driving side of the differential. This unbalance reverses the rotation of the control shaft of the differential, which in turn resets the transmission to high speed required for starting the next coil on the rewinding mandrel.

In operation, any element in the system that tends to change strip tension causes a change in winding power. This change, in turn, is immediately compensated by the rotation (or tendency to rotate) of the controlling shaft in the differential. Hence, the winding mandrel speed is continuously and automatically corrected to maintain constant tension in the strip.

When the correct speed relationships are established in the controller, the system operates automatically for all conditions of operation. In addition, tension in the strip can be adjusted to any value by moving the tension idler on the control belt to increase or decrease the load capacity of the belt to match a desired strip tension.

There are many continuous processing systems that require constant velocity of the material during processing, yet do not require accurate control of the tension in the material. An example of this

Fig. 6 The differential controller has a third shaft that signals the remote actuator when tension in sheet material changes. Coiler power is a secondary-control index.

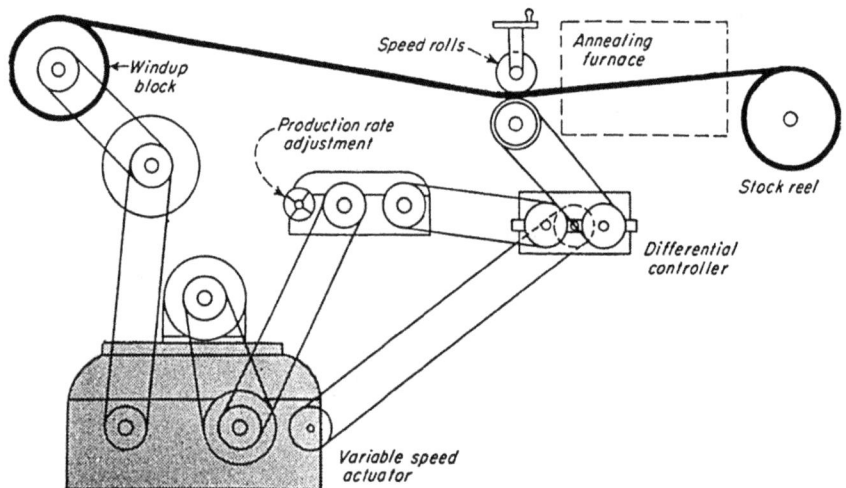

Fig. 7 The movement of wire through the annealing furnace is regulated at constant velocity by continuously retarding the speed of the windup reels to allow for wire build-up.

process is the annealing of wire that is pulled off stock reels through an annealing furnace and then rewound on a windup block.

The wire must be passed through the furnace at a constant rate so that the annealing time is maintained at a fixed value. Because the wire is pulled through the furnace by the wind-up blocks, shown in Fig. 7, its rate of movement through the furnace would increase as the wire builds up on the reels unless a control slows down the reels.

A constant-velocity control that makes use of the wire as a direct indicator measures the speed of the wire to initiate a control action for adjusting the speed of the wind-up reel. In this case, the wire can be contacted directly, and a primary indicator in the form of a contact roll can register any change in speed. The contact roll drives one input shaft of the differential controller. The second input shaft is connected to the driving shaft of

the variable-speed transmission to provide a reference speed. The third, or control, shaft will then rotate when any difference in speed exists between the two input shafts. Thus, if the control shaft is connected to a screw-regulated actuator, an adjustment is obtained for slowing down the wind-up blocks as the coils build up and the wire progresses through the furnace at a constant speed.

434

NINE DRIVES FOR CONTROLLING TENSION

Mechanical, electrical, and hydraulic methods for obtaining controlled tension on winding reels and similar drives, or for driving independent parts of a machine in synchronism.

MECHANICAL DRIVES

A band brake is used on coil winders, insulation winders, and similar machines where maintaining the tension within close limits is not required.

It is simple and economical, but tension will vary considerably. Friction drag at start-up might be several times that which occurs during running because of the difference between the coefficient of friction at the start and the coefficient of sliding friction. Sliding friction will be affected by moisture, foreign matter, and wear of the surfaces.

Capacity is limited by the heat radiating capacity of the brake at the maximum permissible running temperature.

Differential drives can take many different forms, e.g., epicycle spur gears, bevel gear differentials, or worm gear differentials.

The braking device on the ring gear or spider could be a band brake, a fan, an impeller, an electric generator, or an electric drag element, such as a copper disk rotating in a powerful magnetic field. A brake will give a drag or tension that is reasonably constant over a wide speed range. The other braking devices mentioned here will exert a torque that will vary widely with speed, but will be definite for any given speed of the ring gear or spider.

A definite advantage of any differential drive is that maximum driving torque can never exceed the torque developed by the braking device.

Differential gearing can be used to control a variable-speed transmission. If the ring gear and sun gear are to be driven in opposite directions from their respective shafts which are to be held in synchronism, the gear train can be designed so that the spider on which the planetary gears are mounted will not rotate when the shafts are running at the desired relative speeds. If one or the other of the shafts speeds ahead, the spider rotates correspondingly. The spider rotation changes the ratio of the variable-speed transmission unit.

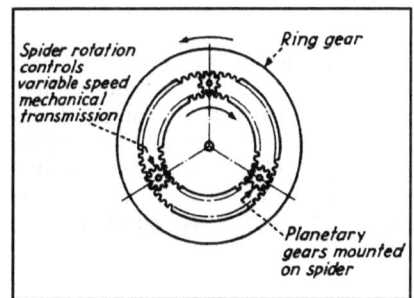

ELECTRICAL DRIVES

The shunt-field rheostat in a DC motor drive can be used to synchronize drives. When connected to a machine for processing paper, cloth, or other sheet material that is passing around a take-up roll, the movement of the take-up roll moves a control arm which is connected to the rheostat. This kind of drive is not suitable for wide changes of speed that exceed about a 2.5 to 1 ratio.

For wide ranges of speed, the rheostat is put in the shunt field of a DC generator that is driven by another motor. The voltage developed by the generator is controlled from zero to full voltage. The generator furnishes the current to the driving motor armature, and the fields of the driving motor are separately excited. Thus, the motor speed is controlled from zero to maximum.

Selsyn motors can directly drive independent units in exact synchronism, provided their inertias are not too great. Regardless of loads and speeds, selsyn motors can be the controlling units. As an example, variable-speed mechanical transmission units with built-in selsyn motors are available for powering constant-tension drives or the synchronous driving of independent units.

HYDRAULIC DRIVES

Hydraulic Control—Tension between successive pairs of rolls, or synchronism between successive units of a machine can be controlled automatically by hydraulic drives. Driving the variable delivery pump from one of the pairs of rolls automatically maintains an approximately constant relative speed between the two units, at all speeds and loads. The variations caused by oil leakage and similar factors are compensated automatically by the idler roll and linkage. They adjust the pilot valve that controls the displacement of the variable delivery pump.

The counterweight on the idler roll is set for the desired tension in the felt, paper, or other material. Increased tension as a result of the high speed of the second pair of rolls depresses the idler roll. The control linkage then moves the pilot valve to decrease pump delivery,

which slows the speed of the second pair of rolls. The reverse operations occur when the tension in the paper decreases, allowing the idler roll to move upwards.

If the material passing through the machine is too weak to operate a mechanical linkage, the desired control can be obtained by photoelectric devices. The hydraulic operation is exactly the same as that described for the hydraulic drives.

A band brake intended to obtain a friction drag will give variable tension. In this hydraulic drive, the winding tension is determined by the difference in torque exerted on the rewinder feed roll and the winding roll. The brake plays no part in establishing the tension.

The constant displacement hydraulic motor and the variable displacement hydraulic motor are connected in series with the variable delivery pump. Thus, the relative speeds of the two hydraulic motors will always remain substantially the same. The displacement of the variable speed motor is then adjusted to an amount slightly greater than the displacement of the constant-speed motor. This tends to give the winding roll a speed slightly greater than the feed roll speed. This determines the tension, because the winding roll cannot go faster than the feed roll. Both are in contact with the paper roll being wound. The pressure in the hydraulic line between the constant and variable displacement pumps will increase in proportion to the winding tension. For any setting of the winding speed controller on the variable delivery hydraulic pump, the motor speeds are generally constant. Thus, the surface speed of winding will remain substantially constant, regardless of the diameter of the roll being wound.

This is a hydraulic drive for fairly constant tension. The variable-delivery, constant-speed pumping unit supplies the oil to two constant displacement motors. One drives the apparatus that carries the fabric through the bath at a constant speed, and the other drives the winder. The two motors are in series, Motor A drives the winding reel, whose diameter increases from about 5 in. when the reel is empty to about 33 in. when the reel is full. Motor A is geared to the reel so that even when the reel is empty, the surface speed of paper travel will be somewhat faster than the mean rate of paper travel established by motor B, driving the apparatus. Only a small amount of oil will be bypassed through the choke located between the pressure and the return line.

When the roll is full, the revolutions per minute of the reel and its driving motor are only about one-seventh of the revolutions per minute when the reel is empty. More oil is forced through the choke when the reel is full because of the increased pressure in the line between the two motors. The pressure in this line increases as the reel diameter increases because the torque resistance encountered by the reel motor will be directly proportional to the reel diameter and

because tension is constant. The larger the diameter of the fabric on the reel, the greater will be the torque exerted by the tension in the fabric. The installation is designed so that the torque developed by the motor driving the reel will be inversely proportional to the revolutions per minute of the reel. Hence, the tension on the fabric will remain fairly constant, regardless of the diameter of the reel. This drive is limited to about 3 hp, and it is relatively inefficient.

LIMIT SWITCHES IN MACHINERY

Limit switches, which confine or restrain the travel or rotation of moving parts within certain predetermined points, are actuated by varying methods. Some of these, such as cams, rollers, push-rods, and traveling nuts, are described and illustrated.

Fig. 1

Spring returns plunger to original position

Spring return of switch must be of sufficient force to overcome friction in system

Guides

Input shaft

Geneva lock

Switch mechanism of cam and roller construction

Cam actuates push-rod from switch

Push-rod from switch follows cam

Spring-return, two-circuit limit switch, stationary mounting

Fig. 2

Rod movable

Cam for roller

One-half of the travel required to actuate the switch

Roller lever

Switch and roller lever mounted stationary

Arrangement designed for use where close adjustments and accuracy of contact operation are important

Spring-return limit switch

Fig. 3

Rod stationary if switch mounted on moving carriage. Rod movable if switch mounted stationary

Adjustable

Stop

Stop

Plunger pushed from left side restores contacts to original position

Plunger pushed from right side operates contacts

Two-circuit, maintained-contact type limit switch mounted stationary, or mounted on moving carriage

Fig. 4

Adjustable

Rod stationary if switch and linkage assembly mounted on moving carriage. Rod movable if switch and linkage mounted stationary

Coarse adjustment

Fine adjustment

Maintained-contact type, push lever operated limit switch

Fixed point

Sleeve

Switch and accompanying linkage mounted stationary, or mounted on moving carriage

Fig. 5

Actuating pin may be used to provide both reverse and forward operation

Rod and actuating pin stationary if switches mounted on moving carriage. Rod and pin movable if switches mounted stationary

Contact is made when stroke or cam lever is moved in a clockwise direction and original contact is restored on return stroke

First switch speeds up or slows down an operation; second switch reverses the motion

Two-circuit maintained contact type limit switches mounted stationary, or mounted on moving carriage

Frame mounted
stationary on
machine

Top of nut arm rides in trough
in frame to prevent nut from turning

Input through gear train,
belt or chain drive

Adjustable stop nuts

Traveling nut

Adjustable
stop nuts

To controlled element
on machine

Screw and nut device
to control travel of
machine by reversing
belt or throwing out
a catch after any speci-
fied number of revolu-
tions.
Travel is adjusted by
stop nuts

Fig. 6

Rod movable

Pin A

Pin B

Prong A

Prong B

Actuating pins A and B
require separate ad-
justments

Two-circuit maintained contact type
limit switch mounted stationary

Front
View

Fig. 7

Pin B

Pin A

Prong A

Prong B

Contact operation
takes place when
fork lever is moved.
Contacts are re-
stored when fork lever
is moved in opposite
direction

Side
View

Movement of the machine lever to the right
operates the contacts of the right-hand
switch, but no contact takes place in the
left-hand switch.
Movement to the left operates the contacts
in the left-hand switch, but no contact takes
place in the right-hand switch

Switch
operation

Switch
operation

Machine
lever

No switch operation

Rod connected
to movable
carriage

Two-circuit, main-
tained contact type
limit switches mounted
stationary

A spring return mechanism
can be used if the weight and
friction of the connecting linkage
does not offset the power of the
return spring

Fig. 8

Cam wheel driven by machine
element

Cam

Rollers

Actuating
lever

When the cam wheel meets
the left-hand roller of the
actuating lever, this roller
rides up on the inclined surface and the limit switch
lever is tripped in a clockwise
direction.
The right-hand roller moves
upwards into the path of
the cam.
When the cam wheel meets
the right-hand roller, the
limit is tripped in a counterclockwise direction

Maintained-contact
type roller lever
limit switch mounted
stationary

Fig. 9

Electrical contact arrangements

All contracts in normal position with limit switch unactuated

SINGLE POLE

Open

Closed

Closed Open

Double throw

3-point contact

TWO POLE

Closed Open

Closed Open

MULTI-CONTACT

Off

POS.	1	2	3	4
R	C	C	O	O
Off	C	C	C	C
L	O	O	C	C

RIGHT
Bar travels in same direction as limit switch lever

WRONG
Bar travels against direction of travel of limit switch

Actuating bar
Roller
Spring return
Stop

Strain on lever and bearing increases wear and friction

RIGHT
Because of angle, on contact cam turns roller; less friction and strain

Spring return

WRONG
Contact too nearly perpendicular to center of roller lever

Position and shape of arm lets roller lever snap back, creating strains

Roller doesn't turn; large force component pushes against bearing, excessive wear

Arm delivers sharp blow

Spring return

PRECISION TYPES

RIGHT
Both motions in same direction; cam contact angle turns roller

WRONG

Operating pin

Cam moves against roller level travel

RIGHT
Change in cam contour just enough to operate switch

Cam clears roller

WRONG
Deep cam causes excessive motion of switch

Cam delivers sharp blow instead of applying actuating force gradually

RIGHT
One-half travel distance

Plunger

WRONG
Harmful horizontal pressure created

Cam to drop out a motion
or repeat timing cycle

Spring return

Off position

Small operating angle assures
drift of machine returns lever
to off position for next angle

Adjustable

Push rod

Limit switch with snap action
maintained type contacts

Correctly shaped and
placed cam

Pin operates
snap action
contacts

Lever

Precision type limit switch

Adjustable operating dogs

Roller

Hand lever for
manual control
during setup

Push rod

Spring return push
type limit switch

Correctly
shaped cam

Centers in line

Roller

Adjustment

Pin operates contact

Precision type limit switch

Actuates snap action contacts
of limit switch which reverses
actuating mechanism

Off
position

Spring return
to off
position

Lever travel

Oscillating
operating lever

On position

Off position

Spring
return

Good position and
shape of cam

Double action
contacts operated
in either right
or left position

Spring return
to center

Switch operation

No switch operation

Contacts not
operated as
actuating
bar returns
to left

Contacts
operated

Spring return
to center

Guide pin

Traveling nut

Roller

Operation of switch
can reverse threaded
shaft or stop a
motion

Precision type limit switch

Machine operated lever to provide
staggered operating sequence

Adjustable

② As lever continues to
move this switch
operates to stop or
reverse the motion

① This switch operates first and
might slow down
an operation

Adjustable

Traveling dog

Operating
screw

Screw limit

Operation
reverses
screw

Push type return limit switch

441

NINE AUTOMATIC SPEED GOVERNORS

Speed governors, designed to maintain the speeds of machines within reasonably constant limits, regardless of loads, depend for their action upon centrifugal force or cam linkages. Other governors depend on pressure differentials and fluid velocities for their actuation.

Primary cylinder and piston
To control valve or switch on machine

Fluid energy from jet pipe enters either passage A or B depending upon position of jet pipe

Heavy fluid energy input from machine enters system and actuates large primary piston according to position of valve gear

Secondary cylinder and piston

Fluid jet pipe actuated laterally by rod from governor mechanism on machine

Small amount of the fluid energy supply entering the system is diverted to movable fluid jet pipe

"D" Valve gear positioned by piston in secondary cylinder

Flexible coupling

Fig. 1   Auxiliary piston governor.

Fixed point

To reciprocating motion on machine

When speed of machine exceeds or falls below desired point fly-balls cause linkage to raise or lower arm

Guide

To control valve or switch

Spur strikes end of rod when aligned by motion of governor and actuates control rod to switch or valve

Fixed point

Fig. 2   Hit-and-miss governor.

Flyball shaft rotated by belt gear or chain drive from machine

Flyball governor positions valve piston in sliding valve cylinder

Controller cylinder and piston

Flexible hose

Flexible hose

Flexible hose

Fluid energy input from machine enters system and actuates controller piston according to position of valve piston

Fixed point pivot

Crosshead

As controller piston moves, it actuates linkage to move sliding valve cylinder in a "follow-up" or "compensated" manner

To controlling element

Fig. 3   Force-compensated regulator.

Fig. 4 Pressure-actuated governor.

Pivot

Fixed point

Control rod actuates as piston rises and falls in cylinder

Setting spring

To control valve or switch

As pneumatic, hydraulic, or steam pressure input from machine rises, it forces piston upward against tension of setting spring

Fluid energy input from machine

Fig. 5 Varying differential governor.

To cylinder slide valve

As input pressure rises beyond desired point, it forces piston upward in cylinder against tension of setting spring

Control rod receives motion from both piston and reciprocating rod. Net movement of control rod is difference of motion of piston rod and reciprocating rod

Setting spring

Pneumatic, hydraulic, or steam pressure input from pump unit driven by engine

To reciprocating motion on machine

Fig. 6 Centrifugal governor.

As ring rises it causes inverted truncated cone to move outward and actuate control rod

Ring

Excessive input through belt, shaft, or chain drive from machine causes flyballs to raise ring

Control rod connected to valve or switch

Spring loaded arm pivots about fixed point

Fig. 7 Constant-volume governor.

Adjusting weight

Fixed point

Float linkage operates control valve

Steam supply

Governor float

Clearance

To turbine

Pressure variations across orifice actuate oil-pressure-operated governor float. Float rises and falls depending on pressure differential

Atmospheric air

To compressor inlet

Pressure drop across orifice proportional to volume of air passing through

Fig. 8 Velocity-type governor (coil spring).

Fuel mixture strikes against governor valve and forces it closed against spring pressure

Tension in spring increases as valve cam turns

Similar in operation to Fig. 9. Engine speed drops when load is applied. Velocity of incoming gases and pressure against valve is reduced. Coil spring opens valve to permit greater fuel flow to meet increased load demand

Increasing force opposes force of incoming fuel mixture because of greater effective radius of cam as valve closes

Fig. 9 Velocity-type governor (cantilever spring).

Carburetor

Vented to atmosphere

Fuel mixture strikes against governor valve and forces it closed against the pressure of the cantilever spring

Cantilever spring

Manifold

Engine speed drops when load is applied and velocity of fuel mixture and pressure against governor valve is reduced. Cantilever spring opens valve to permit greater fuel flow to meet increased load demand

EIGHT SPEED CONTROL DEVICES FOR MECHANISMS

Friction devices, actuated by centrifugal force, automatically keep speed constant regardless of variations of load or driving force.

The weight is counterbalanced by a spring that brakes the shaft when the rotation speed becomes too fast. The braking surface is small.

A sheetmetal brake provides a larger braking surface than in the previous brake. Braking is more uniform, and it generates less heat.

Symmetrical weights give an even braking action when they pivot outward. The entire action can be enclosed in a case.

Weight-actuated levers make this arrangement suitable where high braking moments are required.

Three flat springs carry weights that provide a brake force upon rotation. A speed adjustment can be included.

The typical governor action of swinging weights is applied here. As in the previous brake, adjustment is optional.

The adjustment of the speed at which this device starts to brake is quick and easy. The adjusting nut is locked in place with a setscrew.

A tapered brake drum is another way to provide for varying speed-control. The adjustment is again locked.

CABLE-BRAKING SYSTEM LIMITS DESCENT RATE

This arrangement of capstan and idler sheaves permits the cable contact radius on the capstan to remain constant during cable payout. Cable from a centrifugally braked spool, positioned by three idlers, winds through three capstan grooves before payout. The capstan grooves have equal diameters (see inset), so total payout radius and the spool brake (not shown) keep payout nearly constant.

An improved cable-braking (CB) system that limits the descent rate of a load being dropped has been developed at Caltech for NASA's Jet Propulsion Laboratory. It lowers loads at near constant speed, regardless of the length of cable paid out. By contrast, previous cable-payout systems have lowered loads at speeds that vary with the length of cable paid out. It is expected that the improved CB system will be useful as a braking or shock-absorbing device for many different purposes. For example, it could lower a trapped person from an upper floor of a burning building at a safe speed, capture and slow a moving vehicle, limit the shock of an opening parachute on a parachutist, or supply pallets being dropped.

In previous CB systems of this type, the cable is paid out from a spool on a shaft connected to a centrifugal brake. Because the payout radius on the spool decreases as cord is paid out, the load's descent speed decreases by a corresponding amount. The improved CB system operates at approximately constant payout speed, regardless of the length of cable paid out because the spool radius remains constant.

The improved system, shown in the diagram, includes a cable spool, a three-groove capstan, three idler sheaves, and a centrifugal brake. (The cable spool and centrifugal brake are not shown.) The diagram shows the capstan with grooves of diminishing diameters to make it easier to understand how the system works. However, the actual capstan has three grooves that are equal in diameter. (See inset, right.) The spool stores the cable and, unlike in the previous cable system, does not participate in the primary braking action, but it permits the cable to unwind from it at low tension. The spool is connected to the rest of the system through a constant-torque slip clutch which must slip to pay out the cable.

As the cable from the spool (connected to the constant-torque slip clutch) passes over the first of three idler sheaves, it is directed into the first of three capstan grooves. After the cable travels part way around the circumference of the first capstan groove, it is directed to and passes around the second idler positioned to direct it into the second groove. After traveling a slightly longer distance around the second groove, the cable is transitioned to the third idler directing it into the third and last groove. Finally, after traveling yet another slightly longer distance around the third groove the cable is directed into the payout path. The sum of the wrap distances or radius angles the cable completes around the capstan is large enough to prevent slippage between the cable and the capstan.

The capstan is connected to a shaft which, in turn, is connected to the centrifugal brake. As a result, the effective payout radius, for purposes of braking, is not the varying radius angles of the remaining cable on the spool, but rather the constant radius of the grooves in the capstan. Cable payout speed is determined primarily by the sum of the distances the cable travels in the capstan grooves and the characteristics of the centrifugal brake. Therefore, the payout speed in this system is more nearly constant than in the previous system.

This work was done by Tommaso P. Rivellini, Donald B. Bickler, Bradford Swenson, John Gallon, and Jack Ingle of Caltech for NASA's Jet Propulsion Laboratory.

CHAPTER 16
INSTRUMENTS AND CONTROLS: PNEUMATIC, HYDRAULIC, ELECTRIC, AND ELECTRONIC

TWENTY-FOUR MECHANISMS ACTUATED BY PNEUMATIC OR HYDRAULIC CYLINDERS

Fig. 1 **A cylinder** can be used with a first-class lever.

Fig. 2 **A cylinder** can be used with a second-class lever.

Fig. 3 **A cylinder** can be used with a third-class lever.

Fig. 4 **A cylinder** can be linked up directly to the load.

Fig. 5 **A spring** reduces the thrust at the end of the stroke.

Fig. 6 **The point of application** of force follows the direction of thrust.

Fig. 7 **A cylinder** can be used with a bent lever.

Fig. 8 **A cylinder** can be used with a trammel plate.

Fig. 9 **Two pistons** with fixed strokes position the load in any of four stations.

Fig. 10 **A toggle** can be actuated by the cylinder.

Fig. 11 **The cam** supports the load after the completion of the stroke.

Fig. 12 **Simultaneous thrusts** in two different directions are obtained.

(Note: In place of hydraulic or pneumatic cylinders electrically powered actuators or solenoids can be substituted to perform these functions.)

Fig. 13 **A force** is transmitted by a cable.

Fig. 14 **A force** can be modified by a system of pulleys.

Fig. 15 A force can be modified by wedges.

Fig. 16 **A gear sector** moves the rack perpendicular to the piston stroke.

Fig. 17 **A rack** turns the gear sector.

Fig. 18 **The motion** of a movable rack is twice that of the piston.

Fig. 19 **A torque** applied to the shaft can be transmitted to a distant point.

Fig. 20 **A torque** can also be applied to a shaft by a belt and pulley.

Fig. 21 **A motion** is transmitted to a distant point in the plane of motion.

Fig. 22 **A steep screw nut** produces a rotation of the shaft.

Fig. 23 **A single-sprocket wheel** produces rotation in the plane of motion.

Fig. 24 **A double-sprocket wheel** makes the rotation more nearly continuous.

449

FOOT-CONTROLLED BRAKING SYSTEM

This crane braking system (see figure) operates when the main line switch closes. The full depression of the master-cylinder foot-pedal compresses the brake-setting spring mounted on the hydraulic releasing cylinder. After the setting spring is fully compressed, the hydraulic pressure switch closes, completing the electric circuit and energizing the magnetic check valve. The setting spring remains compressed as long as the magnetic check valve is energized because the check valve traps the fluid in the hydraulic-releasing cylinder. Upon release of the foot peal, the brake lever arm is pulled down by the brake releasing spring, thus releasing the brake shoes.

FIFTEEN TASKS FOR PNEUMATIC POWER

Suction can feed, hold, position, and lift parts, form plastic sheets, sample gases, test for leaks, convey solids, and deaerate liquids. Compressed air can convey materials, atomize and agitate liquids, speed heat transfer, support combustion, and pressurize cable.

FEEDING PARTS

HOLDING PARTS

POSITIONING PARTS

LIFTING PARTS

Hoist

Motor & vacuum pump with fail-safe air valve →

Vacuum cup lifts up to 14 lb/in.³ of cup area

Large piece

"O" ring or gasket

VACUUM FORMING

Plastic sheet to be formed

Radiant heater

To pump

Mold

GAS SAMPLING

To pump

Dust filter or gas analyzer

Atmosphere or processing chamber

LEAK TESTING

To pump

Part to be tested for fit or flatness

Bubbles if leak exists

CONVEYING SOLIDS

To pump

Intercepting trap or chamber

Small parts or granular materials

DEAERATING HIGH-VISCOSITY LIQUIDS

To pump

Low-vacuum test chambers, tanks, containers, etc.

CONVEYING MATERIALS

Plaster, cement, powders etc.

Wet or dry flowable materials

Air from compressor

ATOMIZING LIQUIDS

Air from compressor

AGITATING LIQUIDS

Air

Liquid to be chilled or warmed

Air bubbles agitate cold or hot water

Holes in pipe

AIR COOLING OR HEATING

SUPPORTING COMBUSTION

PRESSURIZING CABLE

TEN APPLICATIONS FOR METAL DIAPHRAGMS AND CAPSULES

Fig. 1

Fig. 2

Fig. 3

A metal diaphragm is usually corrugated (Fig. 1) or formed to some irregular profile. It can be used as a flexible seal for an actuating rod. The capsule (Fig. 2) is an assembly of two diaphragms sealed together at their outer edges, usually by soldering, brazing, or welding. Two or more capsules assembled together are known as a capsular element (Fig. 3). End fittings for the capsules vary according to their function; the "fixed end" is fixed to the equipment. The "free end" moves the related components and linkages. The nested capsule (Fig. 4) requires less space and can be designed to withstand large external overpressures without damage.

Fig. 4

Fig. 5

A **differential pressure gage** (Fig. 5) with opposing capsules can have either single or multicapsular elements. The multicapsular type gives greater movement to the indicator. Capsules give improved linearity over bellows for such applications as pressure-measuring devices. The force exerted by any capsule is equal to the total effective area of the diaphragms (about 40% actual area) multiplied by the pressure exerted on it. Safe pressure is the maximum pressure that can be applied to a diaphragm before hysteresis or set become objectionable.

Fig. 6

A **pressure gage** (Fig. 6) has a capsular element linked to a dial indicator by a three-bar linkage. Such a gage measures pressure or vacuum relative to prevailing atmospheric pressure. If greater angular motion of the indicator is required than can be obtained from the three-bar linkage, a quadrant and gear can be substituted.

Fig. 7

An **absolute pressure gage** (Fig. 7) has an evacuated capsular element inside an enclosure that is connected to the pressure source only. The diaphragm allows the linkage movement from the capsule to pass through a sealed chamber. This arrangement can also be used as a differential pressure gage by making a second pressure connection to the interior of the element.

Fig. 8

An **expansion compensator** (Fig. 8) for oil-filled systems takes up less space when the capsules are nested. In this application, one end of the capsule is open and connected to oil in the system; the other end is sealed. Capsule expansion prevents the internal oil pressure from increasing dangerously from thermal expansion. The capsule is protected by its end cover.

Fig. 9

A **capsule pressure-seal** (Fig. 9) works like a thermometer system except that the bulb is replaced by a pressure-sensitive capsule. The capsule system is filled with a liquid such as silicone oil and is self-compensating for ambient and operating temperatures. When subjected to external pressure changes, the capsule expands or contracts until the internal system pressure just balances the external pressure surrounding the capsule.

Fig. 10

A **force-balanced seal** (Fig. 10) solves the problem, as in the seal of Fig. 9, for keeping corrosive, viscous, or solids-bearing fluids out of the pressure gage. The air pressure on one side of a diaphragm is controlled so as to balance the other side of the diaphragm exactly. The pressure gage is connected to measure this balancing air pressure. The gage, therefore, reads an air pressure that is always exactly equal to the process pressure.

Gage pressure bellows transmitter. The bellows is connected to a cantilever beam with a needle bearing. The beam adopts a different position for every pressure; the transformer output varies with beam position. The bellows are available for ranges from 0–10 in. to 0–200 in. of water for pressure indication or control.

Absolute pressure bellows transmitter. This transmitter is similar to the differential diaphragm transmitter except for addition of a reference bellows which is evacuated and sealed. It can measure negative gage pressures with ranges from 0–50 mm to 0–30 in. of mercury. The reference bellows compensates for variations in atmospheric pressure.

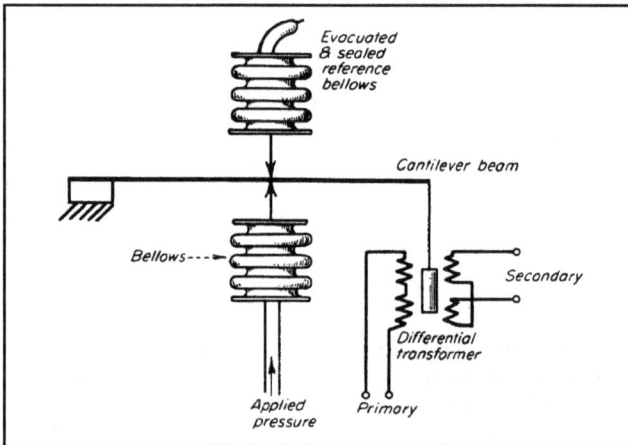

Absolute pressure Bourdon-tube transmitter. This device can indicate or control absolute pressures from 15 to 10,000 psi, depending on tube rating. The reference tube is evacuated and sealed, and compensates for variations in atmospheric pressure by changing the output of the reference differential transformer. The signal output consists of the algebraic sum of the outputs of both the primary and reference differential transformers.

Differential diaphragm pressure transmitter. Differential pressures P_1 and P_2 act on the opposite sides of a sensitive diaphragm and move the diaphragm against the spring load. The diaphragm displacement, spring extension, and transformer core movement are proportional to the difference in pressure. The device can measure differentials as low as 0.005 in. of water. It can be installed as the primary element in a differential pressure flowmeter, or in a boiler windbox for a furnace-draft regulator.

Cantilever load cell. The deflection of a cantilever beam and the displacement of a differential transformer core are proportional to the applied load. The stop prevents damage to the beam in the event of overload. Beams are available for ranges from 0–5 to 0–500 lb, and they can provide precise measurement of either tension or compression forces.

Proving ring. The core of the transmitting transformer, T_1, is fastened to the top of the proving ring, while the windings are stationary. The proving ring and transformer core deflect in proportion to the applied load. The signal output of the balancing transformer, T_2, opposes the output of T_1, so that at the balance point, the null point indicator reads zero. The core of the balancing transformer is actuated by a calibrated micrometer that indicates the proving ring deflection when the differential transformer outputs are equal and balanced.

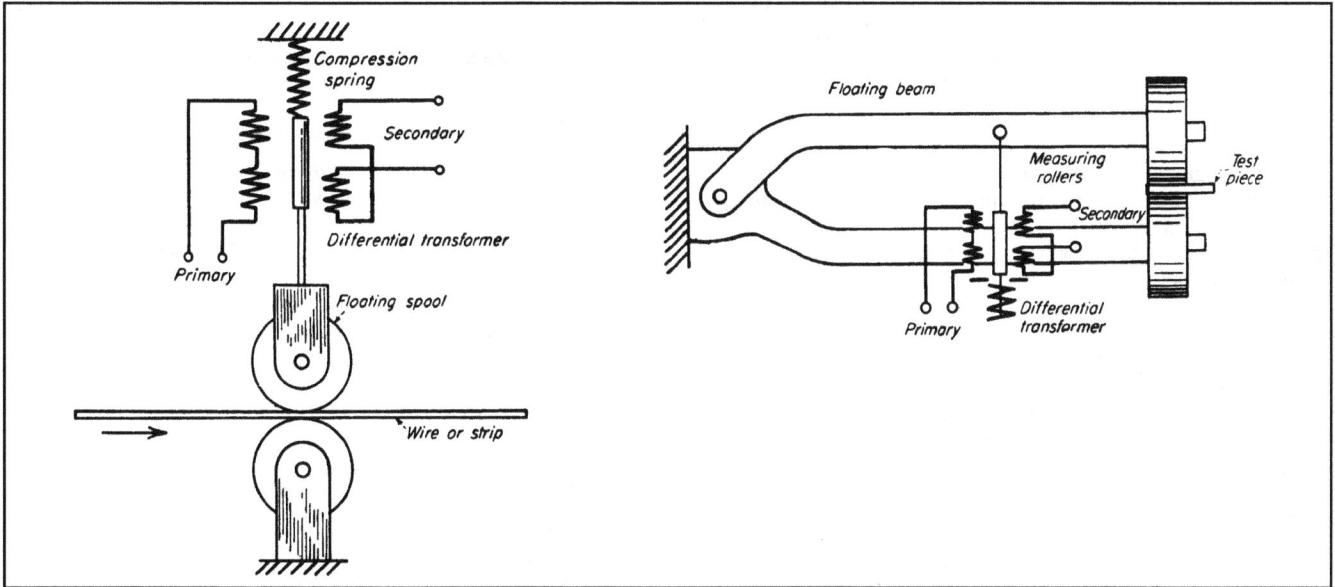

Gaging and calipering. The thickness of a moving wire or strip is gaged by the position of the floating spool and transformer core. If the core is at the null point for proper material thickness, the transformer output phase and magnitude indicate whether the material is too thick or thin and the amount of the error. The signal can be amplified to operate a controller, recorder, or indicator. The device at the right can function as a production caliper or as an accurate micrometer. If the transformer output is fed into a meter indicator with *go* and *no-go* bands, it becomes a convenient device for gaging items.

Flowmeter. The flow area varies as the float rises or falls in the tapered tube. High flows cause the float to rise, and low flows cause it to drop. The differential transformer core follows the float travel and generates an AC signal which is fed into a square-root recorder. A servo can be equipped with a square root cam to read on a linear chart. The transformer output can also be amplified and used to actuate a flow regulating valve so that the flowmeter becomes the primary element in a flow controller. Normally meter accuracy is better than 2%, but its flow range is limited.

Tension control. The loading spring can be adjusted so that when the transformer core is at its null point, the proper tension is maintained in the wire. The amplified output of the transformer is transmitted to some kind of tension-controlling device which increases or reduces the tension in the wire, depending on the phase and magnitude of the applied differential transformer signal.

HIGH-SPEED ELECTRONIC COUNTERS

The electronic counter counts electrical pulses and gives a running display of accumulated pulses at any instant. Because the input is an electrical signal, a transducer is generally required to transform the nonelectrical signal into a usable input for the counter.

With a preset function on the counter, any number can be selected within the count capacity of the device. Once the counter reaches the preset number, it can open or close the relay to control some operation. The counter will either reset automatically or stop. A dual unit permits continuous control over two different count sequence operations. Two sets of predetermining switches are usually mounted on the front panel of the counter, but they can be mounted at a remote location. If two different numbers are programmed into the counter, it will alternately count the two selected numbers. Multiple presets are also available, but at higher cost.

In addition to performing two separate operations, a dual preset can control speeds, as shown in Fig. 1. In the metal shearing operation run at high speed, one preset switch can be used to slow the material down at a given distance before the second preset actuates the shearing. Then both switches automatically reset and start to measure again. The same presets could also be used to alternately shear the material into two different lengths.

One form of measurement well adapted to high-speed counters is the measuring of continuous materials such as wire, rope, paper, textiles, or steel. Fig. 2 shows a coil-winding operation in which a counter stops the machine at a predetermined number of turns of wire.

A second application is shown in Fig. 3. Magazines are counted as they run off a press. A photoelectric pickup senses the alternate light and dark lines formed by the shadows of the folded edges of the magazines. At the predetermined number, a knife edge, actuated by the counter, separates the magazines into equal batches.

A third application is in machine-tool control. A preset counter can be paired with a transducer or pulse generator mounted on the feed mechanism. It could, for example, convert revolutions of screw feed, hence displacement, into pulses to be fed into the counter. A feed of 0.129 in. might represent a count of 129 to the counter.

When preset at that number, the counter could stop, advance, or reverse the feed mechanism.

Fig. 1 **A dual preset function** on a high-speed counter controls the high-speed shearing operation. If the material is to be cut in 10-ft lengths and each pulse of the electromagnetic pickup represents 0.1 ft, the operator presets 100 into the first input channel. The second input is set to 90. When 90 pulses are counted, the second channel slows the material. Then when the counter reaches 100, the first channel actuates the shear. Both channels reset instantaneously and start the next cycle.

Fig. 2 **A coil-winding machine** with electronic counting for measuring length.

Fig. 3 **A magazine counter** has a sharply focused photoelectric pickup that counts magazines. The photocell senses the alternate light and dark lines as the folded edges of the magazines cast their shadows. The counter actuates the knife edge to separate the magazines into equal batches when the predetermined count is reached.

Flexible stationary pin

Guide bearing

1" dia.

Hollow cylindrical permanent magnets

Non-magnetic support

Vertical rotary shaft

Lower guide assembly

Pat. No. 2,305,416

Fig. 1 A suspension.

Stationary shaft

Reel

Non-magnetic materials

Horseshoe magnets attached to housing

Brass disk rotates with reel

Vanes for cooling air

Armature adjusting bushing in stationary housing

Laminated ring armature

Magnets

Section A-A

Pat. No. 2,293,748

Fig. 3 A reel brake.

Magnet

Soft iron shell

Pat. No. 2,242,830

Fig. 5 A crankcase oil drain plug.

Magnet case

Thread

Magnetic tension device

Loom shuttle

Guide pins

Thread

Magnets

Keeper for varying magnetic effect and tension on thread

Fig. 2 Tension devices.

Adjustment for varying thrust

Pat. No. 2,281,204

Fig. 4 An instrument coupling.

Thermostat drive for oil burner control

Rotation holds contacts open or closed, depending on direction of movement

Magnet

Electrical contacts

Leaf spring

Pat. No. 2,305,974

Door jamb

Tapered bolt

Magnet

Lock recess in bolt

Fig. 6 A non-rattling door latch.

Tapered extension of soft steel reduces size of magnet required and shortens air gap before latching

Pat. No. 2,219,186

Fig. 7 A clamp.

Fig. 8 A snap-action switch.

Fig. 9 An instrument holder.

Fig. 10 An escape wheel.

Fig. 11 A pressure release.

Fig. 12 A filter.

FLOWMETER DRIVE

Meter drive pinion in gear train

Armature separate from magnet by thin partition sealing gear train chamber from drive housing

Magnet

Drive shaft

Pat. No. 2,354,563

Metering aperture

Magnets open metering aperture

Bleeder pipe to control valve

Spring

Pressure chamber (or float chamber)

Pat. No. 2,231,158

Fig. 14   A control device.

Keeper

Wood or plastic faces facilitate release of magnets by rolling.

Weld to be tested magnetically

Pat. No. 2,353,550

Fig. 13   A weld tester.

Wood container for 16 magnets arranged circumferentially

16 Magnets circumferentially arranged

Circular wooden turntable

Pat. No. 2,323,857

Fig. 16   A floating advertising display.

Flywheel of gyroscope

Stationary semicircular magnets separated by non-magnetic spacers at ends

Rotary magnet

Large slots for adjustments

Shaft guides flywheel without supporting weight

Stationary magnet support

Pat. No. 2,351,424

Fig. 15   A horizontal-shaft suspension.

Horseshoe magnet

Soft iron bar

Pat. No. 2,306,677

DRAFTING BOARD

Magnet finder for conduit connector to be buried in floor. Small compasses used to locate box

Fig. 17   A finder.

Rotating magnet

Bell rotated by magnetic flux in flange

Pat. No. 2,347,613

Fig. 18   A tachometer.

NINE ELECTRICALLY DRIVEN HAMMERS

The application of controlled impact forces can be as practical in specialized stationary machinery as in the portable electric hammers shown here. These mechanisms have been employed in vibrating concrete forms, nailing machines, and other special machinery. In portable hammers they are efficient in drilling, chiseling, digging, chipping, tamping, riveting, and similar

Fig. 1 **A free-driving throw** of the cam-slotted striker is produced by the eccentric stud roller during contact between points A and B of the slot. This accelerates the striker beyond the tangential speed of the roller for an instant before the striker is picked up for the return stroke.

Fig. 3 **The striker** has no mechanical connection with the reciprocating drive in this hammer.

Fig. 2 **The centrifugal force** of two oppositely rotating weights throws the striker assembly of this hammer. The power connection is maintained by a sliding-splined shaft. The guide, not shown, prevents the rotation of the striker assembly.

Fig. 4 **This combination** of mechanical, pneumatic, and spring action is included in this hammer.

operations where quick, concentrated blows are required. The striker mechanisms illustrated are operated by springs, cams, magnetic force, air and vacuum chambers, and centrifugal force. The drawings show only the striking mechanisms.

Fig. 5 **Two electromagnets** operate this hammer. The weight of the blows can be controlled by varying the electric current in the coils or timing the current reversals by an air-gap adjustment of the contacts.

Fig. 7 **This spring-operated hammer** includes a shaft rotating in a female cam to return the striker.

Fig. 6 **This spring-operated hammer** with a cam and rocker for the return stroke has a screw for adjusting the blow to be imparted.

Fig. 8 **This spring-operated hammer** has two fixed rotating-barrel cams. They return the striker by means of two rollers on the opposite sides of the striker sleeve. Auxiliary springs prevent the striker from hitting the retaining cylinder. A means of rotating the tool, not shown, is also included in this hammer.

Fig. 9 **Two steel balls** rotated in a divided cylinder and steered by an edge cam develop centrifugal force to strike blows against the tool holder. The collar is held clear of the hammer by a compression spring when no tool is in the holder. A second spring cushions the blows when the motor is running, but the tool is not held against the work.

461

Sensitivity or change in deflection for a given temperature change depends upon the combination of meals selected and the dimensions of the bimetal element. Sensitivity increases with the square of the length and inversely with the thickness. The force developed for a given temperature change also depends on the type of bimetal. However, the allowable working load for the thermostatic strip increases with the width and the square of the thickness. Thus, the design of bimetal elements depends upon the relative importance of sensitivity and working load.

Fig. 1   This recording thermometer has a pen that is moved vertically across a revolving chart by a brass-invar bimetal element. To obtain sensitivity, the long movement of the pen requires a long strip of bimetal, which is coiled into a helix to save space. For accuracy, a relatively large cross section gives stiffness, although the large thickness requires increased length to obtain the desired sensitivity.

Fig. 2   Room temperatures in summer and winter are controlled over a wide range by a single, large-diameter coil of brass-invar in this thermometer. To prevent chattering, a small permanent magnet is mounted on each side of the steel contact blade. The magnetic attraction on the blade, which increases inversely with the square of the distance from the magnet, gives a snap action to the contacts.

Fig. 3   This overload relay for large motors passes some of the motor current through a heating coil within the relay. Heat from the coil raises the temperature of a bimetal spiral which rotates a shaft carrying an electrical contact. To withstand the operating temperatures, it includes a heat-resistant bimetal spiral. It is coiled into the spiral form for compactness. Because of the large deflection needed, the spiral is long and thin, whereas the width is made large to provide the required contact pressure.

Heat barriers between the bimetal spiral and the heating coil make the temperature rise of the bimetal spiral closely follow the increase in temperature within the motor. Thus, momentary overloads do not cause sufficient heating to close the contacts. However, a continued overload will, in time, cause the bimetal spiral to rotate the contact arm around to the adjustable stationary contact, causing a relay to shut down the motor.

Fig. 4 Carburetor control. When the engine is cold, a vane in the exhaust passage to the "hot spot" is held open by a bimetal spring against the force of a small counterweight. When the thermostatic spiral is heated by the outside air or by the warm air stream from the radiator, the spring coils up and allows the weight to close the vane. Because high accuracy is not needed, a thin, flexible cross section with a long length provides the desired sensitivity.

Fig. 6 The bimetal element in this time-delay relay protects mercury-vapor rectifiers. This relay closes the voltage circuit to the mercury tube only after the filament has had time to reach its normal operating temperature. To eliminate the effect of changes in room temperature on the length of the contact gap (and therefore the time interval) the stationary contact is carried by a second bimetal strip, similar to the heated element. Barriers of laminated plastic on both sides of the active bimetal strip shield the compensating strip and prevent air currents from affecting the heating rate. The relatively high temperature range allows the use of a straight, thick strip, but the addition of the compensating strip makes accurate timing possible with a short travel.

Fig. 5 Thermostatic relay. A constant current through an electrical heating coil around a straight bimetal strip gives a time-delay action. Because the temperature range is relatively large, high sensitivity is not necessary. Thus, a short, straight strip of bimetal is suitable. Because of its relatively heavy thickness, the strip is sufficiently stiff to close the contact firmly without chattering.

Fig. 7 Oil pressure, engine temperature, and gasoline level are indicated electrically on automobile dashboard instruments whose bimetal element is both the sender and receiver. A grounded contact at the sender completes an electric circuit through heaters around two similar bimetal strips. Because the same current flows around the two bimetal elements, their deflections are the same. But the sender element, when heated, will bend away from the grounded contact until the circuit is broken. Upon cooling, the bimetal element again makes contact and the cycle continues. This allows the bimetal element to follow the movement of the grounded contact. For the oil-pressure gage, the grounded contact is attached to a diaphragm; for the temperature indicator, the contact is carried by another thermostatic bimetal strip; in the gasoline-level device, the contact is shifted by a cam on a shaft rotated by a float. Deflections on the receiving bimetal element are amplified through a linkage that operates a pointer over the scale of the receiving instrument. Because only small deflections are needed, the bimetal element is in the form of a short, stiff strip.

Fig. 8 Oil dashpots in heavy-capacity scales have a thermostatic control to compensate for changes in oil viscosity with temperature. A rectangular orifice in the plunger is covered by a swaged projection on the bimetal element. With a decrease in oil temperature, the oil viscosity increases, tending to increase the damping effect. But the bimetal element deflects upward, enlarging the orifice enough to keep the damping force constant. A wide bimetal strip provides sufficient stiffness so that the orifice will not be altered by the force of the flowing oil.

Fig. 9 Automobile cooling-water temperature is controlled by a self-contained bellows in the thermostat. As in the radiator air valve, the bellows itself is subjected to the temperature to be controlled. As the temperature of the water increases to about 140°F, the valve starts to open; at approximately 180°F, free flow is permitted. At intermediate temperatures, the valve opening is in proportion to the temperature.

Fig. 10 A throttling circulating-water control valve for refrigeration plants has its valve opening vary with the pressure on the bellows. This valve controls the rate of flow of the cooling water through the condenser. A greater amount of water is required when the temperature, and therefore the pressure, increases. The pressure in the condenser is transmitted through a pipe to the valve bellows, thereby adjusting the flow of cooling water. The bronze bellows is protected from contact with the water by a rubber diaphragm.

Fig. 11 An automatic gas-range control has a sealed thermostatic element consisting of a bulb, capillary tube, and bellows. As food is often placed near the bulb, a nontoxic liquid, chlorinated diphenyl, is in the liquid expansion system. The liquid is also non-flammable and has no corrosive effect on the phosphor-bronze bellows. By placing the liquid outside instead of inside the bellows, the working stresses are maximum at normal temperatures when the bellows bottoms on the cup. At elevated working temperatures, the expansion of the liquid compresses the bellows against the action of the extended spring. This, in turn, is adjusted by the knob. Changes in calibration caused by variations in ambient temperature are compensated by making the rocker arm of a bimetal suitable for high-temperature service.

Fig. 12 For electric ranges, this thermostat has the same bellows unit as the gas-type control. But, instead of a throttling action, the thermostat opens and closes the electrical contacts with a snap action. To obtain sufficient force for the snap action, the control requires a temperature difference between *on* and *off* positions. For a control range from room temperature to 550°F, the differential in this instrument is ±10°F. With a smaller control range, the differential is proportionately less. The snap-action switch is made of beryllium copper, giving it high strength, better snap action, and longer life than is obtainable with phosphor bronze. Because of its corrosion resistance, the beryllium-copper blade requires no protective finish.

Fig. 13 For heavy-duty room-temperature controls, this thermostat has a bellows mechanism that develops a high force with small changes in temperature. The bellows is partly filled with liquid butane. At room temperatures this gas exhibits a large change in vapor pressure for small temperature differentials. Snap action of the electrical contact is obtained from a small permanent magnet that pulls the steel contact blade into firm contact when the bellows cools. Because of the firm contact, the device is rated at 20 A for noninductive loads. To avoid chattering or bounce under the impact delivered by the rapid magnetic closing action, small auxiliary contacts are carried on light spring blades. With the large force developed by the bellows, a temperature differential of only 2°F is obtained.

Fig. 14 Snap action in this refrigerator control is obtained from a bowed flat spring. The silver contacts carried on an extended end of the spring open or close rapidly when movement of the bellows actuates the spring. With this snap action, the contacts can control an alternating-current motor as large as $1\frac{1}{2}$ hp without auxiliary relays. Temperature differential is adjusted by changing the spacing between two collars on the bellows shaft passing through the contact spring. For the temperatures needed to freeze ice, the bellows system is partly filled with butane.

Fig. 15 In this refrigerator control, the necessary snap action is obtained from a toggle spring supported from a long arm moved by the bellows. With this form of toggle action, the contact pressure is at a maximum at the instant the contacts start to open. Thermostatic action is obtained from a vapor-filled system. Sulfur dioxide is the fill for typical refrigerating service or methyl chloride where lower temperatures are required. To reduce friction, the bellows makes point-contact with the bellows cup. Operating temperature is adjusted by changing the initial compression in the bellows spring. For resistance to corrosion, levers and blades are made from stainless steel with bronze pin bearings.

Fig. 16 Two bellows units in this thermostatic expansion valve control large refrigeration systems. A removable power bellows unit is operated by vapor pressure in a bulb attached to the evaporator output line. The second bellows serves as a flexible, gas-tight seal for the gas valve. A stainless-steel spring holds the valve closed until opened by pressure transmitted from the thermostatic bellows through a molded push pin.

465

EIGHT TEMPERATURE-REGULATING CONTROLS

Temperature regulators are either on-off or throttling. The characteristics of the process determine which should be used. Within each group, selection of a regulator is governed by the accuracy required, space limitations, simplicity, and cost.

Fig. 1   A bimetallic sensor is simple, compact, and precise. Contacts mounted on low-expansion struts determine slow make-and-break action. A shell contracts or expands with temperature changes, opening or closing the electrical circuit that controls a heating or cooling unit. It is adjustable and resistant to shock and vibration. Its range is 100 to 1500°F, and it responds to a temperature changes of less than 0.5°F.

Fig. 2   This enclosed, disk-type, snap-action control has a fixed operating temperature. It is suitable for unit and space heaters, small hot water heaters, clothes dryers, and other applications requiring non-adjustable temperature control. It is useful where dirt, dust, oil, or corrosive atmospheres are present. It is available with various temperature differentials and with a manual reset. Depending on the model, its temperature setting range is from −10° to 550°F and its minimum differential can be 10, 20, 30, 40, or 50°F.

Fig. 3   This bimetallic unit has a rod with a low coefficient of expansion and a shell with a high coefficient of expansion. A microswitch gives snap action to the electrical control circuit. The current can be large enough to operate a solenoid valve or relay directly. The set point is adjusted by a knob which moves the pivot point of the lever. Its range is −20° to 175°F, and its accuracy is 0.25 to 0.50°.

Fig. 4   This is a bimetallic-actuated, air-piloted control. The expansion of the rod causes an air signal (3 to 15 psi) to be transmitted to a heating or cooling pneumatic valve. The position of the pneumatic valve depends on the amount of air bled through the pilot valve of the control. This produces a throttling type of temperature control as contrasted to the on-off characteristic that is obtained with the three units described previously. Its range is 32 to 600°F, and its accuracy is ±1 to ±3°F, depending on the range.

Fig. 5   This self-contained regulator is actuated by the expansion or contraction of liquid or gas in a temperature-sensitive bulb that is immersed in the medium being controlled. The signal is transmitted from the bulb to a sealed expansion element which opens or closes the ball valve. Its range is 20 to 270°F, and its accuracy is ±1°F. The maximum pressure rating is 100 psi for dead-end service and 200 psi for continuous flow.

Fig. 6 **This remote bulb, nonindicating regulator** has a bellows assembly that operates a flapper. This allows air pressure in the control system to build up or bleed, depending on the position of the changeover link. The unit can be direct- or reverse-acting. A control knob adjusts the setting, and the throttling range adjustment determines the percentages of the control range in which full output pressure (3–15 psi) is obtained. Its range is 0 to 700°F, and its accuracy is about ±0.5% of full scale, depending on the way it is installed.

Fig. 7 **This lever-type pilot valve** is actuated by a temperature-sensitive bulb. The motion of the lever causes the water or steam being controlled to exert pressure on a diaphragm which opens or closes the main valve. Its temperature range is 20 to 270°F, and its accuracy is ±1 to 4°F. It is rated for 5 to 125 psi of steam and 5 to 175 psi of water.

Fig. 8 **These two recording and controlling instruments** have adjustable proportional ranges. In both, air supply is divided by a relay valve. A small part goes through a nozzle and flapper assembly. The main part goes to the control valve. Unit B has an extra bellows for automatic resetting. It was designed for systems with continuously changing control points, and it can be used where both heating and cooling are required in one process. Both A and B are easily changed from direct to reverse acting. Its accuracy is 1% of its temperature range of −40 to 800°F.

SEVEN PHOTOELECTRIC CONTROLS

Typical applications are presented for reducing production costs and increasing operator safeguards by precisely and automatically controlling the feed, transfer, or inspection of products from one process stage to another.

Fig. 1 Automatic weighing and filling. The task is to fill each box with an exact quantity of products, such as screws. An electric feeder vibrates parts through a chute and into a box on a small balance. The photoelectric control is mounted at the rear of the scale. The light beam is restricted to very small dimensions by an optical slit. The control is positioned so that the light is interrupted by a balanced cantilever arm attached to the scale when the proper box weight is reached. The photoelectric control then stops the flow of parts by deenergizing the feeder. Simultaneously, an indexing mechanism is activated to remove the filled box and replace it with an empty one. The completion of indexing reenergizes the feeder, which starts the flow of screws.

Fig. 2 Operator safeguard. Most pressures are operated by a foot pedal that leaves the operator's hands free for loading and unloading. This creates a safety hazard. The use of mechanical gate systems reduces the speed of production. With photoelectric controls, a curtain of light is set up by a multiple series of photoelectric scanners and light sources. When a light beam is broken at any point by the operator's hand, the control energizes a locking mechanism that prevents the punch-press drive from being energized. A circuit or power failure causes the control to function as if the light beam were broken. In addition, the light beam frequently becomes the actuating control because the clutch is released as soon as the operator removes his hand from the die on the press table.

Fig. 3 This apparatus sorts cartons of three different kinds of objects. Because the cartons containing objects differ widely in size, it is not feasible to sort by carton size and shape. A small strip of reflecting tape is put on the cartons by a packer during assembly. On one type of object, the strip is located along one edge of the bottom, and it extends almost to the middle. For the second type, the strip is located along the same edge, but from the middle to the opposite side. No tape is placed on the third type. Cartons are placed on the conveyor so that the tape is at right angles to the direction of travel. Photoelectric controls shown in A "see" the reflecting tape and operate a pusher-bar mechanism shown in B. This pushes the carton onto the proper distribution conveyor. Cartons without tape pass.

Fig. 1

Fig. 2

Fig. 3

Fig. 4

Fig. 4 **This cut-off machine** has a photoelectric control for strip materials that lack sufficient mass to operate a mechanical limit switch satisfactorily. The forward end of the strip breaks the light beam, thus actuating the cut-off operation. The light source and the control are mounted on an adjustable stand at the end of the machine to vary the length of the finished stock.

Fig. 5

Fig. 5 **This heat-treating conveyor** has an electronic timer paired with a photoelectric control to carry parts emerging from a furnace at 2300°F. The conveyor must operate only when a part is placed on it and only for the distance required to reach the next process stage. Parts are ejected onto the conveyor at varying rates. High temperatures caused failures when the mechanical switches were used. Glowing white-hot parts radiate infrared rays that actuate the photoelectric control as soon as a part comes in view. The control operates the conveyor that carries the parts away from the furnace and simultaneously starts the timer. The conveyor is kept running by a timer for the predetermined length of time required to position the part for the next operation.

Fig. 6

Fig. 6 **Jam detector.** Cartons jamming on the conveyor cause losses in production and damage to cartons, products, and conveyors. Detection is accomplished with a photoelectric control that has a timer, as shown in (A). Each time a carton passes the light source, the control beam is broken. That starts the timing interval in the timer. The timing circuit is reset without relay action each time the beam is restored before the preset timing interval has elapsed. If a jam occurs, causing cartons to butt against each other, the light beam cannot reach the control. The timing circuit will then time-out, opening the load circuit. This stops the conveyor motor. By locating the light source at an angle with respect to the conveyor, as shown in (B), the power conveyor can be delayed if cartons are too close to each other but not butting each other.

Fig. 7

Fig. 7 **Automatic inspection.** As steel caps are conveyed to final assembly, they pass an intermediate stage where an assembler inserts an insulation liner into a cap. The inspection point for the missing liners has a reflection-type photoelectric scanner which incorporates both a light source and photosensor with a common lens system to recognize the difference in reflection between the dark liner and the light steel cap instantly. When it detects a cap without a liner, a relay operates an airjet ejector that is controlled by a solenoid valve. The start and duration of the air blast is accurately controlled by a timer so that no other caps are displaced.

LIQUID LEVEL INDICATORS AND CONTROLLERS

Thirteen different systems of operation are shown. Each one represents at least one commercial instrument. Some of them are available in several modified forms.

A diaphragm actuated indicator will work with any kind of liquid, whether it is flowing, turbulent, or carrying solid matter. A recorder can be mounted above or below the level of the tank or reservoir.

A bellows-actuated indicator. Two bellows and a connecting tubing are filled with incompressible fluid. A change in liquid level displaces the transmitting bellows and pointer.

A float-switch controller. When liquid reaches a predetermined level, a float actuates a switch through a horseshoe-shaped arm. A switch can operate the valve or pump.

A bubbler-type recorder measures height *H*. It can be used with all kinds of liquids, including those carrying solids. A small amount of air is bled into a submerged pipe. A gage measures the air pressure that displaces the fluid.

An electrical level controller. The positions of the probes determine the duration of pump operation. When a liquid touches the upper probe, a relay operates and the pump stops. Auxiliary contacts on the lower probe provide a relay-holding current until the liquid level drops below it.

An automotive liquid-level indicator. The indicator and tank unit are connected by a single wire. As the liquid level in the tank increases, brush contact on the tank rheostat moves to the right, introducing an increasing amount of resistance into the circuit that grounds the F coil. The displacement of a needle from its empty mark is proportional to the amount of resistance introduced into this circuit.

A float recorder. The pointer can be attached to a calibrated float tape to give an approximate instantaneous indication of fluid level.

A magnetic liquid-level controller. When the liquid level is normal, the common-to-right leg circuit of the mercury switch is closed. When the liquid drops to a predetermined level, the magnetic piston is drawn below the magnetic field.

A differential pressure system. This system is applicable to liquids under pressure. The measuring element is a mercury manometer. A mechanical or electric meter body can be used. The seal pots protest the meter body.

A direct-reading float gage. This inexpensive, direct-reading gage has a dial calibrated to the tank volume. A comparable gage, in terms of simplicity, has a needle connected through a right-angle arm to the float. As the liquid level drops, the float rotates the arm and the needle.

A pressure gage indicator for open vessels. The pressure of the liquid head is imposed directly upon the actuating element of the pressure gage. A center-line of the actuating element must coincide with the minimum level line if the gage is to read zero when the liquid reaches the minimum level.

A bimetallic indicator. When the tank is empty, contacts in the tank unit just touch. With the switch closed, heaters cause both bimetallic strips to bend. This opens the contacts in the tank, and the bimetals cool, closing the circuit again. The cycle repeats about once per second. As the liquid level increases, the float forces the cam to bend the tank bimetal. This action is similar to that of the float gage, but the current and the needle displacement are increased.

A switch-actuated level controller. This pump is actuated by the switch. The float pivots the magnet so that the upper pole attracts the switch contact. The tank wall serves as the other contact.

APPLICATIONS FOR EXPLOSIVE-CARTRIDGE DEVICES

Cartridge-actuated devices generate a punch that cuts cable and pipe, shears bolts for fast release, and provides emergency thrust.

(a) Percussion fired

(b) Electrically fired (redundant bridgewire, but nonredundant charge)

(c) Electrically fired (redundant wire, redundant ignitor charge)

Cartridge Assemblies

(a) For free-ended pipes

Pipe Cutters

(b) For pipe fixed at both ends

Thruster (provides 6000 lb force for 5.5 in. in 15 milliseconds)

Cable Cutter

(a) Pneumatic-fired

Explosive Bolt

(b) Electrically fired

An emergency hook release lets the loads be jettisoned at any time. The hook is designed to release automatically if it is overloaded.

The pin retracts to release the load or clear a channel for free movement.

This dual valve is designed so that the flow will be started and stopped by the same unit. Firing one squib starts the flow; firing the other squib stops the flow.

Section A-A

A threaded split ring with a helical-spring response holds the ends of the tubes at a joint until the explosive bolt is fired; then it releases instantly.

Quick Disconnector

A tube joint can be separated almost instantaneously by remote control with an explosive bolt and a split threaded ring, in a design developed by James Mayo of NASA's Langley Research Center, Hampton, Virginia.

External threads of the ring mesh with the internal threads of the members that are joined—and they must be separated quickly. The ring has a built-in spring characteristic that will assume a helically wound shape and reduce to a smaller diameter when not laterally constrained. During assembly, it is held to its expanded size by two spring plates whose rims fit into internal grooves machined in the split ring. The plates are fastened together by an explosive bolt and nut.

Upon ignition of the explosive bolt, the plates fly apart from the axial spring tension of the ring. The ring then contracts to its normally smaller diameter, releasing the two structural members.

The tube joint can be made in any size and configuration. The retaining media need not be limited to V-type screw threads.

473

CENTRIFUGAL, PNEUMATIC, HYDRAULIC, AND ELECTRIC GOVERNORS

Centrifugal governors are the most common—they are simple and sensitive and have high output force. There is more published information on centrifugal governors than on all other types combined.

In operation, centrifugal flyweights develop a force proportional to the square of the speed, modified by linkages, as required. In small engines the flyweight movement can actuate the fuel throttle directly. Larger engines require amplifiers or relays. This has led to innumerable combinations of pilot pistons, linear actuators, dashpots, compensators, and gear boxes.

Centrifugal Governors

ACCELERATION GOVERNOR
(steam engine)

CENTRIFUGAL VALVE

TAPERED PLATE

DIFFERENTIAL CENTRIFUGAL

CENTRIFUGAL CONTACTS

CENTRIFUGAL CLUTCH

Pneumatic sensors are the most inexpensive and also the most inaccurate of all speed-measuring and governing components. Nevertheless, they are entirely adequate for many applications. The pressure or velocity of cooling or combustion air is used to measure and govern the speed of the engine.

Pneumatic Governors

CARBURETOR-FLOW VELOCITY
(linkage)

COMPRESSOR PRESSURE
(direct)

FAN-FLOW VELOCITY

More Pneumatic Governors

CENTRIFUGAL SNAP SPRING

Flyweights

Switch

Snap spring

CARBURETOR-FLOW VELOCITY
(cam)

Throttle plate
(positioned by flow)

Flow
(air-gas mixture)

Varying tension holds
speed constant

COMPRESSOR PRESSURE
(differential)

Venturi

Pipeline flow

Speed sensor

Adjustment

Control rod

Fulcrum

Damper

Steam or
air-fuel mixture

Hydraulic Governors

Hydraulic sensors measure the discharge pressure of engine-driven pumps. Pressure is proportional to the square of the speed of most pumps, although some have special impellers with linear pressure-speed characteristics.

Straight vanes are better than curved vanes because the pressure is less affected by the volume flow. Low pressures are preferred over high pressures because fluid friction is less.

Typical applications for these governors include farm tractors with diesel or gasoline engines, larger diesel engines, and small steam turbines.

PUMP PRESSURE
(centrifugal)

Bellows

Spring

Control rod

Engine-driven pump

Control rod

Orifice

Pump

PUMP PRESSURE
(positive displacement)

Fluid-filled

Engine-driven impeller

Output control

VISCOUS DRAG

Electric Governors

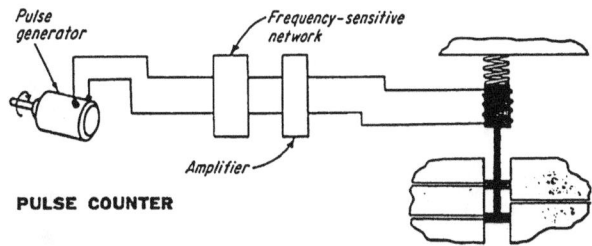

Permanent magnet stator

Calibrated spring

Solenoid

Control valve

Brushes

DC TACHOMETER

Permanent magnet rotor

Rectifier

Ac

AC TACHOMETER

Pulse generator

Frequency-sensitive network

Amplifier

PULSE COUNTER

CHAPTER 17
3D DIGITAL PROTOTYPES AND SIMULATION

INTRODUCTION TO 3D DIGITAL PROTOTYPES AND SIMULATION

Over the past 30 years, engineering drafting has undergone a profound change. This is due to the availability of powerful yet affordable computers combined with increasingly powerful computer-aided design (CAD) software. These advances are expected to continue as more vendors offer ever more versatile software for 3D digital prototyping and virtual simulation. Their more sophisticated software can run on many brands of desktop and notebook computers sold over the Internet or by retail stores rather than being limited to the costly workstations of the past.

Today, designers and engineers have many more choices in software "tools" than ever before that can accelerate and improve their design efforts. Software vendors are offering more features than were available only a few years ago, and they are claiming that their software is easier to learn and use than ever before. Moreover, different brands of software are standardized and compatible making transitions from CAD to computer-aided engineering (CAE) simulation software easier than ever. In addition, most software vendors are offering annual updates.

It is now possible to render 3D digital prototypes or models on a computer screen that can be manipulated and modified in three dimensions, and many different colors are available to make parts of a 3D assembly easier to visualize how they relate. Some digital prototype graphics compare favorably with photography in their ability to illustrate complex details.

Software can now validate the form, fit, and function of a design, minimizing the need to order physical prototypes or models for laboratory stress testing. This saves the expense and time lost building and testing models to determine how they will fare under actual working conditions. Software programs can also be used to initiate 2D concept sketches and convert them to 3D prototypes. Also, 3D models initiated with one brand of software can be imported along with all pertinent design data to different software on a different computer.

Whatever their origin, software permits deconstruction of an assembly into components so that their graphics can be stored in separate files for reference or further modification. Later, the dimensioned geometries of the individual parts can be downloaded and reassembled to form a new digital prototype. That design can then be transferred to simulation software where it can be subjected to a wide range of mechanical stresses and physics phenomena.

Software also permits the designer to make parametric changes in dimensions for the 3D digital prototype including those for length, width, height, wall thickness, fillet radius, and hole diameter. Some software even permits animation of the 3D model for visualization of its function. During this process, it is possible to develop kinematic models of parts and assemblies for further evaluation. After arriving at a validated 3D digital prototype, the revised dimensions and data can be returned to the 2D environment for use in the preparation of production drawings, bills of material, and other essential manufacturing documentation.

Because software permits multiple presentations of a 3D design and its simulation on other computers, both near and far, it permits earlier collaboration between the designer and manufacturing, marketing, and sales groups to facilitate suggestions for real-time improvements. Different workgroups can follow the progress being made on all aspects of the design that relate to their responsibilities. They can also appraise preliminary manufacturing documentation. This feature helps to minimize the need for design changes, but if any are required they can be made faster. They also permit manufacturers to approve designs faster, reduce manufacturing errors and costs, and shorten the time taken to deliver a more innovative product to the market.

Stress analysis can be performed on graphic presentations of individual parts or complete assemblies. They can be subjected to various kinds of physical phenomena, individually or simultaneously. The results of this simulation can be viewed on the computer screen in an understandable format and with appropriate colors. These will permit the designer to confirm the design or make improvements to correct any shortcomings. More importantly, the designer does not need to be an expert in any of the physical stresses being simulated.

Motion simulation software uses the characteristics of the 3D model to identify relevant rigid bodies, generate correct motion joints, and compute dynamic behavior. This helps the designer to understand the behavior of a design, including the position, velocity, and acceleration of moving parts. Different driving loads and torque can be applied to investigate the prototype's performance under a range of different load conditions.

Simulation software also permits the designer to examine the deformation of parts and assemblies under static loads to determine maximum and minimum stresses and deflections. This offers the designer an opportunity to assure that the design meets required safety standards. The study of oscillating modes permits the designer to determine the effects of a range of vibrations as well as the natural frequencies of the parts and assemblies. This information is useful for making the necessary design changes for the reduction of oscillation amplitudes to manageable levels.

Two or more multiphysical stresses can be applied simultaneously to the 3D digital prototype for observation of results. They include the effects of AC or DC current and response to acoustics, heat transfer, radio frequencies, and fluid flow. These simulations permit the designer to substitute alternative materials for manufacturing the end product to evaluate their relative performance for such factors as durability, weight, and cost. The results show up on the computer screen as variations in color of the stressed volumes or surfaces as identified by a magnitude scale for that simulation.

Digital simulations are based on partial differential equations (PDEs), the fundamental equations that describe the laws of science. PDEs are transformed for numerical analysis and solved using the *finite element method* (FEM). Digital simulations have a high correlation with the results obtained by laboratory testing of actual 3D prototypes. This gives the designer a high degree of confidence in the simulation results.

A Short History of Engineering Drawing

For centuries engineering projects were brought to life on paper or parchment with simple drawing tools: pencils, pens, compasses, dividers, triangles, and T squares. Examples can be seen in the exquisite 3D engineering drawings by Leonardo da Vinci. The switch from hand-drawn drawings on paper to those produced by CAD software on a workstation for later printout occurred within the past 30 years. This transition was especially traumatic for many mid-career designers and engineers. It was far more eventful than giving up the slide rule for the multipurpose electronic calculator. In terms of employment, persons

2D drawings, as shown, are still widely used forms of engineering graphics for production planning, tool design, and the manufacture of a product. Many 2D drawings now include 3D wireframe or 3D solid drawings to assist the user's understanding of the intent of the design.

This 3D digital image of a brake assembly permits the designer to use software tools to "slice and dice" it into its component parts, change dimensions, remove excess material (if necessary), and reassemble it into a new or modified prototype while retaining all manufacturing information. *Courtesy of SpaceClaim Corporation*

trained in manual mechanical drawing slowly disappeared from engineering, architectural, and consulting firms to be replaced by a smaller number of persons trained or retrained in CAD.

Before the arrival of CAD, engineering students were required to take courses in engineering drawing and analytical geometry. In the drawing classes, they learned about the conventions and standards of drafting, basics of 2D and perhaps 3D rendering of isometrics and perspectives. The analytical geometry courses covered such subjects as rectangular coordinates, functions and graphs, common geometric forms, rectangular and polar coordinates, and various curves.

Today, engineering colleges typically include CAD in required classroom and laboratory courses as part of the curriculum. The following course descriptions were taken from two catalogs of engineering colleges:

"Provides instruction in the application of computers to the design methods in engineering. Hands-on experience in use of CAD/CAE software packages for geometric modeling, visualization, and drafting. Emphasis on applications to solids and structural problems. Independent design projects are carried out. Course counts as an upper-level design course for mechanical and civil engineering concentrators."

"An introduction to engineering and the concepts of engineering design. Includes an introduction to graphical communication skills used in engineering through the use of sketching and computer-aided design (CAD) on personal computers. The concepts of orthographic and isometric drawings are stressed and extended to include sections and dimensions. The use of spreadsheets in engineering is also included."

Some colleges require that the students purchase a recognized book on CAD as well as a popular brand of CAD software to run on their laptop computers. Learning to master CAD/CAE software takes far more time than learning the fundamentals of manual mechanical drafting, but this was a necessary step even for seasoned professionals if they wanted to stay current in their engineering specialties. Despite what software publishers say, there is little that is intuitive about mastering CAD software because it is revolutionary, but a working knowledge of manual drafting practice is still a prerequisite.

Transition from Board to Screen

The learner's first sight of a professional-level CAD screen can be intimidating because of its seemingly incomprehensible jumble of information on its toolbar menus. The left-side columns and top of the screen toolbars are filled with icons and menus of available "tools" and commands appropriate to the specific software in place. These must be mastered by training and practice, and the student could take years to achieve proficiency. It is also possible to learn CAD in community colleges and technical schools as well as college night sessions. Software venders offer training courses in their own classrooms as well as educational videos and updates of their new or improved products. Tutorials of selected topics are available from software vendors in Internet sessions called *webinars*.

As a reward for learning the software and commands, the designer is relieved of the tedious chores of manual drafting, detailing, and lettering, which were typically turned over to specialized draftspersons. Now digital technology puts the designer in closer touch with the details of the project than ever before, as a result of his or her participation in the CAD and CAE processes. Superior, more accurate and legible drawings are produced in far less time at far lower cost than ever could be accomplished by manual drafting.

The first CAD programs were run on slow and costly computer workstations with only megabytes of memory. By contrast, today's catalog standard off-the-shelf desktop and laptop computers costing less than $1,000 with 3 gigabytes (GB) of memory, 320-GB hard drives, and 18-in. screens can easily run advanced CAD software. Until well into the 20th century, the only practical way to copy a large drawing was with special ultraviolet sensitive paper on a blueprint machine. Today copying is done faster, more accurately, and cheaper with inkjet printers or XY plotters for the larger-sized drawings.

In its earlier days, CAD actually stood for "computer-aided drawing" because that is the only thing the software could do. Today CAD really means computer-aided design, and computer-aided or digital simulation can accurately be called computer-aided engineering. Together these developments have made the traditional trial-and-error product development cycle characterized as *design, build, test, correct the design, analyze, test again*, obsolete.

479

Fluid flow through the 3D image of this fuel injection system can be digitally simulated by software to determine if its internal dimensions and contours are correct for its intended use. If not, design changes can be made easily and rapidly to correct faults without having to test a full-scale physical prototype. *Courtesy of SpaceClaim Corporation*

Form, function, and style changes on this miter saw can be made directly on the computer-screen 3D image. This permits groups to cooperate by offering suggestions for improvements during the design phase; problems can then be anticipated and eliminated before design approval for manufacture. *Courtesy of SpaceClaim Corporation*

CAD Product Features

The value of starting the design as a 3D digital prototype to validate a product design is that 3D line (wireframe) isometric or perspective drawings can then be prepared to enhance the 2D manufacturing drawings. In addition, exploded views of the product can be prepared from the data produced for use in instruction manuals and shop training. This feature also permits the preparation of 2D drawings that are easier for machine operators and field supervisors to understand. It has been reported that this feature has eliminated costly mistakes made in factories, machine shops, and in the field on construction sites.

3D Digital Prototypes vs. Rapid Prototyping

In contrast to physical or rapid prototyping (see Chapter 18, Rapid Prototyping), 3D digital prototyping and system simulation allow a design to be tested as it is being designed, and provide access to its innermost workings at every phase of the design process. (This is difficult or impossible with rapid prototypes.) Moreover, virtual simulation permits the evaluation of different materials and changes that can be made in dimensions and tolerances on the model design and its anticipated performance.

When employed early in the design process, virtual simulation permits the designer to tune and optimize the design, permitting critical insights to be gained before any hardware is built. After the basic design is finalized, simulation can again be employed to verify intended system operation and performance. Parameters can be varied statistically in ways that would be impossible with physical prototypes. To create a system digital model, each subsystem and component of the intended end product needs to have a corresponding digital model. These digital models are then stitched together the same way actual systems are assembled to create a functioning system.

Now that most virtual tests give results that correlate closely with the results of actual physical testing, laboratory testing may no longer be required. This simulation testing is intended to determine how well the end product will withstand real-life environments, and it can be performed on the same computer that was used to design the 3D digital prototype.

However, physical models are still useful for the evaluation of a product's design. One reason may be to demonstrate a design for non-technical marketing and sales personnel to keep them informed about a new product development. It permits comments about the model's style, contours, and color to be made by the design teams as well as non-technical personnel in sales and marketing. These tangible objects can also be useful for advising company management and investors before new product production begins. On the other hand, the end product may be in a category that requires actual laboratory testing and written documentation of results to comply with safety codes or regulations such as those of Underwriters' Laboratories (UL).

With the development of new rapid prototyping systems, prototypes can now be made by additive processes such as "printing" from plastic or powdered metals or by subtractive processes such as computer-controlled milling from solid plastic or metal. It is possible to replicate small quantities of replacement parts or tools for obsolete machines by laser-fusing metal powders. They can be made from the same or comparable metals as those used to make the original machines or tools.

It is important to keep in mind that no rapid prototypes could be made from either additive or subtractive processes without first downloading design and dimensional data from an initial CAD drawing for conversion to software for directing each rapid prototyping system. Obviously, the availability of both digital prototyping and the fabrication of prototypes by rapid prototyping give manufacturers choices that were not previously available.

The Ongoing Role of 2D Drawings

Two dimensional drawings still remain an important medium for the transmission of technical data and engineering information.

This racecar design can be studied with fluid-flow simulation software in a virtual wind tunnel, making it possible to reduce drag and optimize the contours. Simulation software can eliminate the expense and time lost in building and testing a solid model or prototype in an actual wind tunnel. *Courtesy of SpaceClaim Corporation*

This 3D digital prototype image makes it easier for the designer to visualize the internal components of the gear assembly and make any changes necessary to eliminate interference problems before the design is released for manufacture. *Courtesy of SpaceClaim Corporation*

They are cost-effective and widely used graphic presentations for the following engineering specialties:

- *Civil engineers:* Plotting results of surveying, drawing property lines, maps, bridge and highway plans, and recording underground piping. On-site sources of project information for field engineers and construction supervisors.
- *Architects:* Drawing floor plans and elevations of homes, offices, buildings, factories, hotels, etc., and projects from malls to housing developments. On-site sources of project information for construction supervisors.
- *Naval architects:* Drawings of ship layouts, machinery rooms and compartments, recording piping and cable installations, and cutting steel plate shapes. On-site sources of project information for construction supervisors.
- *Mechanical engineers:* Drawings of machines and mechanisms, shop drawings, recording piping, and ductwork. Sources of project information for shop and field installation supervisors.
- *Electrical engineers:* Drawing circuit diagrams, wiring and cabling for homes, buildings, factories, and power plants, transmission lines. On-site sources of project information for field engineers and construction supervisors.
- *Electronic engineers:* Drawing schematic diagrams, integrated circuit manufacturing templates, circuit board component layouts, and installation of wiring and cabling. Sources of information for shop supervisors.

The 2D drawings in these technical specialties can be enhanced by including 3D wireframe views. 2D, 3D wireframe, and exploded view drawings can be used as illustrations in instruction and operation, and maintenance manuals.

Functions of Tools in 3D Digital Prototype Software

Features available in 3D digital prototype software include the following:

- A *pull and move tool* permits the designer to select parts of the model and move them around the computer screen.
- A *combine tool* permits the designer to "slice and dice" parts of the original design so that they can be merged into other parts of the revised design.
- A *fill tool* enables the designer to erase small features and fill in missing gaps on the graphics.

File Types for 3D Digital Prototypes

A complete 3D digital prototype project calls for a wide selection of backup drawings. These include: (1) *part files*, (2) *assembly files*, (3) *presentation files*, and (4) *drawing files*. There is also software for sheet metal parts and weldments.

As implied by their names, part files illustrate components or elements of the prototype assembly, and assembly files show the placement of components that act as single functional units in an assembly. Assembly constraints define the relative positions these components occupy with respect to each other. An example is the axis of a shaft in one part file aligning with a hole in another part file, both components in an assembly. Stored templates usually contain property information such as project data and drawing views. The information stored in a file can be seen by viewing its properties. It is possible to choose from a selection of templates with predefined units. The advantages of using 3D digital modeling software are discussed later in this chapter.

Part Files

Opening a parts file puts the user in the "part environment." Most parts originate as sketches. The parts "tools" permit the manipulation of sketches, features, and components; when combined, they make up parts. A single component can be inserted into assemblies and be constrained in positions they will occupy in the manufactured end product. Multiple parts files can be extracted from a multi-element part.

A sketch is the profile of a feature that can have any geometry (such as a sweep path or axis of rotation) required to define the feature. A part model can have many features. If necessary, solid components in a multi-component part file can share features. Sketch constraints control geometric relationships such as parallel and perpendicular positioning. Dimensions control the size of a part in a method called *parametric modeling*. Constraints or dimensional parameters that control the size and shape of a model can be adjusted and will automatically show the effects of modifications on the computer screen.

Assembly Files

Components that function as a single unit can be placed in an assembly document. Assembly constraints define the relative positions that these components occupy with respect to each other. An example is the axis of a cylinder in a part file aligning with a hole in a different part file. When an assembly file is created or opened, the user is said to be in an "assembly environment." Assembly "tools" manipulate whole subassemblies and assemblies.

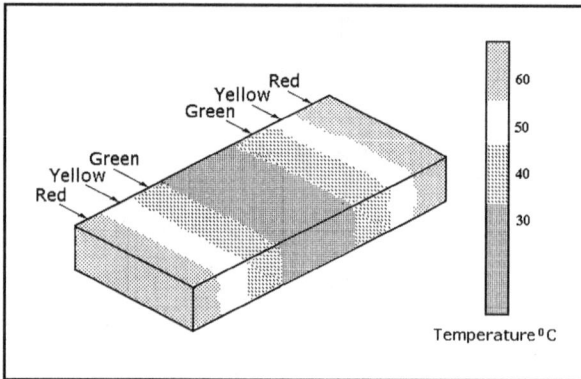

The results of simulated electric current passed through a solid 3D image appear as shaded bands representing electro-thermal heating (red, yellow, and green on the actual image). The temperature values are displayed on the vertical scale (right). Analysis of the magnitude and location of the bands will predict the prototype's ability to tolerate the applied thermal stress without laboratory testing.

Parts that function together can be grouped as a single unit, and this subassembly can be inserted into another assembly. Parts can also be inserted into an assembly or sketch, and part "tools" can be used to create parts of an assembly. While performing these steps, all other components in the assembly are visible on the computer screen. Assembly features that affect multiple components can be created to complete a model. Examples are holes that pass through one or more parts. Assembly features can be used to describe specific manufacturing processes such as post-machining. The assembly browser offers a convenient way to call up these features for editing. It can also be used to edit sketches, features, and constraints, as well as turn visible components on and off the screen.

Presentation Files

Presentation files are multipurpose files that can be used to create exploded view of an assembly for storage in a drawing file or printed out for publication in instruction, operation, and maintenance manuals. They can also be sources for animation that will show step-by-step order of assembly. An animation can display different views of the model that show where visible components are at each step of the assembly process. This animation can then be saved in a separate file.

Drawing Files

After a model has been completed, a drawing of it can be printed out with dimensions and other annotations to document the design. Views of a model can be placed on one or more drawing sheets. A drawing that documents an assembly can contain a parts list that has been automatically created, and item balloons can be added where required.

Computer-Aided Engineering (CAE)

Computer-aided engineering (CAE) involves the engineering and physics functions for simulating physical phenomena. It is to be distinguished from the design and drawing functions of CAD. The CAE functions include the following:

- *Stress analysis* of components and assemblies using finite element analysis (FEA)
- *Thermal flow analysis*, model response to introduced heat
- *Computational fluid dynamics* (CFD), movement of fluids within the model

- *Kinematics*, mechanical event simulation (MES), bending and twisting of elements
- *Analysis tools* for many different industrial processes that contribute to the optimization of product design and process

Simulation Software

Many 3D software vendors have joined with suppliers of simulation and engineering analysis software and bundled their software together. This permits both vendors to offer a complete compatible design-to-analysis package. These packages permit designers and engineers to import most solid 3D model designs created with different brands of software to create a neutral file format that integrates them together. Thus, the designer can create, edit, and prepare 3D models for analysis and verification with detailed stress analysis. It can also be used to test the features of a product's design before any prototypes are built.

Onscreen simulation for 3D solid models is now available because many different partial differential equations (PDEs), the fundamental equations that describe the laws of science, have been incorporated into CAE software in a form suitable for numerical analysis. Virtual tests can be performed on the model as it appears on the computer screen. In mechanical design, for example, simulation software permits linear static stress analysis, drop tests, and fluid flow and analysis.

A designer can perform a structural static or modal analysis of a part or assembly with stress analysis tools. It is possible to apply a force, pressure, bearing load, moment, or body load to vertices, faces, or edges of a model as well as import a motion load from dynamic simulation. It is also possible to apply fixed displacement constraints to the model as well as model various contact conditions between adjacent parts. It is also possible to evaluate the impact of multiple design changes.

The analysis results can be viewed in terms of equivalent stress, minimum and maximum principal stresses, deformation, safety factor, or modal frequency. Features such as gussets, fillets, or ribs can be added or removed. All of these options permit the design to be re-evaluated and updated. It is also possible to animate the model through various stages of deformation, stress, safety factor, and frequencies. Yet another important feature is the ability to generate a complete engineering design report.

Among the benefits of performing a stress analysis in the design phase is the opportunity to determine if the part or assembly is strong enough to withstand expected "real life" loads or vibrations without failing or being bent out of shape. By determining this early-on, the cost of redesign can be small. As a result, the design process is sped up and a better product can be brought to market sooner.

Until recently, computer simulation meant modeling one physical phenomenon at a time such as structural integrity or thermal behavior. However, in the real world, a product must withstand many different kinds of physical phenomena, often applied simultaneously. Today, many different kinds of digital simulation software modules are available at affordable prices to help solve design problems related to these conditions.

In preparing a digital 3D solid model for simulation and analysis, it may be necessary to simplify the design. This is done by removing extraneous features (defeaturing) that would interfere with the analysis, regardless of the objective of simulation. This procedure is normally carried out by applying a mesh to correct the geometry and define specific elements of the geometry. Material is first selected for the manufacture of the end product from a material library, and then constraints or other conditions are set. Finally, the loads to be applied are entered. The results show up graphically as gradations of color representing the magnitude of the physical phenomenon under test, or arrows designating fluid flow.

Stress analysis simulation software is typically an extension of solid 3D CAD software permitting simulation of both solid and sheet-metal parts. *Static analysis* permits the designer to simulate stress, strain, and deformation. *Modal analysis* (or analysis specific to the model or prototype being tested) permits the designer to determine natural frequencies of vibration and mode patterns of the 3D image on the computer screen.

The designer can visualize the affects in 3D as volume plots, create reports for any results, and perform dimensional studies to refine designs. It is useful if the designer has some basic knowledge of stress analysis and the procedures necessary to analyze the loads and constraints that will be imposed on the part or assembly in the real world.

Simulated Stress Analysis

The designer can view the analysis in terms of equivalent stress, minimum and maximum principal stresses, deformation, safety factor, and modal frequency. It is also possible to add or suppress such features as gussets, fillets, or ribs, re-evaluate the design, and update the solution. The model can be animated through various stages of deformation, stress, safety factor, and frequencies and the software can generate a complete and automatic engineering design report. The nine-step process of CAE virtual simulation is summarized later in the chapter.

Early stress analysis of prototypes can show how deforming them would affect the alignment of critical attached components. Deformation can cause forces that would accelerate wear and lead to premature failure. The geometry or form factor of prototypes is an important consideration in the design. It can play a critical role in how the part or assembly behaves when subjected to vibration at its natural frequencies. Sometimes it may be desirable to expose the digital model to a range of frequencies to determine how it withstands each of them, but in other situations it is best to find ways to be sure the model will never be exposed to any of them.

Stress analysis is based on a mathematical representation of a physical model. The factors involved in this representation include the model's material properties, applicable boundary conditions (loads, supports, etc.), contact conditions, and density of the mesh or geometric pattern in which the model's surface is divided. The software adds up the individual behaviors of each element to predict the behavior of the entire model by solving a set of simultaneous algebraic equations. The study of these results is called *post-processing*.

The post-processed results identify both areas of weakness on the model and areas where material is wasted because it bears little or no load. Most evaluation takes place when the results are interpreted. An example is how the numerical output as it relates to the displayed color contours on the model compares with the result expected. If the results make sense based on engineering principles, it can be assumed the design is valid; if not, the cause of the discrepancy must be determined.

Three-dimensional stresses and strains build up in many directions. A common way to express these multidirectional stresses is to summarize them into an *equivalent stress*, also known as the *von Mises stress*. All objects have a stress limit that depends on the material from which it is made. This is expressed as *material yield* or *ultimate strength*. Steel has a yield limit of 40,000 lb/in.^{2}, so any stress above that limit results in some form of permanent deformation. If a steel prototype is not expected to deform permanently, the maximum allowable stress must not exceed that yield limit. This permits a safety factor to be calculated based on that limit.

Advantages of 3D Solid Modeling Software

- 3D designs produced on other CAD systems can be imported and edited directly as if they had originated on the user's computer without concerns about constraints or regeneration failures.
- Existing or imported 3D models can be modified dynamically with a few mouse clicks or be converted into new 3D models by changing their dimensions on the fly.
- Users can switch rapidly through different design scenarios without regeneration failures.
- Difficult to manufacture assemblies can be virtually disassembled on the computer screen into separate component parts and downloaded into separate files for rework apart from the assembly drawings.
- Dimensions or materials can be changed and the record of these changes can be kept in a separate file for further use. Then they can be returned to the assembly file for virtual reassembly. This permits top-down design practices to be carried out easily.
- Parts of an existing 3D solid model's geometry can be reused by cutting and pasting from one design to the next. Even contoured surfaces can be copied and pulled into 3D.
- Geometry of existing 3D models can be combined to create multiple design prototypes rapidly.

Nine-Step Process in CAD/CAE Simulation

---

1. *Determine* the simulation requirements and specify their properties.
2. *Exclude* components not required for simulation.
3. *Assign* materials and run a modal simulation to determine natural frequencies.
4. *Add* constraints.
5. *Add* loads.
6. *Specify* contact conditions (an optional step).
7. *Specify* and preview the mesh (an optional step).
8. *Run* the simulation.
9. *View* and interpret the results.

GLOSSARY OF COMPUTER-AIDED DESIGN TERMS

absolute coordinates: Distances measured from a fixed reference point, such as the origin.

associative dimensioning (AD): A feature in CAD software that automatically updates dimension values in other files when dimensions are changed in one file.

Boolean modeling: A type of *associative dimensioning* 3D modeling technique permitting the user to add or subtract 3D shapes from one prototype to another.

Cartesian coordinates: A rectangular system for locating points on a drawing in which the origin point is 0 and X represents length, Y width, and Z height. The surfaces between them can be designated as the X-Z, X-Y, and Y-Z planes.

composite: A CAD drawing containing stacked layers of drawings with the same orientation and scale, as in the floor plans of a high-rise building or deposition layers in an integrated circuit device.

computer-aided engineering (CAE): Designation for computer software that supports engineering tasks rather than just the presentation of design drawings. CAE tasks include analysis, simulation, planning, manufacturing, and diagnosis.

computer simulation: A method for applying *virtual* physical tests to a 3D digital prototype so the results appear as colored patterns showing the effects of applying stresses such as physical deformation, heat, electric current, or the response of moving fluids. Analysis of the results indicates a prototype's ability to withstand stresses, information that can be fed back for improving the design.

coordinate systems: An application-specific Cartesian system that is user-defined. Loads, constraints, material properties, and post-processing variables are defined by this system.

damping: Dissipation of energy in a vibrating structure. *Viscous damping* occurs when damping is proportional to velocity.

data exchange format (DXF): A standard format or translator for transferring data describing CAD drawings from one brand of software to another.

defeaturing: The removal of irrelevant features from a 3D digital prototype that simplify it to achieve faster and more accurate information from virtual simulation.

digital prototyping: A 3D CAD computer graphic presentation of a prototype design or assembly. It is used to design, iterate, optimize, validate, and visualize a product during its development. It also gives design, engineering, manufacturing, sales, and marketing staff an opportunity to cooperate in suggesting design improvements before the prototype is approved for manufacture.

eigenfrequency analysis: A process for determining the undamped natural frequencies and vibration modes of a prototype.

fatigue analysis: A method for predicting the life of a prototype by showing the effects of cyclic application and removal of loads to identify places where fracture and failure are most likely to occur.

finite element method (FEM)/finite element analysis (FEA): A virtual simulation method based on finding approximate solutions of partial differential equations (PDE) that provides simulation results for a digital prototype displayed on a computer screen. The entire model image is overlaid with a network of small geometrical elements, each having individual behaviors, in a *meshing* process. When summed mathematically, the complete response of the prototype to applied stresses can be viewed.

free vibration: The undamped vibration of a structure after it is displaced from its equilibrium position and released. (See also **eigenfrequency analysis**.)

frequency modes: Results of **modal frequency analysis** for determining the natural resonant frequency of a prototype. *Mode shape* is the displacement shape or distortion of a model when excited at its resonant frequency. This can be destructive and cause the part or assembly to fail.

frequency response analysis: A method for determining the steady-state response of a prototype to harmonic excitation. A frequency-sweep generator can determine many excitation frequencies.

heat transfer analysis: A method for modeling the conductivity or thermal fluid dynamics of a digital prototype. It can consist of a *steady state* or *transient transfer*, a reference to common thermal properties that permit linear heat diffusion.

kinematics: The branch of mechanics that addresses the motion of a body without considering its mass or the forces acting on it.

linear extrusion: A CAD software feature permitting the projection of 2D geometry into a 3D shape along a linear path on a computer screen.

linear material property: The characteristic of the material of a prototype in which the stress is directly proportional to the strain without leaving permanent deformation. It is indicated when the slope of the material's stress-strain curve is constant in the elastic region (measured as the *modulus of elasticity*). Temperature is assumed to have no affect on this material property.

loads: Physical forces that can act on a prototype including tension, compression, shear, torsion, bending, or combinations of these forces.

menu: The display on a CAD computer screen of available modeling "tools" or commands, typically on the left side and upper tool bars. Options can be selected with a mouse or finger tap on the screen.

mesh: A network of lines generated as an overlay on the surface of the model that divide it into smaller elements, each with an individual behavior. When the individual behaviors of each element are added by the simulation software and resolved by a set of simultaneous algebraic equations, the behavior of the entire model can be predicted. [See also **finite element method (FEM)**.]

modal frequency analysis: A method for finding natural frequencies of vibration and mode shapes in prototype designs.

nonlinear models: Prototypes made from materials that will be stressed past their elastic limits. The stress response in the material will depend on the amount of deformation applied.

parametric analysis: A method for determining a prototype's dependence on the variation of a specific parameter or arbitrary constant.

parametric modeling: The use of software that links a 3D image on a CAD computer screen with the data that sets dimensional and positional constraints.

polar coordinates: Elements of a coordinate system that permit the location of points by defining an angle and radial distance from its origin, considered to be the center of a sphere.

polyline: A string of lines that can contain many connected line segments.

post-processing: The evaluation of virtual simulation results as viewed on a 3D prototype image on a computer screen. This last step shows the distribution of stresses, deformations, and other physical effects on the model. These results can show problem areas and places where excess nonfunctional material should be removed because it masks information about how the prototype will react to vibration.

preprocessing: The first step in preparing a 3D digital prototype for the virtual simulation of stresses. It includes *defeaturing* and the application of an appropriate *mesh* to define the finite elements so the computer software can calculate the results.

primitives: Basic elements of a computer graphics display including points, lines, curves, polygons, and alphanumeric characters.

prototype or model: The 3D digital image of a part or assembly displayed on a computer screen for the purpose of design, evaluation, and stress simulation. This contrasts with a solid model made by some form of additive or subtractive prototyping to evaluate style, color, or other characteristic. If it closely replicates the end product, it can be subjected to laboratory stress testing for later evaluation.

prototype drawing: A master drawing or template on the computer screen that includes preset computer defaults permitting it to be reused in other applications.

radial extrusion: A 3D computer software technique for projecting 2D shapes into 3D shapes along a circular path such as projecting a wheel or gear cross-section into a 3D image.

rotational degrees-of-freedom: Degrees of freedom associated with the rotation of a body around an axis. Beams, plates, and shells have rotational degrees-of-freedom.

safety factor/margin of safety: A factor in stress analysis calculated to assure that a part or assembly is strong enough to prevent its failure in service; this is a predetermined design requirement. Safety factor = *ultimate tensile strength/maximum allowable stress.* Margin of safety = *ultimate tensile strength/maximum calculated tensile stress.* A safety factor of 1 means the material is at a yield point. Safety factors between 2 and 4 provide better assurance that the end product will not fail.

solid transfer language (STL): A software data transfer program for converting data from CAD drawings into equipment-specific software for directing either additive or subtractive rapid prototyping equipment or systems. (See Chapter 18, Rapid Prototyping.)

spline: A flexible curve drawn to connect a series of points into a smooth shape on a computer screen.

static stress analysis: An analysis performed after virtual simulation of a 3D digital prototype has been completed. Its materials, loads, and constraints (for the stress or strain to be applied), contact conditions, and mesh preferences have been defined and entered into the computer.

transient analysis: A time-dependent analysis of a digital prototype that involves mass, mass moment of inertia, and damping.

vibrational analysis: Tests a prototype including random vibrations, shock, and impact. Each of these factors can act to reveal the prototype's natural frequency causing it to resonate and fail in service.

von Mises stress/equivalent tensile stress: Conversion of the six stress and strain components that build up in many directions of a 3D solid into a single equivalent stress. A material will start yielding when its von Mises stress reaches a critical value known as the *yield strength.* von Mises stress can predict a yield point of materials under any loading condition from the results of simple tensile tests.

CHAPTER 18
RAPID PROTOTYPING

RAPID PROTOTYPING FOCUSES ON BUILDING FUNCTIONAL PARTS

The opportunity to hold and closely examine a solid model or prototype of a new product before it is manufactured is welcomed by many persons responsible for the success of that product. A model that can be handed around makes it much easier to evaluate a product design. Many find it preferable to viewing a 3D image on a computer screen or even carefully rendered drawings. Designers, production engineers, marketing personnel, and prospective customers can comment on the model in group sessions. Flaws, oversights, and omissions can be detected, and suggestions for improvements can more easily be made.

It is important that problems be discovered early on in the design phase before costly tooling is ordered for mass production. Early identification saves time and can eliminate the higher cost of making design corrections during the production phase. Any corrections that are identified can be made to the initial CAD drawings so they can be reflected in a revised prototype.

In the past, if a 3D model was desired, skilled model-makers built custom models, usually from wood. This costly and time-consuming procedure would be prolonged if the model had to be returned to the shop after evaluation while a corrected model was made. This procedure could delay a new product's time-to-market which is especially serious if it is entering a highly competitive market. Fortunately, with the introduction of computer-aided *rapid prototyping* (RP) some 25 years ago, model or prototype-building has become more popular, faster, and less costly.

Since that time, rapid prototyping or model building has continuously evolved; older processes have been improved or dropped and new ones have been developed. As a result, customers have a wider choice of size, cost, precision in details, color, and materials selection. Models can be made from wax, paper, plastic, ceramics, or metal. The models can be full size or scaled-down versions of the end product, and they can be made from softer material for display or tougher materials to withstand laboratory physical testing.

Rapid prototyping has become a global industry that now includes manufacturers of prototype building equipment, contract prototype building services, and companies that perform both manufacture of equipment and offer building services. The importance of this RP industry can be seen from the growing list of participants in equipment manufacture and building services in the United States, Canada, South and Central America, Europe, and Asia. RP model-making equipment is being sold for in-house use by manufacturers, government and corporate R&D laboratories, and the engineering departments of vocational schools and universities.

The starting point for producing all types of 3D prototypes by RP technologies is the receipt of dimensional data defining the end model in a digital format obtained from CAD drawings. These dimensions are then translated by software into digital building instructions by the selected RP machine to be used in building the 3D model. Prototypes can be built layer by layer by one of the more popular *additive* technologies or sculpted from solid materials by a *subtractive* milling machine. The selection of the process will depend on the end use of the prototype and a number of other factors including cost per unit, desired materials, size or volume, and finishing process.

The first additive RP technologies were developed to produce models from soft materials such as wax or photopolymers because they were intended only for the evaluation of dimensions, form, style, and perhaps color; they did not need to be durable or have long shelf lives. However, as RP technology evolved, new processes were introduced to form models from materials that were strong enough to withstand rough handling and physical stress testing. They could now be made from materials with the same strength and durability as the material intended for manufacturing the end product such as hard plastic, ceramic, or even powdered metal. With the success of these functional prototypes, it was found that similar methods could be used to produce cores or molds for casting with metals or plastic.

As a spinoff of prototype building, some RP technologies have been adapted for short-run manufacture of parts or tools as replacements for machined or cast parts for original equipment no longer in production, made from the same or comparable materials. These can replace worn or broken parts in the field. This is especially important if spare parts are no longer being made for those machines. The cost of retooling to make a small batch of replacement parts would be prohibitive. These RP processes are now called *rapid manufacturing* (RM), *solid free-form fabrication, computer automated manufacturing,* or *layered manufacturing.*

The term "rapid" as it applies to RP technologies is relative. Even the fastest RP fabrication process can take from 3 to 72 hours, depending on the size and complexity of the prototype. Nevertheless, all of these methods are faster than the weeks or even months required to fashion a prototype by traditional hand-crafted carving or machining methods.

RAPID PROTOTYPING STEPS

There are five steps in the production of additive rapid prototypes:

1. *Prepare a CAD drawing of the prototype*: The first step is to provide a 3D drawing file containing the dimensional data necessary to prepare the software for directing the RP build process.

2. *Convert the CAD data to STL format*: Many different CAD software packages are available for preparing the dimensional data for use by an RP system. In the next step, the CAD file is converted into the STL format, which represents a 3D surface as a mesh of planar triangles. (The STL format widely used in many different RP technologies is now the standard in the RP industry.) This file contains information about the shape of the model or prototype in the form of a net of triangles fitted to the model's geometry. Its form is represented precisely by triangles because curved surfaces cannot be digitized. Increasing the number of triangles will reduce the size of the triangles defining the model and improve the accuracy of the representation. But more triangles mean a larger file, which takes more time to convert to build instructions. This, in turn, increases the time required to build the prototype. An early decision must be made about optimum triangle density for defining the precision of the model; this relates directly to the size of the STL file required to obtain that precision.

3. *Convert the STL file into cross-sectional layers*: In this step, pre-processing software converts the STL file into software that will actually instruct the RP system how to build the model in layers. Typically this software permits the size, location, and orientation of the model to be selected. The XYZ coordinates of an RP build envelope are illustrated in Fig. 1. The size of this envelope for most RP processes is typically less than 1 ft^3. The shortest dimension is usually oriented along the Z (height) direction because build time is directly proportional to the number of layers required. Pre-processing software "slices" the STL model into many parallel layers, and the thickness of these layers depends on the RP process selected and the accuracy requirements for the model. RP systems offer "slice" thickness ranging from 0.0024 to 0.5 in. (0.06 to 13 mm). This software can also generate the temporary structures to support features (where needed) such as overhangs, internal cavities, and thin-walled sections. Many RP system manufacturers include their own proprietary pre-processing software, which includes provisions for building these supports for the model while it is being built.

4. *Construct the prototype in layers*: In this step, techniques (to be described later) are used by specific RP systems. Each prototype is built one layer at a time from materials that are principally photopolymers, paper, powdered metal, or powdered ceramic. Most RP systems are essentially autonomous. This means they can operate for hours with little or no human intervention other than occasional supervision of progress until the prototype is finished.

Fig. 1 Build Envelope: The maximum volume of any prototype or model that can be built by a specific RP machine or system. It is stated as length X, width Y, and height Z, in inches, millimeters, or centimeters. Volume is determined by the size of the enclosure or cabinet that contains the equipment and materials for building the prototype; this is influenced by the technology and materials being used. (This applies to enclosures for both the additive and subtractive processes). The Z dimension in the additive process is typically the shortest because cost and time to build a prototype are directly related to the number of "slices" or layers that must be deposited to reach that dimension.

5. *Clean and finish the model*: The final step is *post-processing* (either required or desired) for each RP process. These typically call for removing the prototype from the build system and removing any remaining uncured build material and temporary structural supports. In some systems, the recycling of excess powdered build materials is done at this time. Prototypes made from photosensitive plastic resins usually must undergo a final ultraviolet (UV) curing step in a separate oven to complete the hardening process. Other prototype finishing steps include cleaning and sanding, sealing, painting, or polishing the prototype to improve its appearance and durability. However, prototypes made with some RP processes require additional machining to improve their dimensional accuracy.

Each RP process is focused on one or more market segments such as concept modeling, pre-production prototypes for determining product assembly form and fit, jewelry industry design, and dental structures. The selection of the most appropriate RP process calls for the simultaneous evaluation of such factors as model size, quantities required, complexity, dimensional accuracy, and surface finish requirements. Next is the choice of material and its stability and durability after curing. Some processes offer different colored build material if final painting is to be avoided. Urgency and price are yet other factors to consider. Some RP processes take longer and cost more than others, bearing in mind that all RP processes are faster and cheaper than traditional model-making.

COMMERCIAL RAPID PROTOTYPING CHOICES

According to the Rapid Prototyping Home Page maintained by the University of Utah, there are now at least 17 worldwide sources of additive RP systems selling equipment in the United States. These include six based in the United States, five in Japan, three in Germany, one in Israel, and two in China. Three American firms are recognized as developers of concept modeling systems. In the list of companies worldwide offering subtractive or direct RP milling systems, three are the United States, three are in Germany, and two are in Sweden.

The importance of RP technology can be inferred from the number of RP facilities in universities and research laboratories worldwide: 15 are in American universities and colleges; two are in German universities; and one each are in Spanish, Dutch, and Greek universities. They are also located in laboratories in the United States, England, and Norway.

A commercial RP system is one sold as a complete turn-key package either off-the-shelf or custom-built with standardized components. These packages normally include the manufacturer's proprietary software. Typical customers are industries or laboratories that require enough prototypes to justify purchasing the systems and training technicians to operate them, but some companies buy the packages and train their operators to keep all information about their prototype designs secure and away from competitors.

For those organizations that need only a few prototypes or models per year, there are third-party contract service providers who will produce them in their own facilities. Some of the RP system manufacturers also offer these services in their own shops, using their own equipment. The customer usually provides the requisite CAD data to the service provider.

A measure of the popularity or acceptance of specific commercial RP technologies can be gained from the advertising of contract service providers worldwide. Allowing for possible new entries into the business as well as drop-outs, there are more than 250 service providers in business worldwide. Of those 245 firms reporting, 131 (53 percent) offer Stereolithography (SLA), 45 (18 percent) offer Fused Deposition Modeling (FDM),

42 (17 percent) offer Selective Laser Sintering, (SLS), and 27 (11 percent) offer Laminated Object Manufacturing (LOM).

Many service providers offer two or more of these technologies, while a smaller number of service providers offer technologies other than the four mentioned. These providers include OEMs who offer RP services using their own proprietary systems. The fact that 47, or more than 20 percent, of the service providers are located outside of the United States, predominately in Europe, Canada, Central and South America, and Asia, indicates global adoption of this technology.

Some RP systems called 3D office printers are self-contained autonomous manufacturing units housed in small desktop cabinets suitable for operation within an office environment. They include provisions for either containing or venting any smoke or fumes resulting from their processes. Although model size and choice of materials are limited, these small systems are relatively inexpensive compared to those capable of producing larger prototypes.

Ten recognized commercial additive RP processes require build instructions generated from CAD software and heat to cut or fuse laminations or melt either metal powder or plastic resins to form their prototypes. Stereolithography, the most popular process, as well as Selective Laser Sintering, Laminated-Object Manufacturing, Direct Metal Deposition, and Directed-Light Fabrication use heat from lasers. However, second ranking Fused-Deposition Modeling, as well as Three-Dimensional Printing, Direct-Shell Production Casting, Solid-Ground Curing, and Poly-Jet Matrix 3D Printing use a variety of heating methods including ultraviolet lamps.

The RP original equipment manufacturers (OEMs) mentioned in this chapter were included because of the availability of technical information about their systems. However, this chapter is not intended to be a comprehensive overview of all RP systems now available. Each propriety system included has been identified by the OEM's proprietary name, but their proprietary registered and trade-marked names for computer software and materials being used have not been included.

COMMERCIAL ADDITIVE RP PROCESSES

Stereolithography or SLA

Sterolithography is now known as SLA, the accepted generic abbreviation for the process; it is also known as *3D layering* or *3D printing*. SLA translates a CAD design into a solid 3D prototype or model with a combination of laser, photochemistry, and software technology. The digital data from a CAD drawing is first processed by software, which "slices" the vertical dimensions of the model design into very thin cross sections. This basic build process is shown in Fig. 2. The system is housed in a sealed cabinet to prevent the escape of fumes from the build process.

An SLA platform, which can move up and down like an elevator, is located in a vat filled with 5 to 10 gallons of a clear liquid photopolymer such as urethane acrylate resin. As a first step the computer-controlled platform is lowered below the table surface to a depth equal to the specified layer or "slice" thickness for the model. This permits the platform to be flooded with liquid polymer to that depth. A beam from a low-power solid-state ultraviolet (UV) laser that has been focused by beam-shaping optics traces the outline of the lowest slice of the model on the liquid resin. The beam then scans the traced area to cure or harden the polymer in this step. UV radiation links the molecules of the liquid polymer into chain-like formations. Each layer can be hardened to a depth of 0.0025 to 0.04 in. (0.06 to 0.1 mm).

The platform with the bottom layer is then lowered into the liquid photopolymer vat to the same depth as the thickness of the first layer, and it is recoated with more polymers. The UV laser then repeats its previous steps of tracing and curing this traced area over the bottom layer. This process is repeated layer by layer until the model is complete. A typical SLA prototype can be built within a few hours because the laser beam scans at speeds of up to 350 in./s (890 cm/s). The photopolymer that has not been scanned by the laser remains a liquid. The thinner the resin film (slice thickness), the higher is its resolution. This means that the model will have a more refined finish that requires little or no sanding or polishing. When the model's surface finish is important, layer thicknesses are set for 0.005 in. (0.13 mm) or less.

SLA technology is still the most widely used rapid prototyping technique, but it is limited by the tendency of UV-cured liquid polymer to curl or sag as it cures. This means that models with overhangs or unsupported horizontal sections must be reinforced with supporting structures in the form of walls, gussets, or columns. Without this support, overhanging parts of the model would sag and could break-off before the polymer is fully cured. Instructions for forming these supports are included in the digitized fabrication data entered into the RP system. Each scan of the laser forms support layers where they are necessary while simultaneously forming the layers of the prototype.

When the build process is complete, the SLA prototype is raised from the polymer vat, and the uncured or liquid resin is allowed to drain off. Any excess can be removed manually from the prototype's surfaces. A post-curing step is required because the build process only polymerizes the model to about half of its fully cured strength. This step is done by exposing the whole model to intense overall UV radiation in an enclosed chamber of a separate post-curing apparatus (PCA) unit.

Complete SLA parts have a smoother surface finish than parts formed by the competing SLS process, giving them a more refined appearance. SLA permits fine, accurate detail to be obtained. After all supports are removed, the part can be milled, drilled, bored, or tapped as required. However, sanding or polishing might be needed if paint or a sprayed-on metal coat is desired. SLA is a relatively low cost process. The liquid photopolymers used in the SLA process are similar to those used to form the photoresist masks in semiconductor wafer fabrication.

The SLA process was the first RP technology to gain commercial acceptance, and it still accounts for the largest base of installed systems. 3D Systems Inc., Rock Hill, South Carolina, developed the SLA process (formerly called SL in the industry). 3D Systems manufactures a wide range of SLA equipment.

It offers three versions of its I Pro 3D production systems: I Pro 800, I Pro 9000, and I Pro 9000 XL. All of these systems employ 1450 mW solid-state neodymium lasers and they can build parts with slice thicknesses of 0.002 to 0.006 in. (0.05 to 0.15 mm). The maximum build envelopes for the I Pro 8000 and I Pro 9000 systems are 25.6 × 29.5 × 21.7 in. (650 × 749 × 550 mm) and parts can weigh as much as 165 lb (75 kg). However, the I Pro 9000 XL has a maximum build envelope of 59 × 30 × 22 in. (1499 × 762 × 559 mm) with maximum part weight of 330 lb (150 kg).

3D Systems also offers five versions of its proJet 3D production system of 3D printers. These produce high-definition wax patterns for direct investment casting, high-definition model making, and patterns for dental applications.

Fig. 2 Stereolithography (SLA): A lift platform is immersed in a vat of liquid photopolymer to a depth equal to the prototype's "slice" or layer thickness, and the polymer is leveled by a wiper blade. A solid-state ultraviolet (UV) laser, under computer control, outlines the first (bottom) layer before scanning the outlined area to cure or harden it. The platform is lowered again into the vat to the same depth and outlining and curing are repeated. This sequence is continued until the prototype is built. It is then removed from the unexposed polymer vat, cleaned, and finish-cured by more ultraviolet exposure to complete the hardening process.

Selective Laser Sintering (SLS)

Selective Laser Sintering (SLS), developed at the University of Texas at Austin, is an RP process similar to Stereolithography (SLA). It creates 3D models from plastic, metal, or ceramic powders with heat produced by a carbon dioxide (CO_2) infrared (IR)-emitting laser, as shown in Fig. 3. As in SLA, the model is built in a cylinder with a piston or platform that can move up or

Fig. 3 Selective Laser Sintering (SLS): Plastic powder from the delivery system is pushed up by a delivery piston and spread by roller over a build piston set at a depth below the platform equal to the "slice" or layer thickness of the prototype. The powder layer is than scanned by a computer-controlled CO_2 infrared (IR) laser that melts the powder so it flows and hardens it to form the first (bottom) layer. The piston is lowered again to the same depth and more powder is added, melted, and hardened. This sequence is continued until all layers are bonded and the prototype is complete. It is then removed from the build cylinder and finished as desired. All unbonded plastic powder is recovered and mixed with new powder for use in the next build process.

down like an elevator. The build cylinder is located beside a powder-delivery cylinder filled with preheated powder. Before a piston within the powder-delivery cylinder rises to eject powder onto a table, it is heated further until its temperature is just below its melting point. The powder is then spread by roller over the top of the adjacent build cylinder which is positioned below table level at the depth equal to the specified bottom "slice" thickness.

The laser beam scans the thin layer of powder, controlled by the computerized optical scanning system. The temperature of the powder is raised further until it melts (or sinters) and flows together to form a solid layer in a pattern derived from the STL-formatted CAD data. As in the SLA process, the piston is successively lowered to the slice thickness of each succeeding layer until all are sintered and bonded together to complete the 3D model.

Unsintered powder is brushed away and the part is removed. No final curing is required in the SLS process, but because the sintered slices are porous the completed 3D model has surfaces that can be coarse and powdery. No supports are required in SLS because overhangs and undercuts are supported by the compressed unsintered powder within the build cylinder. The inner and outer porous surfaces of the SLS parts can be smoothed by manual or machine sanding or another melting process. A varnish-like coating can be applied to the model to seal and strengthen it.

An advantage the SLS process has over the SLA process is its ability to form prototypes from many different powdered materials. These include polyamide, glass-filled polyamide, and aluminum-filled polyamide. Polymer-coated metal powder is an alternative. These materials are strong and stable enough to permit the model or prototype to be used in low-stress functional and environmental testing. These thermoplastic-like materials are easier to machine than the photosensitive polymers used in the SLA process. Prototypes made from these materials can also serve as molds or patterns for casting parts.

Inaccuracies in the vertical or Z dimensions of models made by both SLS and SLA can occur because neither process includes milling. But these errors are more likely to occur in the SLS process because of variations in the sintering characteristics of the many materials used in the process.

An SLS system can have as many as five major equipment components: (1) sinter station, (2) build module on wheels for transferring the models between stations, (3) thermal station for preheating the powder delivery station and build cylinder, (4) break-out station for removing the model from the build module, and (5) recycling station for mixing recycled and new powder. These systems also include a nitrogen generator and a new powder storage tank. Customer-installed piping transports new and recycled powder between stations. The SLS sinter station is isolated in a sealed nitrogen-filled enclosure that is maintained at a temperature just below the melting point of the powder. The nitrogen prevents an explosion that could be caused by the rapid oxidation of the build powder.

3D Systems Inc., also offers four SLS systems: two *Sinterstation* HiQs and two *Sinterstation* Pro SLMs.

HiQ systems use CO_2 lasers, 30 W for the standard system and 50 W for the high-speed version. Both of these systems can deposit layers in thicknesses from 0.004 to 0.006 in. (0.1 to 0.15 mm), and both have maximum build envelopes of $15 \times 13 \times 19$ in. ($39 \times 33 \times 50$ cm). They are capable of building complex plastic models and end-use parts as well as complex metal tooling and prototypes. The metal materials available include aluminum and titanium. These systems can also produce complex investment casting patterns as well as functional parts with flexible rubber-like characteristics.

Sinterstation Pro SLM systems are the Pro DM 100 and DM 250 with the SLM designation indicating their ability to produce metal parts. The Pro DM 100 can deposit layers of 0.001 or 0.002 in. (30 or 50 μ) and a maximum part build area of $5 \times 3 \times 3$ in. ($13 \times 8 \times 8$ cm), while the DM 250 can deposit layers of 0.0012 or 0.003 in. (50 or 75 μ) with a maximum part build area of $10 \times 10 \times 10$ in. ($25 \times 25 \times 25$ cm).

EOS GmbH, Electro Optical Systems, Munich, Germany, offers three different classes of machines for laser-sintering plastic, metal, and sand molds.

Formiga P100, Eosint P 390, P 700, P 730, and P 800 are designed to form plastic prototypes. Build size for these systems ranges from $7.9 \times 9.8 \times 13$ in. ($200 \times 250 \times 330$ mm) in the Formiga to $28.6 \times 15 \times 23.9$ in. ($700 \times 380 \times 580$ mm) in the P730. Models P100, Eosint, and P700 have 30- to 50-W lasers, and P730 and P800 series have two 50-W lasers. Laser scan speed is up to 16 to 29 ft/s (5 to 6 m/s), but twice that in models with twin lasers. Variable build layer thickness, depending on the material, is from 0.004 to 0.006 in. (0.1 to 0.15 mm). These systems typically have sealed nitrogen-filled build chambers.

Eosint M 270 performs Direct Metal Laser Sintering (DMLS), and its build envelope maximum is $10 \times 10 \times 9$ in. ($250 \times 250 \times 215$ mm). It has a 200-W ytterbium-fiber laser with a scan speed of up to 23 ft/s (7 m/s). The laser has a variable laser focus diameter of 0.004 to 0.02 in. (0.10 to 0.51 mm), controlled by software as it moves across the powder. Build layer thickness is from 0.0008 to 0.0015 in. (0.02 to 0.04 mm), depending on the material and its application. A nitrogen-filled sealed chamber covers the process except when building in titanium, which requires argon to ensure an impurity-free final prototype.

Eosint S 750, optimized for sintering sand cores and molds for metal casting, has a maximum build envelope of $28.4 \times 15 \times 15$ in. ($720 \times 380 \times 380$ mm).

EOS reports increases in the number of materials suitable for sintering, but the emphasis has been on manufacturing rather than rapid prototyping. Most sintering material is based on PA 12 or PA 11 polyamide. Its fine powder is resistant to most chemicals and, when laser-sintered, produces functional prototypes and manufactured parts with a finish that can withstand high mechanical and chemical loads. In some applications, the polyamide is filled with aluminum, glass, or carbon fiber. The medical and aerospace industries are finding more applications for stainless steel, cobalt chrome and titanium, while demand for maraging steel for making tools is increasing.

Fused Deposition Modeling (FDM)

The Fused Deposition Modeling (FDM) process, shown in Fig. 4, builds prototypes from melted thermoplastic monofilament with a diameter of 0.07 in. (1.78 mm). It is unwound from a spool and passed through an extrusion heater where it is heated to a semiliquid state. The heater, mounted on a computer-controlled XY table, is free to move over a fixtureless platform. The semiliquid filament is fed through a nozzle that extrudes and deposits it in progressive layers from the bottom up. These can range in thickness from 0.002 to 0.030 in. (0.05 to 0.76 mm) with wall thickness of 0.010 to 0.125 in. (0.25 to 3.2 mm). The layers adhere to each other and harden into a solid 3D model. After removal, the completed model can be sanded, painted, drilled, or sealed. This laserless process can form thin-walled, contoured models, functional prototypes, various tools, and molds for investment casting.

The structures needed to support overhanging or fragile structures in FDM modeling must be designed into the STL-formatted CAD data file and deposited as part of the model. These supports can be removed easily in later secondary operations. All functional components of FDM systems are contained within temperature-controlled enclosures.

Stratasys, Inc., Eden Prairie, Minnesota, developed and patented the Fused Deposition Modeling (FDM) process. The company offers three FORTUS 3D production systems: FORTUS 360mc, FORTUS 400mc, and FORTUS 900mc with corresponding envelope volumes in cubic inches. The standard capacity build envelope is $14 \times 10 \times 10$ in. $(36 \times 25 \times 25$ cm) and it contains one build material bay and one support material bay. A large capacity version has a build envelope of $16 \times 14 \times 16$ in. $(41 \times 36 \times 41$ cm), and it contains two build material bays

Fig. 4 Fused Deposition Modeling (FDM): A monofilament of a thermoplastic resin is unwound from a spool and passed through a heater mounted on a computer-controlled XY table. The semisolid filament is extruded from the nozzle to form the first (bottom) layer in a preset pattern by the motion of the XY table as it moves over the platform. The hot plastic binds to each successive layer as it is set down and hardened. This sequence is continued until the prototype is complete. It is then removed from the platform and sanded to improve its finish. This laserless process can form thin-walled, contoured prototypes or molds for investment casting.

and two support material bays. Multiple build speed options are offered: for faster throughput, 0.013 and 0.10 in. (0.330 and 0.254 mm); for finer detail and surface finish, 0.007 in. (0.178 mm); and for the highest feature detail and best surface finish, 0.005 in. (0.127 mm).

Material options are acrylonitrile butadiene styrene (ABS), M30, M30i, and ABSi thermoplastic, polycarbonate thermoplastic (PC), a blend of polycarbonate thermoplastic and ABS thermoplastic (PC-ABS), PC-ISO thermoplastic, ULTEM 9085 thermoplastic, and PPSF/PPSU polyphenylsulfone thermoplastic. These materials melt at temperatures between 180 to 220°F (82 to 104°C).

Stratasys also offers the Dimension 1200es Series 3D printers that use the FDM process to build models in sizes up to $10 \times 10 \times 12$ in. $(25 \times 25 \times 30$ cm). The printers use ABS plus modeling material, a production-grade thermoplastic that is durable enough to perform virtually the same as production parts. Models are printed from the bottom up with precisely deposited layers of modeling and support material. They do not require curing and are ready for support removal right from the printer. Models can be drilled, tapped, sanded, and painted. The 1200es series can produce functional prototypes, molds, patterns, and custom tools and fixtures.

The SST 1200es uses Soluble Support Technology, which dissolves the supports in a water-based solution. This process is recommended for detailed or delicate models. The BST 1200es uses a breakaway support technology that permits the supports to be simply snapped off, revealing the final model. Breakaway supports can be more convenient if there is no ready access to a sink or water supply.

The company also offers two desktop-size 3D printers, the Dimension uPrint and uPrint Plus. The uPrint uses ivory-colored ABS plus, and the uPrint Plus offers a choice of ABS plus colors: ivory, plum, fluorescent yellow, black, red, nectarine, olive green, and gray. As for other FDS 3D printers, the 3D model must be created in CAD software and converted to STL 3D model-printing instructions, including support structures. Build size for both the uPrint and uPrint Plus is $8 \times 6 \times 6$ in. $(20 \times 15 \times 15$ cm). Layer thickness for the uPrint is 0.010 in. (0.254 mm) and for the uPrint Plus it is either 0.010 in. (0.254 mm) or 0.013 in. (0.330 mm). The modeling base must be snapped off and the soluble support material must be dissolved away. The model can then be drilled, machined, sanded, painted, and even chrome plated.

Three-Dimensional Printing (3DP)

The Three-Dimensional Printing (3DP) or the ink-jet printing process, shown in Fig. 5, is similar to Selective Laser Sintering (SLS) except that a multichannel inkjet head and liquid adhesive supply replace the laser. The supply cylinder of the powder delivery system is filled with starch and cellulose powder, which is delivered to the work platform by elevating the powder delivery piston. A roller then distributes a single layer of powder from the supply cylinder to the upper surface of the build piston within the build cylinder that is positioned a distance of one layer thickness below the platform. A multichannel inkjet head sprays a water-based liquid adhesive from the reservoir on the surface of the powder, bonding it in the shape of a horizontal layer or lamination of the model.

In successive steps, the build piston is lowered the same distance while the powder delivery piston pushes up more fresh powder, which the roller spreads over the previous layer on the build piston. This process is repeated until the 3D model is complete. Any loose excess powder is brushed away, and wax is coated on the inner and outer surfaces of the model to improve its strength. The 3DP process was developed at the Three-Dimensional Printing Laboratory at the Massachusetts Institute of Technology (MIT).

Z Corporation, Burlington, Massachusetts, is licensed to use the original MIT process to form 3D models. It offers five 3D

Fig. 5 Three-Dimensional Printing (3DP): Plastic powder from the delivery system is pushed up by the delivery piston and spread by roller over a build piston set below the platform to a depth equal to a "slice" or layer of the prototype. Liquid adhesive from the reservoir is sprayed on the powder by a multichannel inkjet head to form the first (bottom) layer of the prototype. The piston is lowered to the same depth again, another layer of powder is rolled onto the build piston, and more adhesive is sprayed on it to bond the next layer. This sequence is repeated until the hardened prototype is complete. It is then removed and finished as desired.

printers: ZPrinter 310 Plus, Z450, Spectrum Z510, ZPrinter 650, and the ZCast 501.

ZPrinter 310 Plus, for affordable monochrome printing, is capable of build speeds of 2 to 4 layers per minute. It has a maximum envelope size of $8 \times 10 \times 8$ in. ($20 \times 25 \times 20$ cm), and a layer thickness of 0.0035 to 0.008 in. (0.089 to 0.203 mm) is selectable by the user at the time of printing. Materials permitted include high-performance composites, elastomerics, and those for direct and investment casting.

ZPrinter 450, for full color printing, provides build speeds of 2 to 4 layers per minute. It has a maximum build envelope of $8 \times 10 \times 8$ in. ($20 \times 25 \times 20$ cm), and a layer thickness of 0.0035 to 0.004 in. (0.089 to 0.102 mm) can be selected by the user selectable at the time of printing. The material choices include a variety of high-performance composites. It has a resolution of 300×450 dpi, and it can have two print heads, one tricolor or one clear.

Spectrum Z510, for larger size and better resolution and color, is capable of build speeds of two layers per minute. Maximum build envelope size is $10 \times 14 \times 8$ in. ($25 \times 36 \times 20$ cm). Layer thickness of 0.0035 to 0.008 in. (0.089 to 0.203 mm) can be selected by the user at the time of printing. The material options include high-performance composites, elastomerics, and those for direct casting. Its resolution is 600×540 dpi, and it has four print heads.

ZPrinter 650, for the largest models, highest resolution, and best color, is capable of build speeds of two to four layers per minute. It has a maximum build envelope of $10 \times 15 \times 8$ in. ($25 \times 38 \times 20$ cm). Layer thickness of 0.0035 to 0.004 in. (0.089 to 0.102 mm) can be selected by the user at the time of printing. The materials can be selected from high-performance composites. Its resolution is 600×540 dpi, and it has five print heads.

The ZCast 501 Direct Metal Casting process permits cast-metal parts to be poured from molds and cores printed on a 3D printer directly from digital data, eliminating the need for a pattern and core box production step, as in traditional sand casting. Metal is then poured into the 3D printed molds.

Solidscape Inc., Merrimack, New Hampshire, offers four 3D printing systems.

The D76 3D printer is intended for dental applications. It provides a build layer of 0.0010 to 0.0025 in. (25.4 to 63.5 μ), and surface finish of 32 to 63 μ (RMS). Minimum feature size is 0.010 in. (254 μ), and its maximum build envelope is $6 \times 6 \times 4$ in. ($15 \times 15 \times 10$ cm).

The T612 Benchtop and T76 systems are intended for rapid tooling and prototyping of master patterns. Models and patterns built on these systems can be used for design verification and production applications. Solidscape uses its patented thermoplastic build materials, asserting they are well suited for making investment casting patterns. These systems can form deposition layers of 0.0005 to 0.0030 in. (0.013 to 0.08 mm) with build envelopes of $12 \times 6 \times 6$ in. ($30 \times 15 \times 15$ cm) in X, Y, and Z dimensions. Achievable accuracy is stated as ± 0.001 in. (± 0.025 mm) for X, Y, and Z dimensions. Surface finish is 32 to 63 μ (RMS), and minimum feature size is 0.010 in. (0.254 mm).

The R66 is intended for jewelry prototyping. It provides a build envelope of $6 \times 6 \times 4$ in. ($15 \times 15 \times 10$ cm) and build layers of 0.0005 to 0.003 in. (0.013 to 0.076 mm). Achievable accuracy, surface finish, and minimum feature size are equal to those of the T612.

PolyJet Matrix 3D Printing

PolyJet Matrix 3D printing, shown in Fig. 6, permits simultaneous deposition of different types of photopolymer build materials with differing mechanical and physical properties in a single build process. A jetting head assembly, under XY computer control, scans back and forth across the build tray while spraying and simultaneously curing each layer. The assembly includes a stack of eight printing heads, each with 96 nozzles. The synchronized printing heads, filled with both build and support material, are positioned between rows of ultraviolet (UV) lamps at each end that project as flat beams wide enough to span the build tray. Both build and support materials are sprayed in ultra-thin layers onto the build tray, starting with the first (bottom) layer. As the build tray is lowered, successive layers are sprayed on until the

Fig. 6 PolyJet Matrix 3D Printing: This system can spray and cure different photopolymer materials to form layers. The first (bottom) "slice" or layer of a prototype is built as an inkjet housing with eight printing heads, each with 96 nozzles, and is scanned over an elevator-type build tray. Each layer contains both build and support material. Polymer curing or hardening is performed by two ultraviolet (UV) lamps, each projecting a separate flat linear beam, as they are swept over the layer. The build tray then descends to the depth of the next layer and the sequence continues until the prototype is complete. Because the prototype has been fully cured as it was built, it can be handled without additional UV curing.

prototype is complete. Each photopolymer layer, as thin as 16 μ (0.001 in.), is cured by UV immediately after it is sprayed. The results are fully cured prototypes that can be handled immediately without an additional post-curing operation.

Objet Geometries Inc., Billerica, Massachusetts, has PolyJet Matrix 3D printing as its proprietary technology. Objet offers eight systems based on their technology: Connex, six versions of its Eden systems 500V, 300 and 350V, 250V, 260 and 250, and the Alaris 30.

PolyJet prototypes are said to be best suited for applications where accuracy, detail, and surface finish are important and the prototype fits within a build volume of 5 × 5 × 5 in. (13 × 13 × 12 cm). The maximum PolyJet build volume is 19.3 × 15.4 × 7.9 in. (49 × 39 × 20 cm), but Objet believes other RP technologies are more cost-effective for building prototypes of that size.

Directed-Light Fabrication (DLF)

The Directed-Light Fabrication (DLF) process, shown in Fig. 7, uses a neodymium YAG (Nd: YAG) laser to fuse powdered metals for building more durable 3D prototypes than can be made from paper or plastic. Metal powders available are finely milled 300 and 400 series stainless steel, tungsten, nickel aluminides, molybdenum disilicide, copper, and aluminum. This technique is also called *Direct-Metal Fusing, Laser Sintering,* and *Laser Engineered Net Shaping* (LENS).

The laser beam, under computer control, fuses metal powder fed from a nozzle to form dense 3D objects that can have dimensions within thousandths of an inch of design tolerance. DLF technology was developed at the Los Alamos National Laboratory (LANL) in New Mexico.

Optomec Design Company, Albuquerque, New Mexico, offers its LENS additive process based on DLF technology. The work is performed by a turnkey manufacturing unit that includes software, a high-power laser, motion control, and other components. The LENS process builds metal parts directly from CAD designs. It injects metal powder into the focus spot of the laser beam and then "prints" layers of metal to fabricate a prototype from the bottom up. Optomec reports that the LENS process can be used to make short production runs of new metal products or fabricate replacement parts for repairing older or obsolete machines or equipment.

LENS MR-7 system offers a 12 × 12 × 12 in. (31 × 31 × 31 cm) work envelope to make test pieces and smaller components. Ultraviolet radiation from a high-power fiberoptic laser is used to build structures one layer at a time from powdered metals, alloys, ceramics or composites. Sensors gather the data required to allow processing microstructures. Two powder feeders permit adding layers with different powdered metal, ceramic, or composite ingredients.

LENS 750 system is for the rapid manufacturing of metal components or repair of existing parts made from titanium, stainless steel, or Inconel. The system offers a 12 × 12 × 12 in. (31 × 31 × 31 cm) work envelope for smaller parts. As in the LENS MR-7, the 750 uses a laser to build structures one layer at a time directly from powdered metals. Optomec says the process yields fully functional 3D parts with mechanical properties that are equal or superior to cast parts or those machined from solid metal.

LENS 850-R system is for the repair or fabrication of high-value metal components from exotic aerospace materials. As in other LENS systems, the 850-R uses a fiberoptic 500 W or 2 kW laser to build 3D structures from powdered metals that have mechanical properties equal or superior to parts cast or machined from solid metal. It has a build envelope of 35 × 59 × 35 in. (89 × 150 × 89 mm) for the repair of mid- to large-sized metal parts.

Direct Metal Deposition (DMD)

Direct Metal Deposition (DMD) is the proprietary form of DLF developed by The POM (Precision Optical Manufacturing) Group, Inc., Auburn Hills, Michigan. It is an additive 3D, laser-aided technology for fusing powdered-metal, as shown in Fig. 8. DMD produces a completely dense metal product. The laser creates a molten metal pool on the substrate material. Powdered additive materials are then injected in precisely measured amounts and melted to form a tight metallurgical bond.

The DMD systems include process sensors called *Closed-Loop Technology* in their patented equipment because the POM Group says that active process regulation is indispensable for

Fig. 7 **Directed-Light Fabrication (DLF):** Fine metal powder is fed onto an XY work platform that is scanned under computer control beneath a neodymium YAG laser. Heat from the laser beam melts or sinters the metal powder to form a thin "slice" or first (bottom) layer of the prototype. The process is repeated as each layer is deposited and bonded until the prototype is complete. More durable prototypes can be made from metal powder than plastic or paper. The sintered metal prototype or tool is then heat-treated to increase its bond strength. Powdered aluminum, copper, stainless steel, tantalum, and other metals have been sintered to form functional tools or working components.

Fig. 8 **Direct Metal Deposition (DMD):** This additive technology bonds powdered metal to produce dense metal prototypes and parts. A carbon dioxide (CO_2) laser forms a melt pool on the substrate material onto which the powdered additive materials are injected in precisely measured amounts. Then they are melted to form tight metallurgical bonds with the substrate. Three process-sensing cameras are positioned 120° apart in a predetermined configuration to provide closed-loop feedback. If the geometry of the part being formed deviates from the specified geometry, it is detected by the system. The size of the weld pool is also monitored continuously in real time. Process variables such as powder flow rate, CNC velocity, and laser power are adjusted to assure product quality and dimensional stability during the build process.

maintaining consistently high quality deposition of new material layers. If the geometry of the part being formed deviates from the required geometry, it is detected by a closed-loop optical feedback system that continuously monitors, in real time, the size of the weld pool. It then adjusts process variables such as powder flow rate, velocity, and laser power to assure product quality and dimensional stability during part formation.

Melt pool information for DMD layering is monitored by three charge-coupled device (CCD) cameras. The DMD process computer compares the information about the actual geometry with the desired geometry and controls the layering process accordingly. The use of three cameras in an array set apart by 120° controls the DMD process in three dimensions. In setting up this control system, at least one camera must be able to observe the melt pool interference contours. In addition to retaining the designed part's shape and maintaining its dimensional stability, the closed-loop system eliminates the requirement for post-processing.

The POM Group offers two DMD systems: the DMD 105D and the Robotic DMDs.

DMD 105D is a five-axis system that uses a fiberoptic laser to deposit metal alloy powders to form small components by RP. It can also manufacture functional parts and repair damaged or worn industrial equipment. It has a build envelope of $12 \times 12 \times 12$ in. $(31 \times 315 \times 31$ cm$)$. This system has restored critical components including turbine blades, drive shafts, diaphragms, dies, and punches. It deposits complex alloys such as tool steels, Stellite, and Inconel. An evacuated chamber can be added to the system to enclose the toxic fumes emitted while processing alloys such as those that include titanium.

Robotic DMD systems are six-axis industrial robots with an integrated fiberoptic-coupled diode laser. It can deposit complex metal alloy powders to build up worn metal surfaces or make metal repairs on massive industrial equipment. The Model 44R DMD system has a build envelope of 77×84 in. $\times 330°$ $(196 \times 213$ cm $\times 330°)$ and the Model 66R has an envelope of 126×144 in. $\times 360°$ $(320 \times 3665$ cm $\times 360°)$. These Robotic DMD systems bring the repair process to the part rather than bringing the part to the repair process, as in most industrial repair work. This saves time and cost in repairing oil and gas lines, mining equipment, and heavy machines.

Laminated-Object Manufacturing (LOM)

The Laminated-Object Manufacturing (LOM) process, shown in Fig. 9, forms 3D models by cutting, stacking, and bonding successive layers or laminations of paper coated with heat-activated adhesive. It was originally developed by Helisys Inc., but taken over by Cubic Technologies, Carson, California. The carbon dioxide (CO_2) laser beam, directed by an optical system under STL-formatted CAD data control, cuts cross-sectional outlines of each layer of the prototype in the paper. The paper layers are successively bonded to previous layers to form a stack that is the prototype.

The heavy paper forming the bottom layer is unwound from a supply roll and pulled across the vertically movable platform. The laser beam cuts the outline of each layer and cross-hatches the waste material within and around the layer for easy removal after the prototype is completed. The waste paper left over after the first layer has been cut and detached (leaving a large hole) is removed by a take-up roll. The movable platform is lowered and paper that will form the second layer is then pulled from the material supply roll and positioned over the first layer. The laser cutting and cross-hatching steps are repeated and a heated roller applies pressure to bond the adhesive coating on the second layer to the first layer. Then the remaining paper is again removed by the take-up roll.

This procedure is repeated until the final layer is applied and pressed to complete the build process. The platform is then elevated and the excess cross-hatched material, now in stacked segments, is removed exposing the finished 3D prototype. The

Fig. 9 **Laminated Object Manufacturing (LOM):** Adhesive-backed paper is fed across the movable elevator platform, and a computer-controlled carbon dioxide (CO_2) infrared (IR) laser cuts the outline of the first (bottom) layer of the prototype as well as cross-hatch lines on the unused paper. The platform is then lowered and more paper is fed across the first layer. The laser again cuts the outline and cross-hatch lines, and a heated roller bonds the adhesive coating between the layers. This sequence continues until all layers of the prototype have been cut and bonded. The cross-hatched paper is removed, releasing the prototype and the edges of the bonded layers are then sealed to finish the prototype.

LOM process forms large glued monolithic blocks that are more resistant to deformity and cost less to produce than those made from softer material. The prototypes have wood-like finishes that can be sanded or polished before being sealed and painted. These models can be used directly as patterns for investment and sand casting or as forms for making silicone molds. LOM models are generally larger than those that can be made by other RP processes with build dimensions up to $30 \times 20 \times 20$ in. $(76 \times 51 \times 51$ cm$)$.

However, the LOM process has objectionable characteristics which could rule out its use. One is the limitation of its laser's ability to cut through thicker paper, reducing the speed and increasing the cost of prototype building. A second one is the requirement for sealing the edges of the layers to prevent moisture penetration, and a third is that the process should be enclosed in a sealed cabinet with ventilation to dissipate smoke caused as the laser burns through the paper layers.

In addition, the accuracy of the vertical or Z dimension is lower than that for either the SLA or SLS processes because the LOM process lacks a milling step, not practical for reducing the thickness of paper layers. If sheets of adhesive-backed plastic or metal are used to produce prototypes, they must be heat-treated to form more rugged models, but they can be milled to improve vertical accuracy.

Direct-Shell Production Casting (DSPC)

The Direct Shell Production Casting (DSPC) process, shown in Fig. 10, is based on technology developed at MIT and is similar to the 3DP process. Primarily intended for forming molds or shells rather than 3D models, DSPC also begins with a CAD file of the desired shell. Although DSPC is considered to be an RP technology, all actual 3D prototypes must be made by later casting processes.

Two specialized kinds of equipment are used in DSPC: a dedicated computer called a Shell-Design Unit (SDU) and a Shell-Processing Unit (SPU). The CAD file is loaded into the SDU to generate the data needed to define the mold or shell; the resulting

496

Fig. 10 Direct Shell Production Casting (DSPC): Ceramic powder fed from a supply source is smoothed by a roller over the build piston recessed in the build cylinder to a depth equal to the thickness of each "slice" or layer of the mold. An adhesive binder is then sprayed on the powder under computer control to bond the first (bottom) layer. The piston is lowered again a distance equal to layer thickness, more powder is spread over the build piston, and more binder is sprayed to bond the first two layers. This sequence continues until the mold is complete. The bonded "green" ceramic shell is then moved to a furnace for firing to form a durable functional mold. This technology is classed as an RP process because it makes molds faster and cheaper than conventional methods, thus permitting metal castings to be made more rapidly.

SDU software also modifies the original shell dimensions in the CAD file to compensate for ceramic shrinkage. This software can also direct the formation of filets and delete certain features in the shell such as holes or keyways which must be machined in the prototype after it is cast.

The movable platform in DSPC is the piston within the build cylinder. As in other RP technologies, it is lowered to a depth below the rim of the build cylinder equal to the thickness of each layer. Then a thin layer of fine aluminum oxide (alumina) powder is spread by roller over the platform. Next, a fine jet of colloidal silica is sprayed onto the powder to bond it in the shape of a single mold or shell layer. The piston is then lowered for the next layer, and the process is repeated until all layers have been formed to complete the 3D shell.

The excess powder is then removed and the object is furnace fired, converting the bonded powder to monolithic ceramic. After the shell has cooled, it is strong enough to withstand molten metal, and can function like a conventional investment casting mold. After the molten metal has cooled, the ceramic shell and any cores or gating are broken away from the prototype. It can then be finished by any methods suitable for castings. DSPC is a proprietary fabrication process for metal parts that combines the advantages of both casting and machining and eliminates the disadvantages of both processes.

Solid-Ground Curing (SGC)

Solid-Ground Curing (SGC) (or the "solider process") is the multistep in-line process, as shown in Fig. 11. It begins when a photomask for the first layer of the 3D model is generated by the equipment shown at the far left of the figure. An electron gun writes a charge pattern of the photomask on a clear glass plate and opaque toner is transferred electrostatically to the plate, forming a photolithographic mask by a xerographic process. The photomask is then moved to the exposure station

where it is aligned over a work platform and under a collimated UV lamp. The SGC process calls for the work platform to be moved sequentially right and left to complete the process.

Model building begins when the work platform is moved right to the photopolymer resin applicator station where a thin layer of resin is applied to the top surface of the work platform and then wiped to the desired thickness. The platform then moves left to the exposure station. A UV lamp is turned on and a shutter is opened for a few seconds to expose the resin layer to the mask pattern. The UV radiation is intense enough to fully cure the layer, eliminating a secondary curing step.

After curing, the platform is then moved right to the wiper station where the unexposed resin is removed and discarded. The platform continues to move right to the wax applicator station where melted wax is applied and spread into the cavities left by the removal of uncured resin. The platform moves right again to the wax cooling plate where the wax is hardened by pressing it against the plate. Then the platform continues to be moved right to the milling head where both the resin and wax layers are milled to a precise thickness. Then the platform moves left to the resin applicator station again where it is lowered to a depth equal to the thickness of the next layer and more resin is applied.

Meanwhile, the opaque toner has been removed from the glass mask, and a new mask for the next layer is generated on the same plate. The complete platform movement cycle is repeated, and it will continue until the 3D model encased in the wax matrix is completed. (The wax matrix supports any overhangs or undercuts so extra support structures are not needed.) Finally, the prototype is removed from the process equipment and the wax is either melted away or dissolved in a washing chamber similar to a dishwasher. The surface of the 3D model can then be sanded or polished by hand or machine as needed.

The SGC process is similar to *drop on demand inkjet plotting.* This method depends on a dual inkjet subsystem that travels on a precision XY drive carriage and deposits both thermoplastic and wax materials onto the build platform under CAD program control. The drive carriage also energizes a flatbed milling subsystem for obtaining the precise vertical or Z height of each layer and the overall Z height of the prototype by milling off excess material.

Fig. 11 Solid Ground Curing (SGC): A photomask is generated on a glass plate by a xerographic process as the first step. The photomask is then moved to the first station of the work platform where liquid photopolymer is applied and wiped. An ultraviolet (UV) lamp above the photomask illuminates the polymer underneath to define and cure the layer. The platform then moves to the next station where hot wax is applied over the layer to fill in margins and spaces. The wax is hardened by a cooling plate, and at the next station excess polymer and wax is milled off to complete the first "slice" or bottom layer. The first photomask is erased and replaced by a second mask on the same glass plate. The entire sequence is repeated as the platform moves right then left under each workstation until the prototype is complete. Finally, the wax is removed by heating or hot water immersion to release the prototype.

SUBTRACTIVE AND R&D LABORATORY PROCESSES

Desktop Prototyping

Desktop prototyping is a subtractive RP technology offering an alternative to the more common additive RP technologies. As with additive methods, it begins with a 3D CAD drawing that is translated into software to run on a PC computer that will direct the operation of a computer numerically controlled (CNC) milling machine within an enclosed cabinet. This machine is capable of carving or milling relatively small models or prototypes from solid blocks of wax, plastic, or soft metal.

Delft Spline Systems of Utrecht, The Netherlands, offers a self-contained CNC milling machine within a cabinet suitable for desktop prototyping within an office environment. Proprietary software from Delft Spline Systems converts CAD data into instructions for the machine that mills prototypes from blocks of wax, plastic, treated wood, and plastic foam. The company offers milling machines, 3D scanners, and combination machines. The company offers the following milling machines that can be directed from personal computers with data sent to their printer ports.

The CPM-ICP series is intended for CNC machining prototypes from non-metal materials in office locations. The two models are the ICP 3020 with a build volume of 12 × 8 × 4 in. (300 × 200 × 90 mm) and the 4030 with a build volume of 16 × 12 × 6 in. (400 × 300 × 140 mm). However, the maximum height or Z dimension of the prototype being machined is limited 50 to 70 percent by the perpendicular cutting tool projecting down from the top of the build space.

Both the JWX-10 and JWK-30 were developed for creating jewelry designs. Each machines' sizes, speeds, and spindle specifications are suited for creating wax jewelry prototypes. Rotation axes permit modeling rings, the most common type of jewelry. The JWX-10 has a build volume of 6 × 4 × 4 in. (140 × 105 × 105 mm), and the JWX-30 has a build volume of 12 × 8 × 4 in. (305 × 205 × 100 mm).

The MDX-40A is intended for producing rapid prototypes of small consumer products. It is intended to work with soft materials, primarily chemically treated wood as well as wax and plastic foam. It has a build area of 12 × 12 × 4 in. (305 × 305 × 105 mm), and it includes a 100-W, 15,000-rpm tool spindle.

The MDX-540 is a heavy-duty, three-axis CNC milling machine equipped with AC servomotors on the X, Y, and Z axes. These provide high feed rates and smooth motion that can be maintained for each step, slowing down only at sharp corners. It has a build volume of 20 × 16 × 6 in. (500 × 400 × 155 mm), and it can machine many materials including treated wood, engineering plastics, and non-ferrous metals.

The MDX-15 and MDX-20 are combination three-axis CNC milling machines and scanners. As milling machines, they can mill light materials such as plastic foam and wax so their main applications are concept modeling and wax models for jewelry investment casting. Their three axes move over steel bearings and are driven by stepper motors from metal traction wires. Their light-spindle 10-W, 6500-rpm motors are intended for milling only light materials. Their fixed controls can be directed by personal computers with data sent to printer ports. Both models can also perform 2D engraving. The NC files can be derived from most CAM packages. The build volume for the MDX-15 is 6 × 4 × 2 in. (150 × 100 × 60 mm) and for the MDX-20 it is 8 × 6 × 2 in. (200 × 150 × 60 mm). As scanners, both are directed from computers with included software that permits setting the area to be scanned and the resolution (pitch) to be used. The rest is done

automatically, including the conversion of data to a valid CAD geometry (polygon file).

Research and Development in Rapid Prototyping

Some RP technologies are still experimental and have not yet achieved commercial status. Information about this research has been described in patents and announced in news releases to the public by the laboratories performing direct prototype and product manufacturing from metal or ceramics. Two of these experimental technologies are described here: Shape Deposition Manufacturing (SDM) and Mold Shape Deposition Manufacturing (MSDM).

While showing commercial promise, these systems have not been adopted by commercial RP equipment OEMs as have the previous RP technologies described. The reasons relate to equipment cost and lack of a substantial market for them. But, where the resources and space are available, components for this equipment can be purchased off-the-shelf, and facilities for their assembly and use can be organized. The design and construction of these one-of-a kind machines is essentially limited to industrial, academic, and government engineering laboratories.

Shape Deposition Manufacturing (SDM)

The Shape-Deposition Manufacturing (SDM) process was developed at the SDM Laboratory of Carnegie Mellon University's

Fig. 12 Shape Deposition Manufacturing (SDM): Wires made from both primary and sacrificial support metals are fed into separate heaters that melt them to form hot droplets. These are sprayed to form layers at a deposition station (a) starting with the first (bottom) layer. That layer is moved to the shaping station (b) where it is machined under computer control. The layer is then moved to a stress-relief station (c) where it is shot-peened to relieve built-up stress. The layer is then returned to the deposition station (a) for the next layer. The metal droplets retain their heat long enough to remelt the underlying layer on impact to form a metallurgical interlayer bond. All three steps are repeated until the prototype is completed. The sacrificial metal, which supports all undercut features, is removed by acid etching to release the part.

Robotics Institute in Pittsburgh, Pennsylvania. The process, shown in Fig. 12, is a variation of solid free-form (SFF) fabrication. It can produce functional metal prototypes directly from CAD data. Successive layers of hard metal are deposited on a platform without masking for the direct manufacture of rugged functional parts; it is an alternative to conventional manufacturing without the added cost of specialized tooling needed for full-scale production.

The nearly complete structure of primary and sacrificial layers is shown at the deposition station (Fig. 12a). Software derived from CAD drawings of the prototype determines the number and thickness of layers and how they are to be deposited. The first primary metal layer is sprayed as hot metal droplets by an additive process called *microcasting* at the deposition station. This first layer and all of those that follow are deposited slightly oversized so that the outer edges of each layer can be machined to the specified shape and dimensions. After each layer is deposited it is moved to the shaping station (Fig. 12b) where a computer-controlled milling machine or grinder removes excess metal. Next, the work is moved to the stress-relief station (Fig. 12c) where it is shot-peened to relieve stresses that have built up in the layer. The work is then transferred back to the deposition station (Fig. 12a) for the deposition of the next layer of primary metal and the sacrificial metal that will support and protect any overhanging layers. This three-step SDM build cycle is repeated until the prototype is complete.

The metal droplets that form each layer retain their heat long enough to remelt the previous layer on impact and form a strong metal bond between them; shot-peening prevents warping. When the build process is complete, the sacrificial metal is etched away with acid, and any final grinding and polishing is performed. A successful combination of metals in SDM has been stainless steel for the prototype and copper for the sacrificial support layers.

The SDM Laboratory at Carnegie Mellon University (CMU) has investigated many techniques including thermal spraying and plasma or laser welding for depositing high-quality metals before it decided on microcasting. It is a compromise between those two techniques that provides better results than either of the other techniques. It was found that the larger metal droplets (1 to 3 mm diameter) formed by microcasting retain their heat longer than the 50 μm droplets of conventional thermal spraying. SDM can form complex shaped parts rapidly while also permitting both the fabrication of multi-material structures and the embedding of prefabricated components within the parts as they are shaped.

CMU's SDM laboratory has produced custom-made functional mechanical parts, and it has embedded prefabricated mechanical parts, electronic components, electronic circuits, and sensors in the metal layers during the SDM process. It has also made custom tools such as injection molds with internal cooling pipes and metal heatsinks with embedded copper pipes for heat redistribution.

Mold Shape Deposition Manufacturing (MSDM)

Both the Rapid Prototyping Laboratory at Stanford University, Palo Alto, California, and the Robotics Institute at Carnegie Mellon University have Shape-Deposition Manufacturing (SDM) laboratories, but the Stanford Laboratory developed its own version of SDM, called Mold SDM or MSDM. It is also intended for building layered molds for casting ceramics and polymers.

The MSDM process, shown in Fig. 13, uses wax to form the molds. The wax has the same function in MSDM as the sacrificial support metal in SDM; it occupies and supports the mold cavity as it is formed. Water-soluble photopolymer sacrificial support material in MSDM corresponds to the primary metal in

Fig. 13 Mold Shape Deposition Manufacturing (MSDM): The formation of a ceramic part starts at station (a) where wax for casting a mold and water-soluble photopolymer to support the mold cavity are deposited to form the first (bottom) layer. This step is followed by curing or hardening the photopolymer by exposure to ultraviolet lamps. Successive polymer and wax layers are laid down and cured to complete the mold. The polymer support material is removed from the cavity at station (b) by soaking it in water. This leaves a standalone wax structure at station (c) where a gelcasting ceramic slurry is poured into the wax cavity, forming a "green" unfired ceramic mold. The wax is then melted away at station (d), releasing the "green" unfired ceramic mold. Finally, the ceramic mold is furnace-fired at station (e) to harden it. After firing, the vents and sprues are removed from the finished ceramic mold.

SDM deposited to form the finished prototype. It is worth noting that no machining is performed in the MSDM process.

As in other RP technologies, the MSDM process starts with the conversion of data derived from a CAD drawing of the mold into software that determines both the optimum number of layers of wax and sacrificial support material and how they are to be deposited. Layer thickness depends on the complexity and contours of the mold to be built.

The structure is built one layer at a time under computer control. With the exception of the first layer of wax, both mold wax and support photopolymers are deposited in the same layer (Fig. 13a). The polymer is then cured by ultraviolet radiation. The built-up structure is then moved to a station (Fig. 13b) where the sacrificial photopolymer is removed by dissolving it in water, leaving the wax mold cavity empty. The gelcasting ceramic slurry is then poured into the wax mold at station (Fig. 13c) to form the ceramic part. The slurry is allowed to cure into a "green" semi-rigid, clay-like state. Next, the wax mold is melted away, releasing the "green" ceramic part, as shown in (Fig. 13d) for furnace firing. After firing, the last step is the removal of the vents and sprues, as shown in (Fig. 13e). The kinds of materials used in MSDM have been expanded by making parts from a variety of polymer materials, and it has also been used to make pre-assembled mechanisms from both polymer and ceramic materials.

Robocasting Ceramics

An RP method called *robocasting* can form ceramic parts under computer control without molds or machining. Developed at Sandia National Laboratories in Albuquerque, New Mexico, the process can also mix metal with the ceramic slurry to form graded hybrid parts that resist cracking because of different thermal expansion rates. The Sandia engineer who developed the process reports that robocast ceramics are denser than ceramic parts made by other RP processes. He added that the ability to make reliable metal-ceramic parts able to withstand very high temperatures is particularly useful in the manufacture of engine components.

A dense ceramic part, free-formed by robocasting, can be dried and baked in less than 24 hours. Mistakes in the part design or improvements can be made quickly. If a complicated ceramic part is to be made by the standard dry-pressing method, the ceramic powder first must be compacted into a solid form or *billet*, and then the billet must be sculpted into its final shape by costly machining. Intricate ceramic parts are also being made by other techniques. These include slipcasting, gelcasting, and injection molding, but they require the design and manufacture of molds prior to fabrication.

Although still in the laboratory development stage, robocasting holds promise for producing ceramic parts in high quantities. The success of robocasting will depend on the development of ceramic slurries that contain more solid than liquid but are still able to flow. The ceramic slurry is dispensed by a computer-controlled syringe that remains in a fixed position while the platform on which the slurry is deposited moves. The slurry must dry rapidly into a semisolid state so that the next layer can be accepted. The high solid content and the tailored properties of the slurry minimize shrinkage and permit a layer to dry within 10 to 15 seconds of being deposited.

After the part is formed by layering and completely dried, it is sintered by firing to bond the particles at temperatures typically from 1000 to 1700°C for about two hours. Where a design requirement calls for ceramics and metal combined in the same device, joining them together can be difficult because the differences in their thermal expansion rates can cause cracking at the material's interface. Robocasting makes it possible to shift gradually from one material to another, thus spreading the stress evenly to yield a more stable joint. Robocasting also allows the discrete placement of certain materials between the layers that will evaporate or burn away during the sintering process, thereby permitting structures such as cooling channels to be formed within the part.

Resources

The following Web sites provided information included in this chapter:

Carnegie Mellon's The Robotics Institute
www.ri.cmu.edu/

Rapid Prototyping Home Page
http://home.utah.edu/~asn8200/rapid.html

Delft Spline Systems
www.spline.nl

3D Systems Corp.
www.3dsystems.com

EOS Electro Optical Systems
www.eos.info/en

Objet Geometries Inc.
www.objet.com

Optomec Design Company
www.optomec.com

The POM Group Inc.
www.pomgroup.com

Solidscape Inc.
www.solid-scape.com

Soligen Corporation
www.soligen.com

Stratays, Inc.
www.stratasys.com

Z Corporation
www.zcorp.com

CHAPTER 19
NEW DIRECTIONS IN MECHANICAL ENGINEERING

THE ROLE OF MICROTECHNOLOGY IN MECHANICAL ENGINEERING

The role of mechanical engineering has expanded dramatically over the past half century as a result of the introduction of electronics and computer science into this engineering discipline. While the basic physical principals governing mechanics and mechanical design have not changed, engineering practice has. Solid-state digital electronics and microprocessors have obsoleted many traditional instruments and devices: examples include the slide rule, mechanical timers, mechanical numerical displays, and mechanical computing mechanisms. Computers have been responsible for introducing radical changes in mechanical engineering: these include computer-aided design (CAD), computer-aided manufacturing (CAM), computer-based simulation, rapid prototyping (RP), and, of course, the computer's ability to perform complex calculations faster and more accurately than previous calculators.

Together these technologies have increased the kinds of projects that engage mechanical engineers (MEs), and they have introduced new and different educational requirements for entering the profession. While MEs continue to design mechanisms, machines, and mechanical devices, they have recently become participants in such new activities as electronic packaging, the mechanical design of computer hard, CD, and DVD drives, robotics, mechatronics, and microtechnology. They are also making contributions to nanotechnology, a subject which is merging the engineering disciplines with the physical and biological sciences.

MEs now work on megastructures and megamachines the size of aircraft carriers, cruise ships, and deep-ocean oil rigs. By contrast, they are also working on microelectromechanical systems (MEMS), built to micrometer scale (millionths of a meter) that are so small they must be viewed under a microscope. MEMS now include tiny motors, chain drives, torque converters, transmissions, accelerometers, pressure sensors, gyroscopes, and gear reduction units.

Microtechnology Today

MEMS fabrication technology was acquired from the manufacturing processes for large-scale integrated circuits such as microprocessors and memories. This technology calls for sequential masking and chemical etching steps that have made possible the sculpting of dynamic components from multiple layers of silicon. Because MEMS are capable of performing work, they qualify as machines. However, for many reasons both technical and economic, MEMS have yet to live up to the early predictions that they would all soon become mass-produced. However, some MEMS have jumped economic and technical hurdles to achieve success in the marketplace because they created a demand that led to high production levels and, as a result, significant reductions in unit price.

Among the more popular MEMS are acceleration sensors that can detect forces unleashed in vehicular collisions and trigger the deployment of airbags. More than 100 million of these sensors have been installed in motor vehicles worldwide. Analog Devices Inc., a major supplier of these devices, uses its standard, high-volume integrated circuit (IC) manufacturing technology called iMEMS to surface micromachine the precisely patterned sensor structures on a silicon wafer. The company's smallest acceleration sensor, in an aspirin-size $0.2 \times 0.2 \times 0.1$ in. ($5 \times 5 \times 2$ mm) package, performs two-axis motion sensing. A single monolithic chip provides a digital output with low power drain and self-test features.

Analog Devices is offering programmable low-power gyroscopes made with its iMEMS technology. ADIS16250 gyros are complete integrated systems in single compact packages for measuring the angular rate of rotation. Applications include their use for platform stabilization, motion control, navigation, and robotics. Each gyro IC includes two identical MEMS polysilicon sensors called *resonators*. They are operated antiphase to give signals in opposite directions so that a differential output can be obtained. Each resonator contains a mass suspended by springs within an inner dither frame. Other springs suspend the dither frame inside a substrate frame. The inner springs tethering the mass allow it to oscillate only in one direction while the outer springs supporting the dither frame restrict its movement to one direction.

When electrostatically driven at their resonant frequencies, the masses produce Coriolis forces proportional to system movements around a single axis. Capacitors formed by movable sensing fingers on the dither frames interdigitated with fixed fingers on the substrate frame respond to these movements with instantaneous changes in capacitance values directly related to the angular rate of rotation of the gyro. The differential between the instantaneous values of these resonator capacitors is used to measure angular rate. This technique cancels the effects of environmental shock and vibration. The differential signal is fed to a series of on-chip electronic gain and demodulation stages to produce the electrical rate-signal output. ADIS16250 gyros are housed in standard IC packages measuring $0.4 \times 0.4 \times 0.2$ in. ($11 \times 11 \times 5.5$ mm), and they are powered from 4.75 to 5.25 V DC supplies.

Another successful MEMS is the more complex digital micromirror device (DMD) shown in Fig. 1. Invented at Texas Instruments, it is an electromechanical transducer now known as

Fig. 1   Each mirror in a digital micromirror device (DMD) array of mirrors is mounted on a diagonal axis. When switched, it tilts one way to reflect light on a screen to form an "on" pixel or the other way to form a dark "off" pixel. A DMD with more than 2 million mirrors can form high-definition colored images with a resolution of 1920 × 1080.

a DLP chip (for Digital Light Processing); more than five million of these chips have been shipped. With the ability to translate video signals into multicolored video images, DLP chips are now integral components in large-screen digital TVs, projectors for home and business video presentations, and projection systems for displaying movies on theater-size screens.

Each DLP chip contains an array of two million hinge-mounted microscopic mirrors that have been formed on a silicon chip no larger than a finger nail. The micromirrors are sculpted by MEMS photolithographic masking and chemical etching processes. Each mirror measures less than one-fifth the width of a human hair (0.0004 in. or approximately 10 μm). The mirrors are hinged so that they can be deflected or switched from reflecting to nonreflecting positions thousands of times per second. This mechanical DLP chip performs all of the electronic and optical functions necessary to translate the input digital video signals into a sequence of mirror deflections, making it possible for them to form dynamic video images. Each mirror corresponds to a single pixel in a projected or displayed image. When synchronized with a light source and a multicolor projection lens system, the DLP's mirrors reflect full color images onto a screen or other suitable flat surface.

The use of MEMS for direct tire-pressure sensing is another growing market. Millions of indirect automotive tire-pressure monitoring systems have already been installed in high-end vehicles as optional accessories, but the demand for these systems could skyrocket if they are to become requirements for all motor vehicles as a safety feature. However, indirect systems are costly and their mandatory installation on lower-priced vehicles would raise their prices significantly. It is expected that direct monitoring systems, which include MEMS as pressure sensors, will result in lower priced and higher performing tire-pressure monitoring modules.

MEMS sensors would directly measure air pressure in each tire. They would be based on either a piezoresistive effect that would provide a voltage output when stressed by tire pressure changes or capacitive versions whose capacitance values would change with tire pressure. These MEMS would be installed in electronic modules positioned on each wheel rim of a vehicle. The modules might also include temperature and voltage sensors, an accelerometer, a microcontroller chip, a radio transmitter, an antenna, and a battery. The conditioned pressure data signals would first be transmitted wirelessly to a central receiver in each vehicle. The data from the receiver could then be displayed either as a warning light if the tire pressure is too low or continuously as a digital readout of actual pressure. The complete module is expected to weigh between 1.1 and 1.4 oz (30 and 40 grams).

Microtechnology Tomorrow

It has been estimated that about two-thirds of the components in cell phones and radios are filters, and they are mechanical devices. This, according to researchers at Stanford University, suggests that sensors, computers, and communications gear can all be highly miniaturized. The scientists have demonstrated that small, unhardened microscopic radios can transmit and receive messages while in Earth orbit. Conventional radios on satellites must be hardened against radiation, adding to their high cost. However, they could be replaced by large networks of low-cost expendable micron-scale radios. The scientists believe that clusters of expendable radios could be mounted on a spacecraft and, although some might be burned out by exposure to damaging radiation, the survivors could continue to transmit and receive.

Dr. Albert Pisano, chairmen of the Department of Mechanical Engineering at the University of California at Berkeley, believes that a market for tiny transceivers could emerge within 10 years. He also pointed out that most of the components of a typical radio transceiver are passive discrete devices, noting that MEMS devices can replace such discrete components as filters. "Since the MEMS components can be integrated with the transistor electronics, they can reduce the total parts count, the overall size, the cost of the radio," he asserted.

Professor Pisano said MEMS have a role to play in macrotechnology. He noted that with the use of MEMS sensors and actuators, jet engines, for example, could be made more reliable. He explained that heat conductivity can be adjusted and strain can be resolved to very low levels. MEMS sensors are located in jet engines at strategic sites to measure such factors as pressure, strain, vibration, temperature, and acoustic output. According to Dr. Pisano, microstrain gages have over 10,000 times the sensitivity of conventional metal-foil strain gages. MEMS extensometers, he noted, were capable of reading absolute strain with no drift. "With a gage length of 1 mm, this device can resolve strains to help with condition-based maintenance and structural health and monitoring," he reported. The professor went on to say that telemetry systems with sensors and transceivers containing MEMS components could fill a need for improved monitoring of power lines, freeways, and bridges by acting as early warning systems by reporting extreme stresses or strains in structures before they fail.

MEMS devices, he said, could perform the functions of microphones, optical communicators, actuators, uncooled infrared sensors, and inertial measurement units; they could also function as mass data storage devices. Microminiature communicators could also be powered by MEMS-based power generators and energy converters.

The California professor reported that subminiature batteries could be developed for converting thermal energy into electricity. He explained that the use of thermal energy at a small scale is practical, adding that existing tiny rotary engines measuring only microns in diameter would have higher efficiency than existing lithium-ion batteries which must be electrically charged, thus consuming energy. The microminiature thermal motors, powered by methane, could convert the gas into electrical energy. Dr. Pisano agrees that MEMS have not yet reached their early expectations because of the difficulties encountered in manufacturing and packaging them in quantity, but he also believes that progress in nanotechnology will enhance the long-term prospects for MEMS.

MICROMACHINES OPEN A NEW FRONTIER FOR MACHINE DESIGN

The technology for fabricating microelectromechanical systems (MEMS) micron-scale motors, valves, transducers, accelerometers, and other devices, was derived from the proven photolithographic and chemical etching processes used to fabricate silicon integrated circuits. This technology has opened a new field for the mechanical engineer that depends on the application of design rules and manufacturing techniques which differ radically from traditional mechanical engineering practice; they call for the use of materials and chemicals that, until recently, would have been unfamiliar to most mechanical engineers. In MEMS manufacture, silicon replaces steel, brass, aluminum, and other more familiar materials, and chemical etching removes excess material rather than milling, turning, or boring.

MEMS are so small that they can be clearly seen only when viewed under an electron microscope, and the normal laws of physics do not necessarily apply in powering devices this small. For example, MEMS motors are typically driven by electrostatic attraction rather than electromagnetism because when mechanisms are scaled down to this size, electromagnetism is too weak to be effective. Mechanical engineers working on MEMS have ventured into a dimensional realm that was formerly the province of microbiologists, atomic physicists, and microcircuit designers.

Among the more remarkable examples of MEMS made in laboratories are a microminiature electric-powered vehicle that can be parked on a pinhead, electric motors so small that they can easily fit inside the eye of a needle, and pumps and gear trains the size of grains of salt. Far from novelties that serve only to demonstrate the feasibility of the technology, many MEMS are now being mass-produced for automotive, electronic, optical, and biomedical applications. Biomedical researchers, for example, have been developing microminiature medicine dispensers that will reach the site of disease within the human body by traveling through arteries or veins. These natural conduits are also being considered for transporting microscale instruments to precise internal sites where they can perform remote-controlled microsurgery.

The practical MEMS devices now being manufactured in quantity include accelerometers, digital micromirror devices (DMDs), tire-pressure sensors, and modulators. The accelerometers trigger vehicular air bags; the DMDs project television and other video images on large screens; the tire-pressure monitors are included in vehicular tire-pressure monitoring systems; and the modulators convert electronic signals into optical signals in fiberoptic communication systems.

The Microactuators

The rotary micromotor, shown sectioned in Fig. 1, is an example of a microactuator driven by static electricity rather than electric current. Some of the experimental motors made to date have diameters of 0.1 to 0.2 mm and heights of 4 to 6 μm. The rotor, formed with spokes in a "rising sun" pattern, rests on a knife-edge bushing that minimizes frictional contact with the motor's base substrate; this leaves it free to rotate around a central nail-shaped hub which prevents it from being displaced vertically. Insulating slots separate the stator into 20 electrically isolated commutator segments. The conductive inner surfaces of the commutator segments and outer surfaces of the rotor spokes form a rotating capacitor that responds to electrostatic forces produced by

Fig. 1　A cross-section view of a typical micromotor that is driven electrostatically rather than electromagnetically.

voltages applied sequentially to the stator segments; these forces cause the rotor to spin at high speeds.

Rotors of statically powered MEMS motors can reach speeds in excess of 10,000 rpm when driven by excitation voltages of 30 to 40 V. Some of these miniscule motors have been in continuous operation for 150 hours. Researchers in laboratories at the University of California at Berkeley and at Massachusetts Institute of Technology (MIT) have built these tiny machines.

Other successful microactuators that have been built include microvalves and micropumps; a section view of a typical microvalve is shown in Fig. 2. The diaphragm or control element of this microvalve flexes in a direction that is perpendicular to the valve seat in the base substrate. It can be moved by an embedded

Fig. 2　A cross-section view of a typical microvalve. The diaphragm moves perpendicular to its base substrate. Diaphragms can be moved by an embedded piezoelectric film, by electrostatic forces, or by thermal expansion.

Fig. 3 This **linear resonator** consists of a pair of folded beams that are set in vibrational motion in the X direction by an electrostatically driven comb structure. Lateral or Y-direction motion is restrained by the geometry of the folded beams.

piezoelectric film, electrostatic force, or thermal expansion. There are applications for these microminiature valves and similar pumps in biomedical research because they are orders of magnitude smaller than conventional biomedical pumps and valves, and they require very little energy to drive them.

The linear resonator shown in Fig. 3 is another form of microactuator that is driven by electrostatic forces, but its operation depends on different principles than those that apply to the motor. The resonator consists of two major components: a comb drive and a folded-beam structure. The folded-beam structure includes a set of fingers or digits on its left end that is interleaved with a similar set of digits rigidly mounted on a pedestal at the left end of the resonator; together both sets of digits form the comb drive. However, the digits on the folded-beam structure are free to oscillate in the X direction because the structure includes twin-folded beams that are flexible, although the structure itself is rigidly mounted on two anchors. (The pedestal and anchors, grown on the same substrate, are identified by the black lines in the figure.)

The comb drive is set in oscillation electrostatically in the X direction. The flexible folded beams are dimensioned to resonate at a specific frequency when driven by electrostatic charges placed on the comb drive's digits, which act as capacitors. Both folded beams resonate simultaneously, but only in the X direction because motion in the lateral or Y direction is constrained by the geometries of the folded beams.

Microaccelerometers are microactuators that respond to external forces rather than an embedded piezoelectric film, electrostatic forces, or thermal expansion. Three different microminiature capacitive acceleration sensors are shown in Fig. 4. They are identified as (a) cantilever, (b) torsion-bar suspension, and (c) seismic mass suspended from a central pillar. The simple cantilever structure offers the highest sensitivity for a given size of suspension arm. Consequently, it can be made smaller for a desired sensitivity than any of the other configurations.

Capacitive sensing techniques require an AC voltage across the capacitor being measured. This, in turn, produces an electrostatic field which generates an attractive force between the capacitor plates. As the seismic mass deflects, the change in electrostatic field can be measured; with proper signal conditioning, the output will be proportional to acceleration. These accelerometers can operate as switches in either open-loop or closed-loop systems. The selection of the optimum structure for a specific application takes into account such key factors as sensitivity, stability, material fatigue, shock resistance, mass damping, output linearity, and temperature range, but other properties might also

(a)

(b)

(c)

Fig. 4 **Microminiature capacitive acceleration sensors:** (a) cantilever, (b) torsion-bar suspension, and (c) seismic mass suspended from a central pillar.

be considered. MEMS capacitive accelerometers are now widely used in motor vehicles as sensors for detecting collisions and triggering airbag deployment.

Materials

At present, silicon remains the most popular material for fabricating MEMS because it has the most favorable properties of any of the materials tried for fabricating complex microminiature devices. The use of silicon also made it possible to fabricate integrated circuitry for signal processing on the same silicon substrate. However, some MEMS have also included parts made of aluminum and diamonds. The successful design and manufacture of billions of integrated circuits over the past 35 years has resulted in an extensive body of knowledge about the properties of silicon—how it can be grown, how its structure can be altered, how it can be milled chemically, and how slices of silicon can be permanently bonded.

Silicon is a very strong material with a modulus of elasticity that closely matches steel. It exceeds stainless steel in yield strength and aluminum in strength-to-weight ratio. In addition, it exhibits high thermal conductivity and a low thermal expansion coefficient. Because it lacks mechanical hysteresis, it is a near-perfect material for fabricating sensors and transducers. Moreover, silicon's sensitivity to stress, strain, and temperature lends itself to the fabrication of sensors that can easily communicate with electronic circuitry on the same substrate or chip for the transmission of electrical signals. However, silicon must be protected by encapsulation or hermetic sealing because it is susceptible to deterioration if exposed to air or moisture.

In building MEMS, silicon is chemically micromachined or etched into a wide variety of shapes rather than being machined by traditional cutting tools. Silicon, whether in the form of polysilicon or silicon nitride, and aluminum can be etched into many different shapes and contours in batch processes. In the micromachining process, mechanical structures are sculpted from a silicon wafer by selectively etching away sacrificial supporting layers or structures.

Masks produced by the photolithographic methods are used at various stages in the manufacturing process for selective etching of the silicon. They permit the formation of precise contours such as gear teeth, combs, beams, cantilever arms, and seismic masses. The furnace diffusion of chemical "dopants" in gaseous form into the silicon can alter its chemical makeup and change its electrical characteristics. Epitaxy is the process for growing multiple surface layers of material on the basic substrate, and deposition is the process for plating silicon surfaces with materials such as gold, silver, aluminum, or copper.

Power to the Micromachines

In theory, MEMS could be driven by any of four different forces: electromagnetic, thermal expansion, electrostatic, and piezoelectric effect. The choice of actuation method is generally determined by the end use of the device and its performance requirements. As stated earlier, electromagnetic forces are too weak to power MEMS actuators when scaled down to MEMS size. Moreover, thermal expansion is usually ruled out as a driving force for MEMS because of the excessive power required to concentrate enough heat in a small area to drive microminiature parts made of dissimilar materials. This leaves electrostatic force and piezoelectric effect as the only practical forces for powering MEMS.

Electrostatic Forces

Electrostatic force is attractive for actuating MEMS because, unlike magnetic force, it can be scaled down effectively to micro size. The basic requirement for the use of electrostatic forces for driving MEMS is the presence of two electrically conductive surfaces within the MEMS that will act as opposing capacitor plates. The electrostatic force applied is then directly proportional to the product of the square of the voltage across the two plates and inversely proportional to the square of the distance between the plates.

In the MEMS motor shown in Fig. 1, capacitors are formed by the end surfaces of the rotor spokes and the inner walls of the insulated stator segments. To drive this motor, voltage is applied sequentially to the stator segments, switching them on and off. Each rotor spoke is attracted to the nearest powered stator segment as voltage is switched to them in a rotating sequence, causing the whole rotor to spin as it follows the polarity changes. In

this way, the rotor completes a revolution for multiple polarity changes in the stator elements.

The gaps separating the outer ends of the rotor spokes and the inner surfaces of the stator segments act as capacitor plates. However, the gaps are not likely to be uniform because of the microminiature size of this motor. The result is a variation in the electrostatic force with respect to time as the rotor spins, making the electrostatic force a nonlinear function of applied voltage. A MEMS motor can tolerate some variation in electrostatic force, but it can fail or stall for two reasons: if the gap between the opposing surfaces of stator and rotor is not concentric or smooth or if the rotor bearing surfaces are not smooth enough to prevent friction buildup.

The drawback of most surface-micromachined MEMS motors is that both rotor and stator are so thin that their opposing surface areas are too small to provide enough change in capacitance to sustain rotor rotation. One solution to this problem is the use the LIGA process to fabricate the motor because it can form thicker stators and rotors than can be achieved with surface micromachining. (The LIGA process is explained later in this chapter.)

Actuators have also been made in the form of vibrating microstructures with flexible suspensions. An example known as a linear resonator is shown in Fig. 3. It uses a different design to maximize the capacitance change so it can produce motion with larger amplitude by exploiting the advantage of the classical parallel-plate capacitor formula because only attractive forces are generated. The equation for energy stored in an electrostatic comb is:

$$E = \frac{CV^2}{2}$$

where E is the energy stored, C is the capacitance, and V is the voltage across the capacitor.

Surface-micromachined linear resonators with comb drives, as shown in Fig. 3, have multiple fingers or digits. When a voltage is applied, an attractive force is developed between the interleaved digits which then converge. The increase in capacitance is proportional to the number of fingers in the drive: many digits are required to generate enough force to maximize performance. Because the direction of motion of the digits, acting as capacitor plates in the electrostatic comb drive, is parallel along their length, the effective plate area with respect to spacing between the digits' plates remains constant. Consequently, capacitance with respect to the direction of motion is linear, and the induced force in the X direction is directly proportional to the square of the voltage applied across the plates. Comb-drive structures have been driven to deflect by as much as one-quarter of the comb finger length with DC voltages of 20 to 40 V.

The performance of a linear resonator with a comb drive will be degraded if the lateral gaps between the digits are not equal on both sides or if the digits are not straight and parallel. Any of those conditions would cause the digits to diverge at right angles to their intended direction of motion and, as a result, wedge together until the voltage is turned off. However, if they collided with sufficient force they might remain stuck permanently, destroying the linear resonator.

Piezoelectric Films

Microminiature transducers made as rigid beams and diaphragms with cores of polycrystalline zinc-oxide (ZnO) piezoelectric film can change their shape with the application of a voltage. A beam with a central piezoelectric layer of insulated polycrystalline ZnO several micrometers thick is illustrated in Fig. 5. This layer is then insulated on both sides and sandwiched between two conductive electrodes to form a rigid structure. When voltage is applied between the two external electrodes, the piezoelectrically induced stress in the ZnO film causes the structure to deflect.

Fig. 5 A microminiature piezoelectric transducer is made as an insulated layer of polycrystalline zinc oxide (ZnO) sandwiched between two conductive electrodes to form a rigid bimetallic structure.

The converse of the piezoelectric effect can be obtained when the beam functions as a strain sensor; it converts strain applied to the beam into electrical signals that are proportional to the strain.

Bulk Micromachining

Bulk micromachining is a method for fabricating MEMS from a single wafer of silicon by chemically etching away silicon to sculpt the desired microstructure. The bulk silicon is obtained by slicing large, cylindrical, single crystals of purified silicon into thin wafers. Micromachining is done by removing the excess silicon from the wafers with a dry or wet chemical at a removal rate that depends on the orientation of the silicon crystal.

The etching process is controlled by etch masks that define the areas for material removal. The mask, produced by photolithography, is placed over a silicon wafer that has been coated with photoresist, a substance which becomes soluble after being exposed to ultraviolet (UV) energy. The photoresist is then chemically removed only from those areas of the wafer that have been exposed to UV through transparent windows in the mask. Only those areas of the wafer from which the photoresist has been removed to expose bare silicon can be chemically etched. The depth of silicon removal is a function of the time the etching process is allowed to continue.

Bulk micromachining is a process that includes methods for fusion bonding silicon substrates to form precise three-dimensional (3D) structures such as micropumps and microvalves. Two or more etched wafers can be bonded by pressing them together and annealing them to form a permanent 3D microstructure. This is how an internal or reentrant cavity such as the one shown in the cross-section view of a microvalve (Fig. 2) is formed.

Surface Micromachining

Surface micromachining is a manufacturing process that permits multilayer MEMS to be built. Initially developed for fabricating large-scale silicon integrated circuits, it calls for the deposition of multiple layers of various permanent and sacrificial materials on the surface of a silicon wafer in a sequence required to form complex structures. Free-standing movable structures such as microminiature motor rotors, gears, or chains can be sculpted chemically by etching away sacrificial layers. As in bulk micromachining, the process starts with a thin wafer of crystalline silicon which becomes the substrate. Structural material, usually polysilicon, is deposited on underlying layers of sacrificial materials such as silicon dioxide, silicon nitride, or phosphosilicate glass.

The etching of layers of material in surface micromachining is controlled by the same kinds of masks used in bulk micromachining, and the chemical etching techniques are the same as those used in that process. Each structural layer is sculpted to conform to the contours of the many different masks required. Structures that are intended to move or rotate are released from sacrificial layers by highly selective chemical etchants such as hydrofluoric acid. Etching can also remove sacrificial material deposited around the shafts or hubs patterned with the structure. This process provides enough clearance around the hub and between adjacent parts to permit gears or rotors to rotate freely; the hubs are also capped to restrict vertical movement of the gears or rotors as they rotate.

A MEMS motor, such as the one shown in Fig. 1, is surface micromachined in a series of deposition and masking steps in which alternate layers of permanent silicon and sacrificial material are deposited. After the sacrificial material is chemically removed, the motor structure—stator, rotor, and central capped shaft—is complete. The rotor is then free to rotate around the hub. During surface micromachining, electrically conductive pads and conductive paths are deposited by selective plating of gold or other suitable metal on the substrate to connect the power source to the commutator segments of the stator.

MULTILEVEL FABRICATION PERMITS MORE COMPLEX AND FUNCTIONAL MEMS

Researchers at Sandia National Laboratories, Albuquerque, New Mexico, have developed two surface micromachining processes for fabricating multilevel MEMS (microelectromechanical systems) from polysilicon that are more complex and functional than those made from two- and three-level processes. The processes are SUMMiT Technology, a four-level process in which one ground or electrical interconnect plane and three mechanical layers can be micromachined, and SUMMiT V Technology, a similar five-level process except that four mechanical layers can be micromachined. Sandia offers this technology under license agreement to qualified commercial IC producers.

According to Sandia researchers, polycrystalline silicon (also called polysilicon or poly) is an ideal material for making the microscopic mechanical systems. It is stronger than steel, with a strength of 2 to 3 GPa (assuming no surface flaws), whereas steel has a strength of 200 MPa to 1 GPa (depending on how it is processed). Also, polysilicon is extremely flexible, with a maximum strain before fracture of approximately 0.5%, and it does not readily fatigue.

Years of experience in working with polysilicon have been gained by commercial manufacturers of large-scale CMOS integrated circuits chips because it is used to form the gate structures of most CMOS transistors. Consequently, MEMS can be produced in large volumes at low cost in IC manufacturing facilities with standard production equipment and tools. The Sandia researchers report that because of these advantages, polysilicon surface micromachining is being pursued by many MEMS fabrication facilities.

The complexity of MEMS devices made from polysilicon is limited by the number of mechanical layers that can be deposited. For example, the simplest actuating comb drives can be made with one ground or electrical plane and one mechanical layer in a two-level process, but a three-level process with two mechanical layers permits micromachining mechanisms such as gears that rotate on hubs or movable optical mirror arrays. A four-level process such as SUMMiT permits mechanical linkages to be formed that connect actuator drives to gear trains. As a result, it is expected that entirely new kinds of complex and sophisticated micromachines will be fabricated with the five-level process.

According to the Sandia scientists, the primary difficulties encountered in forming the extra polysilicon layers for surface micromachining the more complex devices are residual film stress and device topography. The film stress can cause the mechanical layers to bow from the required flatness. This can cause the mechanism to function poorly or even prevent it from working. The scientists report that this has even been a problem in the fabrication of MEMS with only two mechanical layers.

To surmount the bowing problem, Sandia has developed a proprietary process for holding stress levels to values typically less than 5 MPa, thus permitting the successful fabrication and operation of two meshing gears, whose diameters are as large as 2000 μm.

The intricacies of device topography that make it difficult to pattern and etch successive polysilicon layers restrict the complexities of the devices that can be built successfully. Sandia has minimized that problem by developing a proprietary chemical-mechanical polishing (CMP) process called "planarizing" for forming truly flat top layers on the polysilicon. Because CMP is now so widely used in integrated circuit chip manufacture, it will allow MEMS to be batch fabricated by the SUMMiT processes using standard commercial IC fabrication equipment.

ELECTRON MICROSCOPES: KEY TOOLS IN MICRO- AND NANOTECHNOLOGY

Electron microscopes have long been considered to be essential tools for research in medicine and the biological and physical sciences. However, they have now become essential tools for research and development in materials science and mechanical engineering because of the increasing importance of microtechnology and nanotechnology. There are actually many different types of electron microscopes, and they all work in different ways. The four most familiar versions are the *transmission electron microscope* (TEM), the *scanning electron microscope* (SEM), the *scanning tunneling microscopes* (STM), and the *atomic force microscope* (AFM).

The transmission electron microscope (TEM), Fig. 1, is based on the conventional optical microscope except that electron beams replace light beams. The TEM was the first electron microscope and it was invented in 1931. In common with the optical microscope, specimens for the TEM must be prepared as thin slices that

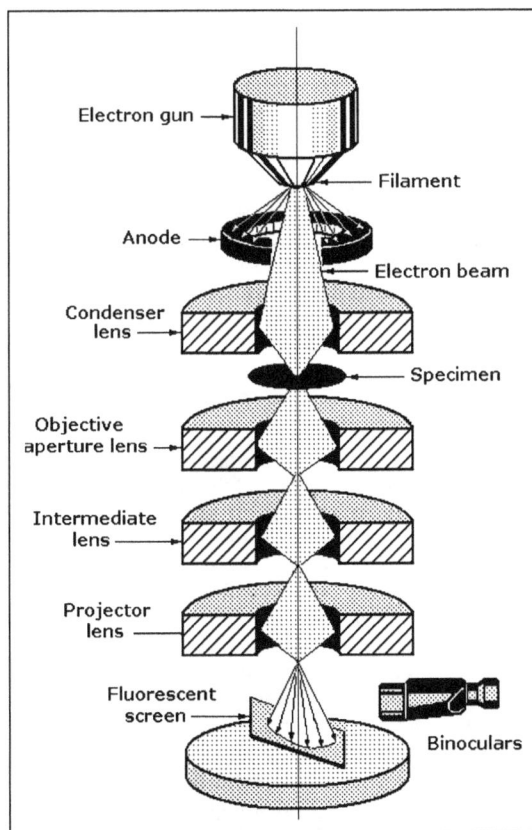

Fig. 1 The transmission electron microscope (TEM) was the first electron microscope to be developed. It works on the same principle as the light transmission microscope except that it focuses a beam of electrons instead of a beam of light on a sample thin enough for the electrons to pass through it, making it possible to "see" through far smaller objects than would be possible with a light transmission microscope. Binoculars permit the image to be seen on a fluorescent screen or the image can be transformed electronically into a video image for display.

will be penetrated by the electrons in the beam. The specimen is placed in the vacuum chamber between the *condensers* and *objective aperture lenses*, electromagnetic coils near the middle of the microscope. The chamber is then pumped down to achieve a vacuum necessary for the transmission of electron beams.

The electrons are emitted by an electron gun, commonly fitted with a tungsten filament cathode as its electron source. The high-voltage electron beam is accelerated by an anode, typically at 50,000 to several million volts with respect to the cathode. The electron beam is focused by electrostatic and electromagnetic lenses. Electrons, normally considered to be particles, can behave like waves just as light waves can behave like particles due to the physical principle of wave-particle duality. The faster the electrons travel, the smaller the waves they form and the more detailed the images they produce. When they reach top maximum speed, some electrons penetrate the specimen and exit out the other side while others are scattered out of the beam.

The electrons merging from the specimen form an image that carries information about the structure of the specimen. This image is then magnified by the objective lens system of the microscope. The information is viewed by projecting the magnified electron image onto a fluorescent viewing screen coated with a phosphor or scintillator material such as zinc sulfide.

The image can be seen directly with binoculars or photographically recorded by exposing a photographic film or plate directly to the electron beam. Alternatively, a high-resolution phosphor may be coupled by means of a lens optical system or a fiberoptic waveguide to the sensor of a charge-coupled device (CCD) camera. The image detected by the CCD can be displayed on a monitor or computer. TEMs are the most powerful electron microscopes because they can effectively magnify the specimen by a million times or more. This permits objects as small as 1 nm to be seen.

The scanning electron microscope (SEM), Fig. 2, similar in many ways to the TEM, was invented four years later in 1935. It was designed to form images of the surfaces of microscopic objects such as microelectromechanical devices (MEMS). As in a TEM, the specimen is first positioned within the chamber (in this case at the bottom) before the chamber is pumped down to obtain a vacuum. The top of a chamber also contains a powerful electron gun that shoots an electron beam down at the specimen.

Unlike the TEM, where electrons of the high-voltage beam carry the image of the specimen, the electron beam of the SEM never carries a complete image of the specimen. A series of electromagnetic lenses moves the beam back and forth, *raster scanning* it slowly and systematically across the specimen's surface. Rather than traveling through the specimen, the electrons in the beam bounce off of it. These secondary electrons reflected from the specimen are directed at a secondary electron detector and the backscattered electrons are directed at a backscatter electron detector. Signals from both of these detectors are combined with signals from the scanning coils to form an image on a remote TV scanner. The varying intensity of these signals forms an image that corresponds to the position of the beam on the specimen when the signals were generated.

Generally, the image resolution of an SEM is about an order of magnitude lower than that of a TEM, limiting definitions of

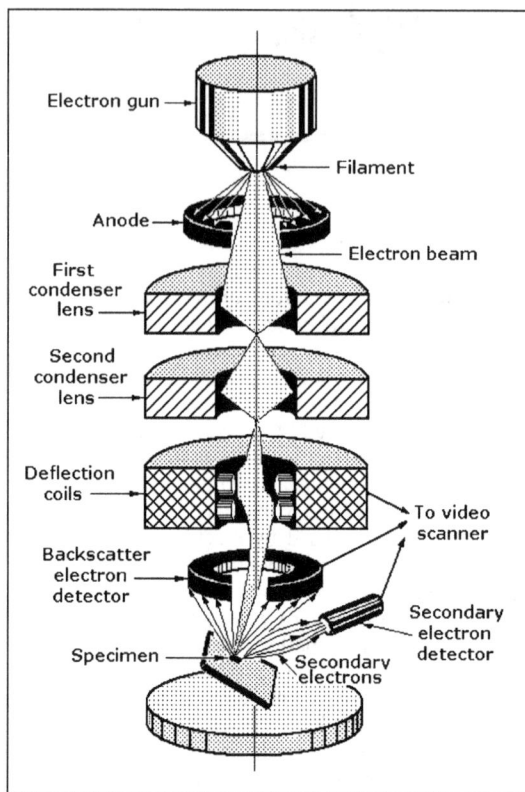

Fig. 2 The scanning electron microscope (SEM) images the specimen by raster scanning it with a beam of electrons. The electrons are reflected as secondary and backscattered electrons from the surface of the specimen rather than passing through it. When signals from the backscattered and secondary electron detectors are collected, they are used to produce signals that contain information about the specimen's surface topography, composition, and electrical conductivity. When combined with raster scanning signals, 3D video images can be produced.

Fig. 3 The scanning tunneling microscope (STM) has a stylus with a fine metal tip that is raster scanned over the sample by a piezoelectric scanner while connected in an electronic feedback circuit. The sample is electrically biased so a small "tunneling current" flows when the tip is close to the sample's surface. This current is amplified and measured to drive a Z piezoelectric electrode that keeps the distance between the tip and the sample constant while X and Y electrodes drive the tip to scan the sample. Signals produced by the electronics form 3D images of the atoms and molecules on the sample's surface.

objects to about 10 nm in size. However, because the SEM image relies on surface processes rather than transmission, it is able to image bulk samples up to many centimeters in size with great depth of field. Consequently, it can produce very sharp 3D images compared to the flat images produced by TEMs. In addition, SEM specimens need less preparation than those for TEMs.

The scanning tunneling microscope (STM), Fig. 3, an electron microscope invented in 1981, works on a principle that differs from those of either the TEM or SEM. It is the first microscope that allowed researchers to "see" objects at the atomic scale. Unlike TEMs, which produce images of the insides of materials, and SEMs, which produce 3D images of specimen surfaces, STMs make detailed 3D images of the atoms or molecules on the surface of objects such as crystals.

Each STM has a conductive metal stylus with an extremely sharp tip formed by a single atom that is scanned over the specimen while it is part of an electronic feedback circuit. As the tip is slowly raster scanned across the surface of an electrically conductive specimen by coordinated X and Y piezoelectric electrodes, it is being held at a constant distance of one atom's diameter of a few Angstroms above the specimen's surface. (An Angstrom is about one-tenth of a nanometer.) When a low bias voltage is applied to the specimen, the rules of quantum mechanics allow electrons to jump or *tunnel* between the specimen's surface and the tip of the stylus.

If the tunneling current exceeds its preset value, the distance between the tip and the sample is decreased, but if it falls

below this value, the feedback increases the distance. The tip is continuously raised and lowered small distances by the Z piezoelectric electrode to keep the tunneling current constant and maintain the distance between the tip and the sample. This enables it to follow even the smallest details of the surface it is scanning. Recording the vertical movement of the tip makes it possible to study the structure of the surface atom by atom. A computer generates a contour map from these profiles of surface scans that can be displayed on a screen. This makes it possible to "see" the individual atoms and molecules that make up the specimen.

The study of surfaces is an important part of physics, with particular applications in semiconductor physics, microelectronics, nanotechnology, and materials science. The STM works best with conducting materials, and it allows for such great precision that individual atoms can be manipulated. This allows for the observation of the results of precise chemical and physical reactions. The STM is a fundamental tool indispensable to the development of nanotechnology.

The atomic force microscope (AFM), Fig. 4, can image almost any kind of solid or organic material, unlike the STM which is limited to imaging conducting or semiconducting surfaces. A laser beam deflection system is used to measure the deflection of a spring called a cantilever with a probe tip at its end while the specimen is moved in a scanning pattern under the tip. The cantilever is made from silicon or silicon nitride and its probe tip has a radius of about 20 nm. (Some of these tips are made from carbon nanotubes.) When the tip is brought into close

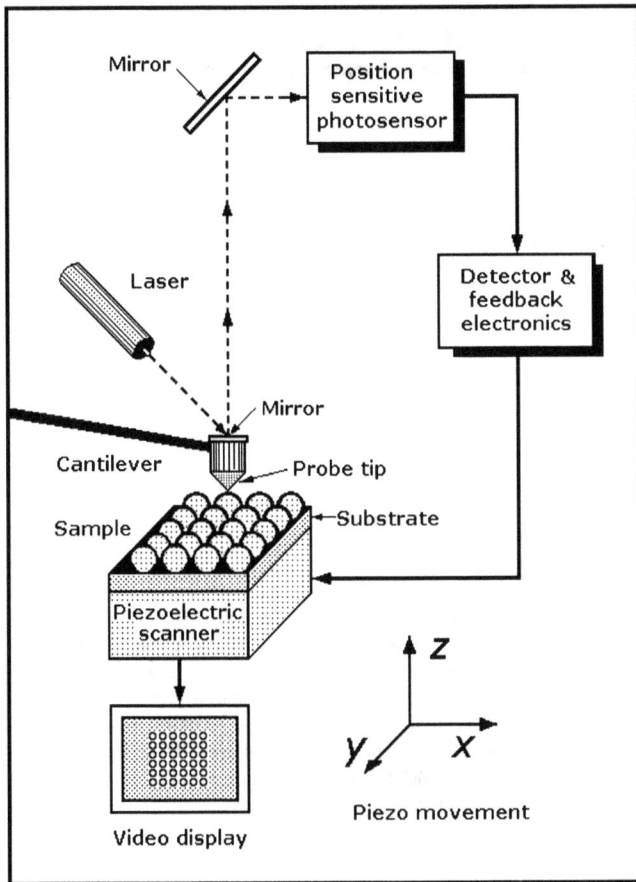

Fig. 4 The atomic force microscope (AFM) can image materials that are neither conductors nor semiconductors. A spring-like cantilever with a probe tip mounted on its end will deflect due to atomic forces existing between the tip and the sample. The sample is raster scanned in an XY direction under the tip by a piezoelectric scanner while a laser-beam based electronic feedback system keeps the tip at a uniform Y distance above the sample. This generates a signal related to the displacement of the cantilever, which can be converted to a video image of the surface topology of the sample.

proximity with a sample surface, any of a wide range of atomic forces existing between the tip and the sample can cause the cantilever to deflect. Typically, tip deflection is measured by laser light that is reflected from a mirror at the end of the cantilever before it is directed into an array of photodiodes called the position-sensitive photosensor.

If the probe tip was scanned at a constant height several nanometers above the sample's surface, it could collide with the surface, causing damage. To prevent this, a feedback circuit that includes a detection and feedback electronics unit is employed to adjust the tip-to-sample distance so a constant force is maintained between the tip and the sample. Variations in tip height are recorded while the sample, mounted on a piezoelectric (PZT) scanner, is scanned in a linear pattern under the tip. The PZT scanner moves the sample up or down in the Z direction to keep the atomic force constant while it scans in the X and Y directions to provide a topographic image of the surface. This image can then be displayed on a computer screen. The ATM is another fundamental tool that is indispensable to research and development in nanotechnology.

GALLERY OF MEMS ELECTRON-MICROSCOPE IMAGES

The Sandia National Laboratories, Albuquerque, New Mexico, have developed a wide range of microelectromechanical systems (MEMS). The scanning electron microscope (SEM) micrographs presented here show the range of these devices, and the captions describe their applications.

Fig. 1   Wedge Stepping Motor: This indexing motor can precisely index other MEMS components such as microgear trains. It can also position gears and index one gear tooth at a time at speeds of more than 200 teeth/s or less than 5 ms/step. An input of two simple input pulse signals will operate it. This motor can index gears in MEMS such as locking devices, counters, and odometers. It was built with Sandia's four-layer SUMMiT technology. Torque and indexing precision increase as the device is scaled up in size.

Fig. 2   Wedge Stepping Motor: A close-up view of one of the teeth of the indexing motor shown in Fig. 1.

Fig. 3   Torque Converter: This modular transmission unit has an overall gear reduction ratio of 12 to 1. It consists of two multilevel gears, one with a gear reduction ratio of 3 to 1 and the other with a ratio of 4 to 1. A coupling gear within the unit permits cascading.

Fig. 4 Torque Converter: By cascading six stages of the modular 12-to-1 transmission units shown in Fig. 3, a 2,985,894-to-1 gear reduction ratio is obtained in a die area of less than 1 mm². The converter can step up or step down.

Fig. 5 Dual-Mass Oscillator: This oscillator uses parallel plate actuation and system dynamics to amplify motion. The 10-mm-long parallel plate actuators on the driving mass produce an amplified motion on the second mass when it is driven by a signal. The actuated mass remains nearly motionless, while the moving mass has an amplitude of approximately 4 µm when driven by a 4-V signal. It was designed to be part of a vibrating gyroscope.

Fig. 6 Rotary Motor: This close-up shows part of a rotary motor that offers advantages over other MEMS actuators. Its operates on linearcomb drive principles, but the combs are bent in a circle to permit unlimited travel. The combs are embedded inside the rotor so that other micromachines can be powered directly from the rotor's perimeter. Built by Sandia's four-level SUMMiT technology, the motor is powered by a lower voltage and produces higher output torque than other MEMS actuators, but it still occupies a very small footprint. It can also operate as a stepper motor for precise positioning applications.

Fig. 7 Comb Drive Actuation: Two sets of comb-drive actuators (not shown) drive a set of linkages (upper right) to a set of rotary gears. The comb-drive actuators drive the linkages 90° out of phase with each other to rotate the small 19-tooth gear at rotational speeds in excess of 300,000 rpm. The operational lifetime of these small devices can exceed 8×10^9 revolutions. The smaller gear drives a larger 57-tooth (1.6-mm-diameter) gear that has been driven as fast as 4800 rpm.

Fig. 8 Micro Transmission: This transmission has sets of small and large gears mounted on the same shaft so that they interlock with other sets of gears to transfer power while providing torque multiplication and speed reduction. Its output gear is coupled to a double-level gear train.

Fig. 9 Microtransmission and Gear Reduction Unit: This mechanism is the same as that in Fig. 8 except that it performs a gear-reduction function. The microengine pinion gear, labeled A in the figure, meshes directly with the large 57-tooth gear, labeled B. A smaller 19-tooth gear, C, is positioned on top of gear B and is linked to B's hub. Because the gears are joined, both make the same number of turns per minute. The small gear essentially transmits the power of the larger gear over a shorter distance to turn the larger 61-tooth gear D. Two of the gear pairs (B and C, D and E) provide 12 times the torque of the engine. A linear rack F, capable of driving an external load, has been added to the final 17-tooth output gear E to provide a speed reduction/torque multiplication ratio of 9.6 to 1.

Fig. 10 Gear-Reduction Units: This micrograph shows the three lower-level gears (A, B, and E) as well as the rack (F) of the system shown in Fig. 9. The large flat area on the lower gear provides a planar surface for the fabrication of the large, upper-level 61-tooth gear (D).

Fig. 11 Microsteam Engine: This is the world's smallest multipiston microsteam engine. Water inside the three compression cylinders is heated by electric current, and when it vaporizes, it pushes the pistons out. Capillary forces then retract the piston once current is removed.

Fig. 12 Microchain Drive: This 50-link microchain drive resembles a bicycle chain and sprocket assembly except that each link of the microchain could rest atop a cross section of human hair with space to spare. The distance on centers between chain links is 50 µm. (The diameter of a human hair is 70 µm.) Because this microchain can rotate many sprockets, a single tiny MEMS motor coupled to the drive shaft of a single sprocket can rotate many other sprockets linked to the microchain. Typically, a single MEMS motor powers each MEMS device. Chain systems, unlike stroke systems, perform oscillatory motions permitting both continuous and intermittent drive translation.

Fig. 13 Detail of Microchain Link and Sprocket: A silicon microchain rather than a silicon microbelt is used in this MEMS because a silicon belt would have spring-like characteristics which would apply too much torque on any sprockets not aligned in the same plane. By contrast, each chain link is capable of ± 52° rotation with respect to the preceding link. This minimizes stress on the support structure. The longest span unsupported by sprockets or bracing is 500 microns, but a microchain tensioner permits longer spans. This multilevel surface-micromachined device was constructed with Sandia's SUMMiT IV and SUMMiT V technology.

MEMS ACTUATORS—THERMAL AND ELECTROSTATIC

Reliable actuators are critical components in most MEMS devices. Well-known for developments in electrostatic comb-drive actuation, Sandia National Laboratories also uses thermal actuators extensively. Some of the common actuators developed for MEMS at Sandia Laboratories are shown here.

V-Shaped Thermal Actuators

The v-shaped thermal actuator shown in Fig. 1 is commonly referred to as a "chevron," or "bent-beam" thermal actuator. It is used in applications requiring high force and reliability. These actuators are based on the constrained thermal expansion of the angled beams (a result of Joule heating when a current is passed through the legs of the actuator), resulting in motion of the center shuttle in the direction shown by the arrow in the figure.

Fig. 1 V-shaped thermal actuator for on-chip, high-force, linear-motion actuation. *Courtesy of Sandia National Laboratories*

Thermal Ratcheting Actuator

The thermal ratcheting actuator (ThRA) shown in Fig. 2 is a V-shaped actuator attached to a ratcheting hub that provides a drive capable of high-force exceeding 30° rotational motion.

Fig. 2 Thermal ratcheting actuator. *Courtesy of Sandia National Laboratories*

Electrostatic Actuators

Torsional Ratcheting Actuator (TRA)

The torsional ratcheting actuator shown in Fig. 3 is used when high-torque rotational motion is needed. Ratcheting allows full unidirectional motion exceeding 360°.

Fig. 3 Torsional ratcheting actuator. *Courtesy of Sandia National Laboratories*

Parallel-Plate Actuation

Electrostatic parallel-plate actuation is used extensively in RF MEMS, oscillators, mirrors, switches, and other applications requiring only small displacements. These actuators can be designed with simple equations.

Comb Drives

The electrostatic "comb drive" shown in Fig. 4 is a common MEMS actuator. It is used in gyroscopes, microengines, resonators, and many other MEMS applications. The force output is generally less than 50 μN, but the linear and highly predictive behavior (very good predictions can be made using simple equations) make it a popular MEMS actuator. Comb drives have been used in microengines by combining the actuator with a rotary transmission.

Fig. 4 Scanning electron microscope view of a typical comb-drive resonator. *Courtesy of Sandia National Laboratories*

MEMS CHIPS BECOME INTEGRATED MICROCONTROL SYSTEMS

The successful integration of MEMS (microelectromechanical systems) on CMOS integrated circuit chips has made it possible to produce "smart" control systems whose size, weight, and power requirements are significantly lower than those for other control systems. MEMS development has previously produced microminiature motors, sensors, gear trains, valves, and other devices that easily fit on a silicon microchip, but difficulties in powering these devices has inhibited their practical applications.

MEMS surface micromachining technology is a spin-off of conventional silicon IC fabrication technology, but fundamental differences in processing steps prevented their successful integration. The objective was to put both the control circuitry and mechanical device on the same substrate. However, the results of recent development work showed that they could be successfully merged.

It has been possible for many years to integrate the transistors, resistors, capacitors, and other electronic components needed for drive, control, and signal processing circuits on a single CMOS silicon chip, and many different MEMS have been formed on separate silicon chips. However, the MEMS required external control and signal-processing circuitry. It was clear that the best way to upgrade MEMS from laboratory curiosities to practical mechanical devices was to integrate them with their control circuitry. The batch fabrication of the electrical and mechanical sections on the same chip would offer the same benefits as other large-scale ICs—increased reliability and performance. Component count could be reduced, wire-bonded connections between the sections could be eliminated, minimizing power-wasting parasitics, and standard IC packaging could replace multichip hybrid packages to reduce product cost.

MEMS sections are fabricated by multilevel polysilicon surface micromachining that permits the formation of such intricate mechanisms as linear comb-drive actuators coupled to gear trains. This technology has produced micromotors, microactuators, microlocks, microsensors, microtransmissions, and micromirrors.

Early attempts to integrate CMOS circuitry with MEMS by forming the electronic circuitry on the silicon wafer before the MEMS devices met with only limited success. The aluminum electrical interconnects required in the CMOS process could not withstand the long, high-temperature annealing cycles needed to relieve stresses built up in the polysilicon mechanical layers of the MEMS. Tungsten interconnects that could withstand those high temperatures were tried, but the performance of the CMOS circuitry was degraded when the heat altered the doping profiles in the transistor junctions.

When the MEMS were formed before the CMOS sections, the thermal problems were eliminated, but the annealing procedure tended to warp the previously flat silicon wafers. Irregularities in the flatness or planarity of the wafer distorted the many photolithographic images needed in the masking steps required in CMOS processing. Any errors in registration can lower attainable resolution and cause circuit malfunction or failure.

Experiments showed that interleaving CMOS and MEMS process steps in a compromise improved yield but limited both the complexity and performance of the resulting system. In other experiments materials such as stacked aluminum and silicon dioxide layers were substituted for polysilicon as the mechanical layers, but the results turned out to be disappointing.

Each of these approaches had some merit for specific applications, but they all resulted in low yields. The researchers persevered in their efforts until they developed a method for embedding the MEMS in a trench below the surface of the silicon wafer

before fabricating the CMOS. This is the procedure that now permits the sections to be built reliably on a single silicon chip.

Sandia's IMEMS Technology

Sandia National Laboratories, Albuquerque, New Mexico, working with the University of California's Berkeley Sensor and Actuator Center (BSAC), developed the unique method for forming the micromechanical section first in a 12-μm-deep "trench" and backfilling that trench with sacrificial silicon dioxide before forming the electronic section. This technique, called Integrated MicroElectroMechanical Systems (IMEMS), overcame the wafer-warping problem. Figure 1 is cross-section view of both sections combined on a single chip.

Fig. 1 A cross-section view of CMOS drive circuitry integrated on the same silicon chip with a microelectromechanical system.

The mechanical polysilicon devices are surface micromachined by methods similar to Sandia's SUMMiT process in the trench, using special photolithography methods. After the trench is filled with the silicon dioxide, the silicon wafer is annealed and that section is "planarized," or etched flat and flush with the rest of the wafer surface, by a process called chemical-mechanical polishing (CMP). After the CMOS section is complete, the sacrificial silicon dioxide in the trench is etched away, leaving the MEMS devices electrically interconnected with the adjacent CMOS circuitry.

Advantages of IMEMS

Sandia spokespersons say the IMEMS process is completely modular, meaning that the planarized wafers can be processed in any facility capable of processing CMOS, bipolar, and combinations of these processes. They add that modularity permits the mechanical devices and electronic circuitry to be optimized independently, making possible the development of high-performance microsystems.

Early Research and Development

Analog Devices Inc. (ADI) was one of the first companies to develop commercial surface-micromachined integrated-circuit accelerometers. ADI developed and marketed these accelerometer chips, demonstrating its capability and verifying commercial demand. Initially ADI built these devices by interleaving, combining, and customizing its internal manufacturing processes to produce the micromechanical devices with the same processes it used to produce monolithic electronic circuitry.

At the same time, researchers at BSAC developed the alternative process for replacing conventional aluminum interconnect layers with tungsten layers to enable the CMOS device to withstand the higher thermal stresses associated with subsequent micromechanical device processing. This process was later superseded by the joint BSAC–Sandia development of IMEMS.

Accelerometers

ADI offered the single-axis ADXL150 and dual-axis ADXL250, and Motorola Inc. offered the XMMAS40GWB. Both of ADI's integrated accelerometers are rated for $\pm 5\,g$ to $\pm 50\,g$. They have been in high-volume production since 1993. The company is now licensed to use Sandia's integrated MEMS/CMOS technology. Motorola is now offering the MMA1201P and MMA2200W single-axis IC accelerometers rated for $\pm 38\,g$.

These accelerometer chips differ in architecture and circuitry, but both work on the same principles. The surface micromachined sensor element is made by depositing polysilicon on a sacrificial oxide layer that is etched away, leaving the suspended sensor element. Figure 2 is a simplified view of the differential-capacitor sensor structure in an ADI accelerometer. It can be seen that two of the capacitor plates are fixed, and the center capacitor plate is on the polysilicon beam that deflects from its rest position in response to acceleration.

Fig. 2 A simplified view of the movement of a polysilicon beam in a surface-micromachined accelerometer moving in response to acceleration. The two fixed plates and one moving plate form a unit cell.

When the center plate deflects, its distance to one fixed plate increases while its distance to the other plate decreases. The change in distance is measured by the on-chip circuitry that converts it to a voltage proportional to acceleration. All of the circuitry, including a switched-capacitor filter needed to drive the sensor and convert the capacitance change to voltage, is on the chip. The only external component required is a decoupling capacitor.

Integrated-circuit accelerometers are now used primarily as airbag-deployment sensors in automobiles, but they are also finding many other applications. For example, they can be used to monitor and record vibration, control appliances, monitor the condition of mechanical bearings, and protect computer hard drives.

Three-Axis Inertial System

When the Defense Advanced Research Projects Agency (DARPA), an agency of the U.S. Department of Defense, initiated a program to develop a solid-state three-axis inertial measurement system, it found that the commercial IC accelerometers were not suitable components for the system it envisioned for two reasons: the accelerometers must be manually aligned and assembled, and this could result in unwanted variations in alignment, and the ICs

lacked on-chip analog-to-digital converters (ADCs), so they could not meet DARPA's critical sensitivity specifications.

To overcome these limitations, BSAC designed a three-axis, force-balanced accelerometer system-on-a-chip for fabrication with Sandia's modular monolithic integration methods. It is said to exhibit an order of magnitude increase in sensitivity over the best commercially available single-axis integrated accelerometers. The Berkeley system also includes clock generation circuitry, a digital output, and photolithographic alignment of sense axes. Thus, the system provides full three-axis inertial measurement, and does not require the manual assembly and alignment of sense axes.

A combined X- and Y-axis rate gyro and a Z-axis rate gyro was also designed by researchers at BSAC. By using IMEMS technology, a full six-axis inertial measurement unit on a single chip was obtained. The 4- by 10-mm system is fabricated on the same silicon substrate as the three-axis accelerometer, and that chip will form the core of a future micro–navigation system. BSAC is teamed with ADI and Sandia Laboratories in this effort, with funds provided by DARPA's Microsystems Technology Office.

Micromechanical Actuators

Micromechanical actuators have not attained the popularity in commercial applications achieved by microminiature accelerometers, valves, and pressure sensors. The two principal drawbacks to their wider application have been their low torque characteristics and the difficulties encountered in coupling actuators to drive circuitry. Sandia has developed devices that can be made by its SUMMiT four-level polysilicon surface-micromachining process, such as the microengine pinion gear driving a 10 to 1 transmission shown in Fig. 3, to improve torque characteristics.

The SUMMiT process includes three mechanical layers of polysilicon in addition to a stationary level for grounding or electrical interconnection. These levels are separated by sacrificial silicon-dioxide layers. A total of eight mask levels are used in this process. An additional friction-reducing layer of silicon nitride is placed between the layers to form bearing surfaces.

If a drive comb, operating at a frequency of about 250,000 rpm, drives a 10-to-1 gear reduction unit, torque is traded off for speed. Torque is increased by a factor of 10 while speed is reduced to about 25,000 rpm. A second 10-to-1 gear reduction would increase torque by a factor of 100 while reducing speed to 2,500 rpm. That gear drives a rack and pinion slider that provides high-force linear motion. This gear train provides a speed-reduction/torque-multiplication ratio of 9.6 to 1.

Fig. 3 This linear-rack gear reduction drive converts the rotational motion of a pinion gear to linear motion to drive a rack. *Courtesy of Sandia National Laboratories*

ALTERNATIVE MATERIALS FOR BUILDING MEMS

Researchers at Sandia National Laboratories, Albuquerque, New Mexico, have developed techniques for fabricating functional microelectromechanical systems (MEMS) components from ceramic and rare earth nanocomposites. Their methods include the formulation of nanocomposite mixtures, micromold fabrication, and micromolding. Sandia Laboratories reports that the fabrication of MEMS from ceramics will increase their robustness, high-temperature inertness, chemical and biological compatibility, magnetic properties, piezoelectric properties, and photochromism.

Silicon and its compounds are now the principal materials used in both the bulk and surface methods for micromachining MEMS. This technology has been derived from the photolithographic and acid etching technology for fabricating large-scale integrated circuits (ICs), permitting MEMS to be made at laboratories and commercial IC wafer fabrication facilities with standard IC manufacturing equipment. One advantage to the use of silicon is that the technology is compatible with complementary metal-oxide semiconductor (CMOS) IC production. As a result, MEMS can be integrated on the same chip with analog or digital ICs for driving, controlling, or communicating with the MEMS in a microminiature system.

Unfortunately, silicon MEMS made by surface micromachining are not intrinsically durable because active mechanical parts such as gears, rotors, or folded beams that have thickness measurable only in microns are subject to fracture. There is an inherent limitation on layer thickness imposed by the deposition and etching process of surface micromachining that has little or no effect on electronic circuitry. Consequently, researchers have been searching for alternative materials such as ceramics and composites to improve the versatility and durability of MEMS without sacrificing IC compatibility.

Sandia Laboratories researchers report that micromolded components can be made free-standing or assembled on substrates. By molding components from nanometer-sized particles, they were able to build them with lateral dimensions of a few microns. Their findings indicated that it is possible to mold micron-size mechanical parts from nanocomposites, and that the process is compatible with existing IC fabrication processes.

Ceramic nanoparticles less that 100 nm in diameter are now readily available. One example is aluminum oxide which offers excellent insulation and wear resistance properties as well as the ability to withstand intense heat. Other nanoparticles suitable for molding micron-sized mechanical components are samarium-cobalt, rare earth, and manganese-iron ferrite because of their magnetic properties.

One way to produce micromolded nanocomposites is to form them in high-aspect ratio micromolds made by the LIGA process. (See *LIGA: An Alternative Method for Making Microminiature Parts* in this chapter.) The molds are formed by x-ray lithography from polymethylmethacrylate (PMMA) on silicon substrates. The PMMA molds are filled with the nanoparticles embedded in a binder in which they are cured. The microparts are then polished flat or "planarized" and removed from the mold.

A variation on the LIGA process involves forming a dovetail-shaped cavity in the silicon substrate to anchor a coating of photoresist which is then patterned by conventional IC photolithographic processes and encased in PMMA. After the PMMA is polished back to reach the top of the photoresist, it is dissolved chemically, leaving a PMMA mold. This mold is filled with the nanoparticles embedded in a binder and cured before the microparts are planarized and removed from the mold.

In other research on the fabrication of more rugged microminiature parts, researchers at the University of California at Los Angeles (formerly working at Pennsylvania State University) developed a process called *microstereolithography*. They succeeded in forming 3D ceramic structures with thicknesses of 50 μm to 1 mm by applying layers of alumina in thicknesses from 10 to 20 μm. Ceramic structures 5.7 $\times$ 10 μm have been produced. The researchers used the method to apply films of lead, zirconate, and titanate greater than 15 μm thick to silicon substrates. The process is a miniaturized version of the rapid prototyping stereolithography process. (See *Stereolithography* in Chap. 18 *Rapid Prototyping*.) An ultraviolet (UV) laser with its beam focused to a 1 to 2 μm width was used to perform the work. This beamwidth contrasts with widths of hundreds of microns used in conventional stereolithography.

Other material alternatives to silicon under consideration in laboratories for the manufacture of MEMS include stainless steel, aluminum, and titanium nanopowders in a binder, as well as bulk titanium. MEMS have also been formed by injection molding, and hot embossing has been used to stamp finely detailed microfeatures on them in thermoplastic. The LIGA process has produced highly detailed microscale tools from a list of metal alloys suitable for fabricating polymeric MEMS. Researchers have also been experimenting with the bulk micromachining of titanium as an alternative to silicon.

Plastic resins including polycarbonate and acrylic are seen as better materials for microfluidic structures because they offer the advantages of biocompatibility and resistance to solvents. A thermoplastic polymer microneedle with a diameter of about 100 μm was produced in the mechanical engineering laboratory at the University of California, Berkeley. It was formed in an aluminum mold on a 30-ton injection-molding machine.

Agilent Technologies, Palo Alto, California, produced a commercial polymeric microfluidic device for separating compounds such as proteins and peptides for analysis. Made from Kapton polyimide thermoplastic film, the microfluidic channels in the device were drilled by a laser, the same process used in making ink-jet printer heads. The device is mounted on a mass spectrometer for performing liquid chromatography.

LIGA: AN ALTERNATIVE METHOD FOR MAKING MICROMINIATURE PARTS

The Sandia National Laboratories, Livermore, California, is using a process called LIGA to form microminiature metal components as an alternative to producing them by the surface micromachining processes used to make microelectromechanical systems (MEMS). LIGA permits the fabrication of larger, thicker, and more durable components with greater height-to-width ratios. They can withstand high pressure and temperature excursions while providing more useful torques than polysilicon MEMS.

The acronym LIGA is derived from the German words for lithography, electroplating, and molding (Lithographie, Galvanoformung, and Abformung), a micromachining process originally developed at the Karlsruhe Nuclear Research Center in Karlsruhe, Germany, in the 1980s. Sandia Labs has produced a wide variety of LIGA microparts, including components for millimotors and miniature stepping motors. It has also made miniature accelerometers, robotic grippers, a heat exchanger, and a mass spectrometer. Sandia is carrying out an ongoing research and development program to improve the LIGA process and form practical microparts for various applications.

In the LIGA process, highly parallel x rays from a synchrotron are focused through a mask containing thin 2D templates of the microparts to be formed. The x rays transfer the patterns to a substrate layered with PMMA (polymethylmethacrylate), a photoresist sensitive to x rays, on a metallized silicon or stainless-steel substrate. When the exposed layer of PMMA (better known as Plexiglas) is developed, the cavities left in the PMMA are the molds in which the microparts will be formed by electroplating. The thickness of the PMMA layer determines the large height-to-width ratio of the finished LIGA microparts. The resulting parts can be functional components or molds for replicating the parts in ceramic or plastic.

The highlights of the LIGA process as illustrated in Fig. 1 are:

(a) An x-ray mask is prepared by a series of plating and lithographic steps. A metallized silicon wafer coated with photoresist is exposed to ultraviolet light through a preliminary mask containing the 2D patterns of the microparts to be produced. Development of the photoresist dissolves the

resist from the plated surface of the wafer, forming the micropart pattern, which is plated in gold to a thickness of 8 to 30 μm. The remaining photoresist is then dissolved to finish the mask.

(b) Target substrate for forming microparts is prepared by solvent-bonding a layer of PMMA to a metallized-silicon or stainless-steel substrate.

(c) PMMA-coated substrate is then exposed to highly collimated parallel x rays from a synchrotron through the mask.

(d) PMMA is then chemically developed to dissolve the exposed areas down to the metallized substrate, etching deep cavities for forming microparts.

(e) Substrate is then electroplated to fill the cavities with metal, forming the microparts. The surface of the substrate is then lapped to finish the exposed surfaces of the microparts to the required height within ±5μm.

(f) Remaining PMMA is dissolved, exposing the 3D microparts, which can be separated from the metallized substrate or allowed to remain attached, depending on their application.

The penetrating power of the x rays from the synchrotron allows structures to be formed that have sharp, well-defined vertical surfaces or sidewalls. The minimum feature size is 20 μm, and microparts can be fabricated with thickness of 100 μm to 3 mm. The sidewall slope is about 1μm/mm. In addition to gold, microparts have been made from nickel, copper, nickel–iron, nickel–cobalt, and bronze.

An example of a miniature machine assembled from parts fabricated by LIGA is an electromagnetically actuated millimotor. With an 8-mm diameter and a height of 3 mm, it includes 20 LIGA parts as well as an EDM-machined permanent magnet and wound coils. The millimotor has run at speeds up to 1600 rpm, and it can provide torque in excess of 1 mN-m. Another example of a miniature machine built from LIGA parts is a size 5 stepper motor able to step in 1.8-deg increments. Both its rotor and stator were made from stacks of 50 laminations, each 1-mm thick.

According to Sandia researchers, the LIGA process is versatile enough to be an alternative to such precision machining methods as wire EDM for making miniature parts. The feature definition, radius, and sidewall texture produced by LIGA are said to be superior to those obtained by any precision metal cutting technique.

In an effort to form LIGA parts with higher aspect ratios, researchers at the University of Wisconsin in Madison teamed with the Brookhaven National Laboratory on Long Island to use the laboratory's 20,000-eV photon source to produce much higher levels of x-ray radiation than are used in other LIGA processes. The higher-energy x rays penetrate into the photoresist to depths of 1 cm or more, and they also pass more easily through the mask. This permitted the Wisconsin team to use thicker and stronger materials to make 4-in.-square masks rather than the standard 1- × 6-cm masks used in standard LIGA. Working with Honeywell, the team developed LIGA optical microswitches.

The primary disadvantage to LIGA is its requirement for a synchrotron or other high-energy sources to image parallel x rays on the PMMA covered substrate. In addition to their limited availability, these sources are expensive to build, install, and operate. Their use adds significantly to the cost of producing LIGA microstructures, especially for commercial applications.

Fig. 1 Steps in fabricating microminiature parts by the LIGA process.

THE ROLE OF NANOTECHNOLOGY IN SCIENCE AND ENGINEERING

Nanotechnology or nanoscience is defined as the science of designing, producing, and viewing structures at the molecular or nanometer scale (billionths of a meter). It has received considerable attention in the media and scientific papers because of its potential applications in the physical and biological sciences as well as mechanical and electronic engineering. However "nano" has become a popular buzz word that grabs attention, and many companies have included nano in their company names for promotional purposes, whether or not they are producing products that include nanoparticles. There has been concern expressed about possible health problems resulting from breathing or accidentally ingesting nanoparticles because their characteristics differ from those of finely ground or powdered forms of the same materials.

Nevertheless, the field shows promise for producing stronger, lighter, and more versatile materials although commercial success in the field has been limited thus far. Researchers in the materials sciences and engineering are, however, exploring the possibilities for creating such practical products as molecular transistors, artificial muscles, super strong aircraft and vehicles. Other possible products are capacitors that can replace batteries, and extremely strong thin films with applications in both electronics and structures. Nanoscale materials are also being used in low-cost, more efficient solar panels.

Nanotechnology concentrates on materials that range in size from 1 to 100 nanometers (nm). One nanometer equals 0.001 micrometer (μm) or, stated in another way, there are one thousand nanometers in a micrometer (micron). The diagram provides a graphic comparison between common organic and inorganic objects in this size range. Although some very small features on silicon microcircuits are only 130 μm wide, they are well outside the size limits that define nanotechnology. Atoms have diameters of about 0.1 nm while molecules have diameters of about 1.0 nm. When objects approach atomic size, the behavior of individual atoms becomes more conspicuous, and the characteristics of common materials near this size range change in unpredictable ways.

At the atomic or quantum level, the boundaries between biology, chemistry, physics, and electronics lose much of their meaning as those disciplines converge. Laboratories that develop promising nanosize inorganic materials are expected to explore the possibilities for creating hybrids of those materials and organic materials. For example, in research work, organic materials have been attached to inorganic carbon nanotubes to create microscopic transistors that can assemble themselves into larger arrays.

Structures in the molecular size range are more likely to break than are larger structures because their imperfections become more significant. Nanotechnology emphasizes the "bottom up" building of structures atom-by-atom as its most effective approach. This concept has caught the attention of researchers in the materials and biological sciences because it has significant implications for chemists and engineers searching for lighter and stronger materials. Nanotechnology could bring about a sea change in mechanical engineering because, unlike electronic systems, many of the tiny mechanisms or mechanical devices with dimensions measurable in nanometers that result will be able to interact directly with the physical world. That could, for example, include nanomotors and nanofluidic pumps. By contrast, electronic and computer engineers focus on devices that generate, transmit, or receive information in the form of electrical impulses.

Experts believe that nanotechnology will have its greatest impact in nanoelectronics, making it possible to produce molecular circuitry that is smaller, faster, and cheaper than silicon circuitry. One technique now being developed is the manufacture of molds from silicon that will permit the mass production of nanoscale circuitry. Photolithographic and chemical-etching processing of silicon will produce molds that will contain the inverse of the structural features needed for "printing" vast quantities of nanoscale transistors.

Elastomeric materials will be poured into the silicon molds, and after curing and removal they will be "rubber stamps" for reproducing large arrays of transistorized circuitry. The stamps, with the necessary positive features for printing the circuitry, will then be coated with an "ink" made from nanomaterials in a suitable binder. By using a technique called "soft lithography," the stamps will imprint fine microscale and nanoscale transistor features on a suitable substrate. The goal of this effort is the production of billions if not trillions of molecular transistors on a chip.

Other approaches being explored for producing large-scale transistor arrays include forming chains of molecules to make nanoprocessors and memory devices. However, it is generally believed that it will take many more years of R&D for molecular computing to become a practical technology, despite promising early results. Hundreds of academic research papers now detail programs in nanotechnology and hundreds of business plans indicate the intention of companies to participate in a commercial nanotechnology business venture. Some are spinoffs of successful academic research on the subject, but many of the research papers and business plans reveal a preoccupation with nanoscale materials rather than devices. Examples of objectives mentioned include the preparation of nanoparticles for coating

Measurement of Objects in Nanometers

One nanometer (nm) = one billionth of a meter

Size range of nanomaterials or nanocrystals

1 to 100 nanometers (nm)

| 0 | 10 | 20 | 30 | 40 | 50 | 60 | 70 | 80 | 90 | 100 |

Optical lithographic structures or viruses

Diameter of proteins in cells = 3 to 20 nm

Electron beam structures = 20 nm

Diameter of a molecule = ~1 nm

Diameter of an atom = ~0.1 nm

1 micrometer (μm) = 1000 nm

Shortest wavelength of visible light = 500 nm

Diameter of a human cell = 5,000 to 200,000 nm

Thickness of human hair = 50,000 to 70,000 nm

Smallest object that can be seen by the unaided human eye = 10,000 nm fabric to make it stain-resistant; semiconductor nanoparticles or quantum dots for biological research; and nanocomposites for making plastics stronger and lighter.

Carbon nanotubes and graphene continue to be subjects for nanotechnology research because of their remarkable strength and unusual electrical properties. (See the article *Carbon: An Engineering Material with a Future* in this chapter.) For example, nanotubes can act either as conductors or semiconductors. Scientists at IBM Watson Research Center believe that every electronic device on a nanochip could be made from nanotubes—sensors, transistors, light emitters, and interconnections. Sheets of self-supporting carbon nanotubes that are stronger than steel have been produced. These sheets are flexible and can be treated to emit light. Scientists at the University of Texas, Dallas (UTD), report that these sheets can act as solar cells to generate electricity from sunlight.

Recently Massachusetts Institute of Technology (MIT) announced research leading to the possible replacement of chemical batteries with capacitors enhanced by nanomaterials. The report pointed out that while batteries produce voltage from a chemical reaction, capacitors store electrical energy between a pair of metal plates. The report noted that the larger the area of capacitor plates and the smaller the space between them, the more energy a capacitor can store. The researchers covered the plates of its capacitor-type batteries with millions of carbon nanotubes, vastly expanding the surface areas of the capacitor plates.

This process is analogous to applying a layer of sponge to a fixed surface to increase the amount of water that can be stored on that surface. Manufacturers of aluminum capacitors etched the surfaces of aluminum foil used as plate material to increase its porosity; this effectively increases its surface area. The MIT researchers say the porosity of capacitor plates coated with conductive nanotubes permits them to store an amount of energy comparable to the capacity of a similar-sized battery. It is expected that these capacitors, when commercially available in perhaps five years, could be charged in minutes or seconds rather than hours. An additional benefit for these capacitor-batteries is that they can be reused indefinitely, thus reducing or eliminating the toxic waste caused by the annual disposal of large numbers of conventional alkaline and lead-acid batteries in landfill.

Researchers at the University of South Carolina have converted simple cotton cloth from which T-shirts are made into a fabric suitable for use in lightweight body armor. The cloth is first dipped into a nanoparticle boron solution and then removed and heated in an oven. The heat changes the cotton fibers into carbon fibers, which react with the boron solution to produce boron-carbide-nanowires. The result is a stiff fabric reinforced with boron carbide that remains flexible enough to be used in the manufacture of bulletproof vests and body armor, replacing the heavy boron-carbide plates now being used. This boron-carbide fabric also has the property of blocking most ultraviolet rays.

Stanford University researchers have used cotton T-shirt fabric dipped in a conductive nanoparticle-infused ink to form batteries and simple capacitors. The resulting conductive textiles, called *eTextiles*, represent a new class of integrated energy storage device. The new eTextiles batteries and capacitors were manufactured by a simple dipping and drying procedure. A strip of fabric is coated with a special ink formula and dehydrated in the oven. The procedure works for the manufacture of batteries or supercapacitors, depending on the contents of the ink. Oxide particles such as $LiCoO_2$ (lithium-cobalt-oxide) are used in the ink for lithium-ion batteries (said to hold up to three times more energy than conventional batteries), and conductive single-walled carbon nanotubes are used in the inks for making the supercapacitors.

CARBON: AN ENGINEERING MATERIAL WITH A FUTURE

It is difficult for most people to believe that a material stronger than steel can be extracted from candle soot, but it is true. The importance of this discovery, made just 25 years ago, has only recently been recognized by scientists and engineers. They found that carbon atoms can arrange themselves into geometries with astonishing properties not previously known to science. The study of atom-to-atom carbon bonding has uncovered dozens of practical applications that could change technology as we know it. Called carbon *allotropes*, both carbon nanotubes and graphene are related to diamonds and graphite, but they are stronger than steel per unit weight and have electrical conductivity that is superior to copper. Amazing as it may seem, it has been found that transistors made from these allotropes are faster and smaller than those made from silicon.

These unusual properties of carbon nanotubes and graphene have yet to be exploited commercially, but they are subjects of intense research and experimentation in academic, government, and industrial laboratories around the world. The reports of experimental work completed thus far suggest that their prospects are unlimited.

By contrast, the properties of carbon bonded with other elements have been well known for centuries. However, carbon stands out among all elements because of its affinity for other atoms and its ability to form so many well-known chemical compounds. But it took curious scientists equipped with modern instruments such as laser spectroscopes and electron microscopes to find out how carbon atoms bond with each other to form different 2D and 3D geometries.

Carbon, the sixth element of the periodic table and fourth most abundant element in the universe, is essential to all life on Earth. Organic chemistry is the study and application of "organic molecules," meaning "carbon-based molecules." This classification applies even where these molecules have no association with living organisms.

A carbon atom, diagrammed in Fig. 1, contains six protons and six neutrons in its nucleus and six electrons in two energy bands or orbits; two are in its first or inner band and four are in the second or outer band. The carbon atom can easily fill its outer energy band by forming four single covalent bonds with other atoms. (The four electron vacancies in the outer band act as "couplings" that latch onto electrons in many other atoms, including those of carbon.)

Ironically, the newly discovered carbon allotropes have been seriously investigated for only about a decade, primarily because the appropriate tools for viewing them were not available. Nevertheless, the simplest carbon and oxygen molecules have long had questionable properties. Carbon monoxide (CO), for one, is a toxic gas capable of asphyxiating people if breathed in high concentrations. This has been known since the introduction of internal combustion engines. It is emitted in the exhaust fumes from hydrocarbon-fueled vehicles, and it has caused countless deaths of persons who have accidentally or deliberately inhaled those fumes while a hydrocarbon-fueled engine was running in a closed space.

By contrast, carbon dioxide (CO_2), best known as the bubbles in carbonated beverages and as a useful agent for extinguishing fires, has been identified as the greenhouse gas principally responsible for causing manmade global warming. Exhaled by humans and animals and absorbed by vegetation, it can,

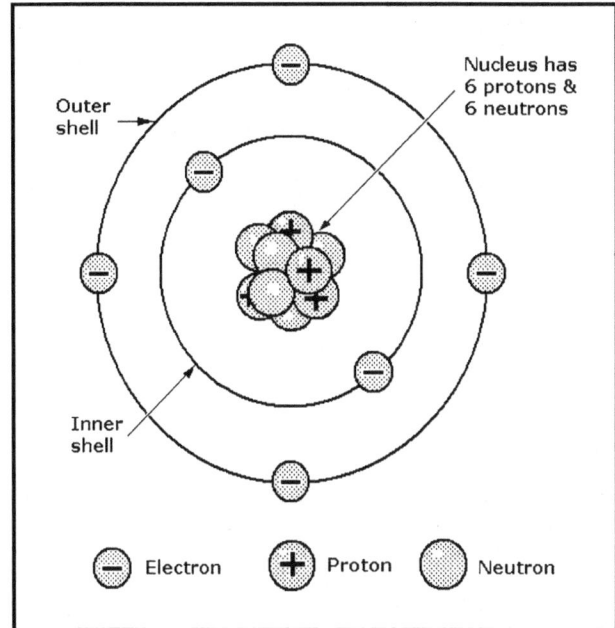

Fig. 1 Carbon is the sixth element in the periodic table and the fourth most abundant in the universe. The carbon atom contains 6 protons, 6 neutrons, and 6 electrons. Two are in the first atomic energy level and four are in the second energy level. Carbon is unique in that it can fill its outer energy level by forming four covalent bonds with other atoms. Carbon forms the basis for organic chemistry and it is essential to all life forms, but it can also form materials that are stronger than steel.

however, also cause death by asphyxiation if inhaled in high concentrations. It does not, however, have the killer reputation of carbon monoxide.

Fullerenes or Buckyballs are the Simplest Nanostructures

Researchers at Rice University, Houston, Texas, discovered and identified fullerenes in 1985 while studying graphite with supersonic jet laser spectroscopy equipment. A laser in the system vaporized bits of graphite into a gas, and after it was cooled to permit the atoms to return to a solid state, they found clusters of atoms in various sizes. The scattered clusters made them easier to investigate and, as a result, they found many contained 60 carbon atoms bonded together in ball shapes.

At first, the scientists were puzzled by how 60 carbon atoms could structure themselves in a stable ball shape, as shown in Fig. 2. Later, upon reflection, they concluded that one carbon atom must lie at each vertex of 12 pentagons and 20 hexagons. Because this arrangement looked like the panels of a soccer ball, they named the molecules "buckminsterfullerenes," later shortened to "buckyballs." This was in recognition of the architect Buckminster Fuller, famous for inventing the geodesic dome that had a similar geometry. Subsequently they found that "buckyballs"

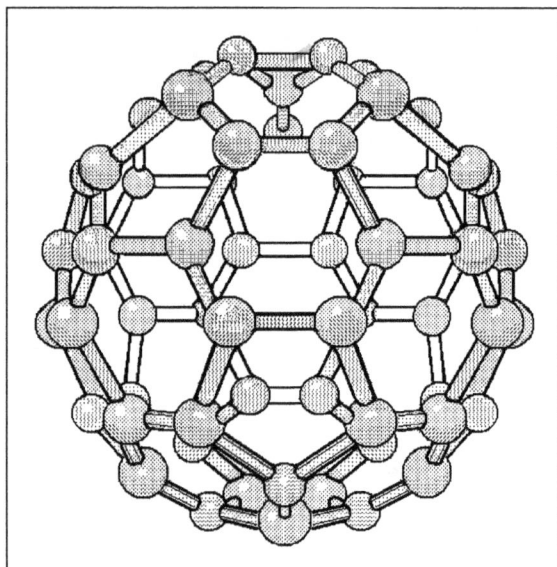

Fig. 2 The Buckminsterfullerene or "buckyball," shown in 3D form, is a carbon molecule that contains 60 atoms. One carbon atom lies at the vertex of 12 pentagons and 20 hexagons arranged like panels of a soccer ball. This configuration has been used to form rigid but light architectural structures.

form naturally in many combustion processes involving the burning of materials containing carbon such as coal. Traces of them were also found in the soot from candle flames.

There is evidence in the scientific literature that both buckyballs and nanotubes were discovered earlier by a Japanese scientist and two researchers in Russia, but no further developments resulted from their observations. However, Rice University's buckeyball findings prompted investigation by other researchers. For example, they led Sumio Iijima, a scientist working at Nippon Electric Company (NEC) in Japan, to discover "needle-like tubes" of carbon atoms in 1991. These configurations are now called "carbon nanotubes." (In later research it was found that some carbon nanotubes were naturally capped at both ends with buckyball hemispheres.)

Carbon Nanotubes Are Prodigies

The carbon nanotube, diagrammed in Fig. 3, has drawn the most attention because of its conspicuous property: great tensile strength. Many might wonder how an almost invisible

Fig. 3 Carbon nanotubes, first discovered in 1991, have the greatest tensile strength of any material known. They can resist one hundred times more strain than structural steel, and are now being used as an engineering material in electronics and medical research. Nanotubes have been spun into yarn, and cloth made from this yarn has been successful in stopping bullets when made into flexible body armor.

strand of carbon, that can only be seen by viewing it with an electron microscope, can resist 100 times more strain than structural steel. However, it is important to keep in mind that these comparisons were made at sub-microscopic scale. Most carbon nanotubes "self-arrange" as miniscule cylindrical cages with the carbon atoms formed in hexagonal geometries like rolled-up chicken wire mesh, but they can also self-arrange in pentagonal geometries.

The diameters of carbon nanotubes can range in size from 1 to 50 nm because their sizes depend on the diameters and thicknesses of the catalyst spots upon which they are grown. But this extremely small size range of their microscopic diameter is better understood when one considers that as many as 50,000 nanotubes must be stacked atop each other to equal the diameter of a human hair.

While nanotube diameters are measured in nanometers, their lengths are typically measured in micrometers (microns) to millimeters. Here length depends on how long the nanotubes are allowed to grow and the conditions of their growth. Carbon nanotubes can form as *single-walled nanotubes* (SWNTs) or *multiwalled nanotubes* (MWNTs), the latter meaning that an inner nanotube is coaxial with an outer nanotube.

A single perfect carbon nanotube is from 10 to 100 times stronger than steel per unit weight. In addition, carbon nanotubes can transmit 1000 times more electrical current than copper, and their thermal conductivity equals that of diamonds. A nanotube's chiral angle (the angle between the axis of its hexagonal pattern and the axis of the tube) determines whether the nanotube will be conductive (metallic) or semiconductive. This chiral angle is set by the way a single sheet of carbon one atom thick (graphene) has been rolled up. It has also been found that nanotubes can function as insulators. However, a defect in the geometrical pattern of the atoms arrayed along its length can cause a carbon nanotube to switch from a semiconductor to a conductor.

Nanotubes can also be formed from atoms and molecules other than carbon. Examples include titanium-dioxide, boron-nitride, and silicon. Nevertheless, none of these nanotubes has the range of properties capable of matching those of carbon nanotubes, which make them suitable for so many different applications. Consequently, the term nanotube now generally implies a carbon nanotube. In addition to buckyballs and nanotubes, carbon atoms can also form into cone shapes called *nanocone*s.

Nanotube Applications Seem Unlimited

Many practical applications for carbon nanotubes are being found such as their use in frictionless nanobearings; nanoelectromechanical systems (NEMS); nanoscale transistors and circuits, and nano-optoelectronic devices. Carbon nanotubes are being used in conductive and high-strength composites, and they are being investigated for various medical applications. The following items drawn from recent laboratory reports and press releases are indicative of the range of recent nanotube research.

Nano radio receiver: Researchers at the University of California, Berkeley, invented a nanotube radio receiver in 2007 based on a single carbon nanotube. It can tune in to a broadcast signal, amplify it, and convert it to an audio signal that can be heard after being sent to an external speaker. This achievement suggests that nanotubes could be used in hearing aids and cell phones small enough to fit within the human ear canal.

Nanotube sonar: Researchers at the University of Texas, Dallas, have found that carbon nanotube sheets excel as underwater sound generators and noise-canceling speakers. These are two highly desirable traits for submarine sonar and military stealth capabilities. The researchers had previously demonstrated that nanotube sheets could produce a wall of sound in air without oscillating like traditional speakers. Later, they showed that the speakers performed as well underwater as they did above water, making them eligible to replace traditional sonar projector arrays.

The team's research found that carbon nanotubes repel water slightly and form a layer of air along their perimeters. Once energized, the thin, light sheets of nanotubes heat and cool quickly, producing a pressure wave in the air around the nanotubes that is perceived as sound. Nanotubes can produce low-frequency sound waves which are effective for sonar scanning of the ocean. They also confirmed that nanotube speakers can cancel noise when tuned to the corresponding frequency.

Nanotube sheets, about 20 μ thick and 99 percent porous, can be fitted on curved surfaces. They can be applied to the outer surfaces of submarines or aircraft to control the boundary layers that these vehicles will move through. The UT scientists also discovered that by periodically heating nanotube sheets to form thin pockets of warm air around vehicles, their speed could be increased because of reduced external friction and turbulence.

Optical waveguides: Researchers at IBM's Watson Research Center demonstrated that certain properties of carbon nanotubes depend on their diameters. Carbon nanotubes with specific diameters can emit light at various wavelengths. This property permits nanotubes to act as optical waveguides by transmitting data along electronic circuits more efficiently than copper wire.

Strong fibers: Scientists at Rice University, Houston, Texas, have been successful in transforming carbon nanotubes into fibers with diameters of human hair. They found that solutions of carbon nanotubes can be formed into strong conductive fibers hundreds of meters long by methods used to make Kevlar fibers. One promising application is the use of electrically conductive nanotube fibers to replace even the finest copper wire conductors in spacecraft instrumentation and prosthetic devices where the weight of copper wires would rule out their use.

Quantum cable: The Rice University scientists have also been working on a prototype of a nanotube-based *quantum cable.* They suggest that power cables made from quantum wires with larger diameters than those made from steel-reinforced aluminum would still be lighter and stronger because their larger diameters would permit a tenfold increase in their capacity for transmitting electric power. Their research found that, at nanometer scale, the unusual properties of quantum physics would permit the wire to carry current without resistance, meaning that electrons could travel down a cable wire made of billions of overlapping nanotubes with almost no energy loss.

Printing ink: Researchers at the University of Illinois, Urbana/Champaign, invented an ink made from carbon nanotubes for printing integrated circuits on flexible substrates. The circuits demonstrated improved performance over circuits printed with conventional semiconducting ink.

Transparent electrodes: Unidym, a Menlo Park, California, company, has developed a roll-to-roll printing technique for rapidly coating sheet plastic with thin films of conductive and transparent carbon nanotube electrodes. These films are intended to replace similar brittle and expensive indium-tin-oxide electrodes printed on transparent plastic panels that enclose the liquid crystals in most flat-panel displays today.

Super capacitors: Scientists at MIT, Cambridge, Massachusetts, developed a capacitor whose plates are coated with nanotubes that dramatically increase capacitor plate surface area. The nanotube coating permits the capacitor to store more electric charge longer than conventional capacitors of comparable size. The MIT researchers, who call these devices *ultracapacitors,* say they are intended to replace miniature batteries. The team believes they will be competitive with alkaline and lithium batteries for powering such electronic devices as cell phones, GPS displays, and mobile radios.

Spring power: Researchers at MIT are also experimenting with nanotube springs, which they say are capable of storing over 1000 times more energy for their weight than steel springs. They point out that this much energy is comparable, pound-for-pound, to that produced by the best lithium-ion batteries, and they are potentially more durable and reliable. Unlike batteries, springs can deliver stored energy either in a rapid, intense burst, or slowly over longer periods. The mechanically stored energy can drive a mechanical load directly, avoiding the need to use less efficient battery-stored electrical energy. Moreover, unlike batteries, energy stored in springs does not slowly leak away over time.

The MIT researchers believe that carbon-nanotube springs could be used for emergency backup power sources that can be left in place for many years. This will eliminate the time and cost of having battery-powered systems tested frequently to verify that they retain full power as well as the need to recharge or replace them if their energy has drained away. They believe that nanotube springs could find their first applications in conventional devices rather just being confined to MEMS-scale devices.

Electron microscope probes: Carbon nanotubes have fine, strong, needle-like properties qualifying them as excellent materials for making electron microscope probe tips. Two different electron microscopes, the *scanning tunneling microscope* (STM), and the *atomic-force microscope* (AFM) include these probe tips. Both microscopes can magnify images of nanomaterials and nanocrystals and manipulate individual atoms in their nanostructures. STMs use the carbon nanotube tips because they conduct electricity, and AFMs use them because their nanoscale tip diameters buckle rather than break when deformed. Both types of probe tips can be modified with chemicals or biological materials for high-resolution imaging. (See *Electron Microscopes: Key Tools in Micro- and Nanotechnology.*)

Fabricating Carbon Nanotubes

Single-walled nanotubes are made in high-pressure reactors at temperatures up to 1000°C. This process forms nanotubes with both diameters and lengths measured in nanometers (nm). They emerge from the reactor in fluffy black streams. Residual iron resulting from reactor processing is removed from the nanotubes by oxidation and treatment with hydrochloric acid, leaving them iron-free. Nanotube fibers are made by dissolving small amounts of nanotubes in strong chlorosulfonic acid to get high nanotube concentrations into solution in liquid crystal form. A piston pushes this liquid through a filter into a hollow needle and out into a bath of ether. This process results in pure nanotube fiber 50 to 100 μm in diameter and many meters long.

Carbon nanotube sheets can be grown at speeds of up to 21 ft/min (6.8 m/min) by rotating trillions of ultra-long nanotubes in a dry-state process per minute for every 0.4 in. (1 cm) of sheet width. A Texas group created sheets of nanotubes that were so thin that an acre of the material would weigh only a quarter of a pound (0.11 kg). In 2005, the group made a sheet 33 ft (10 m) long, and the researchers are working to improve their process.

In addition to being good electrical conductors, the sheets can withstand more than 34,000 lb/in^2 (2390 kg/cm^2) of force without tearing, and they can endure temperatures as high as 840°F (449°C) without losingtheir strength or conductivity. The U.S. Department of Defense is interested in carbon nanotubes for reinforcing composite materials used to make helicopter blades, for manufacturing solar cells, and for use in robotics.

Graphene is Obtained From Graphite

Graphene was first obtained from graphite, an allotrope of carbon. It is actually just a single thin atomic layer of graphite. Its atoms bond weakly to form huge multilayer sheets, and those sheets can slide easily over each other because they are soft and smooth. This form of carbon is best known as the "lead" in pencils. Carbon atoms flake off when graphite is under pressure, as happens when one is writing with a pencil; the marks on paper are actually carbon flakes. One useful property of marks and writing made from carbon flakes is that they are easily erasable.

The atomic structure of isolated, single-layer graphene was studied by transmission electron microscopy (TEM) on sheets of the substance suspended between bars of a metallic grid. Electron diffraction patterns showed the expected hexagonal lattice of graphene, as shown in Fig. 4. It is a sheet of hexagonally shaped carbon rings that look like flat sections of chicken-wire screening. Because graphene is related to the other carbon allotropes, it can be wrapped up into fullerenes, rolled into nanotubes, or stacked to form graphite.

Graphene offers both high electrical conductivity and optical transparency (because it is so thin). This makes it suitable for fabricating transparent sheets of conductive electrodes. These can be used in many applications including computer touchscreens, liquid-crystal displays, organic photovoltaic cells, and organic light-emitting diodes (LEDs). Graphene's mechanical strength and flexibility form better transparent electrodes than widely used indium-tin-oxide electrodes which are brittle. Because graphene films can be deposited from solution over large areas, they show promise as a material for forming radio-frequency transistors. Many applications for graphene parallel those for carbon nanotubes, so they are competitive.

First discovered in 2004 by Andre K. Geim at the University of Manchester, England, graphene has been used to form experimental prototype transistors, memories, and other electronic devices. According to Geim, it is surprisingly easy to make small amounts of graphene simply by writing on paper with a pencil

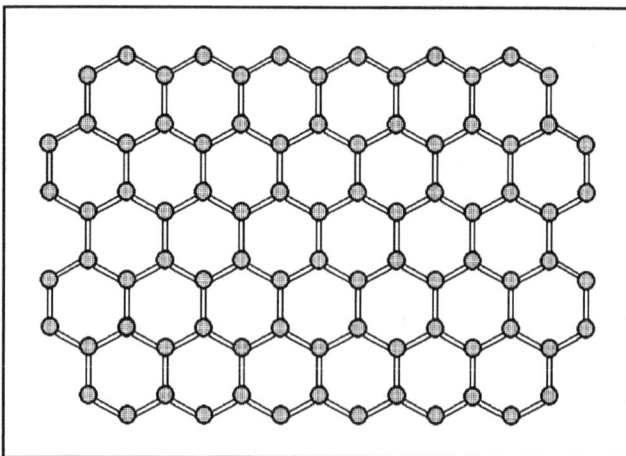

Fig. 4 Graphene is made up of planar sheets of honeycomb (hexagonal) carbon rings just one atom thick. This nanomaterial offers ultrastrength, transparency (because it is so thin), and excellent electron conductivity. A recent significant achievement in nanotechnology was the formation of graphene by chemical vapor deposition (CVD) methods.

because the marks are actually traces of graphene. An early method for obtaining graphene was to abrade graphite to form a powder and then examine this residue with an electron microscope to find suitable samples. The powder was then lifted off the paper with transparent Scotch tape.

Large-Scale Graphene Production

The fabrication of complex graphene-based electronic circuits by lithographic processing will require single-crystal graphene sheets with large areas. Engineers and scientists in many university laboratories as far afield as California, Texas, Georgia, Massachusetts, Pennsylvania, England, and South Korea are trying to devise suitable methods for making industrial quantities of large uniform sheets of pure, single-ply graphene.

Under consideration are several different procedures, but at this time none has advanced far enough for anyone to be accepted as the most successful. For example, scientists at NASA's Jet Propulsion Laboratories in California and MIT have both developed processes for preparing large area (greater than 1 cm²) single-atomic-layer graphene, in 2D crystal form by chemical-vapor deposition (CVD). One process is summarized as follows:

The process calls for two critical ingredients: a substrate with an atomically flat surface suitable for graphene formation and appropriate precursor molecules. These will contain carbon atoms attached to sacrificial atoms in a gaseous form. These molecules will be brought into contact with the nickel substrate surface by allowing them within the CVD furnace. Alternatively, the molecules will be sprayed directly on the substrate, also within the CVD furnace. Conditions of low pressure and elevated temperature are maintained within a furnace, typically in the form of a long tube.

Upon contact with the surface, the precursor molecules will decompose and the sacrificial atoms will detach from the carbon atoms in a gaseous form. The carbon atoms will than fall onto the nickel substrate surface, which acts as a catalyst that helps to form the graphene films. These carbon atoms will diffuse upon the substrate surface where they will encounter and bond to other carbon atoms. If conditions within the furnace are carefully maintained, the surface carbon atoms will arrange themselves to form the single-layer hexagonal structure of graphene desired.

Researchers at Pennsylvania State University (Penn State) have produced 100-mm (4-in.) diameter graphene wafers. They use a high-temperature furnace process called *silicon sublimation*. The thermal process starts with silicon carbide wafers. When placed in a high-temperature furnace, the silicon atoms are driven away from the surface of silicon carbide wafers, leaving graphene with a thickness of one or two layers of carbon atoms.

Researchers at Rice University have learned to make large-area, single-, bi-, or multilayer sheets of high-quality graphene from plain table sugar and other carbon-based substances. This is done in a one-step process at temperatures as low as 800°C (1472°F) in which the carbon-rich source is deposited on a copper or nickel substrate that acts as a catalyst. This compares with the 1000°C (1832°F) required to run a high-temperature furnace. Another carbon-based source being used by scientists is polymethyl methacrylate (PMMA), known commercially as Plexiglas.

Graphene Applications in Electronics

Researchers in Georgia Tech's Atlanta Nanotechnology Research Center have been studying graphene as a replacement for copper for making the fine wire interconnections between transistors and other devices on integrated circuits. They found that graphene *nanoribbons* with widths as narrow as 16 nm have a current-carrying capacity of more than 108 A/cm². Because heat generation is a major cause of the failure of electronic devices, the researchers also measured the ability of the graphene nanostructures to conduct heat away from devices. They determined that graphene nanoribbons have a thermal conductivity of more than 1000 W/m Kelvin for structures less than 20-nm wide.

Penn State researchers have fabricated field-effect transistors (FETS) on their 100-mm silicon carbide graphene wafers and tested their performance. The laboratory has produced wafers that contain more than 22,000 devices and test structures. A further goal of their efforts is an improvement in the speed of electrons in the graphene on the wafers. They say that theoretically electrons could be 100 times faster than their speed in silicon. The Penn State FETs are intended for use in radio frequency (RF) applications.

Diamonds are Different

The molecular structure of a diamond has been well known for years but it differs significantly from those of fullerenes, nanotubes, and graphene, Diamonds are formed when each carbon atom links strongly to four other carbon atoms to form a rigid, compact, 3D, extremely hard structure. Naturally formed, diamonds are the world's most valued gemstones. Even inferior and synthetic diamonds embedded in wheels are hard enough to permit them to cut marble and perform fine grinding and polishing of many different hard materials.

References

Discover Magazine (June 2009) Life after Silicon, pp. 14–15.
Technology Review (Mar/Apr 2009) Nanotube Electronics, p. 14; (May/Jun 2010) Nanotube Fibers, pp. 92–94.
Scientific American (March 2009) Grinding Out Graphene, pp. 20–21; The World's Smallest Radio, pp. 39–45; (Sept 2009) Buckyballs and Nanotubes, p. 82; (Jan 2010) The Next 20 Years of Microchips, p. 86.

NANOACTUATORS BASED ON ELECTROSTATIC FORCES ON DIELECTRICS

Large force-to-mass ratios could be achieved at the nanoscale.
NASA's Jet Propulsion Laboratory, Pasadena, California.

The nanoactuators proposed here would exploit the forces exerted by electric fields on dielectric materials. The term "nanoactuators" as used here includes motors, manipulators, and other active mechanisms that have nanometer dimensions or are designed to manipulate objects that have nanometer dimensions.

The physical principle that explains nanoactuators can be demonstrated by a square parallel plate capacitor with a square dielectric plate inserted part way into the gap between the electrode plates as shown in Fig. 1. The equations for the properties of a parallel-plate capacitor show that the electrostatic field will pull the dielectric slab toward a central position in the gap with a force, F, given by

$$\text{Equation 1: } F = \frac{V^2(1 - 2)a}{2d}$$

where V is the potential applied between the electrode plates, 1 is the permittivity of the dielectric slab, 2 is the permittivity of air, a is the length of an electrode plate, and d is the dimension of the gap between the plates.

In the macroscopic domain, the force F is small, but it becomes more significant in the microscopic domain. Equation 1 states that the force depends on the ratio between the capacitor dimensions but does not depend on the size. In other words, the force remains the same if the capacitor and the dielectric slab are reduced to nanometer dimensions. Nevertheless, the masses of all components are proportional to third power of their linear dimensions. Therefore, the force-to-mass ratio (and, consequently, the acceleration that can be imparted to the dielectric slab) is much larger at the nanoscale than at the macroscopic scale. The proposed actuators would exploit this effect.

A simple linear actuator based on a parallel-plate capacitor is shown in Fig. 2. It is similar to that shown in Fig. 1 except that the upper electrode plate is split into two parts (A and B), and the dielectric slab is slightly longer than either plate A or B. This actuator would be operated in a cycle. During the first half cycle, as shown in Fig. 2a, plate B would be grounded to the lower plate and plate A would be charged to a potential, V, with respect to the lower plate, causing the dielectric slab to be pulled under plate A.

(a)

(b)

Fig. 2 Two stages of a reciprocating actuator. The actuator is driven by electrostatic forces.

During the second half cycle, as shown in Fig. 2b, plate A would be grounded and plate B would be charged to potential V, causing the dielectric slab to be pulled under plate B. The reciprocal motion caused by alternation of the voltages on plates A and B could be used to drive a nanopump.

A rotary motor, shown in Fig. 3, would include a dielectric rotor sandwiched between a top and a bottom plate containing multiple electrodes arranged symmetrically in a circle. Voltages would be applied sequentially to electrode pairs 1 and 1a, then 2 and 2a, then 3 and 3a, attracting the dielectric rotor to sequential positions between the electrode pairs, thus causing it to rotate in a counterclockwise direction.

A micro- or nanomanipulator, shown in Fig. 4, would include lower and upper plates covered by rectangular grids of electrodes—in effect, a rectangular array of nanocapacitors. A dielectric or quasi-dielectric micro- or nanoparticle (a bacterium, virus, or molecule, for example) could be moved from an initial position on the grid to a final position on the grid by applying a potential sequentially to the pairs of electrodes along a path between the initial and final positions.

This work was done by Yu Wang of Caltech for **NASA's Jet Propulsion Laboratory**. *NPO-30747*

Fig. 1 Parallel plate capacitor. The electric field pulls a partially inserted dielectric slab further into a gap.

528

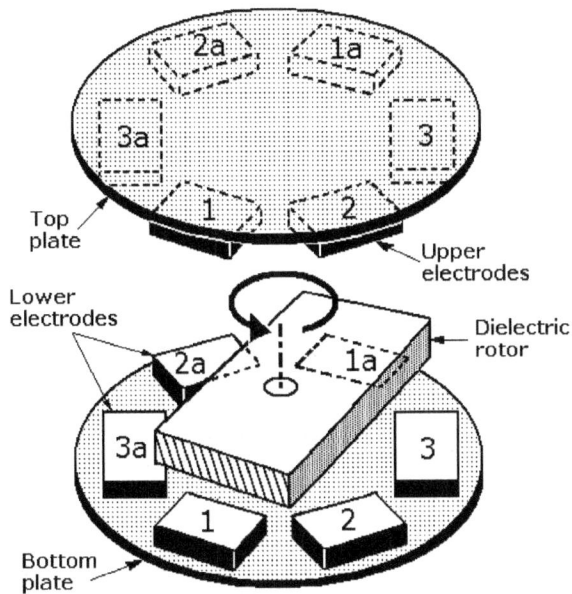

Fig. 3 Exploded view of a micromotor or a nanomotor.

Top plate

Upper electrodes

Lower electrodes

Dielectric rotor

Bottom plate

Manipulated Micro-or Nanoparticle

Fig. 4 A micro or nano capacitor array can pin and move a particle—bacteria, virus, or molecule to any position in the array if a voltage is alternated along the desired path of movement.

THE LUNAR ELECTRIC ROVER: A NEW CONCEPT FOR MOON TRAVEL

NASA had plans to return to the Moon in 2020, but that is unlikely because of existing political and economic conditions. However, evidence of these plans can be seen in the full-scale prototype of an actual *Lunar Electric Rover* (LER) expected to participate in that return. The LER will provide improved range and exploration capabilities over those of previous lunar rovers. The advanced engineering prototype has passed extensive tests of its design concept on Moon-like terrain here on Earth. It has demonstrated that two astronauts can work and drive safely in an air-filled compartment over long lunar distances without having to wear pressurized spacesuits.

The prototype, shown in Fig. 1, is not intended for an actual lunar landing. It is the size of a pickup truck with a pressurized cabin mounted on a 12-wheel chassis. The actual LER will be powered by an advanced battery and will be able to travel 6.2 mph (10 km/h) with a potential range of more than 150 miles in any direction from its lunar base. This lunar "RV camper" will eliminate the need for astronauts to return to base each night. It will be radiation hardened for crew protection and it will set out solar-powered charging stations, communication relays, and scientific packages along its route. The leading specifications are given in the table.

The prototype LER has a removable cabin that includes a sink, toilet, and even a collapsible exercise bicycle. Two seats fold into bunk beds, and curtains can be rolled down from the ceiling to form separate sleeping quarters. The LER contrasts sharply with the LRVs (four-wheeled lunar roving vehicles) used on the last three Apollo Moon missions (15, 16, and 17) of the 1970s. The two astronauts driving in the LRVs had to wear spacesuits because these rovers were completely open to the airless lunar environment. Travel was restricted to about six miles because of time and distance limitations imposed by their oxygen supplies. Moreover, this was the maximum distance from which they could safely walk back to the base station if the rover broke down.

Off-road racing trucks influenced the design of the 12-wheel prototype LER. It has been driven on the rough lava terrain of the Arizona desert for more than 100 miles (160 km). It is expected that the LERs actually sent to the Moon will require little or no maintenance; they will be able to climb easily over rocks and move up slopes as steep as 40° during their anticipated 10-year operational lives.

Five contoured windows at the front of the pressurized cabin give the astronauts a wide-angle view of the lunar landscape as

Fig. 1 The concept Lunar Electric Rover (LER) has two principal modules: a removable pressurized cabin and a 12-wheel chassis. The wheels are paired to form separate motor-driven and steerable units. They allow the LER to move in all directions as well as diagonally. The cabin is living quarters for two astronauts so they can carry out long-distance, long-duration Moon explorations in street clothes. The docking hatch (right) permits astronauts to move from a Moon base station to an LER cabin or between LERs without spacesuits.

Fig. 2 This rear view of the LER shows two spacesuits stored outside attached to box-like suitports. Astronauts climb into their spacesuits from inside the cabin and unplug themselves to walk free in less than 10 minutes without causing a loss of cabin pressure. After completing a mission, they return to the suitports and reverse the suiting procedure. Storing the spacesuits outside prevents Moon dust from contaminating the cabin.

they drive around on missions. The lower middle window contains a clear bubble-shaped dome near the bottom to make it easier for astronauts to observe rocks and other objects of interest more closely by crouching down inside the cabin. The cabin can even be tilted downward to bring surface objects only inches away.

If the astronauts want to look closer at some specific terrain features, they can quickly "suit up" in spacesuits hanging outside the rear bulkhead (shown in Fig. 2) and begin a mission in less than 10 minutes. The large rectangular boxlike backpacks on the spacesuits are "plugged" into recessed hatches in the bulkhead. The astronauts will enter the spacesuits from inside the cabin by climbing through doors on the backpacks that are seamlessly bonded to the spacesuits. This arrangement makes it possible for the astronauts to breathe the air from inside the cabin that keeps the suits permanently inflated while minimizing losses in cabin pressurization. With the spacesuits kept outside, abrasive Moon dust will not be tracked into the cabin accidentally.

When the astronauts are suited up, they close the doors in their backpacks and cabin bulkheads, trapping cabin pressure in both before they "unplug" from the recessed stations. They are then free to walk away from the LER to begin a mission. When the mission is completed, the astronauts reverse the procedure by backing their backpacks into the recessed stations to seal them before opening both doors to climb back into the cabin.

The six pairs of chassis wheels can be moved sideways in a crab-like motion and individually steered to make it easier for the LER to avoid particularly rough terrain in its path. This steering arrangement gives the LER a tight turning radius as well as the ability to drive in any combination of forward and sideways motions.

The versatility of the LER concept can be demonstrated by available options. The astronauts can don spacesuits at a base station and, after removing the LER cabin module, install external controls on the bare chassis converting it into a payload transporter. The astronauts can also elect to drive the bare chassis on missions because the external controls permit driving the rover from standing positions like Roman charioteers. On Moon missions the LER, disassembled into modules, could be sent separately to the Moon, propelled by their own rockets prior to the arrival of the astronauts. When the astronauts land, they can reassemble the LER.

Leading Specifications for Lunar Electric Rover

Rover weight (total)	3.3 ton (3000 kg)
Chassis weight	1.1 ton (1000 kg)
Rover payload	1.1 ton (1000 kg)
Length	15 ft (4.6 m)
Height	10 ft (3.0 m)
Wheels	12 (6 pairs)
Wheel dims (diameter/width)	37/10 in. (94/25 cm)
Crew	Two
Speed (max)	6.2 mph (10 km/h)
Wheel pivot (max)	360°

Resources

The following magazine articles and Web sites provided information included in this chapter:

Magazine Articles

Zach Zorich, "Carbon Nanotubes Burst Out of the Lab." *Discover Magazine*, January 2006, 29.

Erika Jonietz, "Quantum Wires." *Technology Review*, May 2005, 45.

Jia Chen, "Looking Past Silicon to Carbon Nanotubes." *Technology Review*, October 2005, 47.

Web Sites

American Elements
Nanotechnology Information Center
www.americanelements.com/nanotech.html

Mechanical Engineering Magazine
Alan S. Brown, "MEMS across the Valley of Death," April 2006
John De Gaspari, "Beyond Silicon," July 2005
www.memagazine.org/backissues/

Sandia National Laboratories
Microelectromechanical Systems (MEMS)
http://mems.sandia.gov/

Sandia National Laboratories
Liga Technology
http://www.ca.sandia.gov/liga/tech.html

Sandia National Laboratories
Micromachines Image Gallery
http://mems.sandia.gov/scripts/images.asp

University of Texas at Dallas
Carbon Nanotube Sheets
www.utdallas.edu/news/archive/2005/carbon-nanotube-sheets
.html

INDEX

Editor's notes:

1. The number in parentheses in some entries refers to the number of related topics on the listed page(s). For example, an entry *Eccentric cams (7)* indicates seven related captioned figures.
2. Figures and captions in Chaps. 5 to 16 were obtained from many sources that did not uniformly apply words such as devices, drives, mechanisms, and machines. Thus, there may be multiple entries for similar subjects.

544

www.ingramcontent.com/pod-product-compliance
Lightning Source LLC
Chambersburg PA
CBHW060948210326
41598CB00031B/4761